Microbial Photosynthesis

Qiang Wang
Editor

Microbial Photosynthesis

Springer

Editor
Qiang Wang
Key laboratory of Plant Stress Biology, State Key Laboratory
of Cotton Biology, School of Life Sciences
Henan University
Kaifeng, Henan, China

ISBN 978-981-15-3109-5 ISBN 978-981-15-3110-1 (eBook)
https://doi.org/10.1007/978-981-15-3110-1

This Springer imprint is published by the registered company Springer Nature Singapore Pte Ltd.
The registered company address is: 152 Beach Road, #21-01/04 Gateway East, Singapore 189721, Singapore

Preface

As the largest scale chemical reaction, photosynthesis supplies all of the organic carbon and oxygen for life on Earth. It is estimated that more than 50% of the primary production and molecular oxygen on Earth are by the photosynthetic activity of microorganisms.

This book highlights recent breakthroughs in the multidisciplinary areas of microbial photosynthesis. The chapters feature the most recent developments in microbial photosynthesis research, from bacterial to eukaryotic algae, from theoretical biology to structural biology and biophysics. Furthermore, the book also features the latest advancements in artificial photosynthetic. Contributed by leading authorities in photosynthesis research, Microbial Photosynthesis will offer a valuable resource for graduate students and researchers in the field. Furthermore, it will be of great significance for the study of biological evolution and the origin of life.

Kaifeng, Henan, China

Qiang Wang

Contents

Part I
Photosynthesis and Energy Transfer

Molecular Mechanism of Photosynthesis Driven by Red-Shifted Chlorophylls

Artur Sawicki and Min Chen

Abstract Photosynthesis is the process of light-driven production of organic molecules needed as starting components for whole cellular processes and as the energy source. It is carried out by primary producers including land plants, algae, and oxygenic/anoxygenic photosynthetic bacteria. Oxygenic photosynthesis involves two stages: light-dependent reactions generating NADPH and ATP molecules with oxygen as a by-product and light-independent reactions involving utilization of NADPH and ATP as the energy source to convert carbon dioxide into organic molecules. Light-dependent processes require light-absorbing pigments categorized as carotenoids, bilins, and chlorophylls. Chlorophylls and carotenoids are ubiquitous to all photosynthetic organisms, while bilins in phycobiliprotein complexes are specific to cyanobacteria, rhodophytes, glaucophytes, and cryptophytes. Each pigment has several variations, and a particular type of organism has a specific set optimized for light-capture and photosynthetic efficiency in the given environment. The common chlorophyll to each oxygenic photosynthetic organism is chlorophyll *a*, with chlorophyll *b*, *c*, *d*, or *f* synthesized depending on the organism. Chlorophyll *d* (3-formyl-chlorophyll *a*) and chlorophyll *f* (2-formyl-chlorophyll *a*) have far-red long-wavelength absorption features, namely, red-shifted chlorophylls. These chlorophyll molecules enable some cyanobacterial species to grow in shaded environments and establish unique habitats. *Acaryochloris marina* is predominantly a chlorophyll *d*-containing cyanobacterium with minor (≤ 5%) amounts of chlorophyll *a* under all culture conditions, while all chlorophyll *f*-producing cyanobacteria constitute chlorophyll *f* as a minor chlorophyll ~2–15% of the total chlorophyll pool and only produced under far-red light conditions. This chapter will focus on summarizing the current knowledge of the light-dependent processes with particular attention on structures and functions of far-red absorbing pigments and pigment-binding protein complexes.

Keywords *Acaryochloris marina* · Biochemistry of chlorophylls · Chlorophyll *d* · Chlorophyll *f* · Light-harvesting protein complexes · Photosystem

---

A. Sawicki · M. Chen (✉)
School of Life and Environmental Sciences, The University of Sydney,
Sydney, NSW, Australia
e-mail: min.chen@sydney.edu.au

Q. Wang (ed.), *Microbial Photosynthesis*,
https://doi.org/10.1007/978-981-15-3110-1_1

Abbreviations

APC	Allophycocyanin
CBP	Chlorophyll *a*-/*b*-binding protein
Chl	Chlorophyll
CHLF	Chlorophyll *f* synthase
FaRLiP	Far-red light-induced photoacclimation
FRL	Far-red light
IsiA	Iron-stress induced chlorophyll-binding protein
LHC	Light-harvesting protein complex
PAR	Photosynthetically active radiation
PBP	Phycobiliprotein
PBS	Phycobilisome
PC	Phycocyanin
Pcb	Prochlorophyte chlorophyll *a*-/*b*-binding protein complexes
PE	Phycoerythrin
PEC	Phycoerythrocyanin
Pheo *a*	Pheophytin *a*
PS	Photosystem
Q_A	Plastoquinone A
RC	Reaction center

1 General Knowledge of Photosynthesis

Photosynthesis is the conversion of light energy into chemical energy stored in organic molecules. It occurs through two stages: light-dependent and light-independent reactions. Light-dependent processes involve (1) the absorption of light by pigments and energy delivery by antenna/light-harvesting complexes (LHCs), (2) primary electron transfer in reaction centers (RCs), and (3) electron transport processes facilitating the synthesis of NADPH and ATP molecules. Light-independent processes involve utilization of NADPH and ATP for carbon (CO_2) fixation by the Calvin-Benson cycle including carboxylation, reduction, and ribulose 1,5-bisphosphate (RuBP) regeneration. Sucrose is an example of the stable photosynthetic product that is used to power a variety of cellular processes.

The light-gathering antenna system functions to absorb light and to transfer the energy in the light photons to RCs. Various photopigments are used in the antenna system, including chlorophylls (Chls) and carotenoids in integral membrane LHCs, and linear tetrapyrroles bilins complexed with biliproteins organized as a phycobilisome (PBS). The three photopigments have particular absorption wavelength maxima to facilitate photosynthetically active radiation (PAR) between 400 and 700 nm. With the aid of red-shifted Chls (Chl *d* and Chl *f*) and the red-shifted phycobiliproteins (PBPs), some cyanobacteria are able to extend their photosynthetically active spectral region into the long-wavelength region of 700–760 nm, which extends the coverage of the PAR availability in a local habitat (Chen and Blankenship 2011). Increasing the PAR cross section beyond 700 nm has promoted the photoautotrophic growth in conditions where far-red light (FRL) predominates.

Two photosystems (PSs) are essential for conducting sustained oxygenic photosynthesis, PSI and PSII. Each PS has a special pair of Chl molecules otherwise known as primary electron acceptors which consist of Chl *a* and Chl *a′*, which are bound to the RC core peptides of PsaA/PsaB and PsbA/PsbD for PSI and PSII, respectively. Photosynthesis begins with the collective absorption of photons by antenna photopigments, and the collective energy can be transferred from one pigment to another following an energy gradient established by pigments, from higher absorbed light energy to lower absorbed light energy in the RCs. Reaction centers are the location where light-driven charge separation and electron transfer occur and Chls in RC are excited and readily pass electrons to the primary electron acceptor (A) causing a P^+A^- pair (Caffarri et al. 2014). In the unique FRL-absorbing cyanobacterium *Acaryochloris marina* (*A. marina*), Chl *d* replaces functions of Chl *a* in the RCs of PSI and PSII (Mimuro et al. 2004), while in *Prochlorococcus marinus*, Chl *a* is replaced by divinyl Chl *a* (Chl a_2), which coexists with divinyl Chl *b* (Chl b_2) (Partensky et al. 1993). Meanwhile Chl *f* is present in isolated PSI complexes from *Halomicronema hongdechloris* (*H. hongdechloris*) (Li et al. 2018a) and also present in isolated PSI and PSII in *Chroococcidiopsis thermalis* (Nürnberg et al. 2018).

In PSII, the primary electron acceptor is pheophytin *a*, while the secondary electron acceptor is plastoquinone A (Q_A). Each photochemical reaction requires two electrons to be fed into the electron transport chain. The four electrons lost from two water molecules generate one oxygen molecule and four protons (H^+). Similar to PSII, Chl molecules in the RC of PSI are excited, and the separated electrons are readily passed to electron acceptor Chl molecules (A_1), followed by delivery to the next electron acceptors: phylloquinone; FeSx, FeSb, and FeSA; ferredoxin; and $NADP^+$ reductase. The coupling of PSI and PSII via cytochrome b_6f complexes in oxygenic photosynthetic organisms is called noncyclic photophosphorylation. Therefore, PSII and PSI are connected with equal electron currents coupling to production of ATP and NADPH.

Cyclic electron transport involves transfer of electrons from ferredoxin back to cytochrome b_6f complex, plastoquinone, and plastocyanin and onto the RC Chl of PSI. This involves only PSI and couples ATP production without NADPH synthesis (Arnon 1984). Anoxygenic photosynthetic bacteria often only have a cyclic electron transport photosynthetic apparatus. The photosynthetic cyclic electron transport means no direct need for the initial electron donor or terminal electron acceptors. The ATP and NADPH molecules generated from the light-dependent reactions are utilized by the "dark" (light-independent) reactions of the Calvin-Benson cycle which involves the fixing of CO_2 into organic carbon, namely, the production of three-carbon sugar (glyceraldehyde 3-phosphate), and eventually converts into starch or sucrose.

2 Photosynthetic Organisms

Chlorophyll-driven oxygenic photosynthesis occurs in the primary producers: plants, algae, and cyanobacteria. Bacteriochlorophyll-driven anoxygenic photosynthesis occurs only in bacteria under aerobic and anaerobic conditions. Oxygenic photosynthesis facilitated conditions favorable for oxygen-dependent species. Cyanobacteria are the only bacteria that perform oxygenic photosynthesis, which thrive in various environments, including some extreme environmental conditions. For example, cyanobacteria containing FRL-absorbing pigments Chl *d* and Chl *f* can be found from deprived visible light conditions where the FRL is enriched (Zhang et al. 2019; Kühl et al. 2005).

2.1 *Photosynthetic Eukaryotes*

Eukaryotic photosynthesis occurs in a specialized compartment (organelle) called the chloroplast. It is thought chloroplasts originally derived from ancient cyanobacteria through primary endosymbiosis, a process whereby the cyanobacteria was engulfed by a non-photosynthetic eukaryotic organism (a protist) (Archibald 2015). The engulfing event eventually led to the development of the plastid and may have occurred multiple times; however it appears likely that rhodophytes, glaucophytes, and green algae (division chlorophytes or euglenophytes) derived from a common ancestor of primary endosymbiotic origin (Reyes-Prieto et al. 2007). Primary endosymbiotic green algae further evolved into higher plants. Evidence for secondary endosymbiosis of red and green algae exist through the presence of three or four layers of plastid envelope found in the photosynthetic organisms. A common ancestor of engulfed red algae is proposed to have developed into Chl *c*-producing algae, dinoflagellates, diatoms, brown algae, haptophytes, and cryptophytes, while green algae underwent two separate endosymbiotic events forming euglenophytes and chlorarachniophytes.

Algae are unicellular or multicellular organisms with the latter either microscopic or macroscopic, and all species are Chl *a*-dominant. Green algae have Chl *a* and Chl *b*, and an intrinsic light-harvesting system which is structurally similar to that of higher plants compared with other algal divisions. Red algae, glaucophytes, and cryptophytes have PBPs as peripheral light-harvesting systems. Red algae consist of Chl *a* and PBSs consisting of phycoerythrin (PE), phycocyanin (PC), and allophycocyanin (APC). Glaucophytes have PBS consisting of only PC and APC (Chapman 1966; Giddings et al. 1983). Meanwhile cryptophytes contain a Chl *a*/*c* LHC and atypical PBP complexes with unique α-PBP and rhodophyte-like β-PBP subunits, which use either Cr-PE or Cr-PC as bound chromophores. The most striking feature of atypical PBP complexes in cryptophytes is that they are localized in the thylakoid lumen spaces (Glazer and Wedemayer 1995). The remaining Chl *c*-containing algae have a Chl *a*/*c* LHC without PBSs, which include chromophytes (heterokonts and ochrophytes), brown algae, diatoms, dinophytes (dinoflagellates), and haptophytes. Chlorophyll *a*/*c* LHC complexes are structurally similar to Chl *a*/*b* LHC complexes in higher plants, but mostly with different bound carotenoids (Büchel 2019). Higher plants are considered to be the most advanced photosynthetic eukaryotes and have a conserved LHC composed of Chl *a*/*b* with variable carotenoid molecules (Neilson and Durnford 2010).

2.2 *Photosynthetic Prokaryotes*

There are five groups of photosynthetic bacteria: purple, green sulfur, green non-sulfur, heliobacteria, and cyanobacteria. The first four groups are anoxygenic photosynthetic organisms that use infrared-absorbing bacteriochlorophylls in the photosynthetic process (Scheer 1991). Anoxygenic photosynthetic bacteria have only one type of photosynthetic RC, either type I (PSI-like) or type II (PSII-like) RC, which drives only cyclic electron transport surrounding the RC (Allen 2014). Various electron donors use other than water, such as hydrogen, sulfur (H_2S), Fe^{2+}, and reduced organic compounds (Bryant and Frigaard 2006).

Cyanobacteria are oxygenic photosynthetic prokaryotes that use Chl *a* as the main light-absorbing Chl in the PSs, and most of cyanobacteria use PBS as the main part of the antenna system. The photosynthetic oxidation of water provides the electron and the free oxygen molecules like the photosynthesis conducted in higher plants.

2.2.1 Anoxygenic Photosynthetic Prokaryotes

Purple sulfur and non-sulfur bacteria are bacteriochlorophyll *a*- and bacteriochlorophyll *b*-containing organisms that conduct photosynthesis under anaerobic conditions (Frigaard and Dahl 2008; McEwan 1994). Anaerobic conditions trigger the synthesis of additional lipids leading to the invagination of cytoplasmic membrane providing an ultrastructure for the photosynthetic machinery (Madigan and Jung 2009). These intracytoplasmic membranes are comparable to the thylakoid membranes of cyanobacteria. The electron donors are either sulfide, sulfur, or hydrogen; hence no oxygen is produced as a by-product. Some species of purple bacteria are known as aerobic anoxygenic photosynthetic bacteria and carry out photosynthetic processes only under aerobic conditions (Yurkov and Beatty 1998). The purple bacteria use type II RC, which is comparable with PSII of plants, algae, and cyanobacteria. The two RC core subunits are named as L and M with the H subunit stabilizing the dimeric RC complex and enhancing the activity (Allen and Williams 2011). The light-harvesting complex has α- and β-subunits (encoded by *pufA* and *pufB*) forming a ring structure consisting either 9 subunits or the multi-metric complex of 15–16 subunits (Overmann and Garcia-Pichel 2013).

Green sulfur and green non-sulfur bacteria are obligate anoxygenic photosynthetic organisms with specialized extrinsic chlorosomes as the main antenna system thriving in the extreme low-light environments. Chlorosomes contain bacteriochlorophyll *a*/*c*/*d*/*e* and attach to the cytoplasmic membrane via Fenna-Matthews-Olson (FMO) bacteriochlorophyll *a*-binding protein complexes. The ratio of bacteriochlorophylls/protein within chlorosomes is the highest out of any pigment-binding protein complexes reported, providing the advantages for this anoxygenic photosynthetic bacterial group to efficiently capture low-intensity light (Overmann and Garcia-Pichel 2013). Chlorosomes pass the energy to FMO and then to the RC (Hauska et al. 2001; Overmann and Garcia-Pichel 2013). The type I RC resembles that of PSI from oxygenic photosynthetic organisms, with the exception that one polypeptide (dimer) constitutes the structure (Overmann and Garcia-Pichel 2013).

Heliobacteria have a simple type I RC that localized in the cytoplasmic membrane and can only perform photoheterotrophic growth under anaerobic conditions with bacteriochlorophyll *g* as the major photopigment (Tang et al. 2010).

2.2.2 Oxygenic Photosynthetic Prokaryotes (*Cyanobacteria*)

Oxygenic photosynthetic bacteria refer to phylum *Cyanophyta* or *Cyanobacteria* that are sometimes called blue-green algae although they are true prokaryotes with no chloroplast. Cyanobacteria inhabit diverse environments including marine, freshwater, and terrestrial niches. Their morphology varies from coccoid and filamentous to colonial and unicellular with diverse thylakoid membrane architecture (Mareš et al. 2019). Certain species can tolerate high temperatures, periods of desiccation, and high or low light intensities through the ability to produce specific pigments, to modify the chemical structures in order to change the optical properties of the pigment suitable for the given light qualities (Ho et al. 2017b; Rockwell et al. 2016; Gomelsky and Hoff 2011; Kehoe and Gutu 2006).

Cyanobacteria are typically photoautotrophs with some species having the ability to live photoheterotrophically if nutrient conditions are available. Chlorophyll *a* is the major Chl pigment; however, there are some exceptions. *Acaryochloris marina* uses Chl *d* as the major Chl pigment under all light conditions with traces of Chl *a* present and an atypical phycobiliprotein complex (PBP) structure (Chen et al. 2009; Marquardt et al. 2000; Miyashita et al. 1996). All reported *A. marina* spp. phylogenetically belong to a single genus, *Acaryochloris marina* spp. Chlorophyll *f*-producing cyanobacteria phylogenetically are classified into five subsections of cyanobacteria with related morphological differences (Zhang et al. 2019). Chlorophyll *f* extends PAR further into the FRL region, and synthesis of this pigment is induced under PAR conditions up to ~2–15% of the total Chl content along with major Chl, Chl *a* (Zhang et al. 2019; Chen and Blankenship 2011; Chen et al. 2012).

Prochlorophytes are unusual cyanobacteria which synthesize Chl *a* and Chl *b*, and three genera have been identified: *Prochloron*, *Prochlorothrix*, and *Prochlorococcus* (Shih et al. 2013). *Prochloron* is a symbiotic prokaryote associated with marine invertebrates (Lewin 1977), while *Prochlorothrix* is a filamentous free-living genus (Burger-Wiersma et al. 1989; Burger-Wiersma et al. 1986).

Both these genus use Chl *a* and Chl *b* as that in higher plants. *Prochlorococcus marinus* spp. are the smallest free-living unicellular cyanobacteria using Chl a_2 and Chl b_2 instead of the typical Chl *a* and Chl *b* (Partensky et al. 1999; Chisholm et al. 1988). Prochlorophytes have no PBS and instead rely upon membrane-bound intrinsic Chl *a*-/*b*-binding protein complexes (Pcbs) (Chen et al. 2008). These proteins are similar to the core antenna protein of PSII, CP43, and CP47; hence this antenna complex differs from the membrane-bound LHCs of plants and green algae (La Roche et al. 1996).

3 Photopigments

The main light-absorbing photopigments include carotenoids, bilins, and Chls. Bilins capture light energy between approximately 520 and 670 nm that is poorly absorbed by Chls (Fig. 1) (Chen and Scheer 2013). However, Chl molecules are critical for photochemical reactions in the RCs. Together, various photopigments provide the entire PAR between 400 and 700 nm, which is made of 47% of solar photon inputs. The FRL-absorbing Chls (Chl *d* and *f*) extend the PAR beyond 700 nm that is enriched in shade and low-light environments.

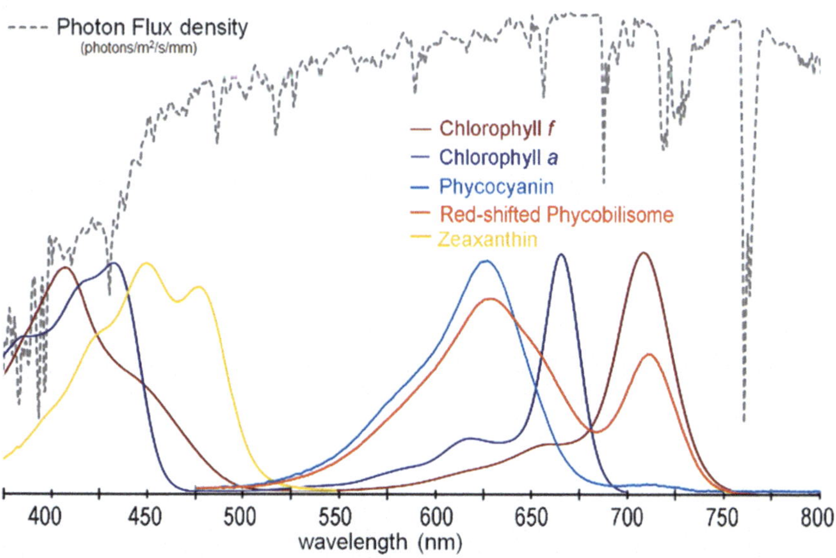

Fig. 1 Solar irradiance and absorption spectrum of some typical common light-absorbing pigments involved in photosynthesis. The spectra of carotenoids (zeaxanthin), chlorophyll *a*, and *f* were recorded in 100% methanol, and the spectra of isolated phycobilisome and red-shifted allophycocyanin were recorded in phosphate buffer. Solar irradiance spectra downloaded from https://www.nrel.gov/grid/solar-resource/spectra-am1.5.html

3.1 Carotenoids

Carotenoids are colored linear tetraterpenoids containing a polyene chain of conjugated double bonds which absorb light mainly in the blue light region, 400–530 nm (Hashimoto et al. 2016). They are involved in light harvesting and also absorption of excessive light energy and quenching of singlet oxygen generated by photosynthetic reactions. Carotenoids may be classified as nonpolar oxygen-lacking structures called carotenes and polar oxygen-containing xanthophylls. There are approximately 30 different carotenoid structures reported in photosynthetic organisms with their own specific pigment composition (Takaichi 2011). Cyanobacteria typically have β-carotene and different xanthophyll structures, such as zeaxanthin, echinenone, myxoxanthophyll (myxol glycosides), and canthaxanthin (Zakar et al. 2016; Takaichi 2011; Takaichi and Mochimaru 2007). In some genus such as *A. marina* spp. and *Prochlorococcus* spp., α-carotene is present instead of β-carotene (Takaichi et al. 2012).

Carotenoid contents and compositions generally change in response to light stress conditions, for example, zeaxanthin concentration is increased under high-light irradiation. In green algae and plants, a rapid epoxidation and de-epoxidation cycle among zeaxanthin, antheraxanthin, and violaxanthin reflects the changed light conditions, which refer as the xanthophyll cycle (Demmig-Adams 1990; Goss and Jakob 2010). Although this cycle is not present in cyanobacteria, zeaxanthin may accumulate in high-light conditions through oxidation of β-carotene by β-carotene hydroxylase (E.C.1.14.13) (Masamoto and Furukawa 1997). Other cyanobacteria produce excess myxoxanthophyll under high-light conditions (Steiger et al. 1999), while a cyanobacteria-specific orange carotenoid protein bound with echinenone is upregulated in high light and can quench the excessive energy generated from the PBS (Schagerl and Müller 2006; Domonkos et al. 2013; Kirilovsky and Kerfeld 2012). The localization of xanthophyll in photosynthetic apparatus has not been determined; however, it plays an important role in photosynthetic reactions. A xanthophyll-less mutant of *Synechocystis* sp. PCC6803 showed a globally negative effect on the photosynthetic apparatus including reduced PBS rod length, monomerized PSI, and irregularly assembled PSII (Tóth et al. 2015).

3.2 Phycobiliprotein Complexes

Bilins are linear tetrapyrroles derived from the oxidative ring opening of heme and are chromophores for bilin-binding PBPs. The bilin-binding site is usually on $C3^1$ of ring A through a cysteine residue of the apoprotein, with some bilins additionally attaching to $C18^1$ of ring D (Fig. 2). In cyanobacteria and some algae (rhodophyte, cryptophyte, glaucophyte), PBS represents an assembly structure of bilin-binding protein complexes (for details, see Sect. 4.4). Allophycocyanin is at the core of PBS, with rods projecting outward as an external antenna, and may be composed of PC,

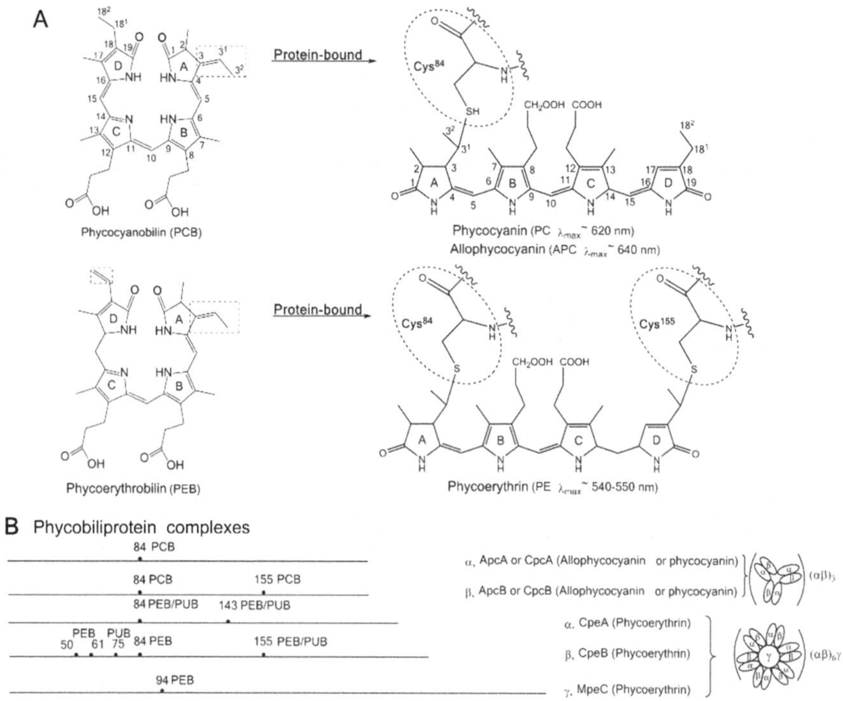

Fig. 2 Chemical structure of PCB and PEB and their association into phycobiliproteins. (**a**) Structure of phycocyanobilin (PCB) and phycoerythrobilin (PEB). The dashed boxes highlight the sites where covalent attachment of bilins occurs with a cysteine residue of the apoprotein. The numbering of carbon atoms is following the IUPAC system. Cys84 and Cys155 represent the relative position of cysteine in apo-phycobiliprotein. (**b**) Primary binding position of bilins on the phycobiliproteins and the quaternary structural model of the phycobiliprotein. The length of the lines represents the approximate relative lengths of the polypeptides

PE, or phycoerythrocyanin (PEC), with PE/PEC at the periphery and PC closer to the core. The basic disc structure of PBP typically has a αβ trimeric $(\alpha\beta)_3$ or $(\alpha\beta)_6\gamma$ structure, and each subunit is structurally related with a variable number of bilin-binding sites. Genes encoding apo-PBP subunits are often clustered together and contain *apc*s, *cpc*s, and *cpe*s, which encode apo-APC, apo-PC, and apo-PE peptides, respectively.

Phycocyanins consist of CpcA/CpcB subunits with PC bound to Cys84 of each subunit and also to Cys84 and Cys155 for the beta-subunit (Fig. 2) (Schirmer et al. 1985). There are various ways which apoproteins bind with bilins: C-PC contains only one bound PC, and R-PCII binds one PCB and two PEB, while R-PCIII binds two PCBs and one PEB (Sidler 1994). C-PC absorbs light in the wavelength of ~610–620 nm (Glazer 1977), while R-PC absorbs at 530–550 and 615 nm (Ong and Glazer 1987). Allophycocyanin has only one PC bilin attached to each Cys84 residue of ApcA and ApcB, and the different chromophore-protein interactions lead to ~30-nm red-shifted absorption maxima (640–670 nm) compared with that of PC

(Brejc et al. 1995). In general, the shorter-wavelength photon-absorbing subunits assembled in PBS surrounding structure and the longer-wavelength photon-absorbing subunits are located in the core of assembled PBSs (Stomp et al. 2007).

Phycoerythrin (PE) contains CpeA/B subunits and the bound PEB (C-PE) (Sonani et al. 2018) or the bound phycourobilin (PUB) and PE (R-PE) (Ritter et al. 1999; Nagy et al. 1985), which covers mainly absorption range of 520–570 nm. There are two PEB-binding sites for the α-subunit and three PEB to each β-subunit in C-PE of the cyanobacterium. PEC is an uncommon PBP present in heterocyst-containing cyanobacteria and is composed of α-subunits (PecA) and β-subunits (PecB) bound to one phycoviolobilin (PVB) at Cys84 and two PCB pigments at sites of Cys82 and Cys153, respectively (Schmidt et al. 2006), and it uses green light in the wavelength range of 570–595 nm (Bryant 1982). Organisms which have PEC do not have PE and vice versa (Bryant 1982).

Linker proteins play important roles for PBS assembly and energy transfers. There are different linkers that are named after their functions: the linkers within rod, L_R, encoded by *cpcC*s; the linker at the rod-core interface, L_{RC}, encoded by *cpcG*s; the linker core-membrane subunits, L_{CM}, encoded by *apcE*; and the rod-capping linker protein, which is encoded by *cpcD* (Six et al. 2007).

3.3 Chlorophylls

Chlorophylls are major photopigments in oxygenic photosynthesis and are structurally described as cyclic tetrapyrroles containing a distinctive five-membered ring (Fig. 3). They absorb light strongly in the blue and red light regions and reflect the green light which explains their inherent green color. There are three possible fates for the absorption of light: (1) emitted as heat or fluorescence through redistribution into atomic vibrations within the molecule (non-photochemical quenching), (2) transfer of light energy to nearby photopigments via resonance, and (3) photochemical reduction/oxidation and the electron transferred to a new molecule. The first process occurs under saturating light conditions and is required to quench the potentially damaging Chl excitation energy which may lead to free radicals. The second and third occur under optimal conditions and lead to photosynthetic light reactions.

There are five types of Chl characterized in oxygenic photosynthetic species, Chl *a*, Chl *b*, Chl *c*, Chl *d*, and Chl *f*, named in the order of discoveries (Fig. 3, Chen 2019). While Chls *a*, *b*, *d*, and *f*, including Chl a_2 and Chl b_2, have a phytyl chain esterified at position C17^3, Chl *c* has the free acid and double bond at C17–C18, which is similar to protochlorophyllide. There are several Chl *c* derivatives with variations in side chains, and Chl c_1 – c_3 are the most common species reported (Fig. 4) (Myśliwa-Kurdziel et al. 2019; Zapata et al. 2006).

The absorbance maxima of Chls depend on the distribution of electron density within the Chl structure, and alterations in the Chl peripheral structure lead to redistribution of the electron density. Chl *a* has Soret absorption maxima of ~433 nm and Q_y absorption at 665 nm in methanol that is taken as a reference point for all other Chl

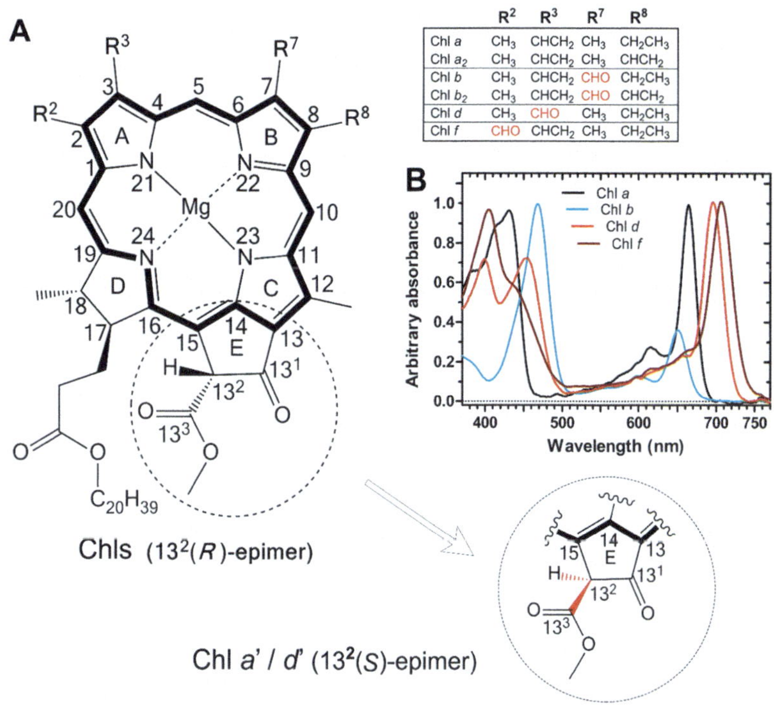

Fig. 3 Structure and spectral properties of chlorophylls present in cyanobacteria. (**a**) Chemical structure of chlorophylls (Chls) *a*, *b*, *d*, and *f* with highlighted 20 π electrons. The different chlorophylls have variations at position C2, C3, C7, and C8 that are summarized in the inserted table. The formyl group substitutions are highlighted in red. The dashed circles highlight the changes from *R*-configuration at position $C13^2$ to the epimer (*S*-stereoisomer) of Chl *a′*/Chl *d′*. (**b**) Absorption spectrum of purified Chl *a*, Chl *b*, Chl *d*, and Chl *f* dissolved in methanol. The absorbance maxima of each of the Chls mentioned here are summarized in Table 2

molecules. Chl *b* has the maxima of Soret and Q_y at 468 and 652 nm, while that of Chl *d* is at 456 and 697 nm and that of Chl *f* at 407 and 707 nm, respectively (Fig. 3b).

Chlorophyll molecules have either singular or dual roles in photosynthesis, namely, (1) accessory light harvesting and energy transfer and (2) photochemical charge separation in the RC (Table 1). Chlorophyll *a* and Chl *d* are involved in all aspects of photosynthesis, and Chl *b* and Chl *c* are strictly light-harvesting pigments, while Chl *f* appears to be light harvesting with an unusual long-lived fluorescent species facilitating uphill energy transfer (Zamzam et al. 2019; Kaucikas et al. 2017).

3.3.1 Chl *a* and Its Spectral Properties

The blue-green-colored Chl *a* molecule is the most ubiquitous pigment in oxygenic photosynthetic organisms, plants, algae, and cyanobacteria, with the exception being Chl a_2 in *Prochlorococcus marinus* that has no detectable Chl *a*

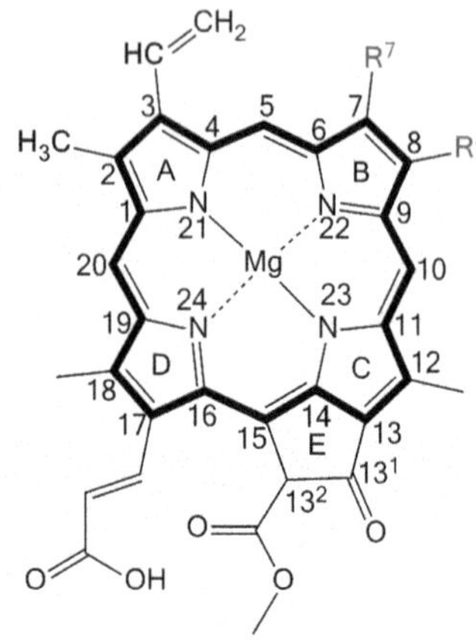

	R^7	R^8	λ_{max} (nm) Soret	Q_y	Solvent
Chl c_1	CH_3	CH_2CH_3	444-5	633-4	Methanol[a]
			444-5	628	Diethyl ether[a]
			442-4	630-1	Acetone[a]
Chl c_2	CH_3	$CHCH_2$	451-2	635	Methanol[a]
			448-9	628-9	Diethyl ether[a]
			444-5	630-1	Acetone[a]
Chl c_3	$COOCH_3$	$CHCH_2$	451	626	Diethyl ether[b]
			452	627	Acetone[b]

Fig. 4 Chemical structure of chlorophyll *c* family. Structure of the three main Chl *c*s found in nature: Chl c_1, Chl c_2, and Chl c_3. The structural differences presented in the inserted table. The respective absorbance maxima are cited from[a] (Jeffrey 1969)[b], (Jeffrey and Wright 1987)

Table 1 The role of chlorophyll molecules in light-dependent processes in photosynthetic organisms

Chlorophylls	Role in photosynthesis	Organisms
Chlorophyll *a*	LH/RC	All oxygenic photosynthetic organisms except *P. marinus* spp. and *A. marina* spp.
Chlorophyll a_2	LH/RC[a]	*P. marinus* spp.
Chlorophyll *b*	LH[b]	Plants, green algae, *Prochlorothrix* spp., *Prochloron* spp., *A. marina* sp. RCC1774
Chlorophyll b_2	LH (PCB)[a,c]	*P. marinus* spp.
Chlorophyll *c*	LH(FCP)[d]	Cryptophyte, dinophytes, haptophytes, some heterokonts, diatoms, Prasinophyceae
Chlorophyll *d*	LH/RC[e]	*A. marina* spp.
Chlorophyll *f*	LH/RC[f,g]	Some cyanobacteria

References: (a) Goericke and Repeta (1992), (b) Scheer (2006), (c) La Roche et al. (1996), (d) Zapata et al. (2000), (e) Loughlin et al. (2013), (f) Ho et al. (2016), and (g) Chen et al. (2019)

(Goericke and Repeta 1992). Chlorophyll *a* is involved in each aspect of light-dependent photosynthetic reactions including light-harvesting processes and photochemical reactions. Different solvents change the Chl spectral maxima to varying degrees (Table 2). The Chl in Chl-binding protein complexes demonstrates a red-shifted absorbance depending on the interaction and influences of surrounding protein structure and residues of amino acid. PsaA/PsaB and PsbA/PsbD are RC subunits and bound Chl *a* P_{700} of PSI and Chl *a* P_{680} of PSII II, respectively.

Table 2 Summary of optical properties of different chlorophyll (Chl) pigments dissolved in selected organic solvents, including their absorbance maxima and molar extinction coefficients (ε)

	λ_{max} (nm)			
	Soret	Q_y	Solvent	ε (L.mol^{-1}.cm^{-1}) × 10^3
Chl *a*	432	665	Methanol[ab]	$\varepsilon_{665.5nm}$ = 70.02[a]; $\varepsilon_{665.2nm}$ = 79.24[b]; $\varepsilon_{431.8nm}$ = 77.05[b]
	428	660	Diethyl ether[b]	
	430	662	Acetone[b]	
Chl a_2	442	668	Aq. methanol*[c]	
	435	660	Diethyl ether[d]	
	438	664	80% acetone[e]	
Chl *b*	469	652	Methanol[b]	$\varepsilon_{652.1nm}$ = 38.87[b]; $\varepsilon_{469.2nm}$ = 105.36[b]
	452	642	Diethyl ether[b]	
	456	645	Acetone[b]	
Chl b_2	478	658	Aq. methanol*[c]	
	460	644	Diethyl ether[d]	
	468	651	80% acetone[e]	
Chl *d*	456	697	Methanol[a]	ε_{697nm} = 63.68[a]; $\varepsilon_{455.5nm}$ = 44.41[a]; ε_{400nm} = 45.74[a]
	446	686	Diethyl ether[a]	
	447	688	Acetone[a]	
Chl *f*	407	707	Methanol[a]	ε_{707nm} = 71.11[a]; $\varepsilon_{406.5nm}$ = 66.92[a]
	396	695	Diethyl ether[a]	
	397	698	Acetone[a]	

References: (a) Li et al. (2012), (b) Lichtenthaler (1987), (c) Goericke and Repeta (1993), (d) Goericke and Repeta (1992), and (e) Shedbalkar and Rebeiz (1992)
*Aq. methanol refers to the HPLC elution buffer which was c.a. 94% methanol in c.a. 31 mM ammonium acetate

3.3.2 Formyl Substitution in Chl *b*, Chl *d*, and Chl *f*

Modifications at C7, C3, and C2 side chains of Chl *a* to formyl (CHO) groups, individually, lead to Chls *b*, *d*, and *f*, respectively (Fig. 3). Chlorophyll *b* is an accessory pigment of plants, green algae, and prochlorophytes (Castenholz 2001; Burger-Wiersma et al. 1986) and recently a Chl *d*-less species of *A. marina* sp. RCC1774 (Partensky et al. 2018). The formyl group substitution at position C7 of Chl *b* shifts the Soret band toward a longer wavelength (469 nm) and the Q_y peak toward the shorter wavelength (652 nm) in comparison with Chl *a*. A formyl group at position C7 in Chl *b* shifts electron densities within the macrocycle, resulting into an increased intensity of the Soret band simultaneously weakening the Q_y dipole moment, resulting in the Soret/Q_y ratio of 2.71 compared with 0.97 in Chl *a* (Hoober et al. 2007).

For Chl *d*, the formyl group substitution at position C3 in Chl *d* strengthens the Q_y dipole moment and results into a red-shifted Q_y band to 697 nm. Chl *f* retains the vinyl group at C3 and has a formyl group replacing the methyl group at C2 position; the Q_y peak of Chl *f* is even further red-shifted to a maximum of 707 nm (Chen et al. 2010). Both Chl *d* and Chl *f* are called longer-wavelength light-absorbing Chls or red-shifted Chls (Chen and Scheer 2013; Chen and Blankenship 2011). Each of these formyl derivatives of Chl *a* is biosynthesized from oxidation, utilizing molecular oxygen as a substrate (Garg et al. 2017; Schliep et al. 2010; Porra et al. 1994).

3.3.3 Diformyl Variants

Chlorophyll *a* oxygenase (CAO) converts Chl *a* and Chlide *a* to Chl *b* and Chlide *b* (Oster et al. 2000). Transformation of the Chl *d*-containing *A. marina* with CAO of *Prochlorothrix hollandica* resulted in the unexpected formation of 7-formyl-Chl *d* or 3,7-diformyl-Chl *a* with absorption maxima at 470 nm and 667 nm in acetone (Tsuchiya et al. 2012a, b). Surprisingly the total amount of Chl *a* in the Δ*CAO*-expressing *A. marina* mutant does not change, but Chl *d* levels in isolated PSII complexes decrease.

The 3,8-diformyl-Chl *a* has not been detected from any natural sources but synthesized in vitro using 3,8-divinyl-Chl *a* as a substrate. Its absorption with the maxima of 484 nm and 674 nm in methanol; hence the Soret and Q_y bands are shifted toward longer wavelengths like that of 3,7-diformyl-Chl *a* with the maxima of 483 and 7-diformyl-Chl *a* 4 and 674 nm in methanol (Loughlin et al. 2014).

3.3.4 Chl *c* Family

Chlorophyll *c* resembles a porphyrin-like structure with 22 π electrons compared with 20 π electrons in other Chl molecules due to double bond between C17 and C18. In Chls c_1, c_2, and c_3, no phytyl chain is present at $C17^3$; however, some Chl *c* pigments are esterified at $C17^3$ with monogalactosyldiacylglycerol (MGDG), which appears to be characteristic of the Prymnesiophyceae class (Zapata et al. 2006). For example, Chl c_2-MGDG was isolated from *Emiliania huxleyi*, which has a blue-shifted Q_y peak at 625–632 nm (Garrido et al. 2000). Conjugation of the entire porphyrin ring in Chl *c* results in weakening of the Q_y dipole strength, resulting in $\varepsilon = 150{,}000\ mol^{-1}.L.cm^{-1}$ at 625–632 nm and approximately 5–13-fold reduced intensity compared with $\varepsilon = 20{,}000\ mol^{-1}.L.cm^{-1}$ at Soret band of 445–454 nm (Scheer 2006). Several Chl *c* structures have been characterized, and up to 11 different types of Chl *c* pigments have been identified (Zapata et al. 2006).

Chl *c* is an accessory pigment present mainly in the algae, including diatoms and dinoflagellates, Prasinophyceae, Xanthophyceae, Phaeophyceae, cryptomonad, and chrysomonad (Cavalier-Smith 2018) (Scheer 2006; Jeffrey 1976; Jeffrey 1972; Jeffrey 1969). The best studied Chl *c*-containing LHC is from diatoms that have specific bound carotenoid, fucoxanthin, called fucoxanthin Chl *a*-/*c*-binding protein (FCP) (Büchel 2019).

In addition Chl *c*-like pigment Mg-divinyl phaeoporphyrin a_5 (MgDVP) or 3,8-divinyl protochlorophyllide is present in some cyanobacteria and functions as light-harvesting pigment (Larkum et al. 1994). There are trace amounts detected in *A. marina* (Schliep et al. 2007; Miyashita et al. 1997). The biosynthesis of Chl *c* most likely is via dehydrogenation of C17[1] of protochlorophyllide (Green 2011; Myśliwa-Kurdziel et al. 2019).

3.3.5 Other Chlorophyll Variants (Including Chemically Modified)

Chlorophyll a_2 and Chl b_2 with a double bond at C8 are present in some photosynthetic organisms including a *Zea mays* mutant ON 8147 (Bazzaz and Brereton 1982), *Prochlorococcus marinus* spp. (Chisholm et al. 1992; Chisholm et al. 1988), and a dinoflagellate (Rodríguez et al. 2016). It has 435 and 660 nm maxima for Chl a_2 and 460 and 644 nm maxima for Chl b_2, slightly red-shifted Soret band and blue-shifted Q_y band compared with that of Chl *a* and Chl *b*, respectively Table 2. Both *Prochlorococcus marinus* spp. and the dinoflagellate are low-light-adapted organisms and have lost the 8-vinyl reductase (DVR, EC 1.3.1.75) gene through evolution (Rodríguez et al. 2016).

The biosynthesis of Chl *b* can be described as two-step reactions, oxygenation and hydroxylation. The intermediate Chl molecule, 7-hydroxymethyl Chl *a*, is generated by oxygenation of the methyl group catalyzed by CAO (EC 1.14.13.122), and this intermediate is not detected in nature (Ito et al. 1996). Currently it is unknown if Chl *d* and Chl *f* biosyntheses proceed via intermediate Chl molecules, such as 3-hydroxymethyl Chl *a* and 2-hydroxymethyl Chl *a*, respectively (Fig. 5). Both putative intermediates have similar spectral feature with more polar retention times using reversed-phase high-performance liquid chromatography compared with that of Chl *a* (Sawicki et al. 2019).

An acidophilic aerobic anoxygenic photosynthetic bacterium consists of zinc bacteriochlorophyll *a* as the major light-harvesting and RC pigment together with bacteriopheophytin *a* (Tomi et al. 2007; Wakao et al. 1996). It appears that the low pH triggers loss of magnesium, which yields zinc-bacteriochlorophyll. No naturally occurring zinc-Chl molecule has been detected in oxygenic photosynthetic organisms.

Two iso-epimer forms of Chl *a* are present in vivo, $13^2(R)$ (Chl *a*) and $13^2(S)$ (Chl *a'*) (Fig. 3). $13^2(R)$ (Chl *a*) is the most common epimer, named as Chl *a*. $13^2(S)$-Chl *a* is the *S*-stereoisomer, named as Chl *a'*, which is found in almost all PSI. The special pair of Chl in PSI is made of Chl *a*/Chl *a'*, which is a feature of one of the two primary electron acceptors (P_A) bound to PsaA of the RC of PSI (Sect. 4.1) (Blankenship 2002). Epimerization of extracted Chl in organic solvent in vitro may occur, particularly, when there are accessible Lewis bases (Hynninen 1991). Epimerization of Chl *a* and *a'* has identical absorbance maxima of 433 nm and 665 nm in methanol, but with slightly changed nonpolar retention time (Loughlin et al. 2015).

Demetallized Chl is called pheophytin, and pheophytin *a* (Pheo *a*) is present as the primary electron acceptor (A_0) of PSII. However, no pheophytin *d* is detected in *A. marina*; even Chl *d* is the main Chl in PSI and PSII. In addition, no pheophytin *b* and *f* are reported to be involved in photochemical reactions (Miyashita et al. 2014).

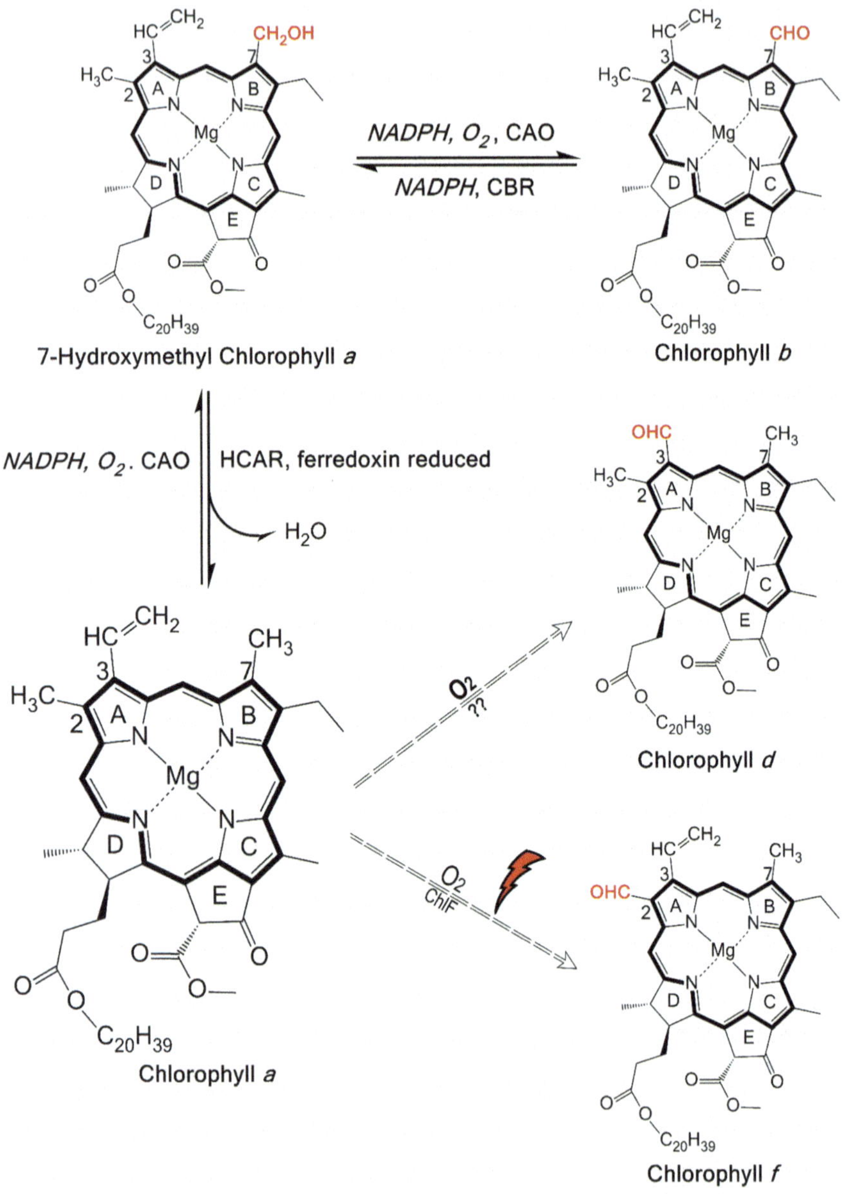

Fig. 5 Oxidation pathway of chlorophyll *a* to produce formyl-substituted chlorophyll. Chlorophyll *b* is synthesized by chlorophyll *a* oxygenase (CAO, EC 1.14.13.122) through the intermediate 7-hydroxymethyl Chl *a*. Oxygen (O_2) and NADPH are required. The Chl *b* cycle is catalyzed by chlorophyll *b* reductase (CBR, EC 1.1.1.294) and 7-hydroxymethyl chlorophyll *a* reductase (HCAR, EC 1.17.7.2). The oxygen atom of formyl group from Chl *b*, *d*, and *f* is derived from molecular oxygen. The question mark represents the uncertainty of Chl *d* synthase. ChlF is proposed the chlorophyll *f* synthase

Pheophytin *a* is also the important breakdown product of Chl degradation, and further breakdown proceeds to numerous breakdown colored or colorless products (Hörtensteiner et al. 2019; Kräutler 2016; Hörtensteiner and Kräutler 2011). Pheophytin *b* is not produced as a breakdown product because Chl *b* is firstly reduced by Chl *b* reductase (CBR, EC 1.1.1.294) (Scheumann et al. 1996) and 7-hydroxymethyl Chl *a* reductase (HCAR, EC 1.17.7.2) (Meguro et al. 2011) into Chl *a* prior to entering the degradation pathway (Christ and Hörtensteiner 2014).

Removal of the phytol chain from Chl is an important Chl degradation, which is catalyzed by chlorophyllase (EC 3.1.1.14) and produces chlorophyllide and phytol molecule. Although not normally prevalent in vivo, in *Synechocystis* PCC6803, N^{15} labeling studies revealed the increased chlorophyllase activity was coordinated with quick turnover of photodamaged PsbD2. Therefore, de novo synthesis of Chl and the recycling of chlorophyllide are avoided under stress conditions (Vavilin and Vermaas 2007).

4 Photopigment-Binding Protein Complexes

All carotenoids, bilins, and Chls are carefully arranged inside the pigment-binding protein complexes. The major Chl-binding protein complexes in oxygenic photosynthetic organisms are PSI and PSII which are associated with either thylakoid membrane-embedded Chl-binding LHCs or extrinsic PBSs. Photosystems are composed of an RC core, surrounding by the inner antenna and associated with either additional intrinsic membrane-bound Chl-binding antenna or extrinsic PBSs. In plants and algae, the intrinsic Chl-binding antenna is three-helical transmembrane Chl *a*-/*b*-binding light-harvesting complexes (LHCs), except for Chl *c*-containing alga, which have Chl *a*-/*c*-binding LHC. There are two main classes of LHC: LHCI and LHCII. LHCI is composed of four Lhca subunits (Lhca1–4) associated with PSI, while LHCII has three LHC subunits (Lhcb1–3) associated with PSII mainly. Multiple LHC subunits associated with PSI or PSII forming supracomplexes provide efficient energy transfer pathways (Neilson and Durnford 2010; Caffarri et al. 2014; Ballottari et al. 2012; Umena et al. 2011; Jordan et al. 2001). Some algae, such as red algae, cryptophytes, and glaucophytes, also have the extrinsic PBSs as the major antenna system, which is commonly arranged in a hemidiscoidal shape (MacColl 1998). There are three main components in assembled PBPs: PE, PC, and APC. In addition, an unusual external soluble antenna complex composed of peridinin-chlorophyll *a* protein (PCP) has been reported from dinophytes (Bautista et al. 1999). How PCP connects with membrane-bound PSs is largely unknown although the route of energy transfer was determined as PCP ➔ LHC ➔ RC (Ogata et al. 1994; Polívka et al. 2008; Bricker and Lo 2015).

Most *Cyanobacteria* have two antenna systems, PBS and Chl-binding antenna systems. The Chl-binding light-harvesting systems are six-helical transmembrane Chl-binding protein complexes, structurally different from the LHCs in eukaryotic photosynthetic organisms, including Pcb and IsiA (Chen et al. 2008; La Roche et al. 1996). Most Pcb and IsiA will bind available Chls to form the functional CBPs. In *Prochlorococcus marina* spp., CBP contains Chls a_2 and b_2; in *A. marina* spp., CBP contains Chl *d* mainly (Chen et al. 2008). Phycobilisomes are efficient at capturing the light in the range of 520–660 nm, complementary with the light absorbed by Chls. There are several nontypical PSB structures reported from various cyanobacteria. *Acaryochloris marina* sp. MIBC11017 has a simple rod structure of PBPs, which tightly associates with the photosynthetic membrane via CpcG linkers (Chen et al. 2009). *Gloeobacter violaceus*, a cyanobacterium lacking thylakoid membranes, has bundle-shaped PBSs attached to the cytoplasmic plasma membrane (Koyama et al. 2006; Guglielmi et al. 1981). According to recent omics studies, PBSs could be remodeled depending on the light conditions (Wiltbank and Kehoe 2019). Under FRL conditions, Chl *f*-producing cyanobacteria remodel the PBS to a red-shifted PBS in order to optimize light capture efficiently (Li et al. 2016; Gan et al. 2014). The red-shifted PBSs are mainly made of APC subunits in *H. hongdechloris* and form the smallest PBS reported up to date (Li et al. 2016). A different mechanism to the red-shifted absorption feature was proposed in *Synechococcus* sp. PCC7335 (Herrera-Salgado et al. 2018; Ho et al. 2017a). Changed chromophore interaction within the PBP subunits and random reorganization of PBP might result in the red-shifted absorption feature of PBS when *Synechococcus* sp. PCC7335 is grown under FRL conditions.

Prochlorophytes use CBP that have sequences homologous to the core antenna protein, CP43 and CP47 of PSII (Bibby et al. 2001b). These proteins form a large antenna-RC supracomplexes, including CBP-PSI and CBP-PSII (Chen et al. 2008; Bibby et al. 2001a, b). Under nutrient limitation or high-light stresses, PBSs rapidly degrade, while PSs are also negatively affected, particularly PSI (Grossman et al. 1998). Different copy numbers of CBP-encoded genes are reported from various *Prochlorococcus* spp. isolated from different niches. The copy numbers of CBP reflect well with the accessible light levels at the eco-location. Multiple copies of CBP-encoded genes are reported from low-light-adapted *Prochlorococcus* sp. SS120, thriving at ~120 m below sea level, in contrast to high-light-adapted *Prochlorococcus* sp. MED4 isolated from the surface of the ocean and have one copy of CBP (Garczarek et al. 2000).

4.1 Photosystem I

Photosystem I from cyanobacteria is typically a trimer and has extrinsic CBP complexes, if present (Jordan et al. 2001; Chen et al. 2008). Each core complex of PSI mainly consists of 12 subunits which includes the RC core comprising PsaA and PsaB subunits surrounded by small transmembrane proteins of PsaF, PsaI, PsaJ, PsaK, PsaL, PsaM, and PsaX and 3 stromal subunits of PsaC, PsaD, and PsaE (Jordan et al. 2001). PsaA and PsaB are approximately 750 amino acids in size, and each contains an RC domain and a core antenna domain at the N-terminal. The six

N-terminal transmembrane helices of PsaA and PsaB show homology to CP43 and CP47 of PSII (Sect. 4.2) (Jordan et al. 2001). The direct interaction between antenna and PSI is determined after isolation of antenna-PSI supracomplexes (Toporik et al. 2019; Ihalainen et al. 2005; Bibby et al. 2001b). Some cyanobacteria show interaction between PBSs and PSI resolved by state-of-the-art cryo-EM technology (Li et al. 2019; Zhang et al. 2017). In addition to P_{700}, PsaA and PsaB host an accessory Chl *a* molecule and an electron acceptor Chl *a* molecule, plus a phylloquinone molecule in the RC domain side. Charge separation occurs when a special pair Chl passes on an electron to the acceptor Chl *a* molecule producing $P_{700}^{+}A_{0A}^{-}$ or $P_{700}^{+}A_{0B}^{-}$ which occurs within ~3.7 ps (Shuvalov et al. 2006). The electron is then transferred to electron carrier of phylloquinone within ~30 ps (Hecks et al. 1994). In *A. marina*, Chl *d* replaced Chl *a* in RCI. The special pair of Chl *d/d′* has a unique absorption maxima of 740 nm and is named as P740 (Hu et al. 1998) (Sect. 5.2).

In Chl *f*-producing cyanobacterium, *H. hongdechloris*, Chl *f* production is induced under FRL conditions, and isolated PSI complexes contain about 8% of Chl *f* from a total of ~110 chlorophylls per isolated PSI from FRL-grown *H. hongdechloris* cells, while the isolated PSI complexes contain only Chl *a* from white light-cultured *H. hongdechloris* cells (Li et al. 2018a). Chl *a/a′* formed P_{700} in isolated PSI from *H. hongdechloris* grown under WL and FR light conditions. The uphill energy transfer was further confirmed by decay fluorescence measures (Schmitt et al. 2018) (Sect. 6.2).

4.2 *Photosystem II*

Photosystem II is a multiple protein subunit complex containing RC and intrinsic core antenna and typically arranges as a dimer (Eaton-Rye and Sobotka 2017). PsbA (D1) and PsbD (D2) are the core subunits in RCII and bind six Chls including the special pair of Chl *a* (P680). CP43 and CP47 are core antenna subunits and bind 13–16 Chls individually. The extra loop of CP43 protein with D1 subunit together forms a binding dock for oxygen evolution center including Mn_4CaO_5 complexes (Umena et al. 2011). There are also small single transmembrane proteins such as PsbE, PsbF, PsbH-N, and PsbX-Z containing a single helix, and PsbO, PsbU, and PsbV are external subunits on the periphery of PSII, which do not have transmembrane domains (Gao et al. 2018). Some small transmembrane proteins (PsbI and PsbM) are involved in dimerization and stability of the structure and function of dimeric PSII (Shi et al. 2012; Guskov et al. 2009). PsbU, PsbV, PsbP, and PsbQ are positioned around D1 and D2 proteins to shield the Mn_4CaO_5 cluster from reductants (Bricker et al. 2012).

The RCII uses Pheo *a* as the primary electron acceptors (A_0) and passes the electrons to the secondary acceptors, A_1 and plastoquinone (Q_A/Q_B). Electrons from Pheo *a* are readily passed onto Q_A by non-heme iron Fe^{2+}. Meanwhile oxidized primary donor Chl is reduced by accepting electrons from the oxidation of water through a redox-active Tyr residue of D1, known as Yz, leading to the formation of

$P_{680}Q_AQ_B^-$ which is a stable charge separation. This process is repeated to produce $P_{680}Q_A^-Q_B^-$ which is stabilized by the uptake of H^+ ions from the stroma to form $P_{680}Q_AQ_BH_2$. Plastoquinol (Q_BH_2) readily diffuses from the Q_B-binding site of PsbD and binds with cytochrome b_6f complex. Cytochrome b_6f complex moves two protons to the lumen, and electrons, which are transferred to the electron carrier of plastocyanin, eventually transfer to PSI.

In Chl *d*-containing *A. marina*, Chl *d* represents more than 95% of total Chls. There are several mechanisms proposed for PSII in *A. marina* due to limited purified PSII complexes. Chen et al. (2005d) reported that Chl *d* is the special pair of Chl in isolated enriched PSII fraction with the proposed absorption maxima of P_{715} nm. Tomo et al. (2007) used FTIR technology to report Chl *d* in isolated PSII with the absorption maximum of 713 nm. In vivo measurements showed a maximum at 724 nm suggesting the primary electron donor is Chl *d* (Mielke et al. 2013) (Sect. 5.3).

4.3 Chlorophyll-Binding Light-Harvesting Protein Complexes (CBPs)

Cyanobacteria use intrinsic membrane-bound CBPs, including prochlorophyte Chl *a-/b*-binding protein (Pcb) and iron-stress induced chlorophyll-binding protein (IsiA) (Chen et al. 2008). The CBPs are six-helical transmembrane Chl-binding proteins, having homologous sequence with CP43 and CP47 of PSII but lacking the extra loop between helix 5 and helix 6 (Chen and Bibby 2005). The CBP proteins are unrelated to LHC proteins from eukaryotic photosynthetic organisms, instead having similarity to CBP with a six-transmembrane helical domain-binding 13–16 Chls (van der Staay and Staehelin 1994). The CBP proteins bind available Chls and form multiple subunit complexes associated with PSs. Under iron-stressed, light-stressed, or oxidatively-stressed conditions, IsiA proteins are upregulated and form 18-mer rings surrounding PSI trimer (Bibby et al. 2001b). Prochlorophytes use Pcb protein forming multiple subunit complexes surrounding PSI and PSII (Bibby et al. 2001b, 2003b).

In addition, there are additional families of CBP found in cyanobacteria, which have one or two helices of transmembrane domain with similarity to the membrane-spanning region of plant chlorophyll *a-/b*-binding proteins (LHCs) (Dolganov et al. 1995). For example, high-light-induced protein (HLIP), found in *Synechocystis* sp. PCC6803, binds Chls and carotenes and associates with PSII to protect D1 protein from degradation under high-light stress conditions (Knoppová et al. 2014).

4.3.1 Inner Antenna Complexes

Constitutively expressed CP43 and CP47 proteins encoded by *psbC* and *psbB* represent the inner antenna domains of PSII, an indispensable subunit of PSII complexes (Eaton-Rye and Putnam-Evans 2005). They function in the same manner as that from PSII system in plants. The N-terminal domain of PsaA and PsaB protein serves as the inner core antenna and has a homologous sequence as CP43 and CP47 (Jordan et al. 2001). The inner core antenna binds 15–30 Chls and directly transfers the captured energy to the core of RC.

4.3.2 Chl-Binding Proteins in Cyanobacteria

In prochlorophytes, several types of Pcbs encoded by different genes, *pcbA–pcbH*, are reported and co-expressed under different culture conditions, including low-light stressed conditions, high-light-stressed conditions, and iron-stressed conditions (Garczarek et al. 2000; La Roche et al. 1996). In *Prochloron didemni*, isolated Pcb-PSII supracomplexes revealed 10Pcb:2PSII complex (Bibby et al. 2003b). *Prochlorococcus* sp. MED4 has a single copy of Pcb-encoded gene, which is constitutively expressed and formed antenna-PS supracomplexes (Bibby et al. 2003a). Since IsiA proteins are prevalent in many cyanobacteria and have high similarity with Pcbs from prochlorophytes, a new name for Chl-binding proteins in cyanobacteria (CBPs) was proposed (Chen et al. 2008). *Acaryochloris marina* uses Chl *d*-binding CBP as its major light-harvesting systems (Chen et al. 2002, 2005c; Chen and Bibby 2005), which will be discussed in details in Sect. 5.1.

4.3.3 Iron-Stress-Induced Chlorophyll-binding Protein A (IsiA)

IsiA protein was firstly reported as Chl-storing sites under iron stressed condition (Burnap et al. 1993). The light-harvesting function of IsiA was confirmed after the structure of IsiA-PSI supracomplexes was solved (Bibby et al. 2001a, b; Boekema et al. 2001). IsiA binds 16–17 Chls and provides additional energy transfer network for PSs (Bibby et al. 2001a; Chen and Bibby 2005; Chen et al. 2005a; Andrizhiyevskaya et al. 2002). The crystal structure of PSI-IsiA revealed an increased surface area for light harvesting of PSI and concentration of Chl *a* molecules to the stromal side of the membrane suggesting an interaction with the remaining PBSs (Toporik et al. 2019).

4.4 Phycobilisomes (PBSs)

Phycobilisomes are the main light-harvesting structure for most cyanobacteria, red algae, and glaucophytes (Overmann and Garcia-Pichel 2013). They are composed of PBPs and linker proteins extending the light-absorbing surface area and particularly the increased light-capture cross section, especially the green-orange light (520–670 nm). Phycobiliproteins consist of apoproteins covalently bound to bilins, mainly at the A ring of bilins through thioether (C-S-C) bonds formed by cysteine residues of the apoprotein and ethylidene of the bilin (Fig. 2) (MacColl 1998).

These bonds are formed and catalyzed by lyase enzymes (Scheer and Zhao 2008). Meanwhile the linker proteins usually do not bind chromophores and function for integrating PBP subunits to PBS. This may involve rod extension (L_R/CpcCs), distal termination of rods (L_{RT}/CpcD), rod and core association (L_{RC}/CpcG), assembly of the APC core (L_c) and attachment of the core membrane linker APC with thylakoid membranes (L_{CM}/ApcE) (Guan et al. 2007). Despite usually not binding bilins, linker proteins influence the absorption maximum of a given PBP through different organizations of PBS (Glazer 1985).

The four different types of bilins in cyanobacteria are PCB (blue), PVB (purple), PEB (pink), and PUB (yellow), in which PCB and PEB are the common chromophores in cyanobacteria. There may be either a single or more bilins bound with various apoproteins, resulting in the different optical properties of APC, PC, PE, and PEC (Fig. 2). The basic structure of PBS is comprised of a αβ heterodimeric, which forms a trimer of heterodimeric disc structure and then interacts with linker proteins to form a cylindrical structure. APC has the lowest light energy (650–660 nm) followed by PC (615–640 nm) and PE (520–575 nm) (Overmann and Garcia-Pichel 2013), so this arrangement facilitates higher light energy captured by PE or PC efficiently to transfer to the lower light energy traps of APC cores (Grossman et al. 1993).

The typical PBS structure is hemidiscoidal that may have bicylindrical, tricylindrical, or pentacylindrical APC core along the thylakoid membranes (Fig. 6a). The two bottom APC core cylinders are bound to the thylakoid, composed of four different APC/linker trimer-like protein complexes, while the top cylinder is composed of four trimer/trimer-like complexes consisting of two different complexes. Each complex is composed of variable APC proteins with or without linker proteins (MacColl 1998). A tricylindrical core structure hosts six PC or PC + PE rods depending on the light conditions. Low light requires maximum rod structures for increased light absorption surface area, while high light has a more compact rod structure; even APC core without the rod structure attached was observed from FRL-adapted *H. hongdechloris* (Fig. 6c) (Li et al. 2016).

The energy transfer funnel was built through the structure of PBS. The APC core has the lowest energy of 650–660 nm, followed by PC absorbing 615–640 nm light, and then the PE using light of 520–575 nm (Overmann and Garcia-Pichel 2013). Phycobilisomes transfer the energy to Chls in PSs, mainly associated with PSII (Chang et al. 2015). A PBS-PSI-PSII supracomplex was recently isolated by pull-down experiments combining with aids of cross-linking agent and proteomic analysis (Liu et al. 2013). An atypical PBS structure lacking APC cores was isolated from *Anabaena* and reported to interact with tetrameric PSI (Watanabe et al. 2014).

5 *Acaryochloris marina*

Acaryochloris marina is the only cyanobacterial genus that uses Chl *d* as the major pigment with the trace amount of Chl *a* of ~1–5% (Loughlin et al. 2013). Chlorophyll *d* was first reported by Manning and Strain (1943) through the identification of a

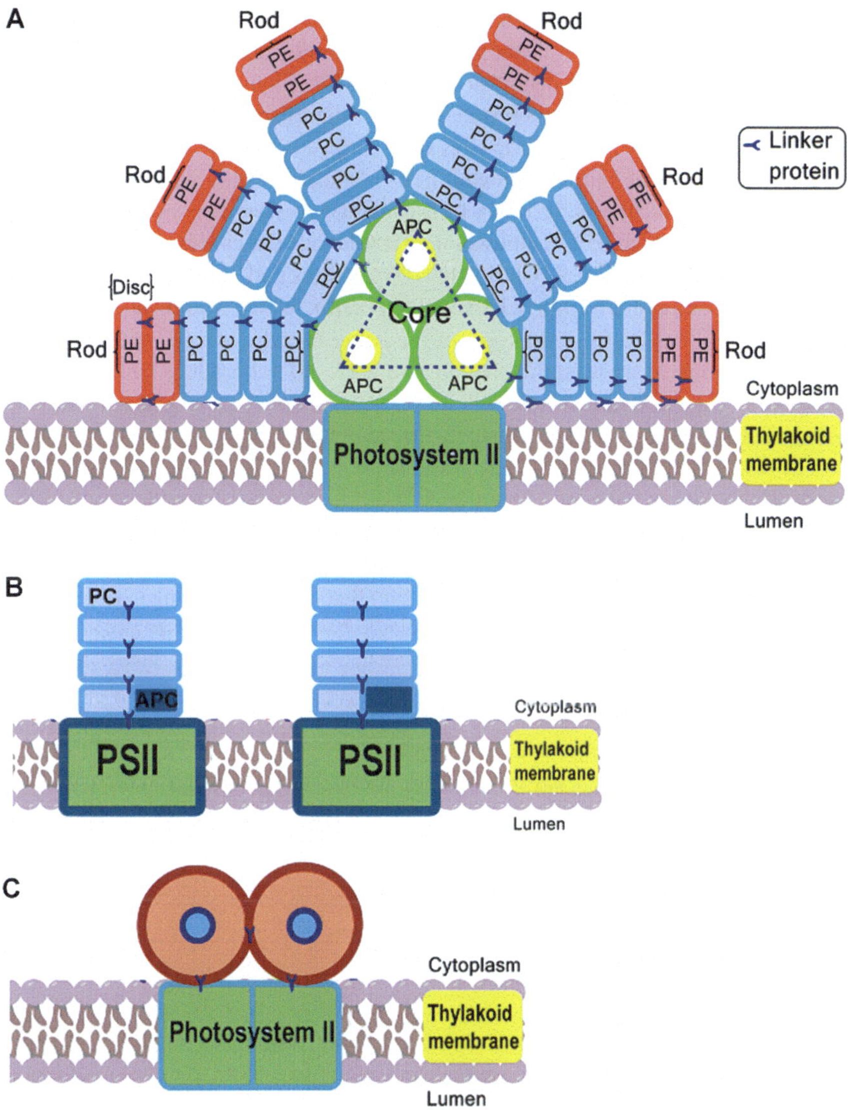

Fig. 6 Structural model of phycobilisomes. (**a**) The typical hemidiscoidal phycobilisome model. Allophycocyanin (APC) builds the core of phycobilisomes. Phycocyanin (PC) and phycoerythrin (PE) assemble the peripheral rods. The structure is held together by various types of linker proteins and interacts with photosystem. (**b**) Atypical phycobiliprotein complexes from *A. marina* sp. MBIC 11017. APC is a minor component at the base of the rod structure. The rod structure directly delivers the energy to the photosystem II. (**c**) The red-shifted APC cores from *H. hongdechloris*, a Chl *f*-producing cyanobacterium. The schematic models were partially drawn with templates from ChemDraw Professional 16.0

new type of Chl that Q_y absorption is at 696 nm; 32 nm shifted to the far-red region. Chlorophyll *d* was named after the order of discoveries, and it was also named as 3-desvinyl-3-formyl-Chl *a* using the revised IUPAC numbering system (Moss 1987) of tetrapyrroles which was updated from the Fischer numbering system of 2-desvinyl-2-formyl-Chl *a* used by Holt and Morley (1959).

Chlorophyll *d* was initially isolated from a sample of the red alga, *Gigartina agardhii* collected from the beach of California, USA (Manning and Strain 1943). Much later, the source of Chl *d* was solely attributed to a cyanobacterium named *Acaryochloris marina* sp. MBIC11017 (Miyashita et al. 1996). *Acaryochloris marina* was discovered from *Lissoclinum patella*, a sea squirt of the family Didemnidae, collected from the coast of the Palau Islands (Miyashita et al. 1996). The same laboratory showed that the red alga itself did not produce Chl *d* and confirmed that Chl *d* was actually extracted from *A. marina* spp. living as endoliths on red algae samples, not from red algae (Murakami et al. 2004).

The widespread nature of *A. marina* is supported by discoveries of *A. marina* spp. from a variety of habitats (Zhang et al. 2019; Li et al. 2013; Larkum et al. 2012; Behrendt et al. 2011; Fleming and Prufert-Bebout 2010; Goh et al. 2008; De Los Ríos et al. 2007; McNamara et al. 2006; Miller et al. 2005). Indeed, *A. marina* grows axenically in the laboratories, suggesting that Chl *d*-containing cyanobacteria occupy both pelagic (open-sea) and hemipelagic habitats from Arctic and Antarctic sea to lake samples and from various sediment samples of bays, lakes, and seas (Kashiyama et al. 2008).

Chlorophyll *d* was also reported with extremely trace amount (~ <0.2%) in some Chl *f*-producing cyanobacteria, such as *Chlorogloeopsis fritschii* (Airs et al. 2014). Since the level is quite low, and considering the relative ease of conversion of vinyl group in Chl *a* to formyl group in Chl *d* in vitro (Loughlin et al. 2014; Oba et al. 2011), the observed Chl *d* could be the results of experimental artifacts.

5.1 Light-Harvesting Systems

There are two antenna systems in *A. marina*: the intrinsic Chl *d*-binding CBP system and extrinsic rod-shaped phycobiliprotein complexes. Both of them are reported to associate with PSI and PSII.

5.1.1 Chl *d*-Binding Light-Harvesting Proteins

Acaryochloris marina uses Chl *d*-binding CBP as the major light-harvesting complex (Chen and Bibby 2005; Chen et al. 2002, 2005b, 2008). There are 11 genes encoding for CBP apoprotein reported from the genome of *A. marina* sp. MBIC11017. According to phylogenetical comparison, three different subgroups of CBP were proposed. An antenna-PSII supracomplex consisting of CBPII proteins was observed in a ratio of eight CBPs per PSII dimer, and associated CBPs with

PSII will increase the surface area of light harvesting to ~300% (Chen et al. 2005b). One of the CBP proteins of *A. marina* (also named PcbC) preferentially associates with PSI (Chen et al. 2005a). Recently, the additional IsiA (CBPIII family) encoded by AM_6003 was reported to upregulate significantly under low iron conditions (Li et al. 2018b)

Three CBP-encoded genes were reported prior to whole genome sequence of *A. marina* MBIC 11017, named as *pcbA*, *pcbB*, and *pcbC* genes, which later changed names as CBPs. The RNAseq comparison revealed that 6 out 11 CBP-encoded genes are upregulated under FRL conditions, while another 2 CBP-encoded genes were upregulated under white light conditions (Hernández-Prieto et al. 2018).

5.1.2 Phycobiliproteins

Acaryochloris marina MBIC11017 has PBPs but does not form a typical phycobilisome structure. Instead a simple rod-shaped structure was composed of predominantly PC with minor components of APC (Marquardt et al. 1997). The rod is composed of four αβ hexameric discs which interact with PSII (Hu et al. 1999). The electron microscope images of *A. marina* cells showed clearly the interaction between rod-structured PBPs and thylakoid membranes. According to the structure model, the rod-shaped PBPs were mainly associated with PSII (Chen et al. 2009), which demonstrated faster energy transfer speed compared with typical PBS structure (Petrasek et al. 2005). Recent structural studies showed seven phycocyanins and one allophycocyanin per rod (Golub et al. 2017). However, the presence of phycobiliproteins is inconsistent within *A. marina* spp. Both *A. marina* sp. CCMEE 5410 and *A. marina* sp. HICR111A strains lack PBP completely (Loughlin et al. 2013). *Acaryochloris marina* sp. HICR111A is the only Chl *d*-containing strain that has the capability to fix dinitrogen (N_2) (Pfreundt et al. 2012).

5.2 *Photosystems*

The genomic sequence information suggests that similar PSI and PSII apparatus are used in *A. marina* because of the homologous sequences of PSI and PSII. The differences of PSs in *A. marina* should mainly reflect the Chl replacements. Chlorophyll *d* is present in both photosystems I and II (Chen et al. 2005d; Hu et al. 1998), with ~140 Chl *d* to 1 Chl *a* in isolated PSI (Hu et al. 1998). The special pair of Chl *d* in PSI has an absorbance maximum of 740 nm that is attributed to Chl *d*/Chl *d′* molecules (Akiyama et al. 2001; Hu et al. 1998). Conflicting conclusions exist as to the nature of the special pair of PSII with an initial report stating Chl *a* as the primary electron donor (Mimuro et al. 2004). Based on the stoichiometry of purified PSII, spectral analysis, and photoreduction of Pheo *a*, it was concluded that the reaction center of *A. marina* consists of Chl *d* as the primary electron donor with a maxima of 713–715 nm and Pheo *a* as the electron acceptor (Chen et al.

2005d; Tomo et al. 2007). Alternatively a Chl *a* + Chl *d* special pair in PSII was proposed at cell levels (Renger and Schlodder 2008; Schlodder et al. 2007). The energetics of photosynthesis driven by Chl *d* is surprisingly comparable with an estimated >0.02 V difference (Tomo et al. 2007; Allakhverdiev et al. 2010); hence the efficiency of photosynthesis is comparable to Chl *a*-driven reactions.

5.3 *Biochemistry of Chlorophyll* d

The aerobic oxidative reduction of a vinyl group into a formyl group is a reaction that may occur with several different types of chemical catalysts, such as organic compounds with a reactive thiol group in the presence of oxygen (Loughlin et al. 2014; Oba et al. 2011) or cerium(IV) ammonium nitrate (Yu et al. 2015). Chlorophyll *a* is readily converted to Chl *d* in aqueous solution if heme, oxygen, and thiols are present (Loughlin et al. 2014).

The chemical reaction converting Chl *a* to Chl *d* involves converting an alkene (C3-vinyl) to a formyl (C3-CHO) group. Two different hypotheses exist as to the identity of the Chl *d* synthase gene. The original idea proposes an oxygenase-type enzyme, with the Rieske 2Fe-2S domain (AM1_5665) (Schliep et al. 2010; Swingley et al. 2008). The $^{18}O_2$ labeling experiments confirmed that Chl *a* and O_2 molecules are precursors of Chl *d* biosynthesis (Schliep et al. 2010). Partensky et al. (2018) isolated a Chl *d*-less *A. marina* strain RCC1774 and deduced that the oxygenase presented by Swingley et al. (2008) has a homolog in both Chl *a*-containing and Chl *d*-dominant *A. marina* species with 74% identity from 99% sequence coverage. Hence two new candidates for Chl *d* synthase were proposed which have no homolog in the Chl *a*-containing strain of *Acaryochloris*: radical SAM-type enzymes (AM1_5023 and/or AM1_5798) (Partensky et al. 2018). The potential of a radical SAM mechanism is evident through recent experiments showing methyl formylation in bacteriochlorophyll *e* biosynthesis by BciD (Thweatt et al. 2017; Harada et al. 2013). However, radical SAM enzymes typically use water as a substrate in anaerobic organisms (Broderick et al. 2014), and since oxygen is the source of the formyl group in chlorophyll *d* (Schliep et al. 2010), it seems unlikely that a radical SAM enzyme is solely responsible.

Some preliminary evidence exists for Chl *d* synthesis from Chl *a*; however, to date the yields are low. The first enzymatic formation of Chl *d* was fortuitously observed following a failed attempt to hydrolyze the phytyl chain of chlorophyll using esterases (Koizumi et al. 2005). When the cysteine protease papain was employed for this same reaction, a small amount of Chl *d* was generated together with chlorophyllide *a*; however serine proteases chymotrypsin and subtilisin showed no activity. Hence there is no clear candidate gene that may fulfill the role of the Chl *d* synthase although several clues exist. Significant problems arise in detecting Chl *d* synthesis due to the spontaneous oxidation of Chl *a* in oxygen by non-specific enzymatic catalysts; hence appropriate controls must be in place.

6 Chl *f*-Producing Cyanobacteria

Chlorophyll *f* production is induced by far-red light-enriched conditions. A far-red light acclimation (FaRLiP) cluster of approximately 20 genes was identified and conserved among all known chlorophyll *f*-producing cyanobacteria, which supported the phenomenon of "Chl *f* is far-red light induced chlorophyll" (Gan et al. 2014). A PsbA paralog was proposed to be directly involved in Chl *f* production and renamed as chlorophyll *f* synthase (ChlF) (Ho et al. 2016).

6.1 Occurrence of Chl f*-Producing Cyanobacteria*

Chlorophyll *f* was firstly isolated from a filamentous, marine cyanobacterium entitled *Halomicronema hongdechloris* (Chen et al. 2010). This pigment is induced under FRL conditions, and the molar ratio of Chl *f*/Chl *a* can reach to maximum of 1:8 under optimized FRL culture conditions (Chen et al. 2012). The ratio of Chl *f*/ Chl *a* may span approximately 1:8–1:35 depending on the species, genus, and culture conditions (Zhang et al. 2019). There are more than 20 different cyanobacterial species reported to have the capability of Chl *f* production under FRL conditions, including 8 new isolates from the sub-forest area in Wuhan, China, and a freshwater Chl *f*-producing cyanobacterium KC1 in Japan (Zhang et al. 2019; Gan et al. 2015; Akutsu et al. 2011).

Since FRL-containing species the production of Chl *f*-producing species have been shown to undergo reorganization of protein components of PSI, PSII and the unique red-shifted PBP components which is enabled by the upregulation of FARLiP clusters (Li et al. 2016, 2018a; Gan et al. 2014). There are a set of paralogous photosystem genes in the FaRLiP cluster, which include the core genes encoding for PSI, PSII, and PBS core (Gan et al. 2015; Chen et al. 2019). It appears the upregulated ChlF synthase is at least partly incompatible with the white light-produced photosynthetic apparatus. Newly synthesized PBSs absorb into the far-red region (Li et al. 2016), which facilitates energy transfer to Chl *a* and Chl *f* in PSs (Schmitt et al. 2018). Chlorophyll *f*-producing cyanobacteria are classified in all five cyanobacterial subsections, including isolates from freshwater and marine environments (Zhang et al. 2019; Akutsu et al. 2011).

6.2 Chl f *and Photosynthetic Reactions*

Chlorophyll *f* is present in PSI and PSII, and rapid energy transfer at a unique wavelength of 748 nm (Chl *f*) from 685 nm (Chl *a*) was observed using time-resolved fluorescence spectroscopy compared with a slower rate under white light

(Tomo et al. 2014). Based on the energy gap required to drive FRL photosynthetic reactions, it is postulated that chlorophyll *f* is located in the A_{-1} position that transfers an electron to the Chl *a*-containing RC of PSI (Zamzam et al. 2019).

6.3 Biochemistry of Chl f

Unlike Chl *d* chemical synthesis, in vitro synthesis of Chl *f* has not been demonstrated chemically. The reaction involves the conversion of a methyl group into a formyl group, which might parallel the chemistry of Chl *b* biosynthesis from Chl *a*. The biosynthesis of Chl *f* occurs through a different mechanism than CAO and was proposed to use a photooxidoreductase ChlF (Ho et al. 2016). PsbA4 is a kind of "super-rogue" D1 paralog with impaired Mn_4CaO_5-binding sites, critical for the water-splitting reaction in PSII (Cardona et al. 2015). However, mutation of the conserved tyrosine residue of ChlF synthase did not cease Chl *f* production, hence the tyrosine residue is not the key redox site contributing to the formation of Chl *f* (Shen et al. 2019).

The most likely substrate of Chl *d* and Chl *f* biosynthesis is Chl *a*, since Chl *a* is rapidly turned over using ^{18}O labeling studies in *A. marina* (Schliep et al. 2010). The same experimental design, ^{18}O labeling of the C2-formyl group of Chl *f*, revealed that the oxygen atom in C2-formyl group in Chl *f* is derived from oxygen molecules and Chl *a* is the precursor of Chl *f* biosynthesis (Garg et al. 2017). Overall the reaction chemistry of Chl *f* synthase remains to be elucidated.

7 Applications of Red-Shifted Chlorophylls

Higher plants are limited by photosynthetic processes driven by Chl *a* and Chl *b* due to the narrow window of PAR. Some cyanobacteria have evolved to produce red-shifted Chls, Chl *d* and Chl *f*, with absorbing and fluorescence maxima in the range of 700–750 nm. Genetically engineering higher plants for utilizing FRL spectral features will potentially increase photosynthetic light-capture efficiency during periods of low light (Wolf and Blankenship 2019). Initial experiments including ChlF synthase in plant biochemistry may prove challenging for the production of appreciable amounts of Chl *f* since (1) optimal Chl *f* production appears to be species-specific even among cyanobacteria, (2) light-harvesting apparatus in FRL-producing organisms (cyanobacteria) is different from eukaryotic LHCI and II, and (3) synthesis of unique paralogous Chl-binding proteins is required for Chl *f* production in cyanobacteria. It is unknown if a single gene is required for Chl *d* synthesis. However, the light-harvesting apparatus in *A. marina* is different from the conserved LHC complexes of plants. Hence considerable genetic engineering involving utilizing cyanobacterial photosynthetic protein complexes may be required for the assembly of a functional Chl *d*- or Chl *f*-containing PS in plants.

Acknowledgments AS and MC gratefully acknowledge the financial support of the Australian Research Council Centre of Excellence for Translational Photosynthesis (CE140100015).

References

Airs, R. L., Temperton, B., Sambles, C., Farnham, G., Skill, S. C., & Llewellyn, C. A. (2014). Chlorophyll *f* and chlorophyll *d* are produced in the cyanobacterium *Chlorogloeopsis fritschii* when cultured under natural light and near-infrared radiation. *FEBS Letters, 588*, 3770–3777.

Akiyama, M., Miyashita, H., Kise, H., Watanabe, T., Miyachi, S., & Kobayashi, M. (2001). Detection of chlorophyll *d′* and pheophytin *a* in a chlorophyll *d*-dominating oxygenic photosynthetic prokaryote *Acaryochloris marina*. *Analytical Sciences, 17*, 205–208.

Akutsu, S., Fujinuma, D., Furukawa, H., Watanabe, T., Ohnishi-Kameyama, M., Ono, H., Ohkubo, S., Miyashita, H., & Kobayashi, M. (2011). Pigment analysis of a chlorophyll *f*-containing cyanobacterium strain KC1 isolated from Lake Biwa. *Photomedicine and Photobiology, 33*, 35–40.

Allakhverdiev, S. I., Tomo, T., Shimada, Y., Kindo, H., Nagao, R., Klimov, V. V., & Mimuro, M. (2010). Redox potential of pheophytin *a* in photosystem II of two cyanobacteria having the different special pair chlorophylls. *Proceedings of the National Academy of Sciences, USA, 107*, 3924–3929.

Allen, J. F. (2014). Origin of oxygenic photosynthesis from anoxygenic type I and type II reaction centers. In J. Golbeck & A. Van Der Est (Eds.), *The biophysics of photosynthesis*. New York: Springer.

Allen, J. P., & Williams, J. C. (2011). The evolutionary pathway from anoxygenic to oxygenic photosynthesis examined by comparison of the properties of photosystem II and bacterial reaction centers. *Photosynthesis Research, 107*, 59–69.

Andrizhiyevskaya, E. G., Schwabe, T. M. E., Germano, M., D'haene, S., Kruip, J., Van Grondelle, R., & Dekker, J. P. (2002). Spectroscopic properties of PSI–IsiA supercomplexes from the cyanobacterium *Synechococcus* PCC 7942. *Biochimica et Biophysica Acta (BBA) – Bioenergetics, 1556*, 265–272.

Archibald, J. O. H. N. M. (2015). Endosymbiosis and eukaryotic cell evolution. *Current Biology, 25*, R911–R921.

Arnon, D. I. (1984). The discovery of photosynthetic phosphorylation. *Trends in Biochemical Sciences, 9*, 258–262.

Ballottari, M., Girardon, J., Dall'osto, L., & Bassi, R. (2012). Evolution and functional properties of photosystem II light harvesting complexes in eukaryotes. *Biochimica et Biophysica Acta (BBA) – Bioenergetics, 1817*, 143–157.

Bautista, J. A., Hiller, R. G., Sharples, F. P., Gosztola, D., Wasielewski, M., & Frank, H. A. (1999). Singlet and triplet energy transfer in the peridinin-chlorophyll *a* protein from *Amphidinium carterae*. *The Journal of Physical Chemistry. A, 103*, 2267–2273.

Bazzaz, M. B., & Brereton, R. G. (1982). 4-Vinyl-4-desethyl chlorophyll *a*: A new naturally occurring chlorophyll. *FEBS Letters, 138*, 104–108.

Behrendt, L., Larkum, A. W. D., Norman, A., Qvortrup, K., Chen, M., Ralph, P., Sørensen, S. J., Trampe, E., & Kühl, M. (2011). Endolithic chlorophyll *d*-containing phototrophs. *The ISME Journal, 5*, 1072–1076.

Bibby, T. S., Nield, J., & Barber, J. (2001a). Iron deficiency induces the formation of an antenna ring around trimeric photosystem I in cyanobacteria. *Nature, 412*, 743–745.

Bibby, T. S., Nield, J., Partensky, F., & Barber, J. (2001b). Antenna ring around photosystem I. *Nature, 413*, 590–590.

Bibby, T. S., Mary, I., Nield, J., Partensky, F., & Barber, J. (2003a). Low-light-adapted *Prochlorococcus* species possess specific antennae for each photosystem. *Nature, 424*, 1051–1054.

Bibby, T. S., Nield, J., Chen, M., Larkum, A. W. D., & Barber, J. (2003b). Structure of a photosystem II supercomplex isolated from *Prochloron didemni* retaining its chlorophyll *a*/*b* light-harvesting system. *Proceedings of the National Academy of Sciences, USA, 100*, 9050–9054.

Blankenship, R. E. (2002). *Molecular mechanisms of photosynthesis*. Oxford: Blackwell Science.

Boekema, E. J., Hifney, A., Yakushevska, A. E., Piotrowski, M., Keegstra, W., Berry, S., Michel, K.-P., Pistorius, & E. K., Kruip, J. (2001). A giant chlorophyll–protein complex induced by iron deficiency in cyanobacteria. *Nature, 412,* 745–748.

Brejc, K., Ficner, R., Huber, R., & Steinbacher, S. (1995). Isolation, crystallization, crystal structure analysis and refinement of allophycocyanin from the cyanobacterium *Spirulina platensis* at 2.3 Å resolution. *Journal of Molecular Biology, 249*, 424–440.

Bricker, W. P., & Lo, C. S. (2015). Efficient pathways of excitation energy transfer from delocalized S2 excitons in the peridinin–chlorophyll *a*–protein complex. *The Journal of Physical Chemistry. B, 119*, 5755–5764.

Bricker, T. M., Roose, J. L., Fagerlund, R. D., Frankel, L. K., & Eaton-Rye, J. J. (2012). The extrinsic proteins of photosystem II. *Biochimica et Biophysica Acta (BBA) – Bioenergetics, 1817*, 121–142.

Broderick, J. B., Duffus, B. R., Duschene, K. S., & Shepard, E. M. (2014). Radical *S*-adenosylmethionine enzymes. *Chemical Reviews, 114*, 4229–4317.

Bryant, D. A. (1982). Phycoerythrocyanin and phycoerythrin: Properties and occurrence in cyanobacteria. *Microbiology, 128*, 835–844.

Bryant, D. A., & Frigaard, N.-U. (2006). Prokaryotic photosynthesis and phototrophy illuminated. *Trends in Microbiology, 14*, 488–496.

Büchel, C. (2019). Light harvesting complexes in chlorophyll *c*-containing algae. *Biochimica et Biophysica Acta (BBA) – Bioenergetics*, In Press. https://doi.org/10.1016/j.bbabio.2019.05.003.

Burger-Wiersma, T., Veenhuis, M., Korthals, H. J., Van De Wiel, C. C. M., & Mur, L. R. (1986). A new prokaryote containing chlorophylls *a* and *b*. *Nature, 320*, 262–264.

Burger-Wiersma, T., Stal, L. J., & Mur, L. R. (1989). *Prochlorthrix hollandica* gen. nov. sp. nov., a filamentous oxygenic photoautotrophic procaryote containing chlorophylls *a* and *b*: Assignment to *Prochlorotrichaceae* fam. nov. and order *Prochlorales* Florenzano, Balloni, and Materassi 1986, with emendation of the ordinal description. *International Journal of Systematic and Evolutionary Microbiology, 39*, 250–257.

Burnap, R. L., Troyan, & T., Sherman, L. A. (1993). The highly abundant chlorophyll-protein complex of iron-deficient Synechococcus sp. PCC7942 (CP43') is encoded by the isiA gene. *Plant Physiology, 103,* 893–902.

Caffarri, S., Tibiletti, T., Jennings, R. C., & Santabarbara, S. (2014). A comparison between plant photosystem I and photosystem II architecture and functioning. *Current Protein & Peptide Science, 15*, 296–331.

Cardona, T., Murray, J. W., & Rutherford, A. W. (2015). Origin and evolution of water oxidation before the last common ancestor of the cyanobacteria. *Molecular Biology and Evolution, 32*, 1310–1328.

Castenholz, R. W. (2001). Oxygenic photosynthetic Bacteria. In D. R. Boone, R. W. Castenholz, & G. Gm (Eds.), *Bergey's manual of systematic bacteriology. Volume 1: The Archaea and the deeply branching and phototrophic Bacteria*. New York: Springer-Verlag.

Cavalier-Smith, T. (2018). Kingdom Chromista and its eight phyla: A new synthesis emphasising periplastid protein targeting, cytoskeletal and periplastid evolution, and ancient divergences. *Protoplasma, 255*, 297–357.

Chang, L., Liu, X., Li, Y., Liu, C.-C., Yang, F., Zhao, J., & Sui, S.-F. (2015). Structural organization of an intact phycobilisome and its association with photosystem II. *Cell Research, 25*, 726–737.

Chapman, D. J. (1966). The pigments of the symbiotic algae (cyanomes) of *Cyanophora paradoxa* and *Glaucocystis nostochinearum* and two Rhodophyceae, *Porphyridium aerugineum* and *Asterocytis ramosa*. *Archiv für Mikrobiologie, 55*, 17–25.

Chen, M. (2019). Chlorophylls *d* and *f*: Synthesis, occurrence, light-harvesting, and pigment organization in chlorophyll-binding protein complexes. *Advances in Botanical Research, 90*, 121–139.

Chen, M., & Bibby, T. S. (2005). Photosynthetic apparatus of antenna-reaction centres supercomplexes in oxyphotobacteria: Insight through significance of Pcb/IsiA proteins. *Photosynthesis Research, 86*, 165–173.
Chen, M., & Blankenship, R. E. (2011). Expanding the solar spectrum used by photosynthesis. *Trends in Plant Science, 16*, 427–431.
Chen, M., & Scheer, H. (2013). Extending the limits of natural photosynthesis and implications for technical light harvesting. *Journal of Porphyrins and Phthalocyanines, 17*, 1–15.
Chen, M., Quinnell, R. G., & Larkum, A. W. D. (2002). The major light-harvesting pigment protein of *Acaryochloris marina*. *FEBS Letters, 514*, 149–152.
Chen, M., Bibby, T. S., Nield, J., Larkum, A., & Barber, J. (2005a). Iron deficiency induces a chlorophyll *d*-binding Pcb antenna system around photosystem I in *Acaryochloris marina*. *Biochimica et Biophysica Acta (BBA) – Bioenergetics, 1708*, 367–374.
Chen, M., Bibby, T. S., Nield, J., Larkum, A. W. D., & Barber, J. (2005b). Structure of a large photosystem II supercomplex from *Acaryochloris marina*. *FEBS Letters, 579*, 1306–1310.
Chen, M., Hiller, R. G., Howe, C. J., & Larkum, A. W. D. (2005c). Unique origin and lateral transfer of prokaryotic chlorophyll-*b* and chlorophyll-*d* light-harvesting systems. *Molecular Biology and Evolution, 22*, 21–28.
Chen, M., Telfer, A., Lin, S., Pascal, A., Larkum, A. W. D., Barber, J., & Blankenship, R. E. (2005d). The nature of the photosystem II reaction centre in the chlorophyll *d*-containing prokaryote, *Acaryochloris marina*. *Photochemical & Photobiological Sciences, 4*, 1060–1064.
Chen, M., Zhang, Y., & Blankenship, R. E. (2008). Nomenclature for membrane-bound light-harvesting complexes of cyanobacteria. *Photosynthesis Research, 95*, 147–154.
Chen, M., Floetenmeyer, M., & Bibby, T. S. (2009). Supramolecular organization of phycobiliproteins in the chlorophyll *d*-containing cyanobacterium *Acaryochloris marina*. *FEBS Letters, 583*, 2535–2539.
Chen, M., Schliep, M., Willows, R. D., Cai, Z.-L., Neilan, B. A., & Scheer, H. (2010). A red-shifted chlorophyll. *Science, 329*, 1318–1319.
Chen, M., Li, Y., Birch, D., & Willows, R. D. (2012). A cyanobacterium that contains chlorophyll *f* – A red-absorbing photopigment. *FEBS Letters, 586*, 3249–3254.
Chen, M., Hernández-Prieto, M. A., Loughlin, P. C., Li, Y., & Willows, R. D. (2019). Genome and proteome of the chlorophyll *f*-producing cyanobacterium *Halomicronema hongdechloris*: adaptative proteomic shifts under different light conditions. *BMC Genomics, 20*, 207.
Chisholm, S. W., Olson, R. J., Zettler, E. R., Goericke, R., Waterbury, J. B., & Welschmeyer, N. A. (1988). A novel free-living prochlorophyte abundant in the oceanic euphotic zone. *Nature, 334*, 340–343.
Chisholm, S. W., Frankel, S. L., Goericke, R., Olson, R. J., Palenik, B., Waterbury, J. B., West-Johnsrud, L., & Zettler, E. R. (1992). *Prochlorococcus marinus* nov. gen. nov. sp.: An oxyphototrophic marine prokaryote containing divinyl chlorophyll *a* and *b*. *Archives of Microbiology, 157*, 297–300.
Christ, B., & Hörtensteiner, S. (2014). Mechanism and significance of chlorophyll breakdown. *Journal of Plant Growth Regulation, 33*, 4–20.
De Los Ríos, A., Grube, M., Sancho, L. G., & Ascaso, C. (2007). Ultrastructural and genetic characteristics of endolithic cyanobacterial biofilms colonizing Antarctic granite rocks. *FEMS Microbiology Ecology, 59*, 386–395.
Demmig-Adams, B. (1990). Carotenoids and photoprotection in plants: A role for the xanthophyll zeaxanthin. *Biochimica et Biophysica Acta (BBA) – Bioenergetics, 1020*, 1–24.
Dolganov, N. A., Bhaya, D., & Grossman, A. R. (1995). Cyanobacterial protein with similarity to the chlorophyll *a*/*b* binding proteins of higher plants: Evolution and regulation. *Proceedings of the National Academy of Sciences, USA, 92*, 636–640.
Domonkos, I., Kis, M., Gombos, Z., & Ughy, B. (2013). Carotenoids, versatile components of oxygenic photosynthesis. *Progress in Lipid Research, 52*, 539–561.
Eaton-Rye, J. J., & Putnam-Evans, C. (2005). The CP47 and CP43 core antenna components. In T. J. Wydrzynski, K. Satoh, & J. A. Freeman (Eds.), *Photosystem II: The light-driven water: Plastoquinone oxidoreductase*. Dordrecht: Springer Netherlands.

Eaton-Rye, J. J., & Sobotka, R. (2017). Editorial: Assembly of the photosystem II membrane-protein complex of oxygenic photosynthesis. *Frontiers in Plant Science, 8*, 1–4.

Fleming, E. D., & Prufert-Bebout, L. (2010). Characterization of cyanobacterial communities from high-elevation lakes in the Bolivian Andes. *Journal of Geophysical Research, 115*.

Frigaard, N.-U., & Dahl, C. (2008). Sulfur metabolism in phototrophic sulfur bacteria. *Advances in Microbial Physiology, 54*, 103–200.

Gan, F., Zhang, S., Rockwell, N. C., Martin, S. S., Lagarias, J. C., & Bryant, D. A. (2014). Extensive remodeling of a cyanobacterial photosynthetic apparatus in far-red light. *Science, 345*, 1312–1317.

Gan, F., Shen, G., & Bryant, D. A. (2015). Occurrence of far-red light photoacclimation (FaRLiP) in diverse cyanobacteria. *Life, 5*, 4–24.

Gao, J., Wang, H., Yuan, Q., & Feng, Y. (2018). Structure and function of the photosystem supercomplexes. *Frontiers in Plant Science, 9*, 357.

Garczarek, L., Hess, W. R., Holtzendorff, J., Van Der Staay, G. W., & Partensky, F. (2000). Multiplication of antenna genes as a major adaptation to low light in a marine prokaryote. *Proceedings of the National Academy of Sciences, USA, 97*, 4098–4101.

Garg, H., Loughlin, P. C., Willows, R. D., & Chen, M. (2017). The $C2^1$-formyl group in chlorophyll *f* originates from molecular oxygen. *Journal of Biological Chemistry, 292*, 19279–19289.

Garrido, J. L., Otero, J., Meaestro, M. A., & Zapata, M. (2000). The main nonpolar chlorophyll *c* from *Emiliania huxleyi* (Prymnesiophyceae) is a chlorophyll c_2-monogalactosyl diacylglyceride ester: A mass spectrometry study. *Journal of Phycology, 36*, 497–505.

Giddings, T. H., Wasmann, C., & Staehelin, L. A. (1983). Structure of the thylakoids and envelope membranes of the cyanelles of *Cyanophora paradoxa*. *Plant Physiology, 71*, 409–419.

Glazer, A. N. (1977). Structure and molecular organization of the photosynthetic accessory pigments of cyanobacteria and red algae. *Molecular and Cellular Biochemistry, 18*, 125–140.

Glazer, A. N. (1985). Light harvesting by phycobilisomes. *Annual Review of Biophysics and Biophysical Chemistry, 14*, 47–77.

Glazer, A. N., & Wedemayer, G. J. (1995). Cryptomonad biliproteins — An evolutionary perspective. *Photosynthesis Research, 46*, 93–105.

Goericke, R., & Repeta, D. J. (1992). The pigments of *Prochlorococcus marinus*: The presence of divinylchlorophyll *a* and *b* in a marine procaryote. *Limnology and Oceanography, 37*, 425–433.

Goericke, R., & Repeta, D. J. (1993). Chlorophylls a and b and divinyl chlorophylls a and b in the open subtropical North Atlantic Ocean. *Marine Ecology Progress Series, 101*, 307–313.

Goh, F., Allen, M. A., Leuko, S., Kawaguchi, T., Decho, A. W., Burns, B. P., & Neilan, B. A. (2008). Determining the specific microbial populations and their spatial distribution within the stromatolite ecosystem of Shark Bay. *The ISME Journal, 3*, 383.

Golub, M., Combet, S., Wieland, D. C. F., Soloviov, D., Kuklin, A., Lokstein, H., Schmitt, F. J., Olliges, R., Hecht, M., Eckert, H. J., & Pieper, J. (2017). Solution structure and excitation energy transfer in phycobiliproteins of *Acaryochloris marina* investigated by small angle scattering. *Biochimica et Biophysica Acta (BBA) – Bioenergetics, 1858*, 318–324.

Gomelsky, M., & Hoff, W. D. (2011). Light helps bacteria make important lifestyle decisions. *Trends in Microbiology, 19*, 441–448.

Goss, R., & Jakob, T. (2010). Regulation and function of xanthophyll cycle-dependent photoprotection in algae. *Photosynthesis Research, 106*, 103–122.

Green, B. R. (2011). After the primary endosymbiosis: An update on the chromalveolate hypothesis and the origins of algae with Chl *c*. *Photosynthesis Research, 107*, 103–115.

Grossman, A. R., Schaefer, M. R., Chiang, G. G., & Collier, J. L. (1993). The phycobilisome, a light-harvesting complex responsive to environmental conditions. *Microbiological Reviews, 57*, 725–749.

Grossman, A. R., Schwarz, R., Bhaya, D., & Dolganov, N. (1998). Phycobilisome degradation and responses of cyanobacteria to nutrient limitation and high light. In G. Garab (Ed.), *Photosynthesis: Mechanisms and effects: Volume I–V: Proceedings of the XIth international congress on photosynthesis, Budapest, Hungary, August 17–22, 1998* (pp. 2853–2858). Dordrecht: Springer.

Guan, X., Qin, S., Zhao, F., Zhang, X., & Tang, X. (2007). Phycobilisomes linker family in cyanobacterial genomes: Divergence and evolution. *International Journal of Biological Sciences, 3*, 434–445.

Guglielmi, G., Cohen-Bazire, G., & Bryant, D. A. (1981). The structure of *Gloeobacter violaceus* and its phycobilisomes. *Archives of Microbiology, 129*, 181–189.

Guskov, A., Kern, J., Gabdulkhakov, A., Broser, M., Zouni, A., & Saenger, W. (2009). Cyanobacterial photosystem II at 2.9-Å resolution and the role of quinones, lipids, channels and chloride. *Nature Structural & Molecular Biology, 16*, 334–342.

Harada, J., Mizoguchi, T., Satoh, S., Tsukatani, Y., Yokono, M., Noguchi, M., Tanaka, A., & Tamiaki, H. (2013). Specific gene *bciD* for C7-methyl oxidation in bacteriochlorophyll *e* biosynthesis of brown-colored green sulfur bacteria. *PLoS One, 8*, e60026.

Hashimoto, H., Uragami, C., & Cogdell, R. J. (2016). Carotenoids and photosynthesis. In C. Stange (Ed.), *Carotenoids in nature: Biosynthesis, regulation and function*. Cham: Springer.

Hauska, G., Schoedl, T., Remigy, H., & Tsiotis, G. (2001). The reaction center of green sulfur bacteria. *Biochimica et Biophysica Acta (BBA) – Bioenergetics, 1507*, 260–277.

Hecks, B., Wulf, K., Breton, J., Leibl, W., & Trissl, H.-W. (1994). Primary charge separation in photosystem I: A two-step electrogenic charge separation connected with $P700^{+}A_{0}^{-}$ and $P700^{+}A_{1}^{-}$ formation. *Biochemistry, 33*, 8619–8624.

Hernández-Prieto, M. A., Li, Y., Postier, B. L., Blankenship, R. E., & Chen, M. (2018). Far-red light promotes biofilm formation in the cyanobacterium *Acaryochloris marina*. *Environmental Microbiology, 20*, 535–545.

Herrera-Salgado, P., Leyva-Castillo, L. E., Ríos-Castro, E., & Gómez-Lojero, C. (2018). Complementary chromatic and far-red photoacclimations in *Synechococcus* ATCC 29403 (PCC 7335). I: The phycobilisomes, a proteomic approach. *Photosynthesis Research, 138*, 39–56.

Ho, M.-Y., Shen, G., Canniffe, D. P., Zhao, C., & Bryant, D. A. (2016). Light-dependent chlorophyll *f* synthase is a highly divergent paralog of PsbA of photosystem II. *Science, 353*, aaf9178.

Ho, M.-Y., Gan, F., Shen, G., & Bryant, D. A. (2017a). Far-red light photoacclimation (FaRLiP) in *Synechococcus* sp. PCC 7335. II.Characterization of phycobiliproteins produced during acclimation to far-red light. *Photosynthesis Research, 131*, 187–202.

Ho, M.-Y., Soulier, N. T., Canniffe, D. P., Shen, G., & Bryant, D. A. (2017b). Light regulation of pigment and photosystem biosynthesis in cyanobacteria. *Current Opinion in Plant Biology, 37*, 24–33.

Holt, A. S., & Morley, H. V. (1959). A proposed structure for chlorophyll *d*. *Canadian Journal of Chemistry, 37*, 507–514.

Hoober, J. K., Eggink, L. L., & Chen, M. (2007). Chlorophylls, ligands and assembly of light-harvesting complexes in chloroplasts. *Photosynthesis Research, 94*, 387–400.

Hörtensteiner, S., & Kräutler, B. (2011). Chlorophyll breakdown in higher plants. *Biochimica et Biophysica Acta (BBA) – Bioenergetics, 1807*, 977–988.

Hörtensteiner, S., Hauenstein, M., & Kräutler, B. (2019). Chlorophyll breakdown—Regulation, biochemistry and phyllobilins as its products. *Advances in Botanical Research, 90*, 213–271.

Hu, Q., Miyashita, H., Iwasaki, I., Kurano, N., Miyachi, S., Iwaki, M., & Itoh, S. (1998). A photosystem I reaction center driven by chlorophyll *d* in oxygenic photosynthesis. *Proceedings of the National Academy of Sciences, USA, 95*, 13319–13323.

Hu, Q., Marquardt, J., Iwasaki, I., Miyashita, H., Kurano, N., Mörschel, E., & Miyachi, S. (1999). Molecular structure, localization and function of biliproteins in the chlorophyll *a*/*d* containing oxygenic photosynthetic prokaryote *Acaryochloris marina*. *Biochimica et Biophysica Acta (BBA) – Bioenergetics, 1412*, 250–261.

Hynninen, P. H. (1991). Chemistry of chlorophylls: Modifications. In H. Scheer (Ed.), *Chlorophylls*. Boca Raton: CRC Press.

Ihalainen, J. A., D'haene, S., Yeremenko, N., Van Roon, H., Arteni, A. A., Boekema, E. J., Van Grondelle, R., Matthijs, H. C. P., & Dekker, J. P. (2005). Aggregates of the chlorophyll-binding protein IsiA (CP43') dissipate energy in cyanobacteria. *Biochemistry, 44*, 10846–10853.

Ito, H., Ohtsuka, T., & Tanaka, A. (1996). Conversion of chlorophyll *b* to chlorophyll *a* via 7-hydroxymethyl chlorophyll. *Journal of Biological Chemistry, 271*, 1475–1479.

Jeffrey, S. W. (1969). Properties of two spectrally different components in chlorophyll *c* preparations. *Biochimica et Biophysica Acta, 177*, 456–467.

Jeffrey, S. W. (1972). Preparation and some properties of crystalline chlorophyll c_1 and c_2 from marine algae. *Biochimica et Biophysica Acta, 279*, 15–33.

Jeffrey, S. W. (1976). The occurrence of chlorophyll c_1 and c_2 in algae. *Journal of Phycology, 12*, 349–354.

Jeffrey, S. W., & Wright, S. W. (1987). A new spectrally distinct component in preparations of chlorophyll *c* from the micro-alga *Emiliania huxleyi* (Prymnesiophycease). *Biochimica et Biophysica Acta (BBA) – Bioenergetics, 894*, 180–188.

Jordan, P., Fromme, P., Witt, H. T., Klukas, O., Saenger, W., & Krauß, N. (2001). Three-dimensional structure of cyanobacterial photosystem I at 2.5 Å resolution. *Nature, 411*, 909–917.

Kashiyama, Y., Miyashita, H., Ohkubo, S., Ogawa, N. O., Chikaraishi, Y., Takano, Y., Suga, H., Toyofuku, T., Nomaki, H., Kitazato, H., Nagata, T., & Ohkouchi, N. (2008). Evidence of global chlorophyll *d*. *Science, 321*, 658–658.

Kaucikas, M., Nürnberg, D., Dorlhiac, G., Rutherford, A. W., & VAN Thor, J. J. (2017). Femtosecond visible transient absorption spectroscopy of chlorophyll *f*-containing photosystem I. *Biophysical Journal, 112*, 234–249.

Kehoe, D. M., & Gutu, A. (2006). Responding to color: The regulation of complementary chromatic adaptation. *Annual Review of Plant Biology, 57*, 127–150.

Kirilovsky, D., & Kerfeld, C. A. (2012). The orange carotenoid protein in photoprotection of photosystem II in cyanobacteria. *Biochimica et Biophysica Acta (BBA) – Bioenergetics, 1817*, 158–166.

Knoppová, J., Sobotka, R., Tichý, M., Yu, J., Konik, P., Halada, P., Nixon, P. J., & Komenda, J. (2014). Discovery of a chlorophyll binding protein complex involved in the early steps of photosystem II assembly in *Synechocystis*. *The Plant Cell, 26*, 1200–1212.

Koizumi, H., Itoh, Y., Hosoda, S., Akiyama, M., Hoshino, T., Shiraiwa, Y., & Kobayashi, M. (2005). Serendipitous discovery of Chl *d* formation from Chl *a* with papain. *Science and Technology of Advanced Materials, 6*, 551–557.

Koyama, K., Tsuchiya, T., Akimoto, S., Yokono, M., Miyashita, H., & Mimuro, M. (2006). New linker proteins in phycobilisomes isolated from the cyanobacterium *Gloeobacter violaceus* PCC 7421. *FEBS Letters, 580*, 3457–3461.

Kräutler, B. (2016). Breakdown of chlorophyll in higher plants--Phyllobilins as abundant, yet hardly visible signs of ripening, senescence, and cell death. *Angewandte Chemie (English Edition), 55*, 4882–4907.

Kühl, M., Chen, M., & Ralph, P. J. (2005). A niche for cyanobacteria containing chlorophyll *d*. *Nature, 433*, 820.

La Roche, J., Van Der Staay, G. W., Partensky, F., Ducret, A., Aebersold, R., Li, R., Golden, S. S., Hiller, R. G., Wrench, P. M., Larkum, A. W., & Green, B. R. (1996). Independent evolution of the prochlorophyte and green plant chlorophyll *a/b* light-harvesting proteins. *Proceedings of the National Academy of Sciences, USA, 93*, 15244–15248.

Larkum, A. W., Scaramuzzi, C., Cox, G. C., Hiller, R. G., & Turner, A. G. (1994). Light-harvesting chlorophyll *c*-like pigment in *Prochloron*. *Proceedings of the National Academy of Sciences, USA, 91*, 679–683.

Larkum, A. W. D., Chen, M., Li, Y., Schliep, M., Trampe, E., West, J., Salih, A., & Kühl, M. (2012). A novel epiphytic chlorophyll *d*-containing cyanobacterium isolated from a mangrove-associated red alga. *Journal of Phycology, 48*, 1320–1327.

Lewin, R. A. (1977). *Prochloron*, type genus of the Prochlorophyta. *Phycologia, 16*, 217–217.

Lichtenthaler, H. K. (1987). Chlorophylls and carotenoids: pigments of photosynthetic biomembranes. *Methods in Enzymology, 148*, 350–382.

Li, Y., Scales, N., Blankenship, R. E., Willows, R. D., & Chen, M. (2012). Extinction coefficient for red-shifted chlorophylls: Chlorophyll d and chlorophyll f. *Biochimica et Biophysica Acta (BBA) - Bioenergetics, 1817*, 1292–1298.

Li, Y., Larkum, A., Schliep, M., Kühl, M., Neilan, B., & Chen, M. (2013). Newly isolated Chl *d*-containing cyanobacteria. In *Photosynthesis research for food, fuel and future – 15th international conference on photosynthesis*. Hangzhou/Heidelberg: Zhejiang University Press/Springer.

Li, Y., Lin, Y., Garvey, C. J., Birch, D., Corkery, R. W., Loughlin, P. C., Scheer, H., Willows, R. D., & Chen, M. (2016). Characterization of red-shifted phycobilisomes isolated from the chlorophyll *f*-containing cyanobacterium *Halomicronema hongdechloris*. *Biochimica et Biophysica Acta (BBA) – Bioenergetics, 1857*, 107–114.

Li, Y., Vella, N., & Chen, M. (2018a). Characterization of isolated photosystem I from *Halomicronema hongdechloris*, a chlorophyll *f*-producing cyanobacterium. *Photosynthetica, 56*, 306–315.

Li, Z.-K., Yin, Y.-C., Zhang, L.-D., Zhang, Z.-C., Dai, G.-Z., Chen, M., & Qiu, B.-S. (2018b). The identification of IsiA proteins binding chlorophyll *d* in the cyanobacterium *Acaryochloris marina*. *Photosynthesis Research, 135*, 165–175.

Li, W., Su, H.-N., Pu, Y., Chen, J., Liu, L.-N., Liu, Q., & Qin, S. (2019). Phycobiliproteins: Molecular structure, production, applications, and prospects. *Biotechnology Advances, 37*, 340–353.

Liu, H., Zhang, H., Niedzwiedzki, D. M., Prado, M., He, G., Gross, M. L., & Blankenship, R. E. (2013). Phycobilisomes supply excitations to both photosystems in a megacomplex in cyanobacteria. *Science, 342*, 1104–1107.

Loughlin, P., Lin, Y., & Chen, M. (2013). Chlorophyll *d* and *Acaryochloris marina*: Current status. *Photosynthesis Research, 116*, 277–293.

Loughlin, P. C., Willows, R. D., & Chen, M. (2014). *In vitro* conversion of vinyl to formyl groups in naturally occurring chlorophylls. *Scientific Reports, 4*, 6069.

Loughlin, P. C., Willows, R. D., & Min, C. (2015). Hydroxylation of the $C13^2$ and C18 carbons of chlorophylls by heme and molecular oxygen. *Journal of Porphyrins and Phthalocyanines, 19*, 1007–1013.

Maccoll, R. (1998). Cyanobacterial phycobilisomes. *Journal of Structural Biology, 124*, 311–334.

Madigan, M. T., & Jung, D. O. (2009). An overview of purple bacteria: Systematics, physiology, and habitats. In C. N. Hunter, F. Daldal, M. C. Thurnauer, & J. T. Beatty (Eds.), *The purple phototrophic bacteria*. Dordrecht: Springer.

Manning, W. M., & Strain, H. H. (1943). Chlorophyll *d*, a green pigment of red algae. *Journal of Biological Chemistry, 151*, 1–19.

Mareš, J., Strunecký, O., Bučinská, L., & Wiedermannová, J. (2019). Evolutionary patterns of thylakoid architecture in cyanobacteria. *Frontiers in Microbiology, 10*, 1–22.

Marquardt, J., Senger, H., Miyashita, H., Miyachi, S., & Mörschel, E. (1997). Isolation and characterization of biliprotein aggregates from *Acaryochloris marina*, a Prochloron-like prokaryote containing mainly chlorophyll *d*. *FEBS Letters, 410*, 428–432.

Marquardt, J., Mörschel, E., Rhiel, E., & Westermann, M. (2000). Ultrastructure of *Acaryochloris marina*, an oxyphotobacterium containing mainly chlorophyll *d*. *Archives of Microbiology, 174*, 181–188.

Masamoto, K., & Furukawa, K.-I. (1997). Accumulation of zeaxanthin in cells of the cyanobacterium, *Synechococcus* sp. strain PCC 7942 grown under high irradiance. *Journal of Plant Physiology, 151*, 257–261.

Mcewan, A. G. (1994). Photosynthetic electron transport and anaerobic metabolism in purple non-sulfur phototrophic bacteria. *Antonie Van Leeuwenhoek, 66*, 151–164.

Mcnamara, C. J., Perry, T. D., Bearce, K. A., Hernandez-Duque, G., & Mitchell, R. (2006). Epilithic and endolithic bacterial communities in limestone from a Maya archaeological site. *Microbial Ecology, 51*, 51–64.

Meguro, M., Ito, H., Takabayashi, A., Tanaka, R., & Tanaka, A. (2011). Identification of the 7-Hydroxymethyl chlorophyll *a* reductase of the chlorophyll cycle in *Arabidopsis*. *The Plant Cell, 23*, 3442–3453.

Mielke, S. P., Kiang, N. Y., Blankenship, R. E., & Mauzerall, D. (2013). Photosystem trap energies and spectrally-dependent energy-storage efficiencies in the Chl *d*-utilizing cyanobacterium, *Acaryochloris marina*. *Biochimica et Biophysica Acta (BBA) – Bioenergetics, 1827*, 255–265.

Miller, S. R., Augustine, S., Olson, T. L., Blankenship, R. E., Selker, J., & Wood, A. M. (2005). Discovery of a free-living chlorophyll *d*-producing cyanobacterium with a hybrid proteobacterial/cyanobacterial small-subunit rRNA gene. *Proceedings of the National Academy of Sciences, USA, 102*, 850–855.

Mimuro, M., Akimoto, S., Gotoh, T., Yokono, M., Akiyama, M., Tsuchiya, T., Miyashita, H., Kobayashi, M., & Yamazaki, I. (2004). Identification of the primary electron donor in PS II of the Chl *d*-dominated cyanobacterium *Acaryochloris marina*. *FEBS Letters, 556*, 95–98.

Miyashita, H., Ikemoto, H., Kurano, N., Adachi, K., Chihara, M., & Miyachi, S. (1996). Chlorophyll *d* as a major pigment. *Nature, 383*, 402.

Miyashita, H., Adachi, K., Kurano, N., Ikemot, H., Chihara, M., & Miyach, S. (1997). Pigment composition of a novel oxygenic photosynthetic prokaryote containing chlorophyll *d* as the major chlorophyll. *Plant & Cell Physiology, 38*, 274–281.

Miyashita, H., Ohkubo, S., Komatsu, H., Sorimachi, Y., Fukayama, D., Fujinuma, D., Akutsu, S., & Kobayashi, M. (2014). Discovery of chlorophyll *d* in *Acaryochloris marina* and chlorophyll *f* in a unicellular cyanobacterium, strain KC1, isolated from Lake Biwa. *Journal of Physical Chemistry & Biophysics, 4*, 149.

Moss, G. P. (1987). Nomenclature of tetrapyrroles (recommendations 1986). *Pure and Applied Chemistry, 59*, 779.

Murakami, A., Miyashita, H., Iseki, M., Adachi, K., & Mimuro, M. (2004). Chlorophyll *d* in an epiphytic cyanobacterium of red algae. *Science, 303*, 1633–1633.

Myśliwa-Kurdziel, B., Latowski, D., & Strzałka, K. (2019). Chapter three – Chlorophylls *c*—Occurrence, synthesis, properties, photosynthetic and evolutionary significance. *Advances in Botanical Research, 90*, 91–119.

Nagy, J. O., Bishop, J. E., Klotz, A. V., Glazer, A. N., & Rapoport, H. (1985). Bilin attachment sites in the alpha, beta, and gamma subunits of R-phycoerythrin. Structural studies on singly and doubly linked phycourobilins. *Journal of Biological Chemistry, 260*, 4864–4868.

Neilson, J. A. D., & Durnford, D. G. (2010). Structural and functional diversification of the light-harvesting complexes in photosynthetic eukaryotes. *Photosynthesis Research, 106*, 57–71.

Nürnberg, D. J., Morton, J., Santabarbara, S., Telfer, A., Joliot, P., Antonaru, L. A., Ruban, A. V., Cardona, T., Krausz, E., Boussac, A., Fantuzzi, A., & Rutherford, A. W. (2018). Photochemistry beyond the red limit in chlorophyll *f*–containing photosystems. *Science, 360*, 1210–1213.

Oba, T., Uda, Y., Matsuda, K., Fukusumi, T., Ito, S., Hiratani, K., & Tamiaki, H. (2011). A mild conversion from 3-vinyl- to 3-formyl-chlorophyll derivatives. *Bioorganic & Medicinal Chemistry Letters, 21*, 2489–2491.

Ogata, T., Kodama, M., Nomura, S., Kobayashi, M., Nozawa, T., Katoh, T., & Mimuro, M. (1994). A novel peridinin—Chlorophyll a protein (PCP) from the marine dinoflagellate *Alexandrium cohorticula*: A high pigment content and plural spectral forms of peridinin and chlorophyll a. *FEBS Letters, 356*, 367–371.

Ong, L. J., & Glazer, A. N. (1987). R-phycocyanin II, a new phycocyanin occurring in marine *Synechococcus* species. Identification of the terminal energy acceptor bilin in phycocyanins. *Journal of Biological Chemistry, 262*, 6323–6327.

Oster, U., Tanaka, R., Tanaka, A., & Rüdiger, W. (2000). Cloning and functional expression of the gene encoding the key enzyme for chlorophyll *b* biosynthesis (CAO) from *Arabidopsis thaliana*. *Plant Journal, 21*, 305–310.

Overmann, J., & Garcia-Pichel, F. (2013). The phototrophic way of life. In E. Rosenberg, E. F. Delong, S. Lory, E. Stackebrandt, & F. Thompson (Eds.), *The prokaryotes: Prokaryotic communities and ecophysiology*. Berlin/Heidelberg: Springer.

Partensky, F., Hoepffner, N., Li, W. K. W., Ulloa, O., & Vaulot, D. (1993). Photoacclimation of *Prochlorococcus* sp. (Prochlorophyta) strains isolated from the North Atlantic and the Mediterranean Sea. *Plant Physiology, 101*, 285–296.

Partensky, F., Hess, W. R., & Vaulot, D. (1999). *Prochlorococcus*, a marine photosynthetic prokaryote of global significance. *Microbiology and Molecular Biology Reviews, 63*, 106–127.

Partensky, F., Six, C., Ratin, M., Garczarek, L., Vaulot, D., Probert, I., Calteau, A., Gourvil, P., Marie, D., Grébert, T., Bouchier, C., Le Panse, S., Gachenot, M., Rodríguez, F., & Garrido,

J. L. (2018). A novel species of the marine cyanobacterium *Acaryochloris* with a unique pigment content and lifestyle. *Scientific Reports, 8*, 9142.

Petrasek, Z., Schmitt, F.-J., Theiss, C., Huyer, J., Chen, M., Larkum, A., Eichler, H. J., Kemnitz, K., & Eckert, H.-J. (2005). Excitation energy transfer from phycobiliprotein to chlorophyll *d* in intact cells of *Acaryochloris marina* studied by time- and wavelength-resolved fluorescence spectroscopy. *Photochemical & Photobiological Sciences, 4*, 1016–1022.

Pfreundt, U., Stal, L. J., Voß, B., & Hess, W. R. (2012). Dinitrogen fixation in a unicellular chlorophyll *d*-containing cyanobacterium. *The ISME Journal, 6*, 1367–1377.

Polívka, T., Pascher, T., & Hiller, R. G. (2008). Energy transfer in the peridinin-chlorophyll protein complex reconstituted with mixed chlorophyll sites. *Biophysical Journal, 94*, 3198–3207.

Porra, R. J., Schäfer, W., Cmiel, E., Katheder, I., & Scheer, H. (1994). The derivation of the formyl-group oxygen of chlorophyll *b* in higher plants from molecular oxygen. *European Journal of Biochemistry, 219*, 671–679.

Renger, T., & Schlodder, E. (2008). The primary electron donor of photosystem II of the cyanobacterium *Acaryochloris marina* is a chlorophyll *d* and the water oxidation is driven by a chlorophyll *a*/chlorophyll *d* heterodimer. *The Journal of Physical Chemistry. B, 112*, 7351–7354.

Reyes-Prieto, A., Weber, A. P. M., & Bhattacharya, D. (2007). The origin and establishment of the plastid in algae and plants. *Annual Review of Genetics, 41*, 147–168.

Ritter, S., Hiller, R. G., Wrench, P. M., Welte, W., & Diederichs, K. (1999). Crystal structure of a phycourobilin-containing phycoerythrin at 1.90-Å resolution. *Journal of Structural Biology, 126*, 86–97.

Rockwell, N. C., Martin, S. S., & Lagarias, J. C. (2016). Identification of cyanobacteriochromes detecting far-red light. *Biochemistry, 55*, 3907–3919.

Rodríguez, F., Garrido, J. L., Sobrino, C., Johnsen, G., Riobó, P., Franco, J., Aamot, I., Ramilo, I., Sanz, N., & Kremp, A. (2016). Divinyl chlorophyll *a* in the marine eukaryotic protist *Alexandrium ostenfeldii* (Dinophyceae). *Environmental Microbiology, 18*, 627–643.

Sawicki, A., Willows, R. D., & Chen, M. (2019). Spectral signatures of five hydroxymethyl chlorophyll *a* derivatives chemically derived from chlorophyll *b* or chlorophyll *f*. *Photosynthesis Research, 140*, 115–127.

Schagerl, M., & Müller, B. (2006). Acclimation of chlorophyll *a* and carotenoid levels to different irradiances in four freshwater cyanobacteria. *Journal of Plant Physiology, 163*, 709–716.

Scheer, H. (1991). Structure and occurrence of chlorophylls. In H. Scheer (Ed.), *Chlorophylls*. Boca Raton: CRC Press.

Scheer, H. (2006). An overview of chlorophylls and bacteriochlorophylls: Biochemistry, biophysics, functions and applications. In B. Grimm, R. J. Porra, W. Rüdiger, & H. Scheer (Eds.), *Chlorophylls and bacteriochlorophylls: Biochemistry, biophysics, functions and applications*. Dordrecht: Springer.

Scheer, H., & Zhao, K. H. (2008). Biliprotein maturation: The chromophore attachment. *Molecular Microbiology, 68*, 263–276.

Scheumann, V., Ito, H., Tanaka, A., Schoch, S., & Rüdiger, W. (1996). Substrate specificity of chlorophyll(ide) *b* reductase in etioplasts of barley (*Hordeurn Vulgare* L.). *European Journal of Biochemistry, 242*, 163–170.

Schirmer, T., Bode, W., Huber, R., Sidler, W., & Zuber, H. (1985). X-ray crystallographic structure of the light-harvesting biliprotein C-phycocyanin from the thermophilic cyanobacterium *Mastigocladus laminosus* and its resemblance to globin structures. *Journal of Molecular Biology, 184*, 257–277.

Schliep, M., Chen, M., Larkum, A., & Quinnell, R. (2007). The function of MgDVP in a chlorophyll *d*-containing organism. In J. F. Allen, E. Gantt, J. H. Golbeck, & B. Osmond (Eds.), *Photosynthesis: Energy from the sun* (pp. 262–263). Dordrecht: Springer.

Schliep, M., Crossett, B., Willows, R. D., & Chen, M. (2010). ^{18}O labeling of chlorophyll *d* in *Acaryochloris marina* reveals that chlorophyll *a* and molecular oxygen are precursors. *Journal of Biological Chemistry, 285*, 28450–28456.

Schlodder, E., Çetin, M., Eckert, H.-J., Schmitt, F.-J., Barber, J., & Telfer, A. (2007). Both chlorophylls *a* and *d* are essential for the photochemistry in photosystem II of the cyanobacteria, *Acaryochloris marina. Biochimica et Biophysica Acta (BBA) – Bioenergetics, 1767*, 589–595.

Schmidt, M., Krasselt, A., & Reuter, W. (2006). Local protein flexibility as a prerequisite for reversible chromophore isomerization in α-phycoerythrocyanin. *Biochimica et Biophysica Acta (BBA) – Proteins and Proteomics, 1764*, 55–62.

Schmitt, F.-J., Campbell, Z. Y., Bui, M. V., Hüls, A., Tomo, T., Chen, M., Maksimov, E. G., Allakhverdiev, S. I., & Friedrich, T. (2018). Photosynthesis supported by a chlorophyll *f*-dependent, entropy-driven uphill energy transfer in *Halomicronema hongdechloris* cells adapted to far-red light. *Photosynthesis Research, 139*, 185–201.

Shedbalkar, V. P., & Rebeiz, C. A. (1992). Chloroplast biogenesis: Determination of the molar extinction coefficients of divinyl chlorophyll a and b and their pheophytins. *Analytical Biochemistry, 207,* 261–266.

Shen, G., Canniffe, D. P., Ho, M.-Y., Kurashov, V., Van Der Est, A., Golbeck, J. H., & Bryant, D. A. (2019). Characterization of chlorophyll *f* synthase heterologously produced in *Synechococcus* sp. PCC 7002. *Photosynthesis Research, 140*, 77–92.

Shi, L.-X., Hall, M., Funk, C., & Schröder, W. P. (2012). Photosystem II, a growing complex: Updates on newly discovered components and low molecular mass proteins. *Biochimica et Biophysica Acta (BBA) – Bioenergetics, 1817*, 13–25.

Shih, P. M., Wu, D., Latifi, A., Axen, S. D., Fewer, D. P., Talla, E., Calteau, A., Cai, F., Tandeau DE Marsac, N., Rippka, R., Herdman, M., Sivonen, K., Coursin, T., Laurent, T., Goodwin, L., Nolan, M., Davenport, K. W., Han, C. S., Rubin, E. M., Eisen, J. A., Woyke, T., Gugger, M., & Kerfeld, C. A. (2013). Improving the coverage of the cyanobacterial phylum using diversity-driven genome sequencing. *Proceedings of the National Academy of Sciences, USA, 110*, 1053–1058.

Shuvalov, V. A., Yakovlev, A. G., Vasileva, L. G., & Ya, S. A. (2006). Primary charge separation between P700∗ and the primary electron acceptor complex A-A_0: A comparison with bacterial reaction centers. In J. H. Golbeck (Ed.), *Photosystem I: The light-driven plastocyanin: Ferredoxin oxidoreductase*. Dordrecht: Springer.

Sidler, W. A. (1994). Phycobilisome and phycobiliprotein structures. In D. A. Bryant (Ed.), *The molecular biology of cyanobacteria. Advances in photosynthesis*. Dordrecht: Springer.

Six, C., Thomas, J.-C., Garczarek, L., Ostrowski, M., Dufresne, A., Blot, N., Scanlan, D. J., & Partensky, F. (2007). Diversity and evolution of phycobilisomes in marine *Synechococcus* spp.: A comparative genomics study. *Genome Biology, 8*, R259–R259.

Sonani, R. R., Roszak, A. W., Ortmann De Percin Northumberland, C., Madamwar, D., & Cogdell, R. J. (2018). An improved crystal structure of C-phycoerythrin from the marine cyanobacterium *Phormidium* sp. A09DM. *Photosynthesis Research, 135*, 65–78.

Steiger, S., Schäfer, L., & Sandmann, G. (1999). High-light-dependent upregulation of carotenoids and their antioxidative properties in the cyanobacterium *Synechocystis* PCC 6803. *Journal of Photochemistry and Photobiology B: Biology, 52*, 14–18.

Stomp, M., Huisman, J., Stal, L. J., & Matthijs, H. C. P. (2007). Colorful niches of phototrophic microorganisms shaped by vibrations of the water molecule. *The ISME Journal, 1*, 271–282.

Swingley, W. D., Chen, M., Cheung, P. C., Conrad, A. L., Dejesa, L. C., Hao, J., Honchak, B. M., Karbach, L. E., Kurdoglu, A., Lahiri, S., Mastrian, S. D., Miyashita, H., Page, L., Ramakrishna, P., Satoh, S., Sattley, W. M., Shimada, Y., Taylor, H. L., Tomo, T., Tsuchiya, T., Wang, Z. T., Raymond, J., Mimuro, M., Blankenship, R. E., & Touchman, J. W. (2008). Niche adaptation and genome expansion in the chlorophyll *d*-producing cyanobacterium *Acaryochloris marina. Proceedings of the National Academy of Sciences, USA, 105*, 2005–2010.

Takaichi, S. (2011). Carotenoids in algae: Distributions, biosyntheses and functions. *Marine Drugs, 9*, 1101–1118.

Takaichi, S., & Mochimaru, M. (2007). Carotenoids and carotenogenesis in cyanobacteria: Unique ketocarotenoids and carotenoid glycosides. *Cellular and Molecular Life Sciences, 64*, 2607–2619.

Takaichi, S., Mochimaru, M., Uchida, H., Murakami, A., Hirose, E., Maoka, T., Tsuchiya, T., & Mimuro, M. (2012). Opposite chilarity of α-carotene in unusual cyanobacteria with unique chlorophylls, *Acaryochloris* and *Prochlorococcus*. *Plant & Cell Physiology, 53*, 1881–1888.

Tang, K.-H., Yue, H., & Blankenship, R. E. (2010). Energy metabolism of *Heliobacterium modesticaldum* during phototrophic and chemotrophic growth. *BMC Microbiology, 10*, 150.

Thweatt, J. L., Ferlez, B. H., Golbeck, J. H., & Bryant, D. A. (2017). BciD is a radical *S*-Adenosyl-l-methionine (SAM) enzyme that completes bacteriochlorophyllide *e* biosynthesis by oxidizing a methyl group into a formyl group at C-7. *Journal of Biological Chemistry, 292*, 1361–1373.

Tomi, T., Shibata, Y., Ikeda, Y., Taniguchi, S., Haik, C., Mataga, N., Shimada, K., & Itoh, S. (2007). Energy and electron transfer in the photosynthetic reaction center complex of *Acidiphilium rubrum* containing Zn-bacteriochlorophyll *a* studied by femtosecond up-conversion spectroscopy. *Biochimica et Biophysica Acta (BBA) – Bioenergetics, 1767*, 22–30.

Tomo, T., Okubo, T., Akimoto, S., Yokono, M., Miyashita, H., Tsuchiya, T., Noguchi, T., & Mimuro, M. (2007). Identification of the special pair of photosystem II in a chlorophyll *d*-dominated cyanobacterium. *Proceedings of the National Academy of Sciences, USA, 104*, 7283–7288.

Tomo, T., Shinoda, T., Chen, M., Allakhverdiev, S. I., & Akimoto, S. (2014). Energy transfer processes in chlorophyll *f*-containing cyanobacteria using time-resolved fluorescence spectroscopy on intact cells. *Biochimica et Biophysica Acta (BBA) – Bioenergetics, 1837*, 1484–1489.

Toporik, H., Li, J., Williams, D., Chiu, P.-L., & Mazor, Y. (2019). The structure of the stress-induced photosystem I–IsiA antenna supercomplex. *Nature Structural & Molecular Biology, 26*, 443–449.

Tóth, T. N., Chukhutsina, V., Domonkos, I., Knoppová, J., Komenda, J., Kis, M., Lénárt, Z., Garab, G., Kovács, L., Gombos, Z., & Van Amerongen, H. (2015). Carotenoids are essential for the assembly of cyanobacterial photosynthetic complexes. *Biochimica et Biophysica Acta (BBA) – Bioenergetics, 1847*, 1153–1165.

Tsuchiya, T., Akimoto, S., Mizoguchi, T., Watabe, K., Kindo, H., Tomo, T., Tamiaki, H., & Mimuro, M. (2012a). Artificially produced [7-formyl]-chlorophyll *d* functions as an antenna pigment in the photosystem II isolated from the chlorophyllide *a* oxygenase-expressing *Acaryochloris marina*. *Biochimica et Biophysica Acta (BBA) – Bioenergetics, 1817*, 1285–1291.

Tsuchiya, T., Mizoguchi, T., Akimoto, S., Tomo, T., Tamiaki, H., & Mimuro, M. (2012b). Metabolic engineering of the Chl *d*-dominated cyanobacterium *Acaryochloris marina*: Production of a novel Chl species by the introduction of the chlorophyllide *a* oxygenase gene. *Plant & Cell Physiology, 53*, 518–527.

Umena, Y., Kawakami, K., Shen, J.-R., & Kamiya, N. (2011). Crystal structure of oxygen-evolving photosystem II at a resolution of 1.9Å. *Nature, 473*, 55–60.

Van Der Staay, G. W., & Staehelin, L. A. (1994). Biochemical characterization of protein composition and protein phosphorylation patterns in stacked and unstacked thylakoid membranes of the prochlorophyte *Prochlorothrix hollandica*. *Journal of Biological Chemistry, 269*, 24834–24844.

Vavilin, D., & Vermaas, W. (2007). Continuous chlorophyll degradation accompanied by chlorophyllide and phytol reutilization for chlorophyll synthesis in *Synechocystis* sp. PCC 6803. *Biochimica et Biophysica Acta, 1767*, 920–929.

Wakao, N., Yokoi, N., Isoyama, N., Hiraishi, A., Shimada, K., Kobayashi, M., Kise, H., Iwaki, M., Itoh, S., & Takaichi, S. (1996). Discovery of natural photosynthesis using Zn-containing bacteriochlorophyll in an aerobic bacterium *Acidiphilium rubrum*. *Plant & Cell Physiology, 37*, 889–893.

Watanabe, M., Semchonok, D. A., Webber-Birungi, M. T., Ehira, S., Kondo, K., Narikawa, R., Ohmori, M., Boekema, E. J., & Ikeuchi, M. (2014). Attachment of phycobilisomes in an antenna–photosystem I supercomplex of cyanobacteria. *Proceedings of the National Academy of Sciences, USA, 111*, 2512–2517.

Wiltbank, L. B., & Kehoe, D. M. (2019). Diverse light responses of cyanobacteria mediated by phytochrome superfamily photoreceptors. *Nature Reviews. Microbiology, 17*, 37–50.

Wolf, B. M., & Blankenship, R. E. (2019). Far-red light acclimation in diverse oxygenic photosynthetic organisms. *Photosynthesis Research, 142*(3), 349–359. https://doi.org/10.1007/s11120-019-00653-6.

Yu, L., Huang, Y., Bai, Z., Zhu, B., Ding, K., Chen, T., Ding, Y., & Wang, Y. (2015). The aerobic oxidation and C=C bond cleavage of styrenes catalyzed by Cerium(IV) ammonium nitrate (CAN). *Journal of the Chinese Chemical Society, 62*, 479–482.

Yurkov, V. V., & Beatty, J. T. (1998). Aerobic anoxygenic phototrophic bacteria. *Microbiology and Molecular Biology Reviews, 62*, 695–724.

Zakar, T., Laczko-Dobos, H., Toth, T. N., & Gombos, Z. (2016). Carotenoids assist in cyanobacterial photosystem II assembly and function. *Frontiers in Plant Science, 7*, 295–295.

Zamzam, N., Kaucikas, M., Nürnberg, D. J., Rutherford, A. W., & Van Thor, J. J. (2019). Femtosecond infrared spectroscopy of chlorophyll *f*-containing photosystem I. *Physical Chemistry Chemical Physics, 21*, 1224–1234.

Zapata, M., Garrido, J. L., & Jeffrey, S. W. (2006). Chlorophyll *c* pigments: Current status. In B. Grimm, R. J. Porra, W. Rüdiger, & H. Scheer (Eds.), *Chlorophylls and bacteriochlorophylls: Biochemistry, biophysics, functions and applications*. Dordrecht: Springer Netherlands.

Zapata, M., Rodríguez, F., & Garrido, J. L. (2000). Separation of chlorophylls and carotenoids from marine phytoplankton: A new HPLC method using a reversed phase C8 column and pyridine-containing mobile phases. *Marine Ecology Progress Series, 195*, 29–45.

Zhang, J., Ma, J., Liu, D., Qin, S., Sun, S., Zhao, J., & Sui, S.-F. (2017). Structure of phycobilisome from the red alga *Griffithsia pacifica*. *Nature, 551*, 57–63.

Zhang, Z.-C., Li, Z.-K., Yin, Y.-C., Li, Y., Jia, Y., Chen, M., & Qiu, B.-S. (2019). Widespread occurrence and unexpected diversity of red-shifted chlorophyll producing cyanobacteria in humid subtropical forest ecosystems. *Environmental Microbiology, 21*, 1497–1510.

Cyanobacterial NDH-1-Photosystem I Supercomplex

Weimin Ma

Abstract Cyanobacterial NDH-1-photosystem I (PSI) supercomplex is localized in the thylakoid membrane, and its formation facilitates cyclic electron transfer around PSI (PSI CET) via NDH-1. Similar NDH-1-PSI supercomplex has also been identified in the majority of higher plants, but not in liverwort. We describe in this book chapter the cyanobacterial NDH-1-PSI supercomplex, focusing on its identification, function, assembly, and evolution from cyanobacteria to higher plants.

Keywords Cyanobacteria · Cyclic electron transfer · NDH-1-PSI supercomplex · Photosynthesis

Abbreviations

CpcG2-PBS	CpcG2-phycobilisome
Fd	ferredoxin
NAI130	an NDH-1 hydrophilic arm assembly intermediate complex with molecular mass of about 130 kDa
NAI300	an NDH-1 hydrophilic arm assembly intermediate complex with molecular mass of about 300 kDa
NDH-CET	NDH-1-mediated cyclic electron transfer around photosystem I
OPS	oxygenic photosynthesis-specific
PSI CET	cyclic electron transport around photosystem I
Synechocystis 6803	*Synechocystis* sp. strain PCC 6803

W. Ma (✉)
Shanghai Key Laboratory of Plant Molecular Sciences, College of Life Sciences, Shanghai Normal University, Shanghai, China
e-mail: wma@shnu.edu.cn

Q. Wang (ed.), *Microbial Photosynthesis*,
https://doi.org/10.1007/978-981-15-3110-1_2

1 Introduction

Photosynthetic NDH-1 is an important branch of a group 4 membrane-bound [NiFe] hydrogenase (Friedrich and Scheide 2000; Peltier et al. 2016) and is identified in a majority of species spanning from cyanobacteria to higher plants (Friedrich et al. 1995). A general function of photosynthetic NDH-1 is to transfer electrons from photoreduced ferredoxin (Fd) by photosystem I (PSI) to plastoquinone (Battchikova et al. 2011a; Yamamoto et al. 2011; Yamamoto and Shikanai 2013; Schuller et al. 2019; Pan et al. 2020; Zhang et al. 2020), thereby balancing the ATP/NADPH ratio required for Calvin-Benson cycle through establishing transmembrane proton motive force and producing additional ATP (Arnon 1971; Kramer and Evans 2011).

Cyanobacterial NDH-1 is an important member of photosynthetic NDH-1 (Peltier et al. 2016). Four types of NDH-1, NDH-1L, NDH-1L', NDH-1MS, and NDH-1MS', have been predicted in a unicellular cyanobacterium *Synechocystis* sp. strain PCC 6803 (hereafter *Synechocystis* 6803) based on results of reverse genetics experiments (Ohkawa et al. 2000; Shibata et al. 2001; Battchikova and Aro 2007). Proteomics and cryo-electron microscopy (cryo-EM) structure studies have confirmed the presence of NDH-1L, NDH-1MS, and NDH-1MS' in either *Synechocystis* 6803 or *Thermosynechococcus elongatus* (Herranen et al. 2004; Zhang et al. 2005; Xu et al. 2008; Wulfhorst et al. 2014; Laughlin et al. 2019; Schuller et al. 2019; Pan et al. 2020; Zhang et al. 2020), although NDH-1L' is not yet identified in cyanobacteria. They all are involved in cyclic electron transfer around PSI (PSI CET) (Bernát et al. 2011) and share the common NDH-1M module that contains 4 oxygenic photosynthesis-specific (OPS) regulatory subunits NdhL, NdhO, NdhS, and NdhV in addition to 11 structural components NdhA to NdhC, NdhE, NdhG to NdhK, NdhM, and NdhN (Ogawa 1992; Prommeenate et al. 2004; Battchikova et al. 2005, 2011a; Arteni et al. 2006; Zhao et al. 2014; Ma and Ogawa 2015; Gao et al. 2016a; He and Mi 2016; He et al. 2016; Peltier et al. 2016), as schematically represented in Fig. 1. To carry out the PSI CET function, the common NDH-1M module has to associate

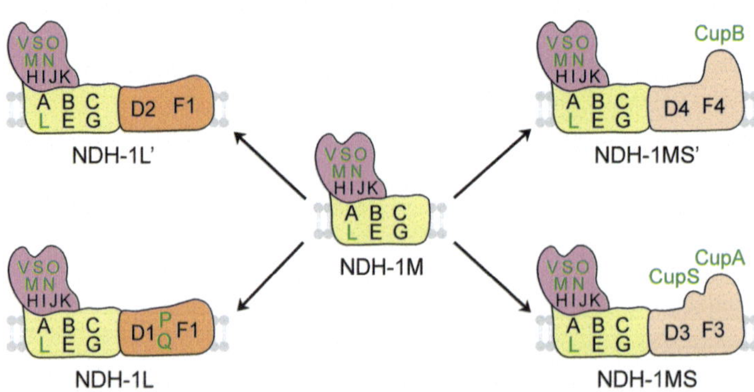

Fig. 1 Subunit compositions of four types of cyanobacterial NDH-1 complexes, all of which share a common NDH-1M module but possess of a specific variable module. These OPS subunits and core subunits of NDH-1 complexes are indicated by green and black letters, respectively

selectively with one of the four variable modules that contain different NdhD and NdhF subunits (Battchikova et al. 2011b; Peltier et al. 2016) (Fig. 1). The NdhD subunit mediates the direct interaction with the NDH-1M module, and one β-carotene molecule may be involved in stabilizing their association (Laughlin et al. 2019; Schuller et al. 2019).

Recently, a NDH-1L-PSI supercomplex has been identified in *Synechocystis* 6803, and its formation facilitates PSI CET via NDH-1L (Gao et al. 2016b). In this book chapter, we describe the cyanobacterial NDH-1-PSI supercomplex, with emphasis on its identification, function, assembly, and evolutional changes.

2 Identification

Under normal growth conditions, the L-shaped NDH-1L complex is a dominant type of NDH-1 in cyanobacteria (Zhang et al. 2004; Laughlin et al. 2019; Schuller et al. 2019). By using the forward genetics together with physiological and biochemical analyses, a NDH-1L-PSI supercomplex has been identified in a unicellular cyanobacterium *Synechocystis* 6803 (Gao et al. 2016b). Further, we found that NDH-1L can interact with PSI to form the supercomplex via a PSI-specific antenna CpcG2-phycobilisome (hereafter CpcG2-PBS) (Fig. 2). As a consequence, deletion of CpcG2, a linker protein for the PSI-specific antenna, completely destabilized the NDH-1L-PSI supercomplex and significantly decreased the accumulated levels of NDH-1L complex through preventing the assembly of CpcG2-PBS antenna (Gao et al. 2016b). In addition, deletion of an OPS regulatory subunit NdhL had little, if any, effect on the accumulation of NDH-1L in the thylakoid membrane (Battchikova et al. 2005) but completely destabilized the NDH-1L-PSI supercomplex (Gao et al. 2016b). It appears plausible that the hydrophilic arm side of NDH-1L is more close to the PSI, as schematically represented in Fig. 2.

When cyanobacterial cells are transferred to environments with either low CO_2 or high light, the L-shaped NDH-1L complex is still constitutively expressed, but a U-shaped NDH-1MS complex is significantly induced (Hihara et al. 2001; Zhang et al. 2004; Arteni et al. 2006). Consequently, NDH-1MS becomes the major type of NDH-1 to reinforce its CO_2 acquisition required for the Calvin-Benson cycle and

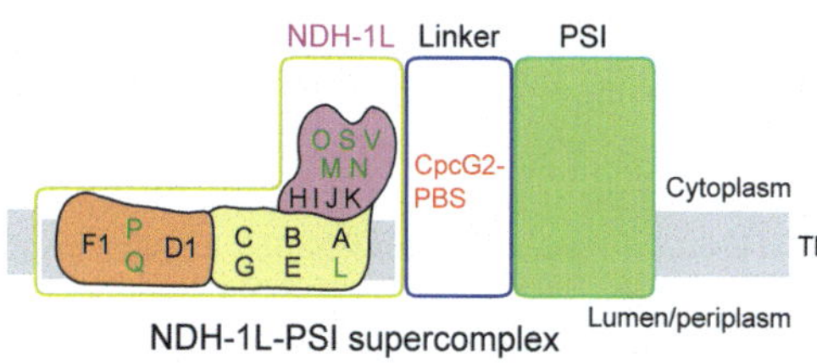

Fig. 2 Schematic model of cyanobacterial NDH-1L-PSI supercomplex. NDH-1L interacts with PSI to form the NDH-1L-PSI supercomplex via a PSI-specific antenna CpcG2-PBS. TM, thylakoid membrane

decrease ROS production (Zhang et al. 2004). Recently, under conditions of low CO_2 or high light, the NDH-1MS-PSI supercomplex is also identified in the cyanobacterium *Synechocystis* 6803 cells using blue-native (BN)-PAGE together with Western blot (unpublished data).

Over the past two decades, studies using the reverse genetics and DNA microarray strategies suggested that NDH-1MS' is constitutively expressed, but NDH-1L' is significantly induced under various environmental stresses (Hihara et al. 2001; Shibata et al. 2001). However, NDH-1MS' and NDH-1L' complexes have not been biochemically isolated yet, possibly because of their low abundance and/or fragile character. Therefore, whether NDH-1MS'-PSI and NDH-1L'-PSI supercomplexes are also present in cyanobacteria remains elusive.

3 Function

About 30 years ago, studies using NDH-1-defective mutants together with physiological analyses indicated that cyanobacterial NDH-1 is involved in a variety of bioenergetic reactions, including respiration, cyclic electron transfer around PSI (PSI CET), and CO_2 acquisition (Ogawa 1991; Mi et al. 1992; Ohkawa et al. 2000). Deletion of the PSI-specific antenna CpcG2-PBS impaired NDH-1-dependent cyclic electron transfer around PSI (NDH-CET) but had little, if any, effect on activities of cellular respiration and CO_2 uptake (Gao et al. 2016b). We thus propose that the formation of the NDH-1-PSI supercomplex in cyanobacteria indeed facilitates PSI CET via NDH-1L.

Recently, studies using cryo-EM and laser flash-induced absorption change strategies demonstrated that photoreduced Fd by PSI in the light can inject electrons into NDH-1L to enable PSI CET (Schuller et al. 2019). Based on this fact and the PSI-specific antenna CpcG2-PBS that simultaneously connects with NDH-1L and PSI complexes (Fig. 2), we speculate that sunlight absorbed by the CpcG2-PBS antenna may be involved in Fd reduction, and then, photoreduced Fd inject electrons into NDH-1L to enable PSI CET. Clearly, formation of NDH-1L-PSI supercomplex provides an important structural basis for efficient operation of PSI CET in cyanobacteria.

More recently, our structural data indicated that on top of the NDH-1L peripheral arm, Fd binds with these two OPS regulatory subunits NdhV and NdhO through electrostatic interactions (Zhang et al. 2020). Our functional data further indicated that NdhV and NdhO positively and negatively PSI CET, respectively, and NdhO is absent in the NDH-1MS complex (Zhang et al. 2020). It is worthy of note that in cyanobacteria, NDH-1L is a dominant type of NDH-1 under normal growth conditions, but NDH-1MS becomes a major type of NDH-1 under conditions of low CO_2 or high light (Hihara et al. 2001; Zhang et al. 2004). Collectively, it appears that operation of these positive and negative regulatory mechanisms under normal growth conditions depends on the NDH-1L-PSI supercomplex and under conditions of either low CO_2 or high light depends on the NDH-1MS-PSI supercomplex.

Therefore, we propose that these positive and negative regulatory mechanisms balance the ATP/NADPH ratio required for the Calvin-Benson cycle and consequently reduce the ROS production against environmental stresses in cyanobacterial natural habitat.

4 Assembly

As schematically represented in Fig. 3, we noticed that the assembly process of the cyanobacterial NDH-1 hydrophilic arm might be involved in a dynamic transition of NAI300 and NAI130 with the aid of several OPS assembly factors, including Ssl3829, Slr1097, and Ssl3451 (Wang et al. 2016; Ran et al. 2019). Two OPS assembly factors Slr1097 and Ssr2781, respectively, interact with NdhI and NdhK, two conserved [4Fe–4S] cluster subunits, and may be involved in their mature process (Dai et al. 2013; unpublished data). A cytoplasmic protein Slr0067 may be also involved in their mature process as scaffold for the assembly of the [4Fe–4S] cluster (Dai et al. 2013; unpublished data). Further, Ssl3829 is considered to be involved in the transition from NAI300 to NAI130 (Wang et al. 2016), and Ssl3451 is removed from NAI130 before the incorporation of the NDH-1 hydrophilic arm into the membrane parts. The NDH-1 hydrophilic arm is almost fully assembled in the cytoplasm and is then incorporated into the thylakoid membrane, probably by interacting with two membrane proteins, NdhA and NdhL (Fig. 3). NdhN and NdhO are likely incorporated into the hydrophilic arm during this process, because they are not being detected in NAI300 and NAI130 (Wang et al. 2016). Although the assembly steps for the hydrophilic arm have been clarified, those for the membrane arm remain elusive. Similarly, the assembly process of NDH-1-PSI supercomplex is

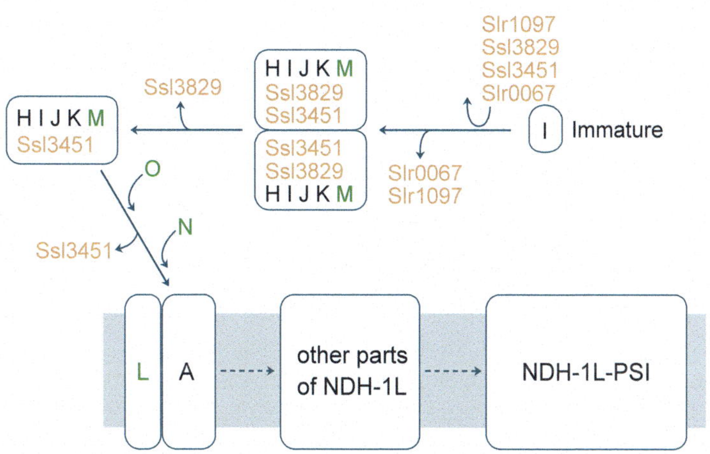

Fig. 3 Schematic assembly model of the NDH-1L-PSI supercomplex in cyanobacteria. These OPS subunits and core subunits of the NDH-1L complex are indicated by green and black letters, respectively. Non-subunit factors required for the assembly process are depicted in orange

even more unclear. For example, little is known regarding the stoichiometry between NDH-1 and CpcG2-PBS-PSI complexes and which non-subunit factors assist in the assembly of the NDH-1 complex with the CpcG2-PBS-PSI complex.

5 Evolutional Change

During the evolution from cyanobacteria to higher plants, there were significant changes in the linker protein in between NDH-1 and PSI and type and assembly of NDH-1-PSI supercomplex, although both cyanobacterial and chloroplastic NDH-1-PSI supercomplexes are involved in NDH-CET (Peng et al. 2009; Gao et al. 2016b). It is now well established that, in higher plants, the chloroplast NDH-1 together with PSI also forms an NDH-1-PSI supercomplex, and its NDH-1/PSI stoichiometry is 1:2 (Peng et al. 2008, 2009; Kouřil et al. 2014). It was demonstrated that two minor proteins of the light-harvesting complex I (LHCI), Lhca5 and Lhca6, are necessary for the supercomplex formation (Peng et al. 2009). A similar NDH-1-PSI supercomplex has also been found in *Physcomitrella patens* that harbors the LHCI protein related to Lhca5 (Alboresi et al. 2008; Neilson and Durnford 2010) but not in *Marchantia polymorpha*, which lacks homologs of Lhca5 or Lhca6 protein (Ueda et al. 2012). Similarly, the homologs of Lhca5 and Lhca6 proteins are also absent in cyanobacteria. However, we found that a PSI-specific antenna CpcG2-PBS can connect simultaneously with NDH-1 and PSI to form the NDH-1-PSI supercomplex in the cyanobacterium *Synechocystis* 6803 (Gao et al. 2016b). This clearly indicates that the linker proteins between NDH-1 and PSI alter during evolution from cyanobacteria to higher plants.

It is well known that in higher plants, only a type of NDH-1-PSI supercomplex is present in *Arabidopsis thaliana*, *Hordeum vulgare*, and *Physcomitrella patens* (Peng et al. 2008, 2009; Kouřil et al. 2014) or is absent in *Marchantia polymorpha* (Ueda et al. 2012). In contrast, multiple types of NDH-1-PSI supercomplex are present in *Synechocystis* 6803, at least including NDH-1L-PSI supercomplex (Gao et al. 2016b) and NDH-1MS-PSI supercomplex (unpublished data). This difference is well consistent with this fact that the activity of NDH-CET in cyanobacteria is much higher than that in higher plants.

We noticed that homologs of certain chloroplast NDH-1 assembly factors are absent in the cyanobacterial genome. For instance, CRR41 is involved in the NDH-1 hydrophilic arm in *Arabidopsis* as a scaffold protein (Peng et al. 2012), but its homolog is missing in the cyanobacterial genome. We also noticed that the roles of certain NDH-1 assembly factors are altered during evolution from cyanobacteria to higher plants. For example, PAM68L in *Arabidopsis* is required for the assembly of the membrane arm part, but its homolog Sll0933 in *Synechocystis* 6803 is not required for NDH-1 assembly (Armbruster et al. 2013). Collectively, the linker proteins between NDH-1 and PSI, together with the type and assembly of NDH-1-PSI supercomplex, change during evolution from cyanobacteria to higher plants, although they originate from the same photosynthetic NDH-1 branch (Peltier et al. 2016).

6 Concluding Remarks

Cyanobacterial NDH-1 is predominantly, if not totally, located in the thylakoid membrane (Ohkawa et al. 2001), accepts electrons from reduced Fd (Battchikova et al. 2011a; Schuller et al. 2019), and is involved in photosynthetic reactions, including NDH-CET and CO_2 acquisition (Ogawa 1991; Mi et al. 1992). Based on these results above, we propose that in addition to PSII, Cytb_6f, and PSI, NDH-1 is the fourth photosynthetic electron transport in the thylakoid membrane. This proposal is corroborated by the cyanobacterial NDH-1-PSI supercomplex formation (Gao et al. 2016b). Further, formation of NDH-1-PSI supercomplex during evolution is closely associated with its photosynthetic reaction, NDH-CET (Gao et al. 2016b). Recently, structural and functional data suggest that NDH-1 constitutes a key portion of the PSI CET pathway—electrons donated by photoreduced Fd are transferred to PQ via the chain of three [4Fe-4S] clusters N6a, N6b, and N2 harbored by NdhI and NdhK subunits (Fd-N6a-N6b-N2-PQ) (Laughlin et al. 2019; Schuller et al. 2019; Pan et al. 2020; Zhang et al. 2020). However, how photosynthetic electrons are transferred from PSI to NDH-1 is still poorly understood. Therefore, obtaining the cryo-EM structure of NDH-1-PSI supercomplex in cyanobacteria will be still necessary to clarify the key transport of electrons in the PSI CET pathway.

Acknowledgments This work was supported by the National Natural Science Foundation of China (31570235 and 31770259) and Science and Technology Commission of Shanghai Municipality (17070502900 and 18DZ2260500).

References

Alboresi, A., Caffarri, S., Nogue, F., Bassi, R., & Morosinotto, T. (2008). *In silico* and biochemical analysis of *Physcomitrella patens* photosynthetic antenna: Identification of subunits which evolved upon land adaptation. *PLoS One, 3*, e2033.

Armbruster, U., Rühle, T., Kreller, R., Strotbek, C., Zühlke, J., Tadini, L., Blunder, T., Hertle, A. P., Qi, Y., Rengstl, B., Nickelsen, J., Frank, W., & Leister, D. (2013). The photosynthesis affected mutant68-like protein evolved from a PSII assembly factor to mediate assembly of the chloroplast NAD(P)H dehydrogenase complex in Arabidopsis. *Plant Cell, 25*, 3926–3943.

Arnon, D. I. (1971). The light reactions of photosynthesis. *Proceedings of the National Academy of Sciences of the United States of America, 68*, 2883–2892.

Arteni, A. A., Zhang, P., Battchikova, N., Ogawa, T., Aro, E. M., & Boekema, E. J. (2006). Structural characterization of NDH-1 complexes of *Thermosynechococcus elongatus* by single particle electron microscopy. *Biochimica et Biophysica Acta, 1757*, 1469–1475.

Battchikova, N., & Aro, E. M. (2007). Cyanobacterial NDH-1 complexes: Multiplicity in function and subunit composition. *Physiologia Plantarum, 131*, 22–32.

Battchikova, N., Zhang, P., Rudd, S., Ogawa, T., & Aro, E. M. (2005). Identification of NdhL and Ssl1690 (NdhO) in NDH-1L and NDH-1M complexes of *Synechocystis* sp. PCC 6803. *The Journal of Biological Chemistry, 280*, 2587–2595.

Battchikova, N., Wei, L., Du, L., Bersanini, L., Aro, E. M., & Ma, W. (2011a). Identification of a novel Ssl0352 protein (NdhS), essential for efficient operation of cyclic electron transport around photosystem I, in NADPH: Plastoquinone oxidoreductase (NDH-1) complexes of *Synechocystis* sp. PCC 6803. *The Journal of Biological Chemistry, 286*, 36992–37001.

Battchikova, N., Eisenhut, M., & Aro, E. M. (2011b). Cyanobacterial NDH-1 complexes: Novel insights and remaining puzzles. *Biochimica et Biophysica Acta, 1807*, 935–944.

Bernát, G., Appel, J., Ogawa, T., & Rögner, M. (2011). Distinct roles of multiple NDH-1 complexes in the cyanobacterial electron transport network as revealed by kinetic analysis of $P700^+$ reduction in various *ndh*-deficient mutants of *Synechocystis* sp. strain PCC6803. *Journal of Bacteriology, 193*, 292–295.

Dai, H., Zhang, L., Zhang, J., Mi, H., Ogawa, T., & Ma, W. (2013). Identification of a cyanobacterial CRR6 protein, Slr1097, required for efficient assembly of NDH-1 complexes in *Synechocystis* sp. PCC 6803. *The Plant Journal, 75*, 858–866.

Friedrich, T., & Scheide, D. (2000). The respiratory complex I of bacteria, archaea and eukarya and its module common with membrane-bound multisubunit hydrogenases. *FEBS Letters, 479*, 1–5.

Friedrich, T., Steinmüller, K., & Weiss, H. (1995). The proton-pumping respiratory complex I of bacteria and mitochondria and its homologue in chloroplasts. *FEBS Letters, 367*, 107–111.

Gao, F., Zhao, J., Wang, X., Qin, S., Wei, L., & Ma, W. (2016a). NdhV is a subunit of NADPH dehydrogenase essential for cyclic electron transport in *Synechocystis* sp. strain PCC 6803. *Plant Physiology, 170*, 752–760.

Gao, F., Zhao, J., Chen, L., Battchikova, N., Ran, Z., Aro, E. M., Ogawa, T., & Ma, W. (2016b). The NDH-1L-PSI supercomplex is important for efficient cyclic electron transport in cyanobacteria. *Plant Physiology, 172*, 1451–1464.

He, Z., & Mi, H. (2016). Functional characterization of the subunits N, H, J, and O of the NAD(P) H dehydrogenase complexes in *Synechocystis* sp. strain PCC 6803. *Plant Physiology, 171*, 1320–1332.

He, Z., Xu, M., Wu, Y., Lv, J., Fu, P., & Mi, H. (2016). NdhM subunit is required for the stability and the function of NAD(P)H dehydrogenase complexes involved in CO_2 uptake in *Synechocystis* sp. strain PCC 6803. *The Journal of Biological Chemistry, 291*, 5902–5912.

Herranen, M., Battchikova, N., Zhang, P., Graf, A., Sirpiö, S., Paakkarinen, V., & Aro, E. M. (2004). Towards functional proteomics of membrane protein complexes in *Synechocystis* sp. PCC 6803. *Plant Physiology, 134*, 470–481.

Hihara, Y., Kamei, A., Kanehisa, M., Kaplan, A., & Ikeuchi, M. (2001). DNA microarray analysis of cyanobacterial gene expression during acclimation to high light. *Plant Cell, 13*, 793–806.

Kouřil, R., Strouhal, O., Nosek, L., Lenobel, R., Chamrád, I., Boekema, E. J., Šebela, M., & Ilík, P. (2014). Structural characterization of a plant photosystem I and NAD(P)H dehydrogenase supercomplex. *The Plant Journal, 77*, 568–576.

Kramer, D. M., & Evans, J. R. (2011). The importance of energy balance in improving photosynthetic productivity. *Plant Physiology, 155*, 70–78.

Laughlin, T. G., Bayne, A. N., Trempe, J. F., Savage, D. F., & Davies, K. M. (2019). Structure of the complex I-like molecule NDH of oxygenic photosynthesis. *Nature, 566*, 411–414.

Ma, W., & Ogawa, T. (2015). Oxygenic photosynthesis-specific subunits of cyanobacterial NADPH dehydrogenases. *IUBMB Life, 67*, 3–8.

Mi, H., Endo, T., Schreiber, U., Ogawa, T., & Asada, K. (1992). Electron donation from cyclic and respiratory flows to the photosynthetic intersystem chain is mediated by pyridine nucleotide dehydrogenase in the cyanobacterium *Synechocystis* PCC 6803. *Plant & Cell Physiology, 33*, 1233–1237.

Neilson, J. A. D., & Durnford, D. G. (2010). Structural and functional diversification of the light-harvesting complexes in photosynthetic eukaryotes. *Photosynthesis Research, 106*, 57–71.

Ogawa, T. (1991). A gene homologous to the subunit-2 gene of NADH dehydrogenase is essential to inorganic carbon transport of *Synechocystis* PCC 6803. *Proceedings of the National Academy of Sciences of the United States of America, 88*, 4275–4279.

Ogawa, T. (1992). Identification and characterization of the *ictA/ndhL* gene product essential to inorganic carbon transport of *Synechocystis* PCC6803. *Plant Physiology, 99*, 1604–1608.

Ohkawa, H., Pakrasi, H. B., & Ogawa, T. (2000). Two types of functionally distinct NAD(P)H dehydrogenases in *Synechocystis* sp. strain PCC6803. *The Journal of Biological Chemistry, 275*, 31630–31634.

Ohkawa, H., Sonoda, M., Shibata, M., & Ogawa, T. (2001). Localization of NAD(P)H dehydrogenase in the cyanobacterium *Synechocystis* sp. strain PCC 6803. *Journal of Bacteriology, 183*, 4938–4939.

Pan, X., Cao, D., Xie, F., Xu, F., Su, X., Mi, H., Zhang, X, & Li, M. (2020). Structural basis for electron transport mechanism of complex I-like photosynthetic NAD(P)H dehydrogenase. *Nature Communications, 11*, 610.

Peltier, G., Aro, E. M., & Shikanai, T. (2016). NDH-1 and NDH-2 plastoquinone reductases in oxygenic photosynthesis. *Annual Review of Plant Biology, 67*, 55–80.

Peng, L., Shimizu, H., & Shikanai, T. (2008). The chloroplast NAD(P)H dehydrogenase complex interacts with photosystem I in *Arabidopsis*. *The Journal of Biological Chemistry, 283*, 34873–34879.

Peng, L., Fukao, Y., Fujiwara, M., Takami, T., & Shikanai, T. (2009). Efficient operation of NAD(P)H dehydrogenase requires the supercomplex formation with photosystem I via minor LHCI in *Arabidopsis*. *Plant Cell, 21*, 3623–3640.

Peng, L., Fukao, Y., Fujiwara, M., & Shikanai, T. (2012). Multistep assembly of chloroplast NADH dehydrogenase-like subcomplex A requires several nucleus-encoded proteins, including CRR41 and CRR42, in Arabidopsis. *Plant Cell, 24*, 202–214.

Prommeenate, P., Lennon, A. M., Markert, C., Hippler, M., & Nixon, P. J. (2004). Subunit composition of NDH-1 complexes of *Synechocystis* sp. PCC 6803: Identification of two new *ndh* gene products with nuclear-encoded homologues in the chloroplast Ndh complex. *The Journal of Biological Chemistry, 279*, 28165–28173.

Ran, Z., Zhao, J., Tong, G., Gao, F., Wei, L., & Ma, W. (2019). Ssl3451 is important for accumulation of NDH-1 assembly intermediates in the cytoplasm of *Synechocystis* sp. strain PCC 6803. *Plant & Cell Physiology, 60*, 1374–1385.

Schuller, J. M., Birrell, J. A., Tanaka, H., Konuma, T., Wulfhorst, H., Cox, N., Schuller, S. K., Thiemann, J., Lubitz, W., Sétif, P., Ikegami, T., Engel, B. D., Kurisu, G., & Nowaczyk, M. M. (2019). Structural adaptations of photosynthetic complex I enable ferredoxin-dependent electron transfer. *Science, 363*, 257–260.

Shibata, M., Ohkawa, H., Kaneko, T., Fukuzawa, H., Tabata, S., Kaplan, A., & Ogawa, T. (2001). Distinct constitutive and low-CO_2-induced CO_2 uptake systems in cyanobacteria: Genes involved and their phylogenetic relationship with homologous genes in other organisms. *Proceedings of the National Academy of Sciences of the United States of America, 98*, 11789–11794.

Ueda, M., Kuniyoshi, T., Yamamoto, H., Sugimoto, K., Ishizaki, K., Kohchi, T., Nishimura, Y., & Shikanai, T. (2012). Composition and physiological function of the chloroplast NADH dehydrogenase-like complex in *Marchantia polymorpha*. *The Plant Journal, 72*, 683–693.

Wang, X., Gao, F., Zhang, J., Zhao, J., Ogawa, T., & Ma, W. (2016). A cytoplasmic protein Ssl3829 is important for NDH-1 hydrophilic arm assembly in *Synechocystis* sp. strain PCC 6803. *Plant Physiology, 171*, 864–877.

Wulfhorst, H., Franken, L. E., Wessinghage, T., Boekema, E. J., & Nowaczyk, M. M. (2014). The 5 kDa protein NdhP is essential for stable NDH-1L assembly in *Thermosynechococcus elongatus*. *PLoS One, 9*, e103584.

Xu, M., Ogawa, T., Pakrasi, H. B., & Mi, H. (2008). Identification and localization of the CupB protein involved in constitutive CO_2 uptake in the cyanobacterium, *Synechocystis* sp. strain PCC 6803. *Plant & Cell Physiology, 49*, 994–997.

Yamamoto, H., & Shikanai, T. (2013). *In planta* mutagenesis of Src homology 3 domain-like fold of NdhS, a ferredoxin-binding subunit of the chloroplast NADH dehydrogenase-like complex in *Arabidopsis*: A conserved Arg-193 plays a critical role in ferredoxin binding. *The Journal of Biological Chemistry, 288*, 36328–36337.

Yamamoto, H., Peng, L., Fukao, Y., & Shikanai, T. (2011). An Src homology 3 domain-like fold protein forms a ferredoxin binding site for the chloroplast NADH dehydrogenase-like complex in *Arabidopsis*. *Plant Cell, 23*, 1480–1493.

Zhang, P., Battchikova, N., Jansen, T., Appel, J., Ogawa, T., & Aro, E. M. (2004). Expression and functional roles of the two distinct NDH-1 complexes and the carbon acquisition complex NdhD3/NdhF3/CupA/Sll1735 in *Synechocystis* sp. PCC 6803. *Plant Cell, 16*, 3326–3340.

Zhang, P., Battchikova, N., Paakkarinen, V., Katoh, H., Iwai, M., Ikeuchi, M., Pakrasi, H. B., Ogawa, T., & Aro, E. M. (2005). Isolation, subunit composition and interaction of the NDH-1 complexes from *Thermosynechococcus elongatus* BP-1. *The Biochemical Journal, 390*, 513–520.

Zhang, C., Shuai, J., Ran, Z., Zhao, J., Wu, Z., Liao, R., Wu, J., Ma, W., & Lei, M. (2020). Structural insights into NDH-1 mediated cyclic electron transfer. *Nature Communications*, *11*, 888.

Zhao, J., Gao, F., Zhang, J., Ogawa, T., & Ma, W. (2014). NdhO, a subunit of NADPH dehydrogenase, destabilizes medium size complex of the enzyme in *Synechocystis* sp. strain PCC 6803. *The Journal of Biological Chemistry, 289*, 26669–26676.

Recent Progress on the LH1-RC Complexes of Purple Photosynthetic Bacteria

Long-Jiang Yu and Fei Ma

Abstract Photosynthetic bacteria have been proven to be excellent model organisms because they own the relatively simplified model systems for us to study the reactions that occurs at the initial stage of photosynthesis, compared with the oxygen-evolving cyanobacteria, algae, and higher plants. In purple bacteria, there are usually two kinds of light-harvesting (LH) complexes, named LH1 and LH2, respectively. LH2 is the peripheral antenna complex, and LH1 is the core antenna complex that surrounds the reaction center (RC) to form the LH1-RC supercomplex. Solar energy is first absorbed by the LH complex and then transferred rapidly and efficiently to the RC, where the charge separation and electron transfer take place. Several high-resolution structures are available for the RC and LH2 for a long time; for LH1-RC complex, its structure was solved and improved to an atomic resolution recently with a thermophilic purple photosynthetic bacterium *Thermochromatium tepidum*. The high-resolution structure provided much more detailed structural information of this supercomplex including the arrangements of protein subunits, pigments, and cofactors; a much more intact RC complex due to the protection of LH1 complex; the detailed coordination of the Ca^{2+} ions in the LH1 that are important for the absorption maximum at 915 nm as well as for the enhanced thermostability; the possible ubiquinone exchange pathway in the closed LH1 ring; and so on. In addition, the dynamic processes involved in this complex were also discussed. All these results greatly advance our understanding on the molecular mechanism of bacterial photosynthesis, which could be essential for designing artificial photoelectronic conversion materials with enhanced performance.

Keywords Charge separation · Energy trapping · Exciton delocalization · LH1-RC · Purple photosynthetic bacteria

L.-J. Yu (✉) · F. Ma (✉)
Photosynthesis Research Center, Key Laboratory of Photobiology, Institute of Botany, Chinese Academy of Sciences, Beijing, China
e-mail: longer@ibcas.ac.cn; fma@ibcas.ac.cn

Q. Wang (ed.), *Microbial Photosynthesis*,
https://doi.org/10.1007/978-981-15-3110-1_3

1 Introduction

Photosynthesis is one of the most fundamentally important reactions on Earth, and it provides the essential oxygen, food, and fossil fuels to survive the living beings. Among these photosynthetic organisms, anoxygenic purple photosynthetic bacteria have been investigated extensively and proved to be excellent model organisms for the study of photosynthesis because of their relatively simple photosynthetic apparatuses including photochemical reaction center (RC) and light-harvesting (LH) complexes. Purple photosynthetic bacteria usually contain two kinds of light-harvesting complexes that are called LH1 and LH2, respectively. LH2 is located around the LH1 and RC, also named "peripheral antenna complex," and is absent in some species, whereas LH1 is located around the RC to form the LH1-RC supercomplex that exists in all purple photosynthetic bacteria. Both LH2 and LH1-RC are integral pigment-membrane proteins located on the photosynthetic membrane to fulfill the photosynthesis. In primary processes of purple bacteria, light energy is first absorbed by the LH complexes and then transferred to the RC by the LH2 → LH1 → RC sequence, where the primary charge separation takes place to initiate the cyclic electron transfer and the subsequent various reactions leading ultimately to the synthesis of ATP. LH1-RC assembly is a "perfect" solar cell, not only due to its high photoelectronic conversion efficiency but also its stability and adaptability to different environmental factors such as low or excess light and low or high temperature. Hence understanding its functioning mechanism is essential for designing artificial photoelectronic conversion materials with enhanced performance.

Almost all RC consists H, M, and L subunits and cytochrome c (Cyt) subunit for some species, and that from *Blastochloris viridis* is the first membrane protein complex whose three-dimension structure was solved by X-ray crystallography (Deisenhofer et al. 1985). The high-resolution structure of LH2 is available from two kinds of photosynthetic bacteria, which showed that it forms a cylinder structure composed of 8–9 pairs of α- and β-subunits (Mcdermott et al. 1995; Koepke et al. 1996). LH1 complex has a similar structure with LH2, but a larger ring composed of 14–17 pairs of α- and β-subunits or α-, β-, and γ-subunits that could encircle RC to form the LH1-RC supercomplex (Roszak et al. 2003; Qian et al. 2013, 2018). Since the structure of RC was determined, a lot of attempts were made to solve the structure of LH1-RC, and several results were reported by using AFM, crystallography, or cryo-EM from different species of bacteria at various resolutions. The overall structures of LH1-RC from different species show similar features that RC is surrounded by a closed/semi-closed circle/elliptical ring; however, there are still some differences that are found among them, such as the number of LH1 subunits associated, the presence or absence of auxiliary small subunits, the metal ion-binding sites, and so on. Among all these reported structures of LH1-RC, the highest resolution is that from *Thermochromatium tepidum* determined by X-ray crystallography at 1.9 Å, the atomic resolution structure provided much detailed information about the organization of protein subunits and pigments, cofactors, as well as the Ca^{2+} ion coordination in this thermophilic complex which is important for its functioning (Yu et al. 2016, 2018a; Niwa et al. 2014). In this chapter, we describe a brief overview of the structural determination of LH1-RC, the structural features, and their functional implications of this complex revealed from its atomic structure; on the other hand, we give an overview of the involved dynamic processes including exciton delocalization and relaxation on the LH1 ring, LH1 → RC energy trapping, and charge separation and electron transfer in RC, which show the complete working scheme of LH1-RC with high efficiency, and a brief introduction is presented regarding self-protection against excess light via transferring energy to carotenoid molecules or via other pathways.

2 Structure of the LH1-RC Complexes

2.1 Overall Structure of LH1-RC Complex

Since the structure of RC from *Blc. viridis* was determined, many structures of LH1-RC were reported by using several different techniques such as atomic force microscopy, cryo-EM, and X-ray crystallography in the last decades. The first crystal structure was reported at a relatively low resolution of 4.8 Å in 2003 from *Rhodopseudomonas palustris*, which showed that the LH1 is an incomplete ring consisting of 15 pairs of the α- and β-subunits and their associated bacteriochlorophylls (Roszak et al. 2003). At the position of the 16th α/β pair, there is a single transmembrane subunit designated W that prevents the complete closure of the ring, and this position is close to the binding pocket of Q_B. The resulted "gap" was proposed to be the exchange channel for quinones/quinols to transport through the LH1 ring (Fig. 1a). LH1-RC from *Rhodobacter sphaeroides* was solved by cryo-EM at 8.5 Å (Qian et al. 2008) and X-ray at 8.0 Å (Qian et al. 2013), respectively, showing that it is a dimer and each monomer consists of 14 pairs of LH1 subunit and a PufX to form the semi-closed ring, and the function of PufX was proposed to facilitate the quinone/quinol exchange (Fig. 1b). With the rapid development of cryo-EM technology, the structure of LH1-RC from *Blc. viridis* was determined by cryo-EM at 2.9 Å (Qian et al. 2018) and that from a filamentous photosynthetic bacterium *Roseiflexus castenholzii* was determined at 4.1 Å resolution (Xin et al. 2018), respectively. The LH1 of *Blc. viridis* contains α-/β-/ũFE;-subunits, and LH1-RC consists of 17 pairs of α-/β-subunits, together with 16 ũFE;-subunits packed between the β-subunits in the ring structure. The lack of the 17th ũFE;-apoprotein created an opening in the ring (Fig. 1c), which was suggested to be like the case of *Rps. palustris* and *Rb. sphaeroides* to provide the space for the translocation of quinones/quinols between the inside and outside of the ring. The complex from *Rf. castenholzii* also showed an elliptical ring; however, the ring is an irregular one composed of 15 pairs of α-/β-subunits, a hypothetical subunit X, and the N-terminal transmembrane helix C-TM from Cyt subunit (Fig. 1d).

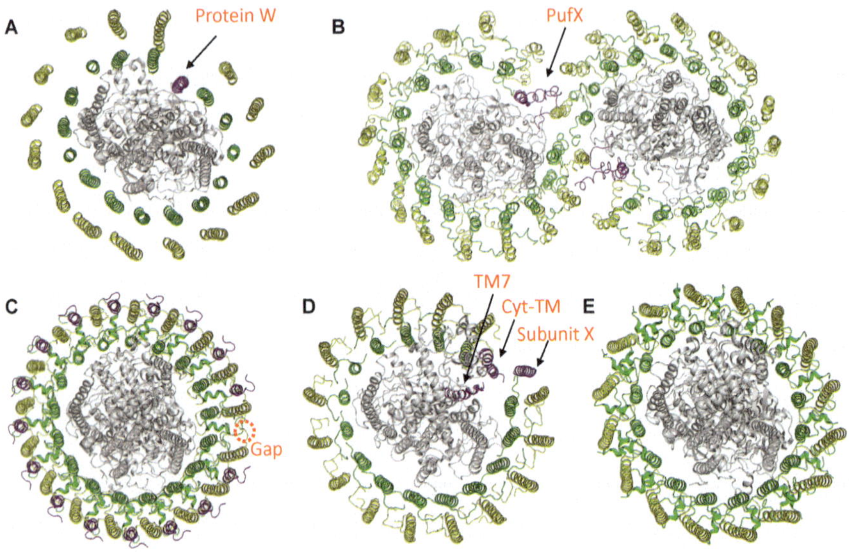

Fig. 1 Comparison of the available LH1-RC structures from photosynthetic bacteria. (**a**) Monomeric LH1-RC complex from *Rps. palustris*. (**b**) Dimeric LH1-RC complex from *Rb. sphaeroides*. (**c**) Monomeric LH1-RC complex from *Blc. viridis*. (**d**) Monomeric LH-RC complex from *Rf. castenholzii*. (**e**) Monomeric LH1-RC from *Tch. tepidum*. Color codes: RC subunits, gray; LH1 α-subunits, green; LH1 β-subunits, yellow; and auxiliary subunits and the LH1 γ-subunits, magenta. Other cofactors including water, lipid, carotenoid, bacteriochlorophyll, and detergent molecules have been omitted for clarity. (The figure is taken from Yu et al. (2018b), copyright by John Wiley & Sons, Inc.)

On the other hand, there was also a completely closed ring structure reported, such as the LH1-RC from *Rhodospirillum rubrum* by cryo-EM at 8.5 Å (Jamieson et al. 2002), whose ring structure consists of 16 pairs of α-/β-subunits. And the crystal structure of LH1-RC from thermophilic photosynthetic bacterium *Tch. tepidum* (Niwa et al. 2014) also showed a completely closed ring consisting of 16 pairs of α-/β-subunits. This structure was first determined at 3.0 Å and improved to 1.9 Å after the further optimization of the crystallization and cryoprotection conditions, and it is among the highest resolution for such a large, multi-subunit membrane protein complex (Yu et al. 2018a) (Fig. 1e). There are totally 36 subunits including Cyt, L, M, and H subunits for RC and 32 α-/β-subunits for LH1, associated with 36 BChls *a*, 16 spirilloxanthin molecules, and 16 Ca^{2+} ions, to form the supercomplex with an overall molecular weight of approximately 400 kDa. In addition, the Sr^{2+}-/Ba^{2+}-substituted structures from *Tch. tepidum* were also determined at 3.0 Å and 3.3 Å (Yu et al. 2016), respectively, which also showed a completely close elliptical ring similar with the Ca^{2+}-bound one. Due to the surrounding of LH1 complex, RC was suggested to be in a more stable state, and many structural details were observed compared with the previously determined isolated RC-only complexes. Owing to the high-resolution structure, some regions in RC which were not observed clearly even in the high-resolution structure could be assigned unambiguously, and flexible loop regions in the N- and C-termini of the LH1 subunits were also traced; in addition, many cofactors including lipids, ubiquinones, and metal ions were identified for the first time.

2.2 Novel Structural Features of the Intact RC Complex

RC complex from *Blc. viridis* is the first membrane protein whose crystal structure was solved, and a large number of RC structures from different species of purple bacteria *Blc. viridis*, *Rb. sphaeroides*, and *Tch. tepidum* have been deposited in the Protein Data Bank since then. Among these structures, *Blc. viridis* (Deisenhofer et al. 1985; Roszak et al. 2012) and *Tch. tepidum* (Nogi et al. 2000) have four subunits, but *Rb. sphaeroides* have three, and it lacks the Cyt subunit (Allen et al. 1988; Xu et al. 2004). Due to the surrounding of LH1, RC complex was suggested to be much closer to its native state, and several large differences were found in the high-resolution structure of LH1-RC supercomplex compared with those isolated RC structures. These differences are mainly located in the Cyt subunit including the N-terminal region (residues 172–196) and a loop region (residues 44–58) of the H subunit (Yu et al. 2018a) (Fig. 2a). It had been reported that the Cyt subunit of the RC from *Blc. viridis* is a lipoprotein and its N-terminal cysteine linked to a diglyceride via a thioether bond (Roszak et al. 2012; Weyer et al. 1987), and this interaction could possibly be broken by the X-ray radiation (Wohri et al. 2009), so it was not confirmed in some other structures. In the high-resolution structure of LH1-RC from *Tch. tepidum*, it was found that the N-terminal cysteine of Cyt subunit was triacylated with *N*-acyl and *S*-diacylglycerol in a manner similar to that of an outer membrane protein (Kulathila et al. 2011), which could anchor the Cyt subunit in the membrane (Fig. 2b). A relatively long loop region (172–196) of Cyt subunit was found to be largely deviated from that of the isolated RCs, and a Ca^{2+} ion previously assigned at a medium resolution located in its vicinity was changed to Mg^{2+} ion based on the X-ray anomalous scattering. With the mediation of Mg^{2+} ion, this loop was closer to the surface of the membrane, and it would not swing in the outer space; it seemed the flexibility of this long loop was reduced to be in the stable state (Fig. 2c, e). At last, the loop region (44–58) of the H subunit was also found to be different with the previously determined RC-only structures. The RC of *Tch. tepidum* was determined at 2.2 Å resolution (Nogi et al. 2000); however, this region was still not observed without the protection of LH1. And in the structure of RC from *Blc. viridis* (Roszak et al. 2012) at 1.92 Å, this loop was also ambiguous due to the reason it located in the crystal lattice contact (Fig. 2d). In the high-resolution structure of LH1-RC (Yu et al. 2018a), this region was traced unambiguously based on the electron density map, and it was similar with the cryo-EM structure of LH1-RC from *Blc. viridis* reported recently (Qian et al. 2018); the slight differences between the two structures may arise from the sequence identity. Both of these results indicate that the RC structure in the LH1-RC is much closer to the native state owing to the surrounding of LH1 and interactions with LH1 complex.

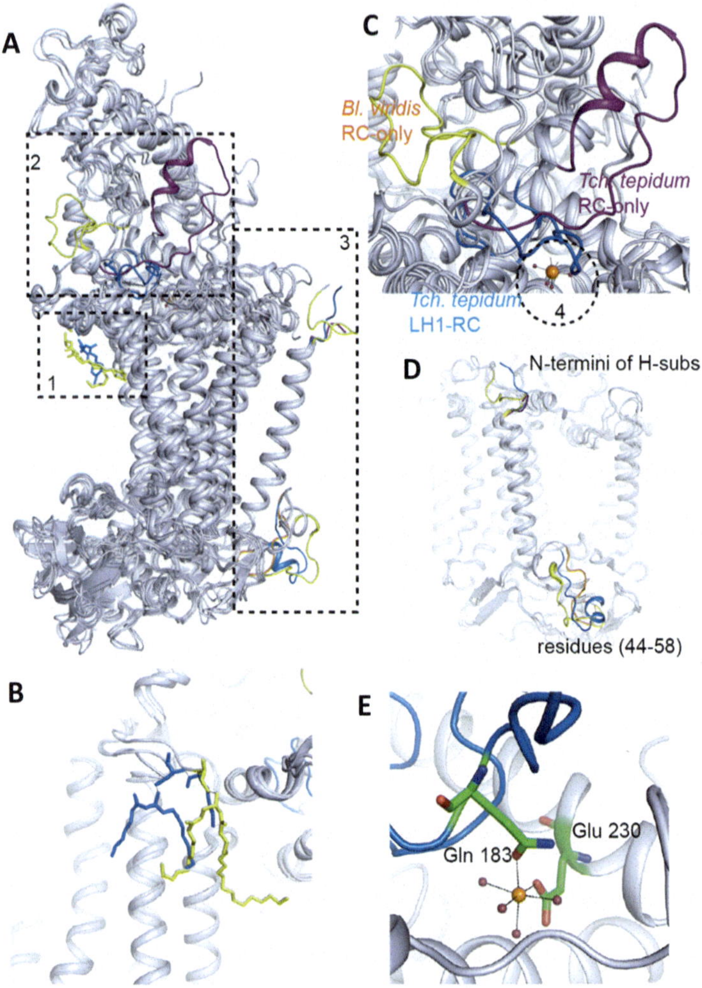

Fig. 2 Novel features found in the intact RC core complex from *Tch. tepidum*. (**a**) Superposition of the RC-only structures from *Tch. tepidum* (PDB code 1EYS, purple), *Rb. sphaeroides* (PDB code 2J8C, orange), *Bl. viridis* (PDB code 3T6E, yellow), and the intact one in the LH1-RC complex (PDB code 5Y5S, marine). Regions with similar structures are colored in gray, whereas the three regions with large differences are boxed and colored differently. Numbers of 1, 2, and 3 in each boxed area are enlarged in panels (**b**), (**c**), and (**d**), respectively. (**b**) Tri-acylation of the Cyt N-terminal cysteine. (**c**) The loop region of Cyt (residues 172–196) including the Mg^{2+}-binding site, which is circled and enlarged in panel (**e**). (**d**) The N-terminal region and residues 44–58 of the H subunit. (**e**) The Mg^{2+}-binding site in the Cyt subunit. (The figure is taken from Yu et al. (2018a), copyright 2018 Nature Publishing Group)

2.3 Potential Exchange Pathway for Quinones

According to the primary reaction in the RC, the secondary quinone electron acceptor Q_B releases from its binding site after accepting two electrons and protonation and then is transferred to the quinone pool to fulfill the electron transport chain. High-resolution crystal structure provides a solid basis for the inspection of ubiquinone exchange and transport in the LH1-RC complex with a closed LH1 ring, which could not be investigated in the RC-only structures. The channels in LH1 are located in the N-terminal ends of transmembrane domain, and both size and shape well fit the head group of the ubiquinone. The LH1 α-polypeptides define the narrowest portion of the channels to form the inner gate, and the similar number of residues in the same region of LH1 β-polypeptides forms the outer gate of the channel (Niwa et al. 2014). Variations were found in the size and shape of channel openings, which may reflect the "breathing motion" of the LH1 complex (Wang-Otomo 2016). Since the LH1 of *Tch. tepidum* is an almost symmetric and completely closed ring, no specific transport route for the quinones/quinols is found. In the high-resolution structure of LH1-RC, another four ubiquinone molecules were assigned between LH1 and RC, and the tail of one ubiquinone was found to be inserted into a pore formed by the α- and β-subunits and BChl of LH1(Yu et al. 2018a) (Fig. 3a, b). This ubiquinone may be in the exchange state between the inside and outside of the ring through this pore. Since the LH1 complex is composed of 16 pairs of α- and

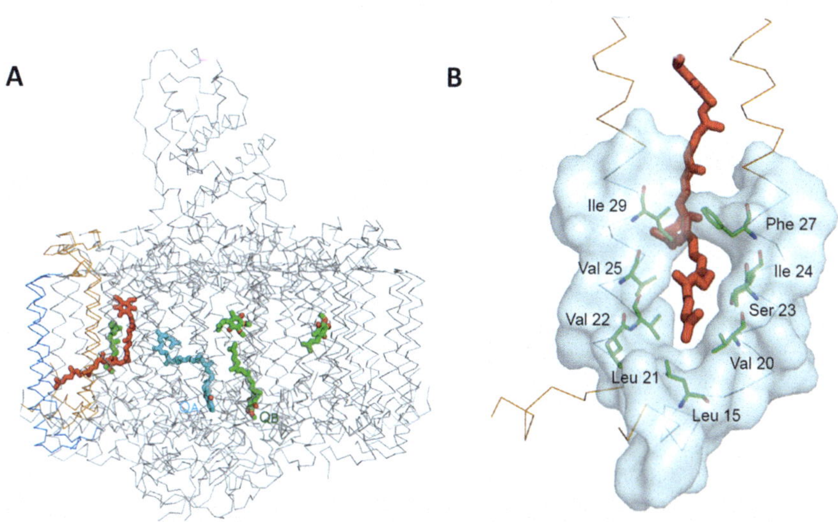

Fig. 3 Distribution of quinone molecules in the LH1-RC. (**a**) Distribution of the menaquinone and ubiquinone molecules in the LH1-RC. Color codes: UQs, green; MQ, cyan; and UQ (one MQ with its tail inserted in the channel between the LH1 α- and β-subunits), red. The two α-subunits and two β-subunits of LH1 that form the channel for the transport of the UQ are shown in orange and blue. (**b**) Quinone exchange channel between the α- and β-subunits of LH1. (The figure is taken from Yu et al. (2018a), copyright 2018 Nature Publishing Group)

β-subunits, there will be 16 potential pores existed for the exchanges of the quinones which will give rise to a high efficiency (Yu et al. 2018b). In the cryo-EM structure of LH1-RC from *Blc. viridis* (Qian et al. 2018), one ubiquinone molecule was found to be located in the position close to LH1, and it was suggested to be a candidate to transport through the pore created by the absence of the 17th γ-subunit. In this semi-closed ring structure, there is only one specific pore, and it may limit the efficiency of the quinone exchange between the inside and outside of the LH1 ring; however, photosynthetic organisms could adopt different strategies to grow photosynthetically.

2.4 Structural Basis for the Redshift and Enhanced Thermostability

Previous investigations have shown that LH1-RC from *Tch. tepidum* has two unique features including LH1 Q_y redshift and higher thermostability compared with its mesophilic counterparts, which has been suggested to be brought about by the binding of Ca^{2+} ions (Kimura et al. 2008, 2009). The Ca^{2+}-binding sites were first suggested to be located in the C-terminal loop region of the LH1 subunits (Yu et al. 2010), and the detailed binding environment of the Ca^{2+} ions was not determined unambiguously at the medium resolution previously (Niwa et al. 2014) due to the flexibility of this region. In the improved high-resolution structure, all ligands involved in the Ca^{2+} coordination were identified clearly based on the electron density map; they include the side chain of α-Asp49; the carbonyl oxygens of α-Trp46, α-Ile51, and (n + 1) β-Trp45; and two water molecules to form a six-coordinate structure (Fig. 4a, b). Around the Ca^{2+}-binding sites, a large amount of interactions were found between the α- and β-subunits, and the neighboring LH1 subunits on the periplasmic side were connected together by the mediation of the Ca^{2+}-binding sites (Fig. 4a, c). The sequence alignment of α-polypeptides shows that there is deletion of the α-43 residue in *Tch. tepidum*, which is strongly related to the Ca^{2+}-binding environment. The insertion of an alanine into this site would disrupt the Ca^{2+} binding, and the LH1 Q_y absorption would blueshift as a result (Nagashima et al. 2017). Spectroscopy measurements also indicate that the binding of Ca^{2+} in LH1 reduces the conformational flexibility and contributes to the redshift of the LH1 Q_y transition (Jakob-Grun et al. 2012; Ma et al. 2015, 2016). In addition, a recent study also showed that the incorporation of Ca^{2+} ions confers a significant redshift (973 nm) as well as thermostability to *Thiorhodovibrio* strain 970, which was cultivated at 22 °C (Imanishi et al. 2019). The comparison showed that Ca^{2+} is important to the unusual redshift of the LH1 Q_y transitions of both strain 970 and *Tch. tepidum*, whereas it confers moderate (*Trv.* strain 970) and major (*Tch. tepidum*) thermal stability on their LH1-RC complexes, respectively. However, the mechanism of this difference is not clear for the lack of detailed structural information of LH1-RC from *Trv.* strain 970.

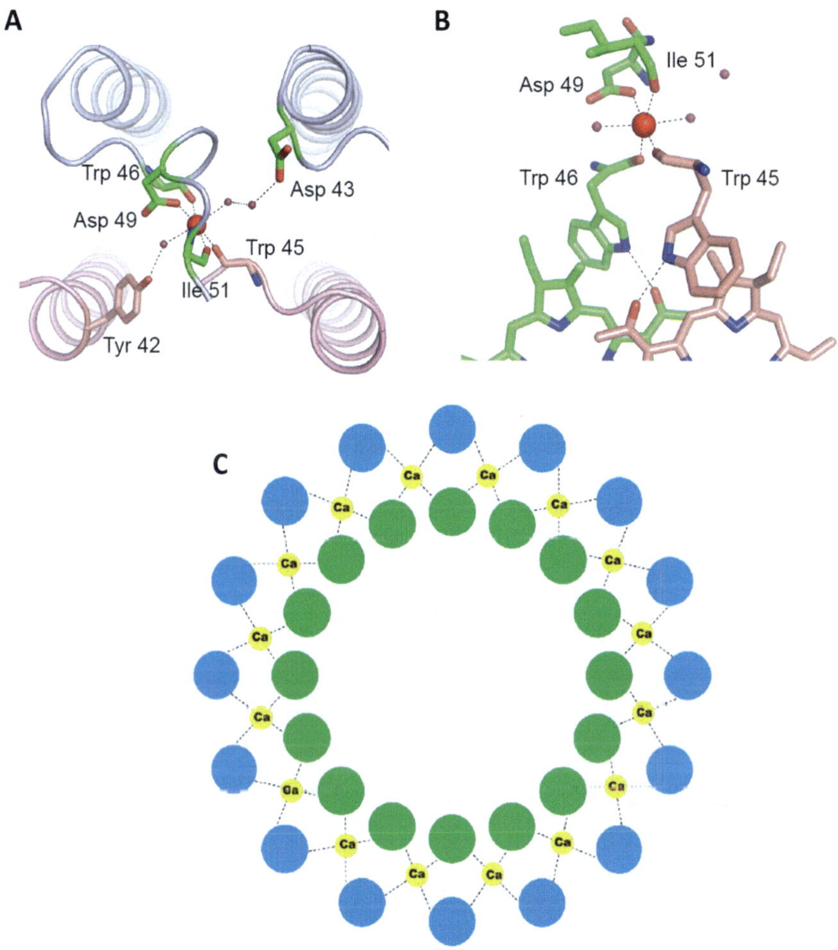

Fig. 4 Calcium-binding sites in the LH1 complex. (**a**) Top view of the Ca^{2+}-binding site among neighboring LH1 subunits. LH1 α-subunits, silver; LH1 β-subunits, pink; Ca^{2+} ion, red spheres. (**b**) A close-up view of the Ca^{2+}-binding site. α-Subunit, green; β-subunits, pink; Ca^{2+} ion, red spheres. (**c**) Binding pattern of 16 Ca^{2+} ions in LH1. α-Subunit, green; β-subunit, blue; Ca^{2+} ion, yellow spheres. (The figure is taken from Yu et al. (2018a), copyright 2018 Nature Publishing Group)

3 Dynamic Process Involved in the LH1-RC

3.1 Exciton Delocalization and Relaxation on the LH1 Ring

The distance between the neighboring BChl *a* molecules is ~9 Å, leading to a considerable nearest-neighbor coupling strength of 250–400 cm^{-1} (Zerlauskiene et al. 2008; Ma et al. 2017a). As a result, excitation is not localized on one BChl *a* molecule; instead, it is coherently delocalized among all BChl *a* molecules on the LH1 ring. Due to the static (inhomogeneity in local site energy) and dynamic (interaction with the phonon bath) disorders induced by slow and fast nuclear motions, the delocalization is not over the whole LH1 ring but over a number of BChl *a* molecules (Zerlauskiene et al. 2008; van Grondelle and Novoderezhkin 2006). These BChl *a* aggregates can be theoretically described with the Frenkel exciton model in combination with modified Redfield theory (van Grondelle and Novoderezhkin 2006; Novoderezhkin et al. 1999). The exciton energies and wave functions are obtained from diagonalization of the exciton Hamiltonian:

$$H = \sum_{n=1}^{N} \left(\varepsilon_n + q_n^{(c)} \right) |n\, n| + \sum_{n \neq m}^{N} V_{nm} \, |n\, m| + H_{ph}, \tag{1}$$

where n is the excitation localized on the nth BChl *a* in LH1 and ε_n is its transition energy. The inhomogeneously distributed ε_n reflecting the static disorder is assumed to be a Gaussian random variable with mean energy ε_0 and width σ. V_{nm} denotes the coupling strength between the nth and the mth BChl *a*. The coupling parameters for neighboring BChl *a* molecules are V_{12} in one heterodimer and V_{23} in adjacent heterodimers. The non-neighboring couplings are calculated with the dipole-dipole interaction, $\frac{1}{4\pi\varepsilon_0}\left[\frac{\vec{\mu}_n \cdot \vec{\mu}_m}{\left|\vec{r}_{nm}\right|^3} - \frac{3\left(\vec{\mu}_n \cdot \vec{r}_{nm}\right)\left(\vec{\mu}_m \cdot \vec{r}_{nm}\right)}{\left|\vec{r}_{nm}\right|^5} \right]$, where $\vec{\alpha}$ and $\vec{r}$ were taken directly from the crystal structures. $q_n^{(c)}$ represents the bath energy fluctuation upon excitation of the nth BChl *a*. H_{ph} is the free phonon Hamiltonian. All relevant information about the dynamic exciton interaction with the bath is contained in the correlation function of this energy fluctuation, the Fourier transform of which is related to the spectral density function (SDF).

Introducing the excitonic states $|k = \sum_{n=1}^{N} c_n^k |n$ with an energy of $E_k = \hbar\omega_k$, the Hamiltonian in Eq. 1 can be diagonalized in the exciton representation:

$$H = \sum_{k=1}^{N} \left(E_k + q_k^{(c)} \right) |k\, k| + H_{ph}, \tag{2}$$

In such a model, the lowest excitonic state gains very little dipole strength; the largest part of the dipole strength is collected in the next degenerate pair of excitonic states, the $k = \pm 1$ states; and all the other states contain no (or very little) dipole strength. $E_{\pm 1}$ is much lower than ε_0 (Fig. 5), responsible for the spectral redshift of the Q_y absorption from 760 nm in organic solvent to 850 nm in LH2 and 880 nm in LH1. Other interactions further modify the electric structure, such as mixing charge-transfer state into the lowest excitonic state (Ma et al. 2015) and hydrogen-bond interaction with environmental amino acids (Sturgis and Robert 1997; Ma et al. 2009). For example, these interactions induce the LH1–Q_y absorption of *Tch. tepidum* redshifted to 915 nm (Ma et al. 2015).

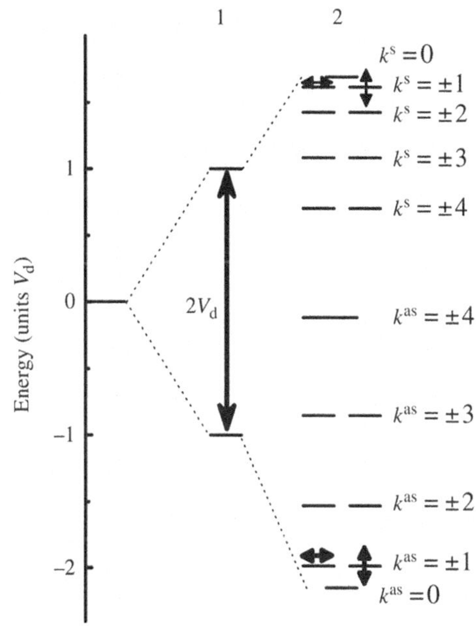

Fig. 5 Energy-level scheme of the excited-state manifold of a dimer (manifold 1) and of the B850 ring of LH2 (manifold 2, can be viewed as a smaller LH1 ring)

Upon excited by absorbing a photon, ultrafast exciton relaxation occurs from higher to lower and finally to the lowest excitonic state. This excitation equilibration process is prompt, as shown by transient absorption technique; it finishes in 200 fs (Visser et al. 1995). Pronounced oscillatory features due to vibrational modes are extensively observed (Monshouwer et al. 1998), so interplay of vibrational coherence and exciton relaxation/migration is taken into account when simulating the excitation dynamics of LH1. Recently, our two-dimensional electronic spectroscopy (2DES) results provide clear evidence that specific vibrational modes coherently link higher and lower excitonic states and this facilitates ultrafast exciton relaxation (Ma et al. 2017b). The prompt excitation equilibration benefits the following energy transfer to RC, which occurs from the lowest excitonic state of LH1.

3.2 *LH1 → RC Energy Trapping*

Excitation on LH1 is transferred to and trapped by RC, a process that follows classical energy hopping mechanism. It is a reversible reaction, with typical time constants for trapping and detrapping of 35~50 and 4~10 ps (Fig. 6a), determined by transient absorption and 2DES (Visscher et al. 1989; Ma et al. 2017c). Despite of the existence of faster detrapping, trapping is still efficient, and its quantum yield is ~95% (Campillo et al. 1977). It is because that P is dominated by another even faster pathway, charge separation. All reversed reactions in the dynamic scheme help to avoid accumulation of excess charges.

There is a noticeable phenomenon that the energy donor, LH1 Q_y, is close to (Visscher et al. 1989) or lower than (Permentier et al. 2000) that of acceptor, a pair of BChl *a* molecules called special pair (P). For example, in *Tch. tepidum*, the excited-state energy of LH1 is 435 cm^{-1} lower than P. To explain this "uphill" energy transfer, a thermally activated barrier-crossing model is employed (Ma et al. 2016; Visscher et al. 1989):

$$k_{\text{trap}} = \left(4\pi 2H_{ab}{}^{2} / h\right)\left(4\pi\lambda RT\right)^{-\frac{1}{2}} \exp\left(-\Delta E^{*} / RT\right) \tag{3}$$

where $\Delta E = E_{Q_y} - E_P$ is the energy difference, calculated from the difference of absorption maxima between LH1 Q_y and P; λ is the total reorganization energy due to the effect of the protein matrix and the solution; $\Delta E^* = (\lambda + \Delta E)^2/4\lambda$ is the activation energy; and $H_{ab} = \hat{\mu}_P \cdot \hat{\mu}_{Q_y} - 3\left(\hat{\mu}_{Q_y} \cdot \hat{R}\right)\left(\hat{\mu}_P \cdot \hat{R}_{Q_yP}\right)$ is the electric state coupling constant. Fitting the temperature dependence of k_{trap} (Fig. 6b) gives the value of λ and H_{ab}. The results show that reorganization energy, reflecting the influence of environmental protein matrix on pigments, could modulate energy transfer rates, and photosynthetic organisms can optimize their structures to adjust different environments.

Other mechanism such as superexchange is also proposed to account for the large energy gap, i.e., P serves as a virtual mediator in a one-step trapping process directly leading to the initial charge separation (Sumi 2004). However, there has been no experimental evidence yet.

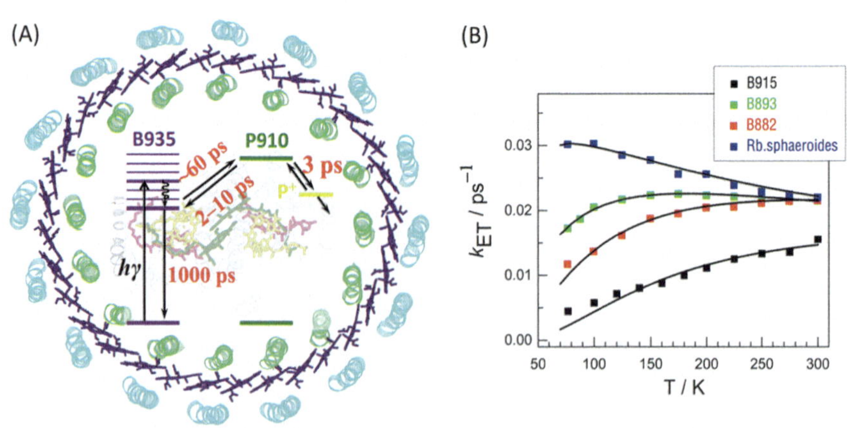

Fig. 6 (**a**) Dynamic scheme of excitation trapping, detrapping, and charge separation in LH1-RC. Copyright American Chemical Society Publications. (**b**) Dependence of the energy transfer rates (k_{ET}) of different LH1-RC complexes on temperature. They give different changing tendency with changing temperature. (The figure is taken from Ma et al. (2016), copyright Elsevier Publications)

3.3 *Charge Separation and Electron Transfer in RC*

Light-powered charge separation is accomplished by bacteriochlorin cofactors: a pair of BChl *a* called P, two monomeric BChl *a* (B), and two bacteriopheophytin *a* (H) molecules arranged in two branches around an axis of quasi-twofold symmetry (Fig. 7a). Charge separation is initiated from the excited state of P (P*), forming a partial intradimer charge-transfer intermediate, $P_A^+P_B^-$. Subsequently an electron is transferred to B_A, H_A, and a ubiquinone (Q_A) sequentially, which make up the "active branch" or "A-branch." At room temperature, the electron transfer processes $P^* \rightarrow P^+B_A^-$, $P^+B_A^- \rightarrow P^+H_A^-$, and $P^+H_A^- \rightarrow P^+Q_A^-$ occur with time constants of ~3 ps, ~1 ps, and ~200 ps, respectively (Kamran et al. 2015). Then within a few milliseconds, Q molecules diffuse and remove electrons from the RC and deliver replacements (Fig. 7b). The quantum efficiency (charges separated per photon absorbed) is near a unity.

This remarkable efficiency has drawn extensive interest. Coherent nuclear motions are found that may facilitate charge separation process in RC (Vos et al. 1993). With 2DES, correlation between electronic/vibronic coherences and efficient charge separation is revealed (Romero et al. 2014; Ma et al. 2018). Very recently, we compare the 2DES of three RC mutants with different charge separation rates and show that various coherences contribute to this efficient process (Ma et al. 2019). Short-lived (~100 fs) electronic coherence connects the initial excited state, P*, and the charge-transfer intermediate, $P_A^+P_B^-$, leading to unidirectional displacement of electron density to establish the charge-transfer state. This $P^* \rightarrow P_A^+P_B^-$ step is associated with a long-lived quasi-resonant vibrational coherence; and another vibrational coherence is associated with stabilizing the primary photoproduct, $P^+B_A^-$. The

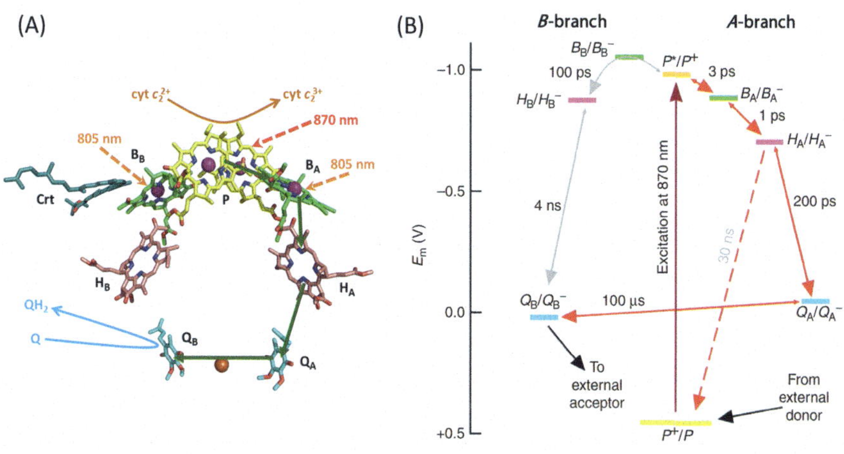

Fig. 7 (**a**) Two branches of cofactors of wild-type RC. (**b**) Midpoint potentials of redox couples involved in A-branch photochemical charge separation (right) and equivalent states involving B-branch cofactors (left). Time constants for each electron transfer step are shown. (The figures are taken from Kamran et al. (2015), copyright Nature Publishing Group)

delicate functioning mechanism could provide insights for new artificial energy conversion systems that exploit coherence phenomena to achieve maximum efficiency.

Excitation energy transfer in RC is also fast, from H over B to P within 200 fs (Jordanides et al. 2001). It is thought that vibronic mixing, i.e., coupling specific vibrational modes to electronic transition, accelerates this process (Paleček et al. 2017).

Another interesting phenomenon is that despite the high level of structural similarity between the two branches of cofactors, only the A-branch is active. Several proposals are advanced to explain this functional symmetry breaking, such as difference in the electronic coupling between cofactors, difference in relative free energies of $P^+B_A^-$ and $P^+B_B^-$, asymmetry in dielectric environments, or asymmetry in protein electrostatic or matrix electric fields (Steffen et al. 1994). A well-accepted point is that the arrangement of protein charges and dipoles create a potential gradient that favors the electron transfer along the A-branch.

3.4 Carotenoid and Photo-Protection

Carotenoid (Car) molecules, another class of pigments in light-harvesting antenna besides BChl *a*, mainly play two roles. Firstly, they capture blue-green light and transfer energy to BChl *a*. Two lowest singlet excited states, S_1 and S_2, and other electronic states such as internal charge-transfer states or vibrational hot excited-state states could transfer singlet excitation to BChl *a* (Polívka and Sundström 2003; Wohlleben et al. 2003) (Fig. 8). The quantum efficiency for energy transfer varies for different Car molecules, ranging from 20% to 90%, depending on electronic structural properties of Car, for example, the number of conjugated C=C double bonds, polar end groups, or twisted configuration (Polívka and Sundström 2003). The light-harvesting pathways of Car in this wavelength region increase the utility of sunlight.

Triplet BChl *a* (^{3}BChl *a*) forms from ^{1}BChl *a* via intersystem crossing in a nanosecond timescale. Long-lived ^{3}BChl *a* is dangerous, since it transfers energy to oxygen and generates harmful singlet O_2 ($^1O_2^*$). The other important role of Car is protecting photosynthetic organisms via quenching ^{3}BChl *a* and/or scavenging singlet $^1O_2$ (Cogdell et al. 2000). Finally, excitation is dissipated by ^{3}Car as heat. Different mechanisms have evolved for photo-protection in higher plants, such as non-photochemical quenching (Ruban et al. 2007).

In bacterial RC, surprisingly, there is only one Car molecule, and it is near the inactive electron transfer branch and cannot proceed triplet energy transfer with B_A or H_A. So, does this mean that Car is not needed for photo-protecting in the electron transfer process? The answer is yes. We recently found that the charge separation and electron transfer processes are reversible reactions and the directions and total rates are modulated by P^+/P potential. When the amount of oxidized P is more than a threshold, charge separation is replaced by charge recombination and formation of

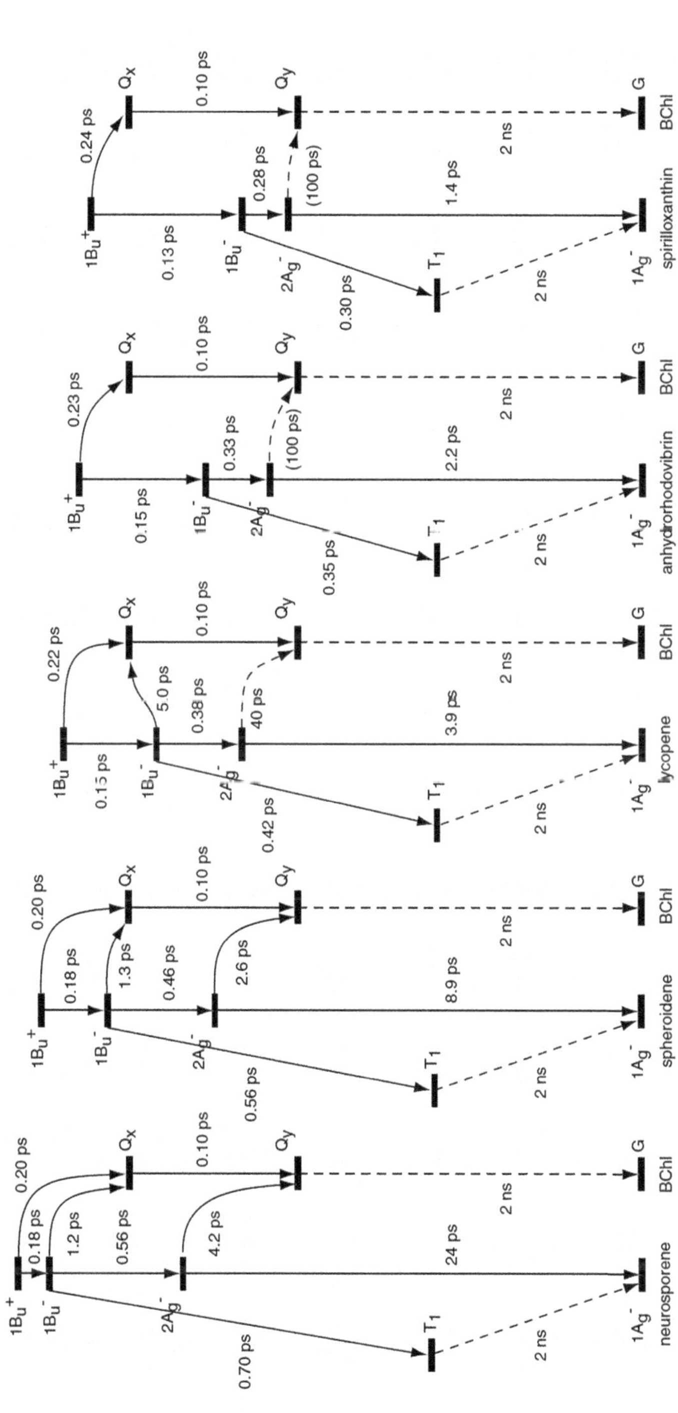

Fig. 8 Relaxation schemes of LH1 binding various Car molecules, including (i) singlet internal conversion in Car and BChl, (ii) singlet-triplet conversion in Car, and (iii) singlet-energy transfer from Car to BChl through different excited states. (This figure is taken from Akahane et al. (2004), copyright Elsevier Publications)

radical pair that dissipates energy to a vibrationally hot ground state (Ma et al. 2017d). This self-regulation mechanism guarantees that no long-lived excited-state bacteriochlorin cofactors accumulate during electron transfer process to produce harmful oxygen species.

4 Concluding Remarks

The recently reported high-resolution structure of the LH1-RC from *Tch. tepidum* provides the first atomic view of this large membrane protein-pigment supercomplex, which revealed the organization of protein subunits, pigment arrangement, an intact RC structure, the exchange channels for the quinones/quinols across the closed LH1 ring, and unique features such as the detailed coordinating pattern of the Ca^{2+} ions which is important for the LH1 Q_y redshift and enhanced thermostability of the supercomplex.

In the natural solar cell LH1-RC, the efficiencies for energy trapping and photoelectronic conversion are remarkably ~95% and ~100%. Meanwhile, it employs various mechanisms to protect from photodamage and adjust to different environments. Further investigations of the unique ecological strategies employed by these extremophiles will provide valuable information for understanding the diversity of photosynthesis. All these are the result of billion-of-years optimization of structures, which will provide an insight to improve artificial photoelectronic materials in the near-infrared region.

Acknowledgments In particular, the authors would like to thank Profs. Zheng-Yu Wang-Otomo, Jian-Ren Shen, Jian-Ping Zhang and Rienk van Grondelle for initiating the related research projects and their general encouragement and helpful hints along the way. This documentation was supported by the National Key R&D Program of China (No. 2019YFA0904600) and the National Natural Science Foundation of China (Project Grant No. 21903086).

References

Akahane, J., Rondonuwu, F. S., Fiedor, L., Watanabe, Y., & Koyama, Y. (2004). Dependence of singlet-energy transfer on the conjugation length of carotenoids reconstituted into the LH1 complex from *Rhodospirillum rubrum* G9. *Chemical Physics Letters, 393*(1–3), 184–191.

Allen, J. P., Feher, G., Yeates, T. O., Komiya, H., & Rees, D. C. (1988). Structure of the reaction center from Rhodobacter sphaeroides R-26: Protein-cofactor (quinones and Fe2+) interactions. *Proceedings of the National Academy of Sciences of the United States of America, 85*(22), 8487–8491.

Campillo, A. J., Hyer, R. C., Monger, T. G., Parson, W. W., & Shapiro, S. L. (1977). Light collection and harvesting processes in bacterial photosynthesis investigated on a picosecond time scale. *Proceedings of the National Academy of Sciences of the United States of American, 74*(5), 1997–2001.

Cogdell, R. J., Howard, T. D., Bittl, R., Schlodder, E., Geisenheimer, I., & Lubitz, W. (2000). How carotenoids protect bacterial photosynthesis. *Philosophical Transactions of the Royal Society B-Biological Sciences, 355*(1402), 1345–1349.

Deisenhofer, J., Epp, O., Miki, K., Huber, R., & Michel, H. (1985). Structure of the protein subunits in the photosynthetic reaction Centre of Rhodopseudomonas viridis at 3A resolution. *Nature, 318*(6047), 618–624.

Imanishi, M., Takenouchi, M., Takaichi, S., Nakagawa, S., Saga, Y., Takenaka, S., Madigan, M. T., Overmann, J., Wang-Otomo, Z.-Y., & Kimura, Y. (2019). A dual role for Ca^{2+} in expanding the spectral diversity and stability of light-harvesting 1 reaction Center Photocomplexes of purple phototrophic Bacteria. *Biochemistry, 58*(25), 2844–2852.

Jakob-Grun, S., Radeck, J., & Braun, P. (2012). Ca(2+)-binding reduces conformational flexibility of RC-LH1 core complex from thermophile Thermochromatium tepidum. *Photosynthesis Research, 111*(1-2), 139–147.

Jamieson, S. J., Wang, P., Qian, P., Kirkland, J. Y., Conroy, M. J., Hunter, C. N., & Bullough, P. A. (2002). Projection structure of the photosynthetic reaction Centre-antenna complex of Rhodospirillum rubrum at 8.5 A resolution. *EMBO Journal, 21*(15), 3927–3935.

Jordanides, X. J., Scholes, G. D., & Fleming, G. R. (2001). The mechanism of energy transfer in the bacterial photosynthetic reaction center. *Journal of Physical Chemistry B, 105*(8), 1652–1669.

Kamran, M., Friebe, V. M., Delgado, J. D., Aartsma, T. J., Frese, R. N., & Jones, M. R. (2015). Demonstration of asymmetric electron conduction in Pseudosymmetrical photosynthetic reaction centre proteins in an electrical circuit. *Nature Communications, 6*, 6530.

Kimura, Y., Hirano, Y., Yu, L. J., Suzuki, H., Kobayashi, M., & Wang, Z. Y. (2008). Calcium ions are involved in the unusual red shift of the light-harvesting 1 Qy transition of the core complex in thermophilic purple sulfur bacterium Thermochromatium tepidum. *The Journal of Biological Chemistry, 283*(20), 13867–13873.

Kimura, Y., Yu, L. J., Hirano, Y., Suzuki, H., & Wang, Z. Y. (2009). Calcium ions are required for the enhanced thermal stability of the light-harvesting-reaction center Core complex from Thermophilic purple sulfur bacterium Thermochromatium tepidum. *The Journal of Biological Chemistry, 284*(1), 93–99.

Koepke, J., Hu, X., Muenke, C., Schulten, K., & Michel, H. (1996). The crystal structure of the light-harvesting complex II (B800-850) from Rhodospirillum molischianum. *Structure, 4*(5), 581–597.

Kulathila, R., Kulathila, R., Indic, M., & van den Berg, B. (2011). Crystal structure of Escherichia coli CusC, the outer membrane component of a heavy metal efflux pump. *PLoS One, 6*(1), e15610.

Ma, F., Kimura, Y., Yu, L.-J., Wang, P., Ai, X.-C., Wang, Z.-Y., & Zhang, J.-P. (2009). Specific Ca^{2+}-binding motif in the LH1 complex from photosynthetic bacterium Thermochromatium tepidum as revealed by optical spectroscopy and structural Modeling. *FEBS Journal, 276*(6), 1739–1749.

Ma, F., Yu, L. J., Wang-Otomo, Z. Y., & van Grondelle, R. (2015). The origin of the unusual Qy red shift in LH1-RC complexes from purple bacteria Thermochromatium tepidum as revealed by stark absorption spectroscopy. *Biochimica et Biophysica Acta, 1847*(12), 1479–1486.

Ma, F., Yu, L. J., Wang-Otomo, Z. Y., & van Grondelle, R. (2016). Temperature dependent LH1→RC energy transfer in purple bacteria Tch. tepidum with shiftable LH1-Q band: A natural system to investigate thermally activated energy transfer in photosynthesis. *Biochimica et Biophysica Acta, 1857*(4), 408–414.

Ma, F., Yu, L.-J., Llansola-Portoles, M. J., Robert, B., Wang-Otomo, Z.-Y., & van Grondelle, R. (2017a). Metal Cations induced αβ-BChl a heterogeneity in LH1 as revealed by temperature-dependent fluorescence splitting. *ChemPhysChem, 18*(16), 2295–2301.

Ma, F., Yu, L.-J., Hendrikx, R., Wang-Otomo, Z.-Y., & van Grondelle, R. (2017b). Excitonic and vibrational coherence in the excitation relaxation process of two LH1 complexes as revealed by two-dimensional electronic spectroscopy. *Journal of Physical Chemistry Letters, 8*(12), 2751–2756.

Ma, F., Yu, L.-J., Hendrikx, R., Wang-Otomo, Z.-Y., & van Grondelle, R. (2017c). Direct observation of energy detrapping in LH1-RC complex by two-dimensional electronic spectroscopy. *Journal of the American Chemical Society, 139*(2), 591–594.

Ma, F., Swainsbury, D. J. K., Jones, M. R., & van Grondelle, R. (2017d). Photoprotection through ultrafast charge recombination in photochemical reaction centres under oxidizing conditions. *Philosophical Transactions of the Royal Society B-Biological Sciences, 372*(1730), 20160378.

Ma, F., Romero, R., Jones, M. R., Novoderezhkin, V. I., & van Grondelle, R. (2018). Vibronic coherence in the charge separation process of the *Rhodobacter sphaeroides* reaction center. *Journal of Physical Chemistry Letters, 9*(8), 1827–1832.

Ma, F., Romero, R., Jones, M. R., Novoderezhkin, V. I., & van Grondelle, R. (2019). Both electronic and vibrational coherences are involved in primary electron transfer in bacterial reaction Center. *Nature Communications, 10*, 933.

Mcdermott, G., Prince, S. M., Freer, A. A., Hawthornthwaitelawless, A. M., Papiz, M. Z., Cogdell, R. J., & Isaacs, N. W. (1995). Crystal structure of an integral membrane light-harvesting complex from photosynthetic bacteria. *Nature, 374*(6522), 517–521.

Monshouwer, R., Baltushka, A., van Mourik, F., & van Grondelle, R. (1998). Time-resolved absorption difference spectroscopy of the LH-1 antenna of *Rhodopseudomonas viridis*. *Journal of Physical Chemistry A, 102*(23), 4360–4371.

Nagashima, K. V. P., Sasaki, M., Hashimoto, K., Takaichi, S., Nagashima, S., Yu, L. J., Abe, Y., Gotou, K., Kawakami, T., Takenouchi, M., Shibuya, Y., Yamaguchi, A., Ohno, T., Shen, J. R., Inoue, K., Madigan, M. T., Kimura, Y., & Wang-Otomo, Z. Y. (2017). Probing structure-function relationships in early events in photosynthesis using a chimeric photocomplex. *Proceedings of the National Academy of Sciences of the United States of America, 114*(41), 10906–10911.

Niwa, S., Yu, L. J., Takeda, K., Hirano, Y., Kawakami, T., Wang-Otomo, Z. Y., & Miki, K. (2014). Structure of the LH1-RC complex from Thermochromatium tepidum at 3.0 A. *Nature, 508*(7495), 228–232.

Nogi, T., Fathir, I., Kobayashi, M., Nozawa, T., & Miki, K. (2000). Crystal structures of photosynthetic reaction center and high-potential iron-sulfur protein from Thermochromatium tepidum: Thermostability and electron transfer. *Proceedings of the National Academy of Sciences of the United States of America, 97*(25), 13561–13566.

Novoderezhkin, V. I., Monshouwer, R., & van Grondelle, R. (1999). Disordered exciton model for the Core light-harvesting antenna of *Rhodopseudomonas viridis*. *Biophysical Journal, 77*(2), 666–681.

Paleček, D., Edlund, P., Westenhoff, S., & Zigmantas, D. (2017). Quantum coherence as a witness of vibronically hot energy transfer in bacterial reaction center. *Science Advances, 3*, e1603141.

Permentier, H. P., Neerken, S., Schmidt, K. A., Overmann, J., & Amesz, J. (2000). Energy transfer and charge separation in the purple non-sulfur bacterium *Roseospirillum parvum*. *Biochimica et Biophysica Acta, 1460*(2-3), 338–345.

Polívka, T., & Sundström, V. (2003). Ultrafast dynamics of carotenoid excited states-from solution to natural and artificial systems. *Chemical Reviews, 104*(4), 2021–2071.

Qian, P., Bullough, P. A., & Hunter, C. N. (2008). Three-dimensional reconstruction of a membrane-bending complex: The RC-LH1-PufX core dimer of Rhodobacter sphaeroides. *The Journal of Biological Chemistry, 283*(20), 14002–14011.

Qian, P., Papiz, M. Z., Jackson, P. J., Brindley, A. A., Ng, I. W., Olsen, J. D., Dickman, M. J., Bullough, P. A., & Hunter, C. N. (2013). Three-dimensional structure of the Rhodobacter sphaeroides RC-LH1-PufX complex: Dimerization and quinone channels promoted by PufX. *Biochemistry, 52*(43), 7575–7585.

Qian, P., Siebert, C. A., Wang, P., Canniffe, D. P., & Hunter, C. N. (2018). Cryo-EM structure of the Blastochloris viridis LH1-RC complex at 2.9 A. *Nature, 556*(7700), 203–208.

Romero, E., Augulis, R., Novoderezhkin, V. I., Ferretti, M., Thieme, J., Zigmantas, D., & van Grondelle, R. (2014). Quantum coherence in photosynthesis for efficient solar energy conversion. *Nature Physics, 10*(9), 676–682.

Roszak, A. W., Howard, T. D., Southall, J., Gardiner, A. T., Law, C. J., Isaacs, N. W., & Cogdell, R. J. (2003). Crystal structure of the RC-LH1 core complex from Rhodopseudomonas palustris. *Science, 302*(5652), 1969–1972.

Roszak, A. W., Moulisova, V., Reksodipuro, A. D., Gardiner, A. T., Fujii, R., Hashimoto, H., Isaacs, N. W., & Cogdell, R. J. (2012). New insights into the structure of the reaction Centre from Blastochloris viridis: Evolution in the laboratory. *The Biochemical Journal, 442*(1), 27–37.

Ruban, A. V., Berera, R., Ilioaia, C., van Stokkum, I. H. M., Kennis, J. T. M., Pascal, A. A., van Amerongen, H., Robert, B., Horton, P., & van Grondelle, R. (2007). Identification of a mechanism of photoprotective energy dissipation in higher plants. *Nature, 450*(7169), 575–578.

Steffen, M. A., Lao, K., & Boxer, S. G. (1994). Dielectric asymmetry in the photosynthetic reaction center. *Science, 264*(5160), 810–816.

Sturgis, J. N., & Robert, B. (1997). Pigment binding-site and electronic properties in light-harvesting proteins of purple bacteria. *Journal of Physical Chemistry B, 101*(37), 7227–7231.

Sumi, H. (2004). Uphill energy trapping by reaction center in bacterial photosynthesis. 2. Unistep charge separation, virtually mediated by special pair, by photoexcitation in place of excitation transfer from the antenna system. *Journal of Physical Chemistry B, 108*(31), 11792–11801.

van Grondelle, R., & Novoderezhkin, V. I. (2006). Energy transfer in photosynthesis: Experimental insights and quantitative models. *Physical Chemistry Chemical Physics, 8*(7), 793–807.

Visscher, K. J., Bergström, H., Sundström, V., Hunter, C. N., & van Grondelle, R. (1989). Antenna BChl-896 to the reaction center in *Rhodospirillum rubrum, Rhodobacter sphaeroides* (W.T. and M21 mutant) from 77 to 177 K, studied by picosecond absorption spectroscopy. *Photosynthesis Research, 22*(3), 211–217.

Visser, H. M., Somsen, O. J. G., van Mourik, F., Lin, S., van Stokkum, I. H. M., & van Grondelle, R. (1995). Direct observation of sub-picosecond equilibration of excitation energy in the light-harvesting antenna of *Rhodospirillum rubrum. Biophysical Journal, 69*(3), 1083–1099.

Vos, M. H., Rappaport, F., Lambry, J.-C., Breton, J., & Martin, J.-L. (1993). Visualization of coherent nuclear motion in a membrane protein by femtosecond spectroscopy. *Nature, 363*(6427), 320–325.

Wang-Otomo, Z.-Y. (2016). Recent understanding on the photosystem of purple photosynthetic bacteria. In *Solar to chemical energy conversion* (pp. 379–390). Cham: Springer.

Weyer, K. A., Schafer, W., Lottspeich, F., & Michel, H. (1987). The cytochrome subunit of the photosynthetic reaction center from Rhodopseudomonas viridis is a lipoprotein. *Biochemistry, 26*(10), 2909–2914.

Wohlleben, W., Buckup, T., Herek, J. L., Cogdell, R. J., & Motzkus, M. (2003). Multichannel carotenoid deactivation in photosynthetic light harvesting as identified by an evolutionary target analysis. *Biophysical Journal, 85*(1), 442–450.

Wohri, A. B., Wahlgren, W. Y., Malmerberg, E., Johansson, L. C., Neutze, R., & Katona, G. (2009). Lipidic sponge phase crystal structure of a photosynthetic reaction center reveals lipids on the protein surface. *Biochemistry, 48*(41), 9831–9838.

Xin, Y., Shi, Y., Niu, T., Wang, Q., Niu, W., Huang, X., Ding, W., Yang, L., Blankenship, R. E., Xu, X., & Sun, F. (2018). Cryo-EM structure of the RC-LH core complex from an early branching photosynthetic prokaryote. *Nature Communications, 9*(1), 1568.

Xu, Q., Axelrod, H. L., Abresch, E. C., Paddock, M. L., Okamura, M. Y., & Feher, G. (2004). X-ray structure determination of three mutants of the bacterial photosynthetic reaction centers from Rb. sphaeroides; altered proton transfer pathways. *Structure, 12*(4), 703–715.

Yu, L. J., Kato, S., & Wang, Z. Y. (2010). Examination of the putative Ca^{2+}-binding site in the light-harvesting complex 1 of thermophilic purple sulfur bacterium Thermochromatium tepidum. *Photosynthesis Research, 106*(3), 215–220.

Yu, L. J., Kawakami, T., Kimura, Y., & Wang-Otomo, Z. Y. (2016). Structural basis for the unusual Qy red-shift and enhanced thermostability of the LH1 complex from Thermochromatium tepidum. *Biochemistry, 55*(47), 6495–6504.

Yu, L. J., Suga, M., Wang-Otomo, Z. Y., & Shen, J. R. (2018a). Structure of photosynthetic LH1-RC supercomplex at 1.9 A resolution. *Nature, 556*(7700), 209–213.

Yu, L. J., Suga, M., Wang-Otomo, Z. Y., & Shen, J. R. (2018b). Novel features of LH1-RC from Thermochromatium tepidum revealed from its atomic resolution structure. *The FEBS Journal, 285*(23), 4359–4366.

Zerlauskiene, O., Trinkunas, G., Gall, A., Robert, B., Urboniene, V., & Valkunas, L. (2008). Static and dynamic protein impact on electronic properties of light-harvesting complex LH2. *Journal of Physical Chemistry B, 112*(49), 15883–15892.

Composition, Organisation and Function of Purple Photosynthetic Machinery

Leanne C. Miller, David S. Martin, Lu-Ning Liu, and Daniel P. Canniffe

Abstract Purple photosynthetic bacteria are metabolically versatile, anoxygenic phototrophs that produce bacteriochlorophylls *a* or *b* and display a wide range of metabolic lifestyles, which are reflected in their diverse range of habitats. Under oxic conditions, energy is derived from aerobic respiration, and the synthesis of photosynthetic pigments, and pigment-binding proteins, is repressed; when conditions shift to anoxic, the ultrastructure of the cytoplasmic membrane of purple bacteria changes, invaginations forming towards the inside of the cell creating intracytoplasmic membranes (ICMs). The photosynthetic machinery for the light-dependent reactions is housed in these ICMs. Light-harvesting (LH) antenna complexes absorb light energy, which is then transferred through a network of pigment–protein complexes, eventually promoting charge separation in the reaction centre, ultimately resulting in the formation of a proton motive force, used to drive ATP synthesis. ATP is then used to reduce inorganic carbon into organic compounds in the Calvin-Benson-Bassham cycle. Purple bacteria have evolved highly intricate assemblies of pigment–protein complexes to efficiently carry out this process. The assembly and organisation of these complexes differ among different species. In this chapter, we will discuss the structure, function and organisation of the photosynthetic components and the mechanisms underlying the photosynthetic process.

Keywords Antenna · Atomic force microscopy · ATP synthase · Bacteriochlorophyll · Bacteriopheophytin · Calvin-Benson-Bassham cycle · Carotenoid · Core complex · Cytochrome · Excitation · Intracellular membrane · Light-harvesting · Photosynthesis · Pigment · Proton motive force · Purple bacteria · Quinone · Reaction centre

---

L. C. Miller · D. P. Canniffe (✉)
Institute of Integrative Biology, University of Liverpool, Liverpool, UK
e-mail: d.canniffe@liverpool.ac.uk

D. S. Martin
Department of Physics, University of Liverpool, Liverpool, UK

L.-N. Liu
Institute of Integrative Biology, University of Liverpool, Liverpool, UK

College of Marine Life Sciences, Ocean University of China, Qingdao, China

Q. Wang (ed.), *Microbial Photosynthesis*,
https://doi.org/10.1007/978-981-15-3110-1_4

1 General Introduction

Photosynthesis is a globally important, fundamental process by which plants, algae and some bacteria convert sunlight into chemical energy. The process relies on a set of highly-ordered, multicomponent assemblies of pigment–protein complexes, located within invaginated membranes; the organisation and dynamics of these complexes are vital for efficient photosynthesis. Although the overall objective of photosynthesis across different organisms is similar—the production of metabolites for the cell—the multiprotein assemblies and compartments that perform this differ among different organisms.

Unlike the chlorophyll-containing plants, algae and cyanobacteria that perform oxygenic photosynthesis, purple bacteria are anoxygenic phototrophs that use bacteriochlorophyll (BChl) *a* or *b* (Fig. 1), as well as carotenoid pigments that are ubiquitous in photosynthesis. The differences in structure between the bacteriochlorophyll and chlorophyll pigments result in varied absorption properties of phototrophic organisms (Fig. 1). Purple phototrophic bacteria are *Proteobacteria*, the most metabolically versatile of the seven phyla of photosynthetic prokaryotes: *Chlorobi*, *Acidobacteria*, *Firmicutes*, *Cyanobacteria*, *Proteobacteria*, *Chloroflexi* and *Gemmatimonadetes* (Gupta and Khadka 2016). Purple bacteria display a wide range of metabolic lifestyles with the capability to be photoautotrophic, photoheterotrophic, chemoautotrophic and mixotrophic, capable of fermentation and aerobic or anaerobic respiration, along with many different metabolic pathways for energy generation and carbon and sulfur metabolism (Imhoff et al. 2005). This is reflected in the diverse habitats in which they grow, including lakes, ponds, estuaries, marine environments, microbial mats, sewage and waste lagoons, as well as extreme niches including hot, cold, acidic, alkaline and hypersaline environments (Madigan 2003). Purple bacteria include both purple sulphur and purple non-sulphur bacteria, classically dependent on their ability to metabolise and assimilate sulphur compounds (Frigaard and Dahl 2008), although more contemporary studies have demonstrated that all purple bacteria are capable of sulphur metabolism to an extent (Hansen and Gemerden 1972). They can use a variety of reductants as electron donors, including H_2S, other sulphur-containing compounds (such as cyst(e)ine and thiosulfate) or H_2 (Truper and Fischer 1982; Brune 2004); some species can use Fe^{2+} iron as an electron donor (Ehrenreich and Widdel 1994). Table 1 shows the basic characteristics of various species of purple bacteria.

Anoxygenic photosynthesis is a simpler form of the process carried out by oxygenic organisms, which makes purple bacteria ideal model systems for dissecting the physiology, biochemistry and molecular biology of photosynthesis. Under oxic conditions, energy is derived from aerobic respiration, and the synthesis of photosynthetic pigments and pigment-binding proteins is repressed. As conditions change through microoxic to anoxic, the ultrastructure of the cytoplasmic membrane forms invaginations towards the inside of the cell called intracytoplasmic membranes (ICMs) (Hunter et al. 2008); these can take the form of vesicles, tubes or lamellae. Figure 2 shows a variety of different ICMs from different species.

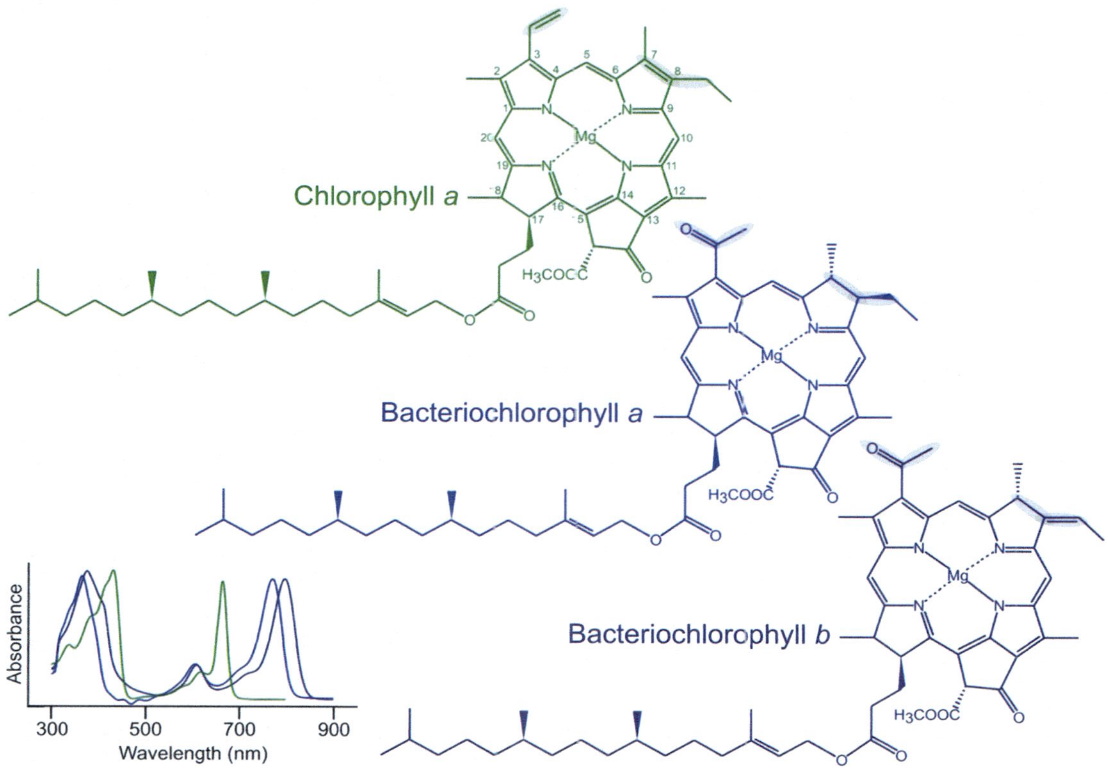

Fig. 1 Structures of (bacterio)chlorophylls. Chemical structures of common (bacterio)chlorophylls. The carbon numbering system displayed on chlorophyll *a* is that approved by the International Union of Pure and Applied Chemistry (IUPAC). Structural differences between these pigments are highlighted in gray. Absorption spectra of these pigments in methanol are displayed (inset)

Table 1 Basic characteristics of various model purple bacterial photosynthetic organisms. AAP, aerobic anoxygenic phototroph; BChl, bacteriochlorophyll

Taxonomy	Phylogeny/class	Family	Genus	Abbreviation	AAP	BChl	Morphology
Purple non-sulfur bacteria	*Alphaproteobacteria*	*Rhodobacteraceae*	*Rhodobacter*	*Rba.*	–	*a*	Rods
		Bradyrhizobiaceae	*Rhodoblastus*	*Rbl.*	–	*a*	Budding rods
			Rhodopseudomonas	*Rps.*	–	*a*	Budding rods
		Rhodospirillaceae	*Rhodospirillum*	*Rsp.*	–	*a*	Spirilla
			Phaeospirillum	*Phs.*	–	*a*	Spirilla
		Hyphomicrobiaceae	*Blastochloris*	*Blc.*	–	*b*	Budding rods
		Acetobacteraceae	*Roseococcus*	*Rsc.*	✓	*a*	Cocci
			Acidiphilium	*Acp.*	✓	*a*	Rods
		Sphingomonadales	*Erythromicrobium*	*Erm.*	✓	*a*	Rods, branched
			Erythromonas	*Emn.*	✓	*a*	Ovoid
	Betaproteobacteria	*Burkholderiales*	*Rubrivivax*	*Rbv.*	–	*a*	Rods, curved rods
Purple sulfur Bacteria	*Gammaproteobacteria*	*Chromatiaceae*	*Allochromatium*	*Alc.*	–	*a*	Rods
			Thermochromatium	*Tch.*	–	*a*	Rods

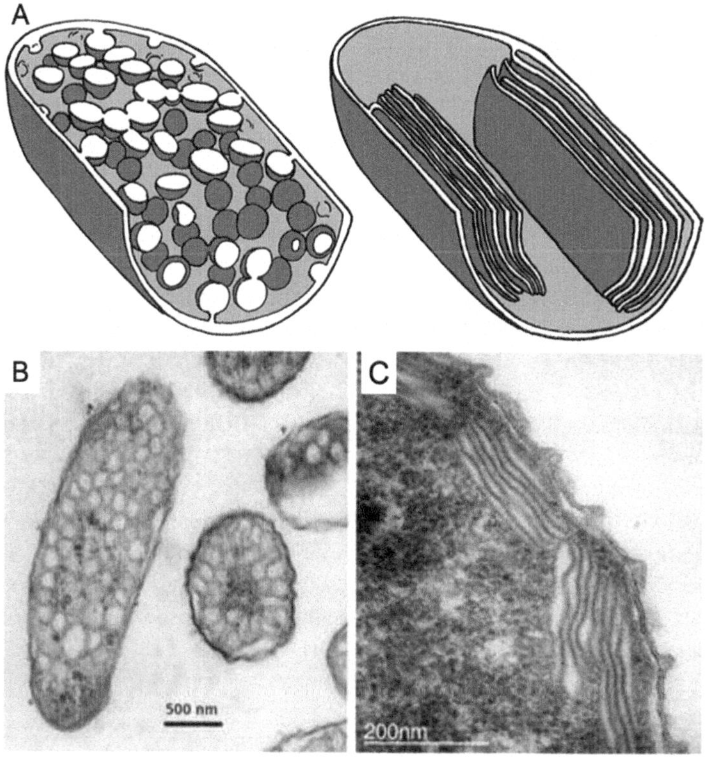

Fig. 2 Intracellular membranes of purple bacteria. (**a**) Schematic illustration of vesicular and lamellar ICM architectures (LaSarre et al. 2018). (**b**) Transmission electron micrograph of *Rhodospirillum* (*Rsp.*) *rubrum* cells, ICM vesicles are visible throughout the cytoplasm (Scheuring and Sturgis 2009). (**c**) TEM of photosynthetic membranes of *Phs. molischianum* cells containing stacked lamellar-type ICM continuous with the cytoplasmic membrane (Scheuring and Sturgis 2009)

Photosynthesis occurs in two stages: the light-dependent reactions, followed by the light-independent reactions. The light-dependent reactions occur in the ICM, where the photosynthetic apparatus is housed. The pigments within light-harvesting (LH) antenna complexes absorb photons of light as resonance energy. The derived excitation energy is then transferred through a network of pigments in the LH complexes towards the reaction centre (RC), where a 'special pair' of BChl pigments are housed. In the RC, charge separation occurs leading to the release of excited electrons, which are then passed down an electron transfer chain to RC-bound quinone electron carriers. This quinone is fully reduced to quinol after two electron transfers, at which point it is released from the RC and diffuses to cytochrome bc_1 (cyt bc_1). The quinol is then oxidised, releasing two protons into the periplasm of the cell; the electrons are passed to the soluble cytochrome c_2 (cyt c_2) and eventually back to the special pair in 'cyclic electron flow'. This results in an electrochemical gradient between the cytoplasm and the periplasm, generating a proton motive force (pmf) utilised by ATP synthase (ATPase) in the production of ATP (Hu et al. 1998) (Fig. 3). In the light-independent stage, the energy produced in the light-dependent reactions is used to fix CO_2 in the Calvin-Benison-Bassham cycle (see details below) (Tabita 2004; Blankenship 2014).

Aerobic anoxygenic phototrophs (AAPs) are a unique purple bacterial functional group that contain fully functional photosynthetic apparatus; however, this is assembled and operative under oxic conditions. Their photosynthetic machinery has various extensive modifications, including different peripheral antennas, and some organisms using zinc-chelated BChls in place of the more common magnesium-containing pigments (Wakao et al. 1996). AAPs have a much lower number of photosynthetic complexes per cell and a huge abundance of carotenoids (Rathgeber et al. 2004). Interestingly, although they are incapable of photoautrophy and rely on heterotrophy for 80% of their cellular metabolism, phototrophy can double organic carbon assimilatory efficiency (Kolber et al. 2001). BChl synthesis in these organisms is inhibited by sunlight (Aagaard and Sistrom 1972). Examples of these organisms can be seen in Table 1.

The majority of the genetic information needed to build the photosynthetic apparatus in purple bacteria is clustered into large (40–50 kbp) groups of genes called the photosynthesis gene cluster (PGC). The precise organisation of these genes within the cluster is highly variable. This cluster enables the precise control of expression levels and spatial proximity of the components (Alberti et al. 2004). The discovery of a phototrophic member of the Gemmatimonadetes bacteria, carrying a PGC of

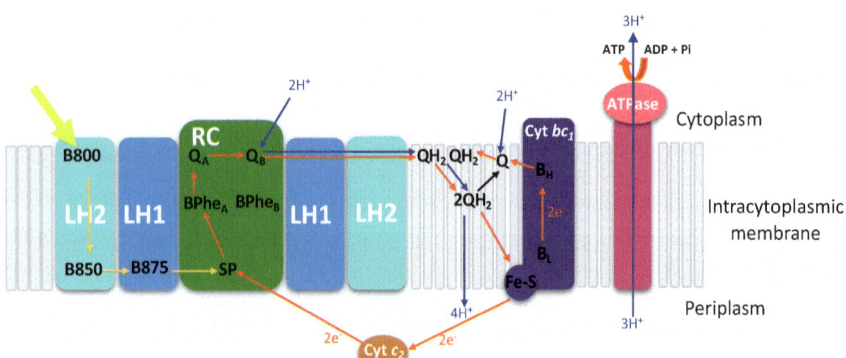

Fig. 3 Schematic of the light-dependent reactions of purple bacterial photosynthesis. Light energy is first absorbed by B800 in LH2; this is then transferred to lower energy BChls until it reaches and excites the special pair BChls in the RC. An electron is then passed along a series of electron donors to Q_B, and after a second charge separation, Q_B is fully reduced to quinol; the electrons are then passed to cyt bc_1 and cyt c_2 which re-reduces the special pair. This creates an electrochemical gradient across the membrane, which is used by ATPase to create ATP. Note the exact locations of cyt bc_1 and ATPase are not known. Yellow arrows, excitation transfer; red arrows, electron transfer; blue arrows, proton transfer; *BChl* bacteriochlorophyll; *LH* light harvesting; *RC* reaction centre; *SP* special pair; *BPhe* bacteriopheophytin; *Q* quinone; *QH_2* quinol; *cyt* cytochrome, *ATPase* ATP synthase

purple bacterial origin, demonstrates that the organisation of these genes into this superoperon permitted horizontal gene transfer between distant bacterial phyla (Zeng and Koblizek 2017).

2 Structural Components

2.1 *Peripheral Antenna Complexes*

The RC itself is a poor antenna, harvesting insufficient photons to support rapid growth via phototrophy. Purple bacteria have thus evolved specialised antenna complexes to increase the cross-sectional area available for light harvesting and maximise photon capture. The energy absorbed by the antenna complexes is then passed to the RC. Purple bacteria typically contain two types of antenna complexes: the 'core' light-harvesting complex 1 (LH1) intimately associates with the RC, encircling it to form the 'core complex' and RC–LH1 complex, found in all purple bacterial species. There are also peripheral antennas, known as LH2, which are separated from the RC by the LH1 ring. These antenna complexes are embedded in the ICM of purple bacteria, and bind the light-absorbing BChl and carotenoid pigments non-covalently.

2.1.1 Light-Harvesting Complex 2

The peripheral LH2 complexes funnel excitation energy to the RC via the core LH1 complexes (discussed in detail later). The LH2 complexes have been well studied across different species; all exist as octa- or nonameric quaternary structures. The crystal structures of LH2 from *Rhodoblastus* (*Rbl.*) *acidophilus* (Fig. 4a) (McDermott et al. 1995; Prince et al. 1997; Papiz et al. 2003) and *Phaeospirillum* (*Phs.*) *molischianum* (Fig. 4b) (Koepke et al. 1996) have been solved to 2.0 Å and 2.4 Å resolution, respectively. Both structures adopt the same modular principle: oligomers consisting of numerous pairs of α- and β-polypeptides, known as the αβ-heterodimer, forming double-ring structures with their associated pigments. The α- and β-polypeptides of the two species share 26% and 31% sequence identity, respectively.

The LH2 complex of *Rbl. acidophilus* is formed of a nonameric ring structure of 9 inner α- and 9 outer β-polypeptide chains (Fig. 4a), with ring diameters of 36 Å and 68 Å, respectively; each polypeptide crosses the membrane once, via an α-helical transmembrane domain (McDermott et al. 1995). The LH2 structure acts as a scaffold for the associated pigments, including carotenoids and two populations of BChls. One BChl population in the *Rbl. acidophilus* LH2 contains nine pairs of overlapping pigments held perpendicular to the plane of the membrane via histidine residues in the α- and β-polypeptides (Prince et al. 1997); the short distances between these paired BChls (~9 Å) provide strong excitonic coupling between the

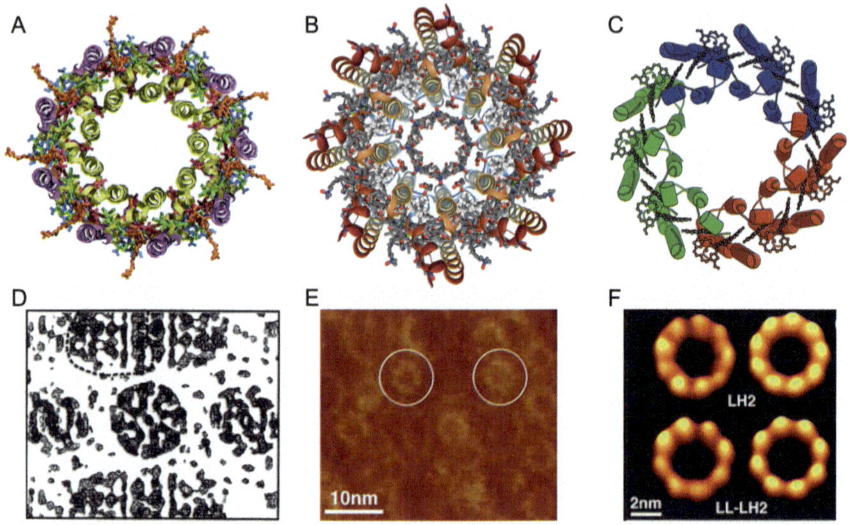

Fig. 4 Structures of different forms of light-harvesting antenna complexes in purple bacteria. (**a**) Structure of the LH2 complex from *Rbl. acidophila 10,050*. Cartoon representation of the nonameric B800-B850 LH2 complex viewed from above. The α- and β-polypeptides are represented in light-green and purple, respectively. The BChls α-B850, β-B850 and B800 are represented in red, green and blue, respectively, and the carotenoid is represented in orange (Papiz et al. 2003). (**b**) Crystal structure of the LH2 complex from *Phs. molischianum*. The octameric B800-B850 LH2 complex viewed from above with the N-termini pointing upwards. The α- and β-polypeptides are represented in orange and red, respectively. The BChl *a* molecules are in grey (Koepke et al. 1996). (**c**) Structure of the LH3 complex from *Rbl. acidophilus* 7050. Schematic representation of the nonameric LH3 complex from *Rbl. acidophilus* 7050 (McLuskey et al. 2001). (**d**) Electron densities obtained from *Rps. palustris* LH4 X-ray data, revealing an octameric structure (Hartigan et al. 2002). (**e**) AFM analysis of peripheral antenna complexes from *Rps. palustris* grown under low light; the white circles represent LH4 complexes (Scheuring et al. 2006). (**f**) Non-symmetrised (left) and symmetrised averages (right) of nonameric LH2 and octameric LH4 (LL-LH2) from *Rps. palustris* grown in low light (Scheuring et al. 2006)

pigments, shifting the absorption maximum to ~850 nm; this population is known as B850 (after their corresponding absorption maxima). The complex also contains 9 B800 BChl molecules inserted between the β-polypeptides, parallel to the membrane plane. The central Mg atom of these BChls are coordinated by methionine residues in the α-polypeptide (Papiz et al. 2003); these pigments are not tightly coupled and so can be considered 'monomeric', thus contribute a band at ~800 nm, close to the absorption maximum of the pigment in solution. Finally, a carotenoid (rhodopin glucoside in the *Rbl. acidophilus* LH2) threads the space between the α- and β-pairs. Together, these pigments contribute to the structural and functional integrity of LH2 (Lang and Hunter 1994).

The overall structure of the B800–850 LH2 complex from *Phs. molischianum* is similar to that of *Rbl. acidophilus* (McDermott et al. 1995; Koepke et al. 1996), but with some key differences. The ring structure of this complex is made up of 8 αβ-pairs (Fig. 4b), thus possessing 3 BChls and 1 carotenoid fewer than that from *Rbl. acidophilus*, as well as reduced diameters of the inner (31 Å) and outer rings (62 Å) (Koepke et al. 1996). The 8 B800 molecules are coordinated by aspartate rather than methionine residues of α-polypeptides almost parallel to the membrane plane, with a distance of 2.45 Å. The acetyl carbonyl groups of BChl were also shown to interact with tryptophan residues in the α- and β-polypeptides; these residues were shown to be important in stabilising the BChl–LH1 interaction (Davis et al. 1997; Kehoe et al. 1998). The 8 carotenoids found in the complex are lycopene pigments stretched between the B800 and B850 molecules (Koepke et al. 1996). The crystal structure of *Phs. molischianum* LH2, along with mutational studies (Olsen et al. 1997), indicated that His residues also form a hydrogen bond with the carbonyl group of the coordinated BChl. The difference in oligomerisation seen between the two LH2 proteins was suggested to be due to the differences in the interaction angle between subunits: 40° for nonameric rings and 45° for octomeric rings (Janosi et al. 2006); these angles are believed to be determined by the surface interactions in the transmembrane regions (Janosi et al. 2006). The intermolecular forces underlying the ring shape of the LH2 complex were shown to influence the structural and functional integrity of the LH2 complex in *Pararhodospirillum* (*Psp.*) *photometricum* (previously *Rhodospirillum*) (Liu et al. 2011a, b).

2.1.2 Light-Harvesting Complexes 3 and 4

Under stress conditions, such as low light or low temperature, absorption bands of several purple bacteria vary due to the expression of atypical peripheral light-harvesting complexes, known as LH3 and LH4 (Evans et al. 1990; Mcluskey et al. 2001; Hartigan et al. 2002; Niedzwiedzki et al. 2011). These are spectral variants of LH2, functioning to increase the size of the photosynthetic unit and broaden the spectral range of solar energy captured by the cell.

Both the *Rbl. acidophilus* 7050 and 7750 strains contain LH3 complexes (Fig. 4c) (Cogdell et al. 1983; Angerhofer et al. 1986). In the 7050 strain, LH3 complexes always coexist with LH2 complexes; however, low-light conditions can also induce the expression of LH3 (Cogdell et al. 1983; Angerhofer et al. 1986). LH2 was almost completely replaced with LH3 complexes in the 7750 strain under low-light conditions (Gardiner et al. 1993). The LH3 complexes absorb light at 800–820 nm, a blue shift from the 800–850 nm absorption seen in LH2. The overall structure of the LH3 complex from *Rbl. acidophilus* 7050 (Mcluskey et al. 2001) is analogous to LH2 from *Rbl. acidophilus* 10,050 (McDermott et al. 1995), with the exception of reorientation of acetyl groups at the C3 position of B820 BChl in LH3 relative to the membrane plane; this causes a change in the hydrogen bonding patterns between the protein and the coupled BChl *a* molecules, resulting in the change in spectral properties of the apoprotein (Mcluskey et al. 2001).

Rhodopseudomonas (Rps.) palustris predominantly contains LH4 complexes in low-light conditions (Tharia et al. 1999). This complex, also referred to as the low-light LH2 B800 complex, is composed of eight αβ-polypeptide pairs; each pair contains four BChl molecules, distinct from the three BChls in LH2 complexes (Hartigan et al. 2002); electron microscopy (EM) and atomic force microscopy (AFM) studies have confirmed its octameric structure (Scheuring et al. 2006) (Fig. 4d, e).

The majority of purple phototrophic bacteria that use BChl *a* employ a peripheral antenna, but there are some exceptions; *Rhodospirillium (Rsp.) rubrum* and AAPs within the Eryth-Citro clde lack the genes that encode LH2 (and LH3 and LH4) (Munk et al. 2011; Zheng et al. 2011). To date, peripheral antennas have not been found in any BChl *b*-containing phototrophs; one possible explanation being that these antennas evolved in an ancestor using BChl *a*, after divergence of the two pigment pathways.

2.2 The Core Complex of Purple Bacterial Photosynthesis

Once light is harvested by LH2, the excitation energy is then passed to the BChl molecules housed in the LH1 antenna proteins and subsequently to the special pair BChls in the RC, enabling charge separation. Across all species, the LH1 complex surrounds the RC forming a ring; this RC–LH1 supercomplex, central to purple bacterial photosynthesis, is known as the core complex. As mentioned above, some purple bacteria, e.g. BChl *a*-containing *Rsp. rubrum* and BChl *b*-containing *Blastochloris (Blc.) viridis*, exclusively use the RC–LH1 core complex for both light harvesting and photochemical functions. The aggregation states of the RC–LH1 core complexes amongst different species are diverse (Fig. 5), with both monomeric and dimeric cores being found in the membranes.

2.2.1 Light-Harvesting Complex 1

The LH1 complex is composed of a ring of α- and β-polypeptides forming a heterodimer together with their associated pigment molecules: BChl *a* or *b* and carotenoids. These generally short, 40–70 amino acid residue, α-helical antenna polypeptides contain three domains: polar C- and N-terminal domains lying either side of a hydrophobic transmembrane domain (Brunisholz and Zuber 1992; Zuber and Cogdell 2004). A conserved histidine residue acts as the ligand to the central magnesium atom of the BChl (Qian 2017); in *Blc. viridis*, the C3 acetyl groups of BChl *b* hydrogen bond to tryptophan residues in LH1 (Qian et al. 2018), adopting an in-plane conformation similar to coordination in the LH2 of *Rbl.* acidophilus.

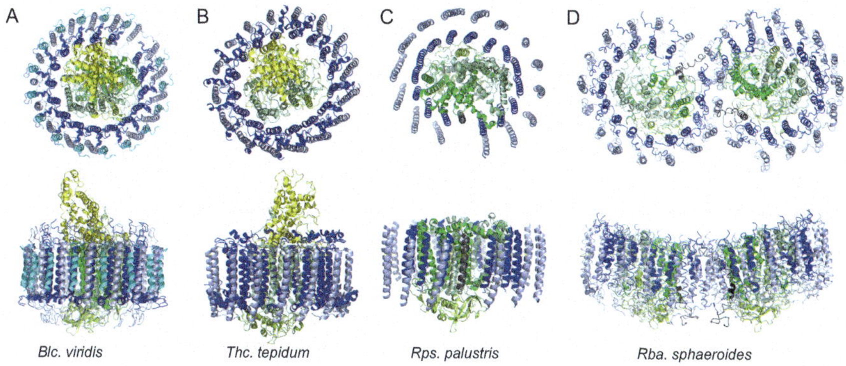

Fig. 5 Diverse structures of RC–LH1 complexes from different purple bacterial species. Cartoon representations viewed perpendicular to the membrane plane (top) and from the membrane plane (bottom) of (**a**) RC–LH1 from *Blc. viridis* (PDB: 6ET5) (Qian et al. 2018), (**b**) RC–LH1 of *Thc. tepidum* (5Y5S) (Yu et al. 2018), (**c**) RC–LH1-PufW complex of *Rps. palustris* (PDB: 1PYH) (Roszak et al. 2003), (**d**) RC–LH1-PufX dimer of *Rba. sphaeroides* (PDB: 4V9G) (Qian et al. 2013). The 4Hcyt, L, M and H subunits of the RC are represented in yellow, bright green, pale green and lime green, respectively; α -, β- and γ-chains of LH1 are shown in blue, light blue and cyan, respectively. PufX and protein W are represented in black and dark grey, respectively

2.2.2 The Photochemical Reaction Centre

The RC, home of the light-induced charge separation across the intracytoplasmic membrane of *Blc. viridis*, was the first membrane protein structure to be solved at high resolution (Deisenhofer et al. 1985; Deisenhofer and Michel 1988). The RC consists of a common, basic structure of three transmembrane polypeptides, the L, M and H subunits (Fig. 6b), and their associated pigments, similar in all species of purple photosynthetic bacteria. L and M subunits each contain five transmembrane helices, and the LM heterodimer displays pseudo twofold axis symmetry, perpendicular to the membrane plane. These subunits possess 25–30% sequence identity. L and M act as a scaffold for the arrangement of cofactors: 4 BChls, 2 bacteriopheophytin (BPhe) pigments, 1 non-haem iron, 2 quinones and 1 carotenoid. The cofactors are also arranged with a pseudo twofold symmetry and form two separate branches, A and B, within the M and L subunits, respectively (Fig. 6c). Electron transfer has been shown to occur through branch A (Kirmaier et al. 1985; Bylina and Youvan 1988; Kellogg et al. 1989). The H subunit contains only 1 transmembrane helix, anchoring the subunit to the membrane, a cytoplasmic globular domain docks to L and M. The RC structure of *Blc. viridis* revealed an extra periplasmic tetrahaem cytochrome subunit (4Hcyt) attached to the membrane via a covalently attached fatty acid molecule (Deisenhofer et al. 1985). The 4Hcyt houses four haem groups covalently attached to cysteine residues, arranged in a high-low-high-low redox potential pattern (Fig. 6c), functioning to re-reduce the special pair BChl (Ortega et al. 1999). The conserved structure of RC is demonstrated by the crystal structures of *Rhodobacter (Rba.) sphaeroides* (Fig. 6d) (Allen and Holmes 1986;

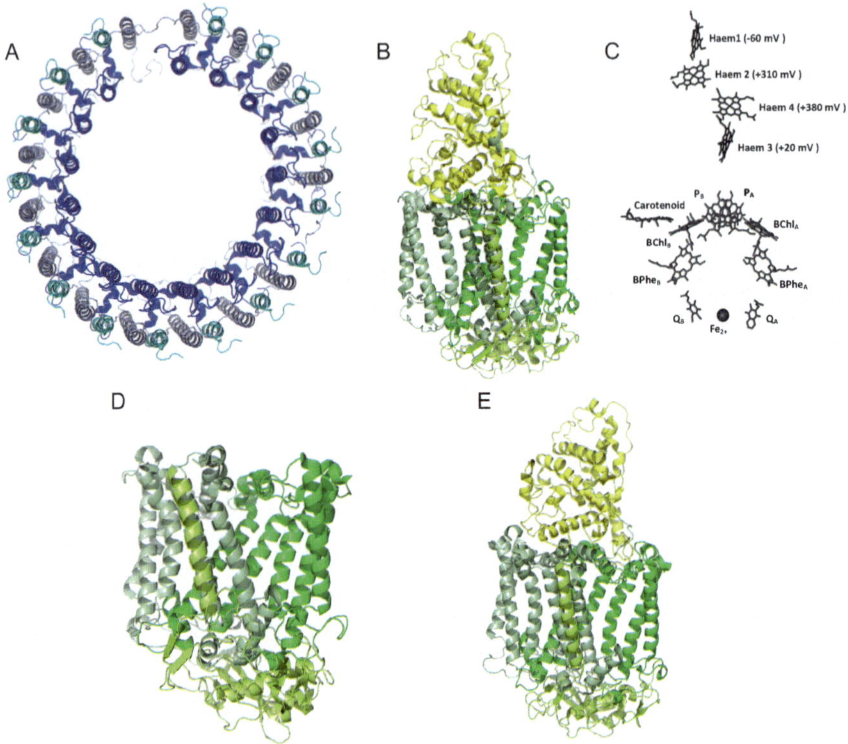

Fig. 6 Structures of LH1 and RC complexes. Cartoon representation of (**a**) LH1 from *Blc. viridis* (PDB: 6ET5) (Qian et al. 2018) from the periplasmic side of the membrane and (**b**) RC from *Blc. viridis* (PDB: 6ET5) (Qian et al. 2018) from the membrane plane. The RC resides in the centre of the LH1 complex; this has been removed from the structure. (**c**) Arrangement of cofactors in the RC from *Blc. viridis*; cofactors are represented as sticks. The redox potentials of the haems are labelled. (**d**) Crystal structure of the RC from *Rba. sphaeroides* (PDB: 2RCR) (Camara-Artigas et al. 2002). (**e**) Crystal structure of the RC from *Thc. tepidum* (PDB: 5Y5S) (Yu et al. 2018). The α-, β- and γ-chains of LH1 are shown in blue, light blue and cyan, respectively. The 4Hcyt, L, M and H subunits are represented in yellow, bright green, pale green and limon, respectively

Chang et al. 1986) and *Thermochromatium (Thc.) tepidum* (Fig. 6e) (Nogi et al. 2005) RCs, both of which are very similar to that of *Blc. viridis* (Qian et al. 2018); however, *Rba. sphaeroides* (Allen and Holmes 1986; Chang et al. 1986) lacks the 4Hcyt subunit and both contain BChl *a* rather than BChl *b*.

2.2.2.1 Quinones

As well as being the home of charge separation, the RC also functions to produce fully reduced quinol (QH_2) to drive electron transfers through cyt bc_1, ultimately resulting in the formation of an electrochemical gradient driving ATPase. The RCs of purple bacteria consist of two acceptor quinones that act in series. The primary quinone, Q_A, is bound tightly to the RC, and cycles between oxidised quinone and singly reduced semiquinone. The secondary quinone, Q_B, is reduced twice by Q_A to form a doubly reduced, fully protonated QH_2. The electron donor for these reactions is cyt c_2, so the RC can be thought of as a 'cytochrome c_2/ubiquinone' photo-oxidoreductase enzyme. After two turnovers of the RC, QH_2 is released from the Q_B site and replaced by oxidised quinone to complete the quinone cycle. In the majority of purple bacteria, both Q_A and Q_B are ubiquinone (UQ) molecules, but in *Blc. viridis*, *Allochromatium vinosum* and *Thc. tepidum*, Q_A is a menaquinone (MQ). The X-ray crystal structure of the RC from *Rba. sphaeroides* revealed hydrogen bonding to Q_A (UQ-10) through carbonyl oxygens, a histidine residue and an amide backbone; a similar hydrogen bonding pattern was seen in Q_A from *Blc. viridis* (MQ-9) (Gast et al. 1985; Shopes and Wraight 1985). The acceptor quinones are symmetrically positioned about a non-haem iron atom coordinated by four histidines, and a glutamate residue, $Q_{A,}$ is bound in the L subunit and Q_B within the M subunit (Deisenhofer and Michel 1988).

2.2.3 Additional Core Complex Components

2.2.3.1 PufX

The core complex of many species within the *Rhodobacter* genus contains an extra transmembrane polypeptide, PufX (Youvan et al. 1984; Sheng and Hearst 1986; DeHoff et al. 1988; Holden-Dye et al. 2008). PufX creates a break in the LH1 ring, keeping it in an 'open' conformation to facilitate rapid quinone/quinol exchange between the RC and cyt bc_1 (Klug et al. 1988; Farchaus et al. 1990; Lilburn et al. 1992). PufX represents an α-helical structure consisting of 34 residues (Tunnicliffe et al. 2006); its N-terminus is largely unstructured, followed by a structured helical domain positioned to ensure basic residues are at the membrane interface to fulfil a membrane-anchoring role (Parkes-Loach et al. 2000). Studies on *Rba. capsulatus* and *Rba. sphaeroides* indicated that deletion of the *pufX* gene led to the inability of these strains to grow phototrophically (Klug et al. 1988; Farchaus et al. 1990; Lilburn et al. 1992). For a history of PufX, see the review by Holden-Dye et al. (Holden-Dye et al. 2008).

2.2.3.2 Protein W

The X-ray crystal structure of the core complex from *Rps. palustris* revealed that the RC was surrounded by an incomplete elliptical ring of 15 αβ-polypeptide pairs. Electron density maps suggested that a single transmembrane helix, named protein W, prevented closure of the LH1 ring (Fig. 5c) (Roszak et al. 2003). The break in the LH1 ring created by protein W is positioned adjacent to the secondary electron acceptor quinone (Q_B) binding site, as is the case for PufX, which suggested a similar potential role in quinone/quinol exchange with cyt bc_1. The identity of protein W remained unknown for many years, until proteomic analysis of purified core complexes from the sequenced *Rps. palustris* strain CGA009 identified a polypeptide (RPA4402) that was present in the preparation (Jackson et al. 2018). Biochemical and electron microscopic analysis indicated that RPA4402/protein W is only present in ~10% of the purified core complexes, and deletion of the encoding gene had no measurable effect on phototrophic growth; thus, the precise role of protein W remains enigmatic (Jackson et al. 2018).

2.2.3.3 The Gamma Subunit

An additional LH1 subunit was also identified in core complex preparations of *Blc. viridis* (Jay et al. 1983); unlike protein W and PufX, this γ-polypeptide was found to be in apparent equimolar ratio with the LH1 α- and β-polypeptides (Brunisholz et al. 1985). Subsequently, the structure of the core complex from this organism revealed the location of γ, packing between β-polypeptides on the outside of the LH1 ring (Qian et al. 2018). The α:β:γ ratio was found to be 17:17:16, the 'missing' γ-subunit creating the channel for quinone diffusion, converse to the proposed roles for PufX and protein W, whose presence at a single location around the LH1 ring is believed to create this channel. An additional role of the γ-polypeptide in increasing the packing of LH1 BChls was also proposed; the resulting shorter intra-subunit Mg-Mg distances increase excitonic coupling, making the *Blc. viridis* RC–LH1 complex the most red-shifted photosynthetic complexes described to date (Qian et al. 2018).

2.2.4 Architectures of Core Complexes

The RC–LH1 complexes possess different forms across different purple photosynthetic bacteria (Fig. 5). Table 2 details the RC–LH1 composition for several species of purple bacteria.

Early, low-resolution electron microscopy studies on membranes from *Blc. viridis* and *Halorhodospira* (*Hlr.*) *halochloris*, followed by single particle analysis on purified core complexes from *Phs. molischianum*, indicated that the common structure of RC–LH1 complexes consisted of a circular LH1 antenna surrounding the RC (Stark et al. 1984; Engelhardt et al. 1986; Boonstra et al. 1994). Later studies employing atomic force microscopy (AFM) to interrogate the topography of membranes from purple bacteria (Liu and Scheuring 2013; Liu 2016) achieved sufficient resolution to assign the number of αβ-polypeptide pairs in the LH1 antenna from *Blc. viridis*, *Rsp. rubrum*, *Psp. photometricum* and *Phs. molischianum*; all considered the RC to be enclosed by 16 αβ dimers, with those from *Blc. viridis* (Fig. 7a) and *Rsp. rubrum* interpreted to be elliptical in shape (Fotiadis et al. 2003; Scheuring et al. 2003a, 2004a; Gonçalves et al. 2005). This elliptical shape was reinforced with the 4.8 Å crystal structure of the *Rps. palustris* core complex, although the LH1 ring was found to contain the W polypeptide in place of one of the 16 αβ pairs (Roszak et al. 2012).

Table 2 Composition of RC–LH1 complexes in purple bacteria. Species with '2x' subunits form dimeric complexes

	RC components				LH1 components		Additional peptides	
Species	4HCyt	L	M	H	αβ	αβγ	PufX	Protein W
Blc. viridis (PDB: 6ET5) (Qian et al. 2018)	✓	✓	✓	✓	1	16	–	–
Rba. sphaeroides(PDB: 4V9G) (Qian et al. 2013)	–	2x	2x	2x	28	–	2	–
Rps. palustris (PDB:1PYH) (Roszak et al. 2003)	–	✓	✓	✓	15	–	–	1
Rsp. rubrum (Jamieson et al. 2002)	–	✓	✓	✓	16	–	–	–
Thc. tepidum (PDB: 5Y5S) (Yu et al. 2018)	✓	✓	✓	✓	16	–	–	–
Rba. blasticus (Scheuring et al. 2005a)	–	2x	2x	2x	26	–	2	–
Rba. veldkampii (Busselez et al. 2007)	–	✓	✓	✓	15	–	1	–

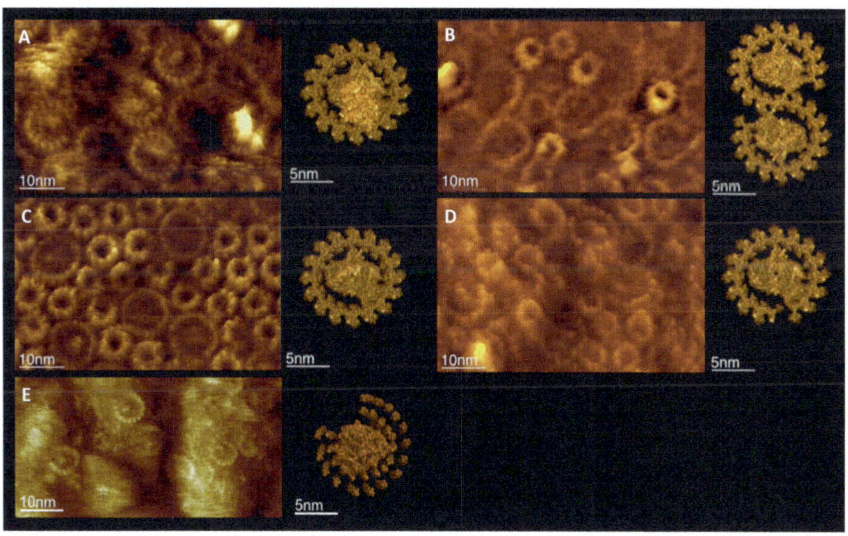

Fig. 7 Diversity of core-complex architectures among purple photosynthetic bacteria species. Core complexes in their native membranes (left) and the models of individual core complexes (right) for (**a**) *Blc. viridis* ($LH1_{16}$-$RC_{L,M,H}$-4Hcyt) (Scheuring et al. 2003a), (**b**) *Rba. blasticus* ($(PufX_2$-$LH1_{13}$-$RC_{L,M,H})_2$) (Scheuring et al. 2005a), (**c**) *Psp. photometricum* ($LH1_{16}$-$RC_{L,M,H}$) (Scheuring et al. 2004a), (**d**) *Rps. palustris* (W-$LH1_{15}$-$RC_{L,M,H}$) (Scheuring et al. 2006) and (**e**) *Rba. veldkampii* (Liu et al. 2011a, b)

The 'gap' in the elliptical ring created by protein W was also identified by AFM (Fig. 7d) (Scheuring et al. 2006). The contemporary, high-resolution structures of the RC–LH1 complexes from *Tch. tepidum* and *Blc. viridis* (Fig. 7a), consist of 16 and 17 αβ heterodimers, respectively; the latter also containing 16 γ-subunits, illustrating that monomeric core complexes do not have a uniform architecture.

This lack of uniformity is compounded by the presence of dimeric core complexes in species of *Rhodobacter*. Cryo-EM of negatively stained membranes from an LH2 mutant of *Rba. sphaeroides* (the mutation causing the switch from its native vesicular to tubular membranes) revealed highly ordered core complexes in S-shaped conformations, believed to be comprised of two RCs, each surrounded by a C-shaped LH1 (Jungas et al. 1999). Sucrose density gradients of solubilised membranes from *Rba. sphaeroides* also revealed that monomeric and dimeric core complexes could be isolated in which PufX was detected in a 1:1 ratio with the RC, but when *pufX* was deleted, only monomeric RC–LH1 could be isolated, indicating that PufX promotes dimerisation of the core complex (Francia et al. 1999). AFM analysis of membranes from *Rba. sphaeroides* and *Rba. blasticus* revealed that the majority of core complexes exist in dimeric form in the native membranes of these strains (Fig. 7b) (Bahatyrova et al. 2004; Scheuring et al. 2005a). A subsequent structure of the dimeric core from *Rba. sphaeroides* revealed that each RC was surrounded by 14 $\alpha\beta$ heterodimers, with each PufX adjacent to the Q_B site and interacting with the RC-H subunit via its N-terminal extension, and both N- and C-termini of PufX promoting dimerisation by interacting with a β polypeptide in the other half of the dimer (Qian et al. 2013). Interestingly, the core complex of *Rba. veldkampii* was found to contain a PufX polypeptide, but sedimentation and single particle analysis, and subsequent AFM analysis of its native membrane, demonstrated the complex exists solely as monomeric cores in this organism (Fig. 7e) (Gubellini et al. 2006; Liu et al. 2011a, b). More recently, a systematic analysis of additional *Rhodobacter* species demonstrated that dimeric core complexes were also found in *Rba. azotoformans* and *Rba. changlensis*, but those of *Rba. capsulatus* and *Rba. vinaykumarii* were monomeric, yet all contained PufX (Crouch and Jones 2012). This study also indicated that there was no clear motif within the amino acid sequences of PufX from the tested strains that could be identified as key for dimerisation; further work is required to clarify the biochemical basis of this morphological difference.

2.3 *Cofactors and Pigments*

2.3.1 Carotenoids

Carotenoids have several essential functions within photosynthetic systems:

1. They are accessory pigments functioning in the collection and absorption of light energy and transferring this to BChl molecules.
2. They function in the photoprotection process, rapidly quenching triplet excited states of BChl and preventing them from reacting with oxygen and forming the highly reactive and damaging excited singlet state of oxygen. If singlet oxygen is formed, carotenoids can also quench this (Blankenship 2014).

3. Carotenoids also play an important role in the assembly of LH complexes; in order to reversibly dissociate LH complexes, removal of carotenoids from the complex is required (Loach and Parkes-Loach 1995).

In LH2 complexes, carotenoids extend between the subunits (McDermott et al. 1995; Koepke et al. 1996; Prince et al. 1997), stabilising the complex and interacting with the BChls.

Structures of several carotenoids present in purple bacteria can be found in Fig. 8. All of these carotenoids are extended molecules with delocalised π electron systems. A number of common characteristics exist including, but not limited to, the presence of tertiary hydroxy and methoxy groups at C-1, keto groups at C-2 and frequent double bonds in the C-3,4 position (Takaichi 2000).

2.3.2 Bacteriochlorophylls

The main structural cofactors and LH pigments of photosynthetic antennae of purple bacteria are BChls, which also play a key role in photochemistry in the RC; energy is harvested and transferred from BChls within the LH complexes to the RC for charge separation at the special pair. Many different types of BChl exist within phototrophic prokaryotes, but only two types exist in purple photosynthetic bacteria, BChl *a* and BChl *b* (Fig. 1). BChl *a* is the most common pigment found in purple bacteria, permitting absorption of wavelengths between 800 and 970 nm when assembled into LH complexes, shifting the Qy absorption band of the pigment in solution into the near infrared (Imanishi et al. 2019) (Table 1). BChl *b* is only found in a few species; it differs from BChl *a* by the presence of an exocyclic double bond at C-8 in ring B, known as an ethylidene substituent (Fig. 1). BChl *b* permits absorption between 960 and ~1050 nm, the longest wavelength absorbance band of any chlorophyll-type pigment. The use of BChls *a* or *b* allows purple bacteria to absorb wavelengths of light outside the range of the chlorophylls used by plants and cyanobacteria.

2.3.3 Bacteriopheophytins

BPhe are easier to reduce than BChls and hence function as electron acceptors within the sequence of electron carriers in places where BChls are not active (Fajer et al. 1975). They are metal-free BChls, lacking the central Mg^{2+} present in BChls. They are predominantly recognised as breakdown products, due to their formation during the degradation of BChl. However, small amounts of BPhe are essential components of RCs, which are produced in specific, yet unknown, pathways. In charge separation, BPhe (termed H_A) function as the first electron acceptor subsequent to the excitation of the special pair of BChls (Zinth and Wachtveitl 2005; Jones 2009).

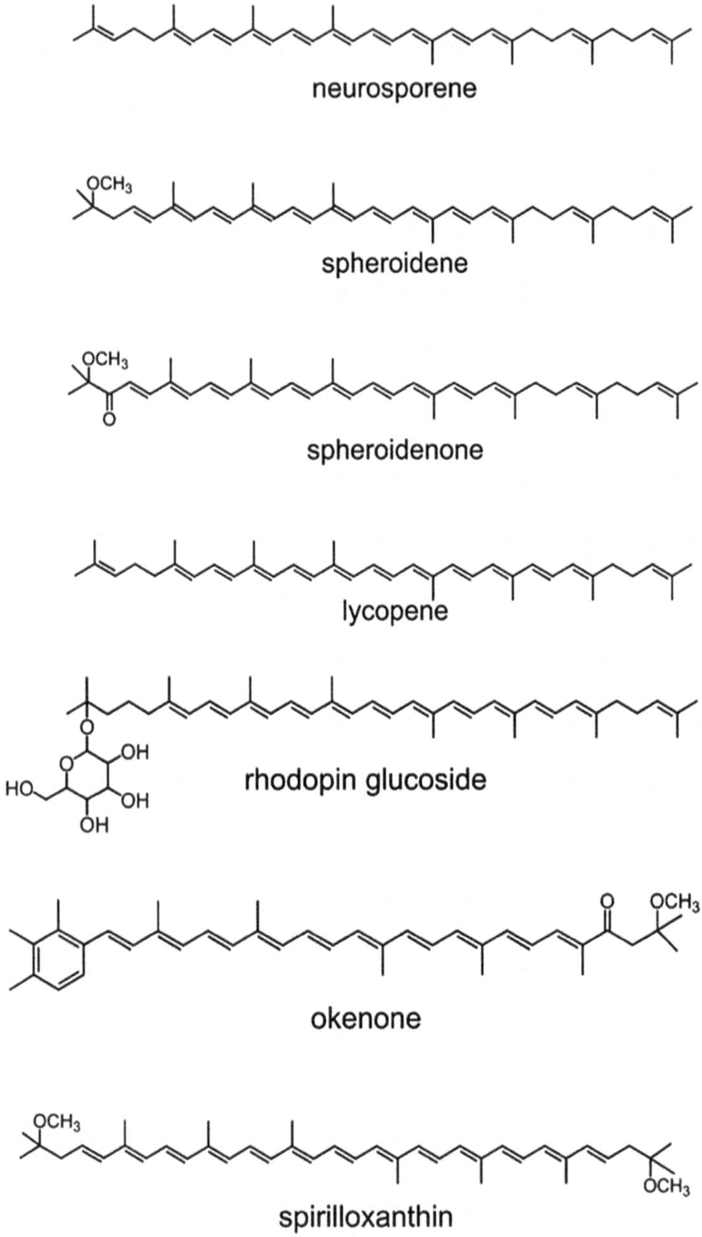

Fig. 8 Structures of common carotenoids in LH complexes of purple photosynthetic bacteria

2.4 Cofactor-Cofactor and Protein-Protein Interactions

A large, overlapping array of BChl molecules is formed by dense packing of LH1 and LH2 proteins in the photosynthetic membranes. This ensures that absorbed energy and the excited state are both quickly delocalised among the array and thus can be easily transferred to other LH complexes and eventually to the RC. Oligomerisation of B820 is required for the formation of LH1 and LH2 complexes (Miller et al. 1987; Chang et al. 1990). B820 subunits readily associate with the corresponding LH apoproteins to form LH1 and LH2 complexes with dependence on the details of cofactor-protein and protein-protein interactions. The heterodimeric B820 complex is stabilised by the interaction of the N-terminal regions of α- and β-polypeptides (Parkes-Loach et al. 2004). Several studies have demonstrated that fine-tuning of the absorption maximum of the associated LH complex may be due to evolution of the α-polypeptide (Todd et al. 1998). Dissociation and reassociation experiments revealed that under in vitro conditions, B820 reassociates to form LH1 via several intermediate species, reflecting the different numbers of subunits that associate during growth of the complex and eventual formation of the ring. Once this structure is formed, some rearrangements of the protein segments occur on a slower timescale (Miller et al. 1987; Pandit et al. 2003).

2.5 Assembly of Complexes

Assembly factors LhaA and PucC are important, but not essential, in the specific assembly of LH1 and LH2 complexes in vivo, respectively. These factors form oligomers at sites of initiation of membrane invagination. LhaA makes associations with RCs, BChl synthase, the protein translocase subunit YajC and the membrane protein insertase YidC, aiding coordination of pigment delivery, co-translational insertion of the LH polypeptides and the folding and assembly of these form functioning complexes (Mothersole et al. 2016). Reconstitution experiments demonstrated that LH1 subunits possess the ability to reassociate with RCs to form functional, native-like core complexes, in the absence of LhaA (Bustamante and Loach 1994). The RC–LH1 may follow a similar pattern of assembly in vivo. As is seen in *Rba. sphaeroides* (Pugh et al. 1998), the RC is likely to form first, allowing the LH1 subunits to subsequently encircle it. This theory suggests that the LH1 subunits are static within the complete circle; however, these subunits may in reality be flexible (Jamieson et al. 2002; Bahatyrova et al. 2004). The forces underlying the stability of the B820 subunit may be minimal, allowing the subunit to readily dissociate; this would provide a possible pathway for quinones on release from the Q_B binding site (Hunter et al. 2008).

2.6 Spectroscopic Properties of Light-Harvesting Complexes

Understanding the spectroscopic properties of LH complexes is fundamental as they are the key factors in directing the flow of excitation energy towards the RC. Although many antenna proteins contain the same BChl molecules, their absorption spectra can differ. BChl *a* and BChl *b* (Fig. 1) molecules of LH1 complexes absorb at 870–890 nm and near 1000 nm, respectively, whereas the BChl *a* molecules of LH2 complexes absorb at 800 and 850 nm. Additionally, a number of strains have been shown to synthesise modified antenna proteins absorbing at unusual wavelengths, for instance *Merichromatium purpuratum* (830 nm) (Cogdell et al. 1990) and *Roseococcus thiosulfatophilus* (856 nm) (Gall et al. 1999). The arrangement of BChl molecules within LH complexes could modify the spectroscopic properties of LH complexes. 24–32 and 16–18 BChl molecules associate with LH1 and LH2, respectively, through van der Waals contacts, leading to the red shifting of the absorption properties of the complexes (van Grondelle et al. 1994), compared to the BChls in solution.

The absorption spectra of LH3 and LH4 complexes are shifted to be shorter than 820 nm. The primary sequence of LH proteins between complexes and species differs; changes in the amino acid sequence in the protein environment of BChls can affect the hydrogen-bonding state and the dielectric properties of the protein environment (Fowler et al. 1994, 1997; Gall et al. 1997); this explains the part of the blue shift observed. Recent experiments showed that a change in the hydrogen bonding network affected the Q_y transition of BChls but also indirectly affected the role of charge-transfer states (Nottoli et al. 2018), leading to a new explanation of the blue shift seen in LH3 and LH4 complexes.

Homogenous LH proteins do not necessarily share the same electronic properties (van Mourik et al. 1992); slight differences in the conformation of the immediate environment of the BChls or dynamic fluctuations in the system, known as the static and dynamic disorder, respectively, lead to changes in the local physicochemical properties of the BChl and heterogeneity in the electronic properties. This disorder in the structure changes the properties of otherwise chemically identical BChls, which results in the localisation of excited states onto a subset of BChls within the interacting group (Hunter et al. 2008).

2.7 Cytochrome bc_1

Cyt bc_1 is a multisubunit membrane-bound enzyme and an essential component of the energy transduction machinery in purple photosynthetic bacteria (Fig. 9) (Berry et al. 2000, 2004; Crofts 2004). The enzyme functions to electronically connect the quinone pool to the downstream electron transport chain (ETC) and translocate protons across the membrane helping to produce the pmf (Crofts 2004). The cyt bc_1 oxidises QH_2 and donates the electron to the membrane-soluble electron carrier

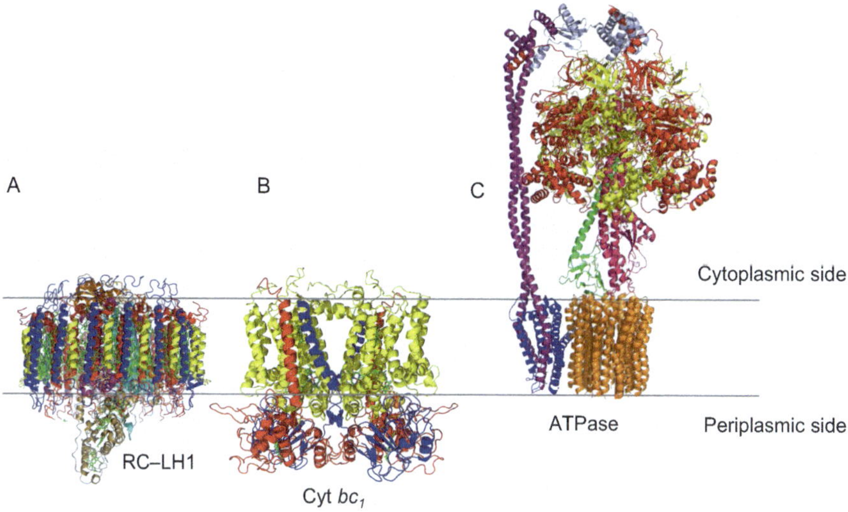

Fig. 9 Structures of cytochrome bc_1 and ATP synthase. (**a**) Structure of RC–LH1 from *Blc. viridis* (Qian et al. 2018) shown for comparison, colour scheme identical to Fig. 5. (**b**) Structure of cyt bc_1 from *Rba. capsulatus* (PDB: 1ZRT) (Berry et al. 2000). The cyt *b*, cyt c_1 and Rieske subunits are represented in yellow, red and blue, respectively. (**c**) Structure of the ATPase from *Bacillus* PS3 (PDB: 6 N30) (Guo et al. 2019). The c-ring, A, B, α, β, δ, γ and ε subunits are represented in orange, blue, purple, red, yellow, light blue, pink and green, respectively. The grey lines represent the intracytoplasmic membrane

cyt c_2 (Jenney and Daldal 1993). Therefore, this enzyme can be thought of as a 'ubiquinol/cytochrome c_2 oxidoreductase', operating in the opposite direction to the RC and completes the cyclic electron flow utilised by purple bacterial photosynthesis.

The overall structure of cyt bc_1 (Fig. 9b) consists of an intertwined homodimer with two monomers organised about a twofold molecular axis; each monomer is comprised of three subunits: a high-potential [2Fe–2S] cluster-containing subunit, known as the Rieske protein (Rieske et al. 1964), an integral membrane cyt *b* subunit and a *c*-type cytochrome subunit, known as cyt c_1, in which the haem cofactor is covalently attached to the polypeptide (Berry et al. 2000; Darrouzet et al. 2004). Cyt *b* consists of two *b*-type haems located on the positive (*p*) and negative (*n*) sides of the membrane. The central core of the monomer is formed by ten transmembrane helices, one from the Rieske protein, one from cyt c_1 and eight from cyt *b*. The large protrusion on the *p* side of the membrane consists mainly of the hydrophilic parts of the Rieske protein and the cyt c_1 subunits, whereas the *n* side is seemingly devoid of proteins (Fig. 9) (Berry et al. 2004). Each monomer contains two Q/QH_2 binding sites referred to as Q_o (QH_2 oxidation, H^+ output) and Q_i (Q reduction, H^+ input) sites; the Q_o site is at the interface of the Rieske protein and cyt *b* on the *p* side of the membrane, and the Q_i site is in the cyt *b* subunit, closer to the *n* side of the membrane (Sarıbaş et al. 1998; Berry et al. 2000).

2.8 ATP Synthase

The electrochemical gradient produced by the actions of the photosynthetic machinery is used by F_0F_1-ATPase, a membrane protein that synthesises ATP from ADP and inorganic phosphate (Pi). The enzyme contains two distinct portions of multi-subunit complexes; a hydrophobic proton-translocating F_0 proton channel is embedded within the membrane, consisting of three subunits, A, B and C, and a hydrophilic, enzymatic F_1 portion, consisting of α, β, ε and γ subunits, protrudes from the plane of the membrane by ~100 Å (Fig. 9) (Guo et al. 2019). Three αβ heterodimers in F_1 form a hexamer, with a long central cavity filled with the elongated portion of the γ-subunit comprising an N-terminal coiled-coil structure and an α-helical domain close to the C-terminus; this protrudes ~30 Å into the stalked region. A short α-helix of the γ-subunit is at a 45° incline to the coiled-coil domain at the bottom of F_1. The γ-subunit makes up a coupling domain, coupling ATP hydrolysis with ion pumping (Capaldi et al. 1996; Junge et al. 1997).

The αβ heterodimer contains six nucleotide-binding sites within the six clefts between adjacent subunits; only three of these participate in catalysis (Weber et al. 1994, 1995); the role of the other three clefts is not known but they are believed to play a regulatory role (Milgrom et al. 1991; Jault and Allison 1993; Jault et al. 1995; Matsui et al. 1997). The catalytic sites are mainly formed by the β subunit. Although they are all identical, at any given time, they have different conformations and affinities to nucleotides (Kayalar et al. 1977). Nucleotide binding to a particular site triggers the release of a nucleotide from another site; this is intimately linked to the rotation of the γ-subunit within the αβ heterodimer (Senior et al. 2002; Weber and Senior 2003).

2.9 Cytochrome c_2

The water-soluble *c*-type haem protein cyt c_2 serves as an electron donor to the RC, rapidly reducing the oxidised special pair created during the charge separation step due to its optimisation for rapid association and dissociation (Tiede and Dutton 1993). The haem group of cyt c_2 is covalently linked to the protein through thioether linkages; one edge of this haem group is exposed to solvent, which is important for reduction of the RC (Axelrod et al. 1994) (Fig. 10). Many positively charged Lys and Arg residues surround the solvent-exposed haem edge; these interact electrostatically with a cluster of negatively charged residues on the surface of the RC, positioning the cyt c_2 in the centre of the periplasmic surface of the RC, so that the solvent-exposed haem edge is directly over the BChl dimer, the primary donor, for rapid electron transfer (Rosen et al. 1983; Caffrey et al. 1992; Drepper et al. 1997). This binding region exhibits close van der Waals contacts important for both binding and electron transfer. A crystal structure of the cyt c_2:RC complex from *Rba. sphaeroides* has been determined, confirming this binding

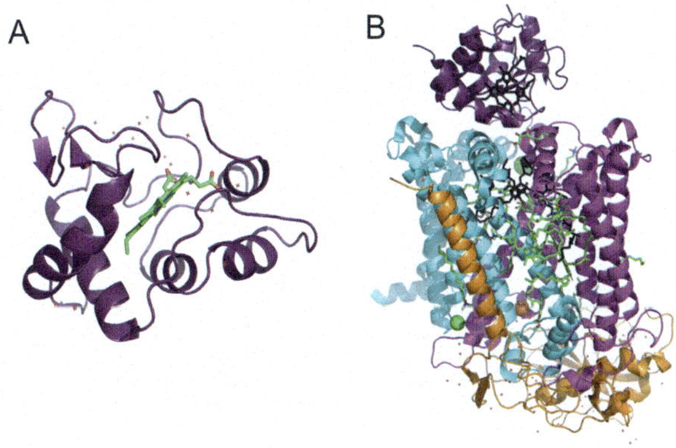

Fig. 10 Structure of cytochrome c_2. (**a**) Crystal structure of cyt c_2 from *Rba. sphaeroides* (PDB: 1CXC) (Axelrod et al. 1994); the haem group is represented in green. (**b**) Crystal structure of *Rba. sphaeroides* cyt c_2:RC (PDB: 1L9B) (Axelrod et al. 2002). Cyt c_2 is represented in purple, the L, M and H subunits of the RC are represented in magenta, cyan and orange, respectively. The haem group in cyt c_2 and the special pair BChl in RC are represented in black to highlight their positioning in the complex

position (Axelrod et al. 2002). The position of the haem is similar to that found in the 4Hcyt subunit of the RC of *Blc. viridis* (Fig. 6b) (Deisenhofer and Michel 1988). Binding of the cyt c_2 to RC leads to a slight conformational change creating a kink in the polypeptide chain of the RC, indicating a rigidly fixed-specific binding site of the cyt c_2 on the RC (Axelrod et al. 2002).

3 Organisation and Assembly of Photosynthetic Membranes

Knowledge of the static structures of individual membrane proteins is insufficient to fully understand their function and physiological coordination. An appreciation of in situ assembly and distribution of the proteins within their native membrane, and analysis under near-native conditions, are fundamental. In addition, it is increasingly accepted that structural heterogeneity exists within these proteins, between proteins from different species and between distinct gene products within a species, and between individual complexes. In addition to studying the structures of individual membrane proteins, AFM has matured to be a powerful technique to visualise the organisation of different membrane proteins in the native environment, enabling visualisation of the macromolecular structures in biological membranes and the dynamics of adaptation to different conditions (Scheuring et al. 2005b, 2007). AFM imaging of photosynthetic membranes from various purple bacteria have shown a variety of different organisational patterns of the photosynthetic apparatus, ranging from highly ordered to less ordered arrangements of proteins (Liu and Scheuring 2013; Liu 2016).

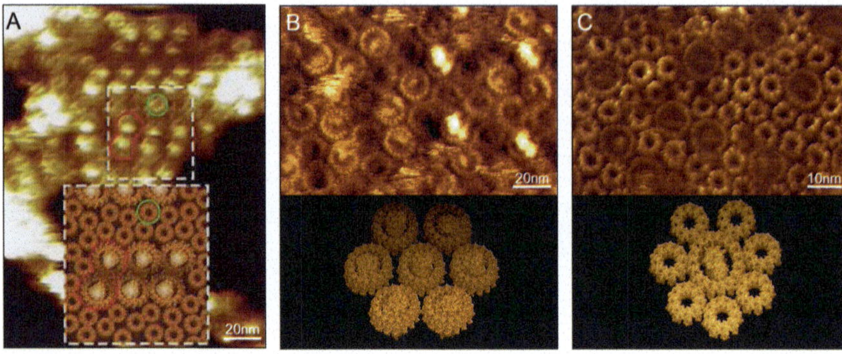

Fig. 11 Molecular resolution AFM topographs of purple photosynthetic apparatus. (**a**) Protein organisation in the vesicular ICMs of *Rba. sphaeroides* (Bahatyrova et al. 2004). (**b**) Protein organisation in the flat ICM of *Blc. viridis* (Scheuring et al. 2003b). (**c**) Protein organisation in the stacked ICMs of *Psp. photometricum* (Scheuring et al. 2004a)

The simplest purple bacterial photosynthetic membrane is that of *Blc. viridis* (Scheuring et al. 2003a; Qian et al. 2018), which contains simplified photosynthetic units lacking LH2 complexes. Small chromatophores were isolated from *Blc. viridis* and analysed using AFM (Scheuring et al. 2003a), revealing a highly organised architecture composed of hexagonally dispersed single RCs, each surrounded by a closed ellipsoid of LH1 subunits (Fig. 11b), similar to the cryo-EM structure (Qian et al. 2018). After removing the 4Hcyt subunit from the RC, AFM images revealed that the underlying RC-L and -M subunits adopt an asymmetric topography, whereby orientation of the long axis of LH1 coincides with this, representing an important constraint for energy transfer (Scheuring et al. 2003b). When the entire RC is removed, the LH1 units adopt a closed circular structure of ~100 Å diameter. This change reflects the flexibility of the LH1 heterotrimer assembly and the strong and specific interactions between the RC and LH1 components. This flexible motion could promote a 'breathing' motion to facilitate quine/quinol exchange, similar to that proposed for the LH1 of *Thc. tepidum* (Niwa et al. 2014).

The presence of LH2 complexes leads to a slightly more complex photosynthetic unit (PSU) arrangement; nevertheless, AFM images of the photosynthetic membrane of wild-type *Rba. sphaeroides* revealed a relatively ordered, interconnected network of LH2 and dimeric RC–LH1-PufX core complexes, coexisting with ordered LH2-only domains (Fig. 11a). The RC–LH1-PufX dimers form linear arrays, composed of pairs of elliptical structures with a protruding protein in the centre (Fig. 11a), interconnected with rows of peripheral antenna LH2 complexes (Bahatyrova et al. 2004). This physical continuity between the complexes ensures a highly efficient transfer of excitation energy from the peripheral LH2 to the LH1 complexes and onto an active RC (Bahatyrova et al. 2004; Scheuring et al. 2005a). The LH1 complexes are positioned to serve as a hub for the collection of excitons from LH2; the excitons can migrate along a series of dimers until they reach an open RC (Bahatyrova et al. 2004). LH2 was found to preferentially hexagonally pack within convex vesicular regions of the membrane (Scheuring et al. 2014). Cyt bc_1 was postulated to localise in disordered areas adjacent to RC–LH1 complexes; cyt bc_1 is outnumbered by RC–LH1 by a 3:1 ratio, placing them out of direct contact with the core complexes (Cartron et al. 2014; Scheuring et al. 2014). Mutants of *Rba. sphaeroides* lacking PufX had a completely reorganised ICM with ordered arrays of monomeric RC–LH1 complexes from which LH2 is largely excluded and segregated to the curved membrane domains (Frese et al. 2004). The interconnection of RC–LH1-PufX arrays interspersed with rows of LH2 complexes was also prevented in the mutant, suggesting that PufX plays a long-range organisational role within the membrane (Frese et al. 2004).

Despite the similarity in the structures of the RC–LH1-PufX complex in *Rba. blasticus* (Scheuring et al. 2005a) and *Rba. sphaeroides* (Jungas et al. 1999; Qian et al. 2008), AFM images revealed that the *Rba. blasticuus* RC–LH1-PufX complexes did not form rows that were seen in *Rba. sphaeroides* (Bahatyrova et al. 2004). Instead individual S-shaped dimeric core complexes are interspersed with nonameric LH2 rings (Fig. 7d) (Scheuring et al. 2005a).

Other LH2 containing species displayed much less regular arrangements of their PSUs. *Psp. photometricum* ICMs, consisting of stacked, flattened thylakoid-like discs, contain a mixed, disordered arrangement of monomeric RC–LH1 complexes and circular nonameric LH2 rings (Fig. 11c) (Scheuring et al. 2004a, b), and inter-complex clustering was visualised (Scheuring et al. 2004b; Scheuring and Sturgis 2005; Liu et al. 2009). The membranes are heterogeneously organised with RC–LH1-rich domains separate from hexagonally packed, almost crystalline LH2 domains. Some of the core complexes were completely surrounded by 7 LH2 complexes, whereas others made multiple contacts with other core complexes, increasing the probability that excitons will find an open RC, enhancing rapid energy trapping as is seen in the more organised *Rba. sphaeroides* membranes (Bahatyrova et al. 2004). This organisation provides efficient movement of excitons towards the RC whilst opening the ring of LH2 complexes to ensure all LH2 are in contact with one another. A heterogeneous variability in ring sizes of LH2 was also observed including octamers, nonamers and decamers, representing a natural variability in ring sizes, suggesting a possible strategy for broadening absorption to maximise the collection of near-IR radiation and optimisation of the packaging of photosynthetic components within the membrane (Scheuring et al. 2004b). Although cyt bc_1 was demonstrated to be in the membrane preparations, it was not identified in AFM images (Scheuring et al. 2004a).

Similarly, the ICMs of *Phs. molischianum* also consist of thylakoid-like structures giving rise to flat membrane sheets. The LH2 and core complexes are also arranged similarly to that in *Psp. photometricum*; LH complexes are segregated into two structurally different domains: one consists of a mixture of core complexes and a small amount of octameric LH2 complexes, and the other consists of paracrystalline, hexagonally packed octameric LH2 rings (Gonçalves et al. 2005). The crystal structure of the LH2 complex from *Phs. molischianum* (Koepke et al. 1996) enabled the exact pigment distances between LH2 complexes to be deduced (Hu et al. 2002); intercomplex distances between central Mg^{2+} atoms are ~16.3–28.3 Å, enabling efficient energy transfer.

The membranes of *Rps. palustris* consist of a complex ICM structure made of infoldings of the cytoplasmic membrane forming regular bundles of stacked and flattened thylakoid-like membrane sacs. AFM images of *Rps. palustris* membranes were obtained under differing light conditions, reflecting the versatility of photosynthetic apparatus in chromatic adaptation (Fig. 7f) (Scheuring et al. 2006). Under low light, mixed domains were commonly observed, along with paracrystalline domains of peripheral LH2 complexes, as is seen for other purple bacteria containing lamellar ICM structures. The mixed domains are likely to arise from the proximal layers within the lamellar ICM folds, whereas the LH2-only regions arise from distal layers formed subsequently; this is supported by the assembly sequence of complexes in developing ICMs (Koblízek et al. 2005). Under high light, regions of crystalline RC–LH1 core complexes were observed, with a few LH2 complexes. This arrangement of randomly ordered RCs would be expected to facilitate efficient energy trapping under high photon fluxes. Chromatic adaptation also resulted in the modification of LH2 ring sizes and absorption spectra (Scheuring et al. 2006). AFM images also confirmed the structures of the RC–LH1 complex, consisting of LH1 rings composed of 15 heterodimers interrupted by a gap, presumed to be the location of protein W (Scheuring et al. 2006).

3.1 Common Features of the Photosynthetic Membranes

Although the organisations of photosynthetic membranes in differing species vary, there are a number of recurring features. All the species contain membranes densely clustered with photosynthetic complexes, ensuring efficient excitation energy capture and transfer between antennae and ultimately to the RC for charge separation. In addition, the organisation of photosynthetic apparatus is never entirely random; clustering of RC–LH1 and formation of LH2 domains are often seen. Although the locations of cyt bc_1 and ATPase in most species remain enigmatic, the location of *Rba. sphaeroides* cyt bc_1 has been proposed (Cartron et al. 2014; Scheuring et al. 2014), and AFM on intact chromatophores from the same organism revealed a high protrusion from the membrane surface, speculated to be ATPase (Kumar et al. 2016).

3.2 Functional Importance of Photosynthetic Membrane Organisation

The architectural variability of the photosynthetic membranes permits the systems to work efficiently in their native environments under various conditions. It is difficult to simultaneously satisfy the paradox of the organisation of the photosynthetic apparatus; these include, firstly, the expectation that the LH system should completely surround the RC for maximum efficiency and to minimise the distances for energy to travel between complexes. Next, the core complexes should be connected in order to avoid energy loss from closed RCs by allowing energy transfer between core complexes (Bahatyrova et al. 2004; Scheuring et al. 2005a). And thirdly, the cyt bc_1 complex is expected to be in close proximity to the RC to close the electron circuit by quinol diffusion.

The first problem lies with the location of cyt bc_1 and the release of quinone/quinol. For *Rba. sphaeroides* and *Rps. palustris* whereby the RC–LH1 complexes contain PufX and protein W, respectively, the diffusion of quinone is fairly simple due to the gap produced by these proteins creating a possible pathway (Roszak et al. 2003; Bahatyrova et al. 2004; Qian et al. 2005). In other species, however, the ring of LH1 was observed to be closed; therefore, quinone transfer across the LH1 wall would depend on the dynamics of the process by which the integrity of the ring is breached. Circular and elliptical closed forms of LH1 were observed in AFM images of *Rsp. rubrum* 2D crystals, suggesting the structural flexibility of LH1 that facilitates the transfer of quinone through molecular motions (Jamieson et al. 2002; Siebert et al. 2004). The missing 17th gamma subunit of *Blc. viridis* RC–LH1 (Qian et al. 2018) also provides a path for this diffusion, in this case through the absence of an additional subunit, rather than the presence of one.

Once the quinone is released from the RC, another problem arises in its rapid diffusion through a crowded membrane to cyt bc_1 complexes. A lipid environment was found to be created around the core complexes of *Psp. photometricum*, due to a size mismatch of the LH2 and core complexes, providing a potential long-range pathway for quinones to travel (Liu et al. 2009). However, it was revealed that static paracrystalline LH2 complexes in the photosynthetic membrane, with significantly restricted diffusion, provided no space for quinones to diffuse between them (Scheuring et al. 2006).

4 Energy Transfer

4.1 Transfer of Excitation Energy

The knowledge of the structures and spatial arrangement of photosynthetic complexes and association with cofactors provide a basis for understanding energy transfer performed by the photosynthetic system. Energy is collected by antenna systems via the funnel concept; the pigments of the most peripheral antenna complexes, e.g. LH2, absorb shorter wavelengths than the pigments within RC. Photons of shorter wavelengths are higher in energy than longer wavelength photons and thus ensure 'downhill' energy flow: LH2 → LH1 → RC. A certain amount of energy is lost as heat during each transfer, providing some irreversibility and resulting in the funnelling of excitation energy into the RC. This funnel model is enabled by the spatial and energetic organisation of the antenna pigments. This concept also applies within LH complexes in the energy transfer from carotenoids to BChls; however, some energy transfers occur 'uphill', e.g. between LH1 and the RC in *Rba. sphaeroides* and *Blc. viridis* (Sumi 2002).

Higher-energy photons are first absorbed by carotenoids in LH2, or in LH1 in species lacking LH2; carotenoids absorb short wavelengths, 400–550 nm, of light and subsequently transfer the energy to both Q_x and Q_y states of B800 in LH2. Upon absorption of light, carotenoids are promoted from the ground state, S_0 to the S_2 state (transition to S_1 state is symmetry-forbidden); the molecule then quickly relaxes to the S_1 state in less than 300 fs. The lifetime of the S_1 state varies from 300 ps to ~1 ps, depending on the conjugation length of the carotenoid (Sundström 2004). Once the excitation energy is passed from S_1, or sometimes S_2 (Polívka and Frank 2010), to B800, it is quickly transferred to B850 in ~1 ps by Förster resonance energy transfer (FRET). The excitation energy is subsequently delocalised over the ring of tightly coupled B850 pigments, i.e. it hops rapidly around the ring, within 100 fs (van Oijen et al. 1999). The excited state is thus equally probable to transfer from any site within the ring, implying that no fixed arrangement of LH2 and LH1 complexes is required for efficient transfer (Cogdell et al. 1999). If the energy does not reach another LH complex within 1 ns, it will decay. Energy from B850 is subsequently transferred to the B875 pigments in ~3 ps, and, again, it is delocalised over the tightly coupled B875 pigments. The relatively large distance between B875 and the RC makes the transfer between these pigments the slowest step, occurring in ~35–50 ps, which is essentially irreversible (Pullerits and Sundström 1996; Sundstrom and Grondelle 1999).

4.2 Charge Separation in the RC

The transfer of excitation energy to the RC induces the separation and stabilisation of charge across the photosynthetic membrane. A BChl dimer, known as the special pair (P), acts as the primary electron donor and is excited to P∗; this excited state then transfers an electron to a BPhe molecule via a transient radical pair state; BPhe then subsequently transfers an electron to Q_A. A *c*-type cytochrome, or the 4HCyt subunit in certain species, reduces oxidised P^+, whilst Q_A^- transfers an electron to Q_B. At this point, an electron has been transferred across the membrane, and the primary reactants are ready for the next electron transfer.

4.3 Electron Transfer in Cytochrome c_2

In species that do not contain the 4Hcyt subunit, cyt c_2 re-reduces P^+. *C*-type cytochromes contain covalently attached haem groups; the haem group is located in the crevice of cyt c_2, with one edge exposed to the outside of the protein (Axelrod et al. 2002). This exposed edge is surrounded by a group of positively charged lysine residues which interact with the negative charges of the reaction partner forming a tightly bound complex. Cyt c_2 thus binds the RC at a specific site with the haem optimally positioned for electron transfer (Fig. 9) (Axelrod et al. 2009). The oxidation kinetics of cyt c_2 are multiphasic; when bound to the RC, the oxidation occurs within ~1 μs; however when the reduced cyt c_2 is unbound prior to electron transfer, the phase is slower (Mathis et al. 1994). The cytochrome needs to diffuse to the RC, and dock in place ready for oxidation; this results in the slow phase.

In species that contain the 4Hcyt subunit, the cytochrome is permanently positioned for optimal electron transfer. After flash excitation, the haem located closest to P is oxidised within a few hundred ns; electron transfer then occurs down the chain of haems within 4Hcyt re-reducing the first haem in μs. The 4Hcyt is then reduced by cyt c_2 (Menin et al. 1998; Myllykallio et al. 1998). At this stage, protons have been taken up from the cytoplasmic side of the membrane, and cyt c_2 has been oxidised in the periplasmic side.

4.4 *Modified Q Cycle*

Cyt bc_1, located in the ICM (Xia et al. 1997; Zhang et al. 1998; Berry et al. 2000; Crofts 2004), mediates the completion of cyclic electron transfer in the modified Q cycle (Fig. 3) (Crofts 2004). Ubiquinone (UQ) binds the Q_o site (see Sect. 2.3 for details on the structure of cyt bc_1) of cyt *b*. UQ then transfers the electron to the Rieske subunit, leaving ubisemiquinone in the Q_o site, and a proton is released to the periplasmic side of the membrane. Next, ubisemiquinone transfers a second electron to the lower redox potential haem and subsequently to the higher redox haem; a second proton is also released into the periplasmic side of the membrane. The Rieske subunit then undergoes a large amplitude motion, moving from the cyt *b* to the cyt c_1 subunit, displacing the Fe-S centre. UQ next binds to the Q_i site and is reduced to the semiquinone by the higher redox haem of cyt *b*. Oxidised UQ is then dissociated from the Q_o site, and the Rieske Fe-S centre reduces cyt c_1, which in turn reduces cyt c_2. The Rieske subunit then moves back to its original position. At this stage, one quinol has been oxidised, an electron has traversed the full electron transport chain, and another has reduced the oxidised UQ to the semiquinone state for the following turnover. A new UQ binds to the Q_o site and transfers its electron to the Rieske subunit, which then subsequently passes it to cyt c_1 and then cyt c_2 along one branch and then the two haem groups from cyt *b* on the other branch. The electron on the higher redox potential haem then reduces semiquinone to quinol, taking two protons from the cytoplasmic side. Finally, the reduced quinol dissociates from the complex (Crofts 2005).

Two turnovers of cyt bc_1 result in the transfer of two electrons to cyt c_2, two UQ oxidised to quinone form and one UQ reduced to hydroquinone quinol form and four protons transferred from the cytoplasmic to the periplasmic side of the membrane. Overall, this electron flow gives rise to pmf across the membrane which is subsequently utilised by ATPase.

4.5 Proton Translocation and ATP Synthase

ATPase exploits the proton motive force created by the photosynthetic electron transfer chain to produce ATP from ADP and P_i. Specifically, the F_0 portion of the enzyme uses the proton motive force to generate a torque, which is then used by the F_1 portion in the synthesis of ATP (Noji et al. 1997; Wada et al. 1999; Pänke et al. 2000; Kinosita et al. 2004; Junge et al. 2009) (Fig. 8). ATPase operates via the binding change mechanism (Boyer 1993). The major energy-requiring step is the simultaneous binding of substrates to, and release of products from, the ATP catalytic sites (Boyer et al. 1973; Kayalar et al. 1977). The rotation of subunits extending through the ATPase complex couples these affinity changes to proton transport. Three ATP catalytic sites located in the β-subunit of the complex sequentially alternate between open, loose and tight binding sites. As the γ-subunit rotates, it acts as a camshaft alternatively distorting the β-subunit, leading to the cycling between the three binding sites (Hunter et al. 2008). Rotation of the γ-subunit in the centre of the F_1 portion deforms the surrounding catalytic subunits, giving rise to the three binding sites (Cross 2000). Rotation of the c subunit relative to the α subunit, however, is required for completion of the proton pathway (Vik and Antonio 1994; Duncan et al. 1995; Hatch et al. 1995; Junge et al. 1996). The energy from this conformational change is transduced into the ATP phosphoanhydride bond. Entrance and exit channels for protons are contained within the F_0 subunit, but there is no connection between the two channels (Junge et al. 1997). An essential glutamic acid residue in a transmembrane helices within the entrance channel is protonated by residing protons; this conformational change ratchets the c subunit with respect to the α subunit (Rastogi and Girvin 1999; Junge et al. 2009; Pogoryelov et al. 2009). The c ring rotates; in turn each σ subunit picks up a proton and releases it through the exit channel. Therefore, during one complete cycle, a proton is transported through each c subunit, and each of the three β-subunits goes through open, loose and tight conformations, each producing and releasing a molecule of ATP.

5 Calvin-Benson-Bassham Cycle

The final stage in purple bacterial photosynthesis is the use of the energy generated in the steps discussed to drive the conversion of inorganic carbon into organic compounds. Purple photosynthetic bacteria contain the most diverse metabolism of carbon compounds. Both purple non-sulphur and sulphur bacteria assimilate CO_2 into 3-phosphoglycerate (3-PG) and reduce CO_2 mainly via the Calvin-Benson-Bassham (CBB) cycle (Fig. 12) (Benson 2002; Tabita 2004). The CBB cycle is a complex series of reactions, which can be broken down into three phases: carboxylation, reduction and regeneration (Heldt and Piechulla 2011).

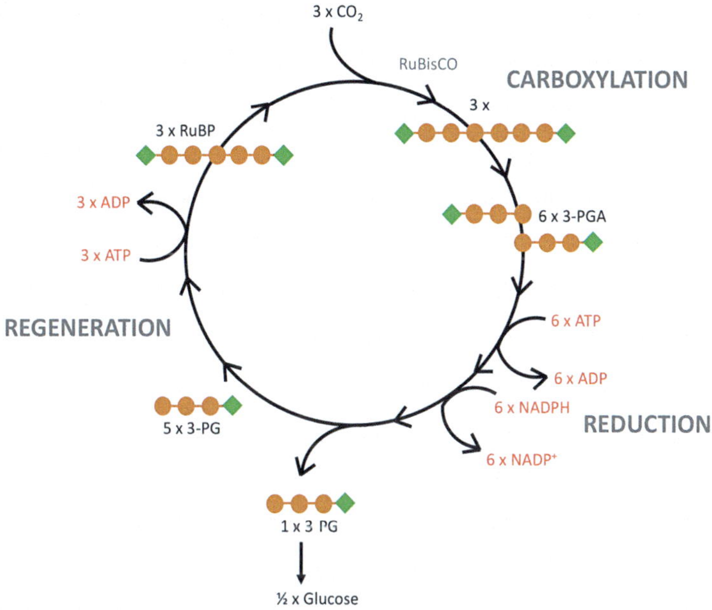

Fig. 12 Schematic of the Calvin-Benson-Bassham (CBB) cycle. The CBB cycle occurs in three main stages; RuBisCO is first used in carboxylation to produce six molecules of 3-PGA from the incorporation of inorganic carbon (CO_2) and RuBP. ATP and NADPH are then utilised to phosphorylate and reduce 3-PGA to produce 3-PG in the reduction stage. Finally, in the regeneration stage, 5 3-PG molecules are recycled to regenerate RuBP, and one molecule of 3-PG is used to make glucose; two cycles of the cycle are required to make a molecule of glucose. Orange circle = carbon, green diamond = phosphate

Carboxylation is the first stage of the cycle and utilises ribulose-bisphosphate carboxylase/oxygenase (RuBisCO) to incorporate carbon from CO_2 into an organic molecule. Ribulose-1,5-bisphosphate (RuBP) is attached to a molecule of CO_2 to form a six-carbon compound; this spontaneously breaks down into two molecules of 3-phosphoglyceric acid (3-PGA). The next stage of the cycle is reduction (Heldt and Piechulla 2011); 3-PGA molecules are converted to 3-PG by phosphorylation. First ATP donates a phosphate group to each of the two 3-PGA molecules producing 1,3-bisphosphoglycerate. This is then reduced by NADPH releasing a phosphate group and producing 3-PG. The final stage is regeneration, in which some of the 3-PG molecules are used to make glucose, while others are recycled to regenerate the RuBP acceptor.

AAPs are obligate photoheterotrophs; no autotrophic carbon assimilation pathways have been reported (Yurkov and Beatty 1998; Fuchs et al. 2007; Swingley et al. 2007). AAPs and some anaerobic anoxygenic phototrophs have been reported to grow heterotrophically on acetate using the ethylmalonyl-CoA pathway and the oxidative glyoxylate cycle (Tang et al. 2011; De Meur et al. 2018).

References

Aagaard, J., & Sistrom, W. R. (1972). Control of synthesis of reaction center bacteriochlorophyll in photosynthetic bacteria. *Photochemistry and Photobiology, 15*(2), 209–225. https://doi.org/10.1111/j.1751-1097.1972.tb06240.x.

Alberti, M., Burke, D. H., & Hearst, J. (2004). *Advances in photosynthesis. Anoxygenic photosynthetic Bacteria*. Available at: https://link.springer.com/content/pdf/10.1007%2F0-306-47954-0.pdf. Accessed 29 Aug 2019.

Allen, J. F., & Holmes, N. G. (1986). A general model for regulation of photosynthetic unit function by protein phosphorylation. *FEBS Letters*. https://doi.org/10.1016/0014-5793(86)80682-2.

Angerhofer, A., Cogdell, R. J., & Hipkins, M. F. (1986). A spectral characterisation of the light-harvesting pigment–protein complexes from Rhodopseudomonas acidophila. *Biochimica et Biophysica Acta (BBA) – Bioenergetics, 848*(3), 333–341. https://doi.org/10.1016/0005-2728(86)90208-2.

Axelrod, H. L., et al. (1994). Crystallization and X-ray structure determination of cytochrome c2 from Rhodobacter sphaeroides in three crystal forms. *Acta Crystallographica Section D Biological Crystallography*. International Union of Crystallography (IUCr), *50*(4), 596–602. https://doi.org/10.1107/S0907444994001319.

Axelrod, H. L., et al. (2002). X-ray structure determination of the cytochrome c2: Reaction center electron transfer complex from Rhodobacter sphaeroides. *Journal of Molecular Biology*. Academic Press, *319*(2), 501–515. https://doi.org/10.1016/S0022-2836(02)00168-7.

Axelrod, H., Miyashita, O., & Okamura, M. (2009). Structure and function of the cytochrome c 2: Reaction center complex from Rhodobacter sphaeroides. In (323–336). Dordrecht: Springer. https://doi.org/10.1007/978-1-4020-8815-5_17.

Bahatyrova, S., et al. (2004). The native architecture of a photosynthetic membrane. *Nature*. Nature Publishing Group, *430*(7003), 1058–1062. https://doi.org/10.1038/nature02823.

Benson, A. A. (2002). P aving the P ath. *Annual Review of Plant Biology*. Annual Reviews 4139 El Camino Way, P.O. Box 10139, Palo Alto, CA 94303-0139, USA, *53*(1), 1–25. https://doi.org/10.1146/annurev.arplant.53.091201.142547.

Berry, E. A., et al. (2000). *Structure and function of cytochrome bc complexes*. Available at: http://www.life.illinois.edu/crofts/pdf_files/ARB_review.pdf. Accessed 30 Aug 2019.

Berry, E. A., et al. (2004). X-Ray structure of rhodobacter capsulatus cytochrome bc $_1$: Comparison with its mitochondrial and chloroplast counterparts. *Photosynthesis Research, 81*(3), 251–275. https://doi.org/10.1023/B:PRES.0000036888.18223.0e.

Blankenship, R. E. (2014). *Molecular mechanisms of photosynthesis Robert E. Blankenship* (2nd ed.). St Louis: Wiley Blackwell.

Boonstra, A. F., Germeroth, L., & Boekema, E. J. (1994). Structure of the light harvesting antenna from Rhodospirillum molischianum studied by electron microscopy. *Biochimica et Biophysica Acta (BBA) – Bioenergetics*. Elsevier, *1184*(2–3), 227–234. https://doi.org/10.1016/0005-2728(94)90227-5.

Boyer, P. D. (1993). The binding change mechanism for ATP synthase — Some probabilities and possibilities. *Biochimica et Biophysica Acta (BBA) – Bioenergetics*. Elsevier, *1140*(3), 215–250. https://doi.org/10.1016/0005-2728(93)90063-L.

Boyer, P. D., Cross, R. L., & Momsen, W. (1973). A new concept for energy coupling in oxidative phosphorylation based on a molecular explanation of the oxygen exchange reactions. *Proceedings of the National Academy of Sciences of the United States of America*. National Academy of Sciences, *70*(10), 2837–2839. https://doi.org/10.1073/pnas.70.10.2837.

Brune, D. (2004) *Advances in photosynthesis. Anoxygenic photosynthetic Bacteria*. Available at: https://link.springer.com/content/pdf/10.1007%2F0-306-47954-0.pdf. Accessed 29 Aug 2019.

Brunisholz, R. A., & Zuber, H. (1992). Structure, function and organization of antenna polypeptides and antenna complexes from the three families of Rhodospirillaneae. *Journal of Photochemistry and Photobiology, B: Biology*. https://doi.org/10.1016/1011-1344(92)87010-7.

Brunisholz, R. A., et al. (1985). The light-harvesting polypeptides of Rhodopseudomonas viridis. The complete amino-acid sequences of B1015-alpha, B1015-beta and B1015-gamma. *Biological Chemistry Hoppe-Seyler, 366*(1), 87–98. https://doi.org/10.1515/BCHM3.1985.366.1.87.

Busselez, J., et al. (2007). Structural basis for the PufX-mediated dimerization of bacterial photosynthetic Core complexes. *Structure, 15*(12), 1674–1683. https://doi.org/10.1016/j.str.2007.09.026.

Bustamante, P. L., & Loach, P. A. (1994). Reconstitution of a functional photosynthetic receptor complex with isolated subunits of Core light-harvesting complex and reaction centers. *Biochemistry, 33*(45), 13329–13339. https://doi.org/10.1021/bi00249a020.

Bylina, E. J., & Youvan, D. C. (1988). Directed mutations affecting spectroscopic and electron transfer properties of the primary donor in the photosynthetic reaction center. *Proceedings of the National Academy of Sciences of the United States of America*. National Academy of Sciences, *85*(19), 7226–7230. https://doi.org/10.1073/pnas.85.19.7226.

Caffrey, M., et al. (1992). Cytochrome c2 mutants of Rhodobacter capsulatus. *Archives of Biochemistry and Biophysics, 292*(2), 419–426. https://doi.org/10.1016/0003-9861(92)90011-k.

Camara-Artigas, A., Brune, D., & Allen, J. P. (2002). Interactions between lipids and bacterial reaction centers determined by protein crystallography. *Proceedings of the National Academy of Sciences of the United States of America, 99*(17), 11055–11060. https://doi.org/10.1073/pnas.162368399.

Capaldi, R. A., et al. (1996). Structural changes in the γ and ε subunits of the Escherichia coli F1F0-type ATPase during energy coupling. *Journal of Bioenergetics and Biomembranes*. Kluwer Academic Publishers-Plenum Publishers, *28*(5), 397–401. https://doi.org/10.1007/BF02113980.

Cartron, M. L., et al. (2014). Integration of energy and electron transfer processes in the photosynthetic membrane of Rhodobacter sphaeroides. *Biochimica et Biophysica Acta (BBA) – Bioenergetics, 1837*(10), 1769–1780. https://doi.org/10.1016/j.bbabio.2014.02.003.

Chang, C.-H., et al. (1986). *Structure of Rhodopseudomonas sphaeroides R-26 reaction center*.

Chang, M. C., et al. (1990). Spectroscopic characterization of the light-harvesting complex of Rhodospirillum rubrum and its structural subunit. *Biochemistry*. American Chemical Society, *29*(2), 421–429. https://doi.org/10.1021/bi00454a017.

Cogdell, R. J., et al. (1983). The isolation and partial characterisation of the light-harvesting pigment–protein complement of Rhodopseudomonas acidophila. *BBA – Bioenergetics, 722*(3), 427–435. https://doi.org/10.1016/0005-2728(83)90058-0.

Cogdell, R. J., et al. (1990). Isolation and characterisation of an unusual antenna complex from the marine purple sulphur photosynthetic bacterium Chromatium purpuratum BN5500. *Biochimica et Biophysica Acta (BBA) – Bioenergetics*. Elsevier, *1019*(3), 239–244. https://doi.org/10.1016/0005-2728(90)90199-E.

Cogdell, R. J., et al. (1999). How photosynthetic bacteria harvest solar energy. *Journal of bacteriology*. American Society for Microbiology (ASM), *181*(13), 3869–3879. Available at: http://www.ncbi.nlm.nih.gov/pubmed/10383951. Accessed 6 Sept 2019.

Crofts, A. R. (2004). The cytochrome BC 1 complex: Function in the context of structure. *Annual Review of Physiology, 66*, 689–733. https://doi.org/10.1146/annurev.physiol.66.032102.150251.

Crofts, A. R. (2005). The Q-cycle — A personal perspective. In *Discoveries in photosynthesis* (pp. 479–499). Berlin/Heidelberg: Springer-Verlag. https://doi.org/10.1007/1-4020-3324-9_46.

Cross, R. L. (2000). The rotary binding change mechanism of ATP synthases. *Biochimica et Biophysica Acta (BBA) – Bioenergetics*. Elsevier, *1458*(2–3), 270–275. https://doi.org/10.1016/S0005-2728(00)00079-7.

Crouch, L. I., & Jones, M. R. (2012). Cross-species investigation of the functions of the Rhodobacter PufX polypeptide and the composition of the RC–LH1 core complex. *Biochimica et Biophysica Acta (BBA) – Bioenergetics, 1817*(2), 336–352. https://doi.org/10.1016/j.bbabio.2011.10.009.

Darrouzet, E., Cooley, J. W., & Daldal, F. (2004). The cytochrome bc $_1$ complex and its homologue the b $_6$ f complex: Similarities and differences. *Photosynthesis Research, 79*(1), 25–44. https://doi.org/10.1023/B:PRES.0000011926.47778.4e.

Davis, C. M., et al. (1997). Evaluation of structure-function relationships in the core light- harvesting complex of photosynthetic bacteria by reconstitution with mutant polypeptides. *Biochemistry*. American Chemical Society, *36*(12), 3671–3679. https://doi.org/10.1021/bi962386p.

De Meur, Q., et al. (2018). Genetic plasticity and ethylmalonyl coenzyme a pathway during acetate assimilation in Rhodospirillum rubrum S1H under photoheterotrophic conditions. *Applied and Environmental Microbiology*. American Society for Microbiology, *84*(3), e02038-17. https://doi.org/10.1128/AEM.02038-17.

DeHoff, B. S., et al. (1988). In vivo analysis of puf operon expression in Rhodobacter sphaeroides after deletion of a putative intercistronic transcription terminator. *Journal of Bacteriology, 170*(10), 4681–4692. https://doi.org/10.1128/jb.170.10.4681-4692.1988.

Deisenhofer, J., & Michel, H. (1988). *The photosynthetic reaction center from the purple bacterium rhodopseudomonas viridis*. Available at: http://science.sciencemag.org/. Accessed 4 July 2019.

Deisenhofer, J. et al. (1985). Structure of the protein subunits in the photosynthetic reaction center of. *Nature* Available at: https://www.nature.com/articles/318618a0.pdf. Accessed 3 July 2019.

Drepper, F., Dorlet, P. A., & Mathis, P. (1997). Cross-linked electron transfer complex between cytochrome c2 and the photosynthetic reaction center of Rhodobacter sphaeroides†. *American Chemical Society*. https://doi.org/10.1021/BI961350U.

Duncan, T. M., et al. (1995). Rotation of subunits during catalysis by Escherichia coli F1-ATPase. *Proceedings of the National Academy of Sciences of the United States of America*. National Academy of Sciences, *92*(24), 10964–10968. https://doi.org/10.1073/pnas.92.24.10964.

Ehrenreich, A., & Widdel, F. (1994). *Anaerobic oxidation of ferrous Iron by purple bacteria, a new type of phototrophic metabolism, applied and environmental microbiology*. Available at: https://www.ncbi.nlm.nih.gov/pmc/articles/PMC202013/pdf/aem00029-0311.pdf. Accessed 29 Aug 2019.

Engelhardt, H., Engel, A., & Baumeister, W. (1986). Stoichiometric model of the photosynthetic unit of Ectothiorhodospira halochloris. *Proceedings of the National Academy of Sciences, 83*(23), 8972–8976. https://doi.org/10.1073/pnas.83.23.8972.

Evans, M. B., Hawthornthwaite, A. M., & Cogdell, R. J. (1990). Isolation and characterisation of the different B800–850 light-harvesting complexes from low- and high-light grown cells of Rhodopseudomonas palustris, strain 2.1.6. *Biochimica et Biophysica Acta (BBA) – Bioenergetics, 1016*(1), 71–76. https://doi.org/10.1016/0005-2728(90)90008-R.

Fajer, J. et al. (1975). Primary charge separation in bacterial photosynthesis: Oxidized chlorophylls and reduced pheophytin. *Proceedings of the National Academy of Sciences*. National Academy of Sciences, *72*(12), 4956–4960. https://doi.org/10.1073/PNAS.72.12.4956.

Farchaus, J. W., Gruenberg, H., & Oesterhelt, D. (1990) Complementation of a reaction Center-deficient Rhodobacter sphaeroides pufLMX deletion strain in trans with pufBALM does not restore the photosynthesis-positive phenotype. *Journal of Bacteriology*. Available at: https://www.ncbi.nlm.nih.gov/pmc/articles/PMC208526/pdf/jbacter01044-0475.pdf. Accessed 29 Aug 2019.

Fotiadis, D. et al. (2003) *Structural analysis of the RC–LH1 photosynthetic core complex of Rhodospirillum rubrum using atomic force microscopy downloaded from*. JBC Papers in Press. Available at: http://www.jbc.org/. Accessed 20 Feb 2019.

Fowler, G. J. S., et al. (1994). Blue shifts in bacteriochlorophyll absorbance correlate with changed hydrogen bonding patterns in light-harvesting 2 mutants of Rhodobacter sphaeroides with alterations at α-Tyr-44 and α-Tyr-45: Figure 2. *Biochemical Journal, 299*(3), 695–700. https://doi.org/10.1042/bj2990695.

Fowler, G. J. S., et al. (1997). The role of βArg_{-10} in the B800 Bacteriochlorophyll and carotenoid pigment environment within the light-harvesting LH2 complex of *Rhodobacter sphaeroides* †. *Biochemistry, 36*(37), 11282–11291. https://doi.org/10.1021/bi9626315.

Francia, F., et al. (1999). The reaction Center–LH1 antenna complex of Rhodobacter sphaeroides contains one PufX molecule which is involved in dimerization of this complex. *American Chemical Society*. https://doi.org/10.1021/BI982891H.

Frese, R. N., et al. (2004). The long-range organization of a native photosynthetic membrane. *Proceedings of the National Academy of Sciences of the United States of America, 101*(52), 17994–17999. https://doi.org/10.1073/pnas.0407295102.

Frigaard, N. U., & Dahl, C. (2008). Sulfur metabolism in phototrophic Sulfur bacteria. *Advances in Microbial Physiology.* https://doi.org/10.1016/S0065-2911(08)00002-7.

Fuchs, B. M., et al. (2007). Characterization of a marine gammaproteobacterium capable of aerobic anoxygenic photosynthesis. *Proceedings of the National Academy of Sciences, 104*(8), 2891–2896. https://doi.org/10.1073/pnas.0608046104.

Gall, A., et al. (1997). Influence of the protein binding site on the absorption properties of the monomeric Bacteriochlorophyll in Rhodobacter sphaeroides LH2 complex†. *American Chemical Society.* https://doi.org/10.1021/BI9717237.

Gall, A., et al. (1999). Certain species of the Proteobacteria possess unusual bacteriochlorophyll environments in their light-harvesting proteins. *Biospectroscopy, 5*(6), 338–345. https://doi.org/10.1002/(SICI)1520-6343(1999)5:6<338::AID-BSPY3>3.0.CO;2-D.

Gardiner, A. T., Cogdell, R. J., & Takaichi, S. (1993). *The effect of growth conditions on the light-harvesting apparatus in Rhodopseudomonas acidophila, Photosynthesis Research*. Kluwer Academic Publishers. Available at: https://link.springer.com/content/pdf/10.1007%2FBF00146415.pdf. Accessed 9 July 2019.

Gast, P., et al. (1985). Determination of the amount and the type of quinones present in single crystals from reaction center protein from the photosynthetic bacterium Rhodopseudomonas viridis. *FEBS Letters.* https://doi.org/10.1016/0014-5793(85)80544-5.

Gonçalves, R. P., et al. (2005). Architecture of the native photosynthetic apparatus of Phaeospirillum molischianum. *Journal of Structural Biology, 152*(3), 221–228. https://doi.org/10.1016/j.jsb.2005.10.002.

Gubellini, F., et al. (2006). Functional and structural analysis of the photosynthetic apparatus of *Rhodobacter veldkampii* †. *Biochemistry, 45*(35), 10512–10520. https://doi.org/10.1021/bi0610000.

Guo, H., Suzuki, T., & Rubinstein, J. L. (2019). Structure of a bacterial ATP synthase. *eLife, 8*. https://doi.org/10.7554/eLife.43128.

Gupta, R. S., & Khadka, B. (2016). Evidence for the presence of key chlorophyll-biosynthesis-related proteins in the genus Rubrobacter (Phylum Actinobacteria) and its implications for the evolution and origin of photosynthesis. *Photosynthesis Research, 127*(2), 201–218. https://doi.org/10.1007/s11120-015-0177-y.

Hansen, T. A., & Gemerden, H. (1972). Sulfide utilization by purple nonsulfur bacteria. *Archiv für Mikrobiologie*. Springer, *86*(1), 49–56. https://doi.org/10.1007/BF00412399.

Hartigan, N., et al. (2002). The 7.5-Å electron density and spectroscopic properties of a novel low-light B800 LH2 from Rhodopseudomonas palustris. *Biophysical Journal.* https://doi.org/10.1016/S0006-3495(02)75456-8.

Hatch, L. P., Cox, G. B., & Howitt, S. M. (1995). The essential arginine residue at position 210 in the alpha subunit of the Escherichia coli ATP synthase can be transferred to position 252 with partial retention of activity. *The Journal of biological chemistry*. American Society for Biochemistry and Molecular Biology, *270*(49), 29407–29412. https://doi.org/10.1074/jbc.270.49.29407.

Heldt, H.-W., & Piechulla, B. (2011). *Plant biochemistry*. Academic.

Holden-Dye, K., Crouch, L. I., & Jones, M. R. (2008). Structure, function and interactions of the PufX protein. *Biochimica et Biophysica Acta, 1777*, 613–630. https://doi.org/10.1016/j.bbabio.2008.04.015.

Hu, X., et al. (1998). Architecture and mechanism of the light-harvesting apparatus of purple bacteria. *Computational Biomolecular Science, 95*, 5935–5941.

Hu, X., et al. (2002). Photosynthetic apparatus of purple bacteria. *Quarterly Reviews of Biophysics*. Cambridge University Press, *35*, 1–62. https://doi.org/10.1017/S0033583501003754.

Hunter, C., et al. (2008). *The purple phototrophic bacteria*. Dordrecht: Springer.

Imanishi, M., et al. (2019). A dual role for ca $^{2+}$ in expanding the spectral diversity and stability of light-harvesting 1 reaction Center Photocomplexes of purple phototrophic Bacteria.

Biochemistry. American Chemical Society, *58*(25), 2844–2852. https://doi.org/10.1021/acs.biochem.9b00351.

Imhoff, J. F., Hiraishi, A., & Süling, J. (2005) *Anoxygenic phototrophic purple bacteria, Bergey's manual® of systematic bacteriology*. https://doi.org/10.1007/0-387-28021-9_15.

Jackson, P. J., et al. (2018). Identification of protein W, the elusive sixth subunit of the Rhodopseudomonas palustris reaction center-light harvesting 1 core complex. *Biochimica et Biophysica Acta (BBA) – Bioenergetics*. Elsevier, *1859*(2), 119–128. https://doi.org/10.1016/j.bbabio.2017.11.001.

Jamieson, S. J., et al. (2002). Projection structure of the photosynthetic reaction centre-antenna complex of Rhodospirillum rubrum at 8.5?? resolution. *EMBO Journal*. https://doi.org/10.1093/emboj/cdf410.

Janosi, L., et al. (2006). Influence of subunit structure on the oligomerization state of light-harvesting complexes: A free energy calculation study. *Chemical Physics*. https://doi.org/10.1016/j.chemphys.2005.08.038.

Jault, J., & Allison, W. (1993). *Slow binding of ATP to noncatalytic nucleotide binding sites which accelerates catalysis is responsible for apparent negative cooperativity exhibit... – PubMed – NCBI, The Journal of Biological Chemistry*. Available at: https://www.ncbi.nlm.nih.gov/pubmed/8420930. Accessed 30 Aug 2019.

Jault, J.-M. et al. (1995). *Complex of the Fi-ATPase from thermophilic Bacillus PS 3 containing the CX.D261N substitution fails to dissociate inhibitory MgADP from a catalytic site when ATP binds to noncatalytic sites1&quot, Biochemistry*. Available at: https://pubs.acs.org/sharingguidelines. Accessed 30 Aug 2019.

Jay, F., Lambillotte, M., & Mühlethaler, K. (1983). Localisation of Rhodopseudomonas viridis reaction centre and light harvesting proteins using ferritin-antibody labelling. *European Journal of Cell Biology*, *30*(1), 1–8. Available at: http://www.ncbi.nlm.nih.gov/pubmed/6189715. Accessed 18 Sept 2019.

Jenney, F. E., & Daldal, F. (1993). A novel membrane-associated c-type cytochrome, cyt cy, can mediate the photosynthetic growth of Rhodobacter capsulatus and Rhodobacter sphaeroides. *The EMBO Journal*. European Molecular Biology Organization, *12*(4), 1283–1292. https://doi.org/10.1002/J.1460-2075.1993.TB05773.X.

Jones, M. R. (2009). The petite purple photosynthetic powerpack. *Biochemical Society Transactions, 37*(2), 400–407. https://doi.org/10.1042/BST0370400.

Jungas, C., et al. (1999). Supramolecular organization of the photosynthetic apparatus of Rhodobacter sphaeroides. *The EMBO Journal, 18*(3), 534–542.

Junge, W., Sabber, D., & Engelbrecht, S. (1996). ATP-synthesis. Rotatory catalysis by F-ATPase: Real-time recording of intersubunit rotation. *Berichte der Bunsengesellschaft für physikalische Chemie*. John Wiley & Sons, Ltd, *100*(12), 2014–2019. https://doi.org/10.1002/bbpc.19961001215.

Junge, W., Lill, H., & Engelbrecht, S. (1997). ATP synthase: An electrochemical ransducer with rotatory mechanics *Trends in Biochemical Sciences*. Elsevier Current Trends, *22*(11), 420–423. https://doi.org/10.1016/S0968-0004(97)01129-8.

Junge, W., Sielaff, H., & Engelbrecht, S. (2009). Torque generation and elastic power transmission in the rotary FOF1-ATPase. *Nature, 459*(7245), 364–370. https://doi.org/10.1038/nature08145.

Kayalar, C., et al. (1977). An alternating site sequence for oxidative phosphorylation suggested by measurement of substrate binding patterns and exchange reaction inhibitions∗. *The Journal of Biological Chemistry*. Available at: http://www.jbc.org/. Accessed 30 Aug 2019.

Kehoe, J. W., et al. (1998). Reconstitution of Core light-harvesting complexes of photosynthetic Bacteria using chemically synthesized polypeptides. 2. Determination of structural features that stabilize complex formation and their implications for the structure of the subunit complex †. *Biochemistry, 37*(10), 3418–3428. https://doi.org/10.1021/bi9722709.

Kellogg, E. C., et al. (1989). Measurement of the extent of electron transfer to the bacteriopheophytin in the M-subunit in reaction centers of Rhodopseudomonas viridis. *Photosynthesis Research*. Kluwer Academic Publishers, *22*(1), 47–59. https://doi.org/10.1007/BF00114766.

Kinosita, K., Adachi, K., & Itoh, H. (2004). Rotation of F_1 -ATPase: How an ATP-driven molecular machine may work. *Annual Review of Biophysics and Biomolecular Structure, 33*(1), 245–268. https://doi.org/10.1146/annurev.biophys.33.110502.132716.

Kirmaier, C., Holten, D., & Parson, W. W. (1985). Picosecond-photodichroism studies of the transient states in Rhodopseudomonas sphaeroides reaction centers at 5 K. effects of electron transfer on the six bacteriochlorin pigments. *Biochimica et Biophysica Acta (BBA) – Bioenergetics*. Elsevier, *810*(1), 49–61. https://doi.org/10.1016/0005-2728(85)90205-1.

Klug, G. et al. (1988). Pleiotropic effects of localized Rhodobacter capsulatus puf operon deletions on production of light-absorbing pigment–protein complexes. *Journal of Bacteriology*. Available at: https://www.ncbi.nlm.nih.gov/pmc/articles/PMC211687/pdf/jbacter00190-0422.pdf. Accessed 29 Aug 2019.

Koblížek, M., et al. (2005). Sequential assembly of photosynthetic units in Rhodobacter sphaeroides as revealed by fast repetition rate analysis of variable bacteriochlorophyll a fluorescence. *Biochimica et Biophysica Acta (BBA) – Bioenergetics, 1706*(3), 220–231. https://doi.org/10.1016/j.bbabio.2004.11.004.

Koepke, J., et al. (1996). The crystal structure of the light-harvesting complex II (B800-850) from Rhodospirillum molischianum. *Structure, 4*(5), 581–597. https://doi.org/10.1016/S0969-2126(96)00063-9.

Kolber, Z. S. et al. (2001). Contribution of aerobic photoheterotrophic bacteria to the carbon cycle in the ocean. *Science (New York, N.Y.)*. American Association for the Advancement of Science, *292*(5526), 2492–2495. https://doi.org/10.1126/science.1059707.

Kumar, S., et al. (2016). Direct imaging of protein organization in an intact bacterial organelle using high-resolution atomic force microscopy. https://doi.org/10.1021/acsnano.6b05647.

Lang, H. P., & Hunter, C. N. (1994). The relationship between carotenoid biosynthesis and the assembly of the light-harvesting LH2 complex in Rhodobacter sphaeroides. *Biochemical Journal, 298*(1), 197–205. https://doi.org/10.1042/bj2980197.

LaSarre, B., et al. (2018). Restricted localization of photosynthetic intracytoplasmic membranes (ICMs) in multiple genera of purple nonsulfur bacteria. *mBio, 9*(4), 780–798. https://doi.org/10.1128/mBio.00780-18.

Lilburn, T. G., et al. (1992). Pleiotropic effects of pufX gene deletion on the structure and function of the photosynthetic apparatus of Rhodobacter capsulatus. *Biochimica et Biophysica Acta (BBA) – Bioenergetics, 1100*(2), 160–170. https://doi.org/10.1016/0005-2728(92)90077-F.

Liu, L.-N. (2016). Distribution and dynamics of electron transport complexes in cyanobacterial thylakoid membranes. *Biochimica et Biophysica Acta*. Elsevier, *1857*(3), 256–265. https://doi.org/10.1016/j.bbabio.2015.11.010.

Liu, L.-N., & Scheuring, S. (2013). Investigation of photosynthetic membrane structure using atomic force microscopy. *Trends in Plant Science, 18*, 277–286. https://doi.org/10.1016/j.tplants.2013.03.001.

Liu, L.-N., et al. (2009). Quinone pathways in entire photosynthetic Chromatophores of Rhodospirillum photometricum. *Journal of Molecular Biology, 393*, 27–35. https://doi.org/10.1016/j.jmb.2009.07.044.

Liu, L.-N., et al. (2011a). Forces guiding assembly of light-harvesting complex 2 in native membranes. *Proceedings of the National Academy of Sciences, 108*(23), 9455–9459. https://doi.org/10.1073/pnas.1004205108.

Liu, L. N., Sturgis, J. N., & Scheuring, S. (2011b). Native architecture of the photosynthetic membrane from Rhodobacter veldkampii. *Journal of Structural Biology*. https://doi.org/10.1016/j.jsb.2010.08.010.

Loach, P. A., & Parkes-Loach, P. S. (1995). Structure-function relationships in Core Light-Harvesting Complexes (LHI) as determined by characterization of the structural subunit and by reconstitution experiments. In *Anoxygenic photosynthetic bacteria* (pp. 437–471). Dordrecht: Kluwer Academic Publishers. https://doi.org/10.1007/0-306-47954-0_21.

Madigan, M. T. (2003). *Anoxygenic phototrophic bacteria from extreme environments, Photosynthesis Research*. Available at: http://www.life.illinois.edu/govindjee/Part2/14_Madigan.pdf. Accessed 29 Aug 2019.

Mathis, P., Ortega, J. M., & Venturoli, G. (1994). Interaction between cytochrome c and the photosynthetic reaction center of purple bacteria: Behaviour at low temperature. *Biochimie, 76*(6), 569–579. https://doi.org/10.1016/0300-9084(94)90181-3.

Matsui, T., et al. (1997) *Catalytic activity of the 3 3 complex of F 1-ATPase without Noncatalytic Nucleotide Binding Site**. Available at: http://www-jbc.stanford.edu/jbc/. Accessed 30 Aug 2019.

McDermott, G., et al. (1995). Crystal structure of an integral membrane light-harvesting complex from photosynthetic bacteria. *Nature*. https://doi.org/10.1038/374517a0.

Mcluskey, K., et al. (2001). The crystallographic structure of the B800-820 LH3 light-harvesting complex from the purple bacteria Rhodopseudomonas acidophila strain 7050 †. https://doi.org/10.1021/bi010309a.

Menin, L., et al. (1998). Role of HiPIP as electron donor to the RC-bound cytochrome in photosynthetic purple bacteria. *Photosynthesis Research*. Kluwer Academic Publishers, *55*(2/3), 343–348. https://doi.org/10.1023/A:1005989900756.

Milgrom, Y. M., Ehler, L. L., & Boyer, P. D. (1991). The characteristics and effect on catalysis of nucleotide binding to noncatalytic sites of chloroplast F1-ATPase. *The Journal of Biological Chemistry, 266*(18), 11551–11558. Available at: http://www.ncbi.nlm.nih.gov/pubmed/1828802. Accessed 30 Aug 2019.

Miller, J. F., et al. (1987). Isolation and characterization of a subunit form of the light-harvesting complex of Rhodospirillum rubrum. *Biochemistry*. American Chemical Society, *26*(16), 5055–5062. https://doi.org/10.1021/bi00390a026.

Mothersole, D. J., et al. (2016). PucC and LhaA direct efficient assembly of the light-harvesting complexes in Rhodobacter sphaeroides. *Molecular Microbiology*. https://doi.org/10.1111/mmi.13235.

Munk, A. C., et al. (2011). Complete genome sequence of Rhodospirillum rubrum type strain (S1T). *Standards in Genomic Sciences, 4*(3), 293–302. https://doi.org/10.4056/sigs.1804360.

Myllykallio, H., et al. (1998). Membrane-anchored cytochrome c_y mediated microsecond time range Electron transfer from the cytochrome bc_1 complex to the reaction Center in *Rhodobacter capsulatus* †. *Biochemistry, 37*(16), 5501–5510. https://doi.org/10.1021/bi973123d.

Niedzwiedzki, D. M., et al. (2011). Energy transfer in an LH4-like light harvesting complex from the aerobic purple photosynthetic bacterium Roseobacter denitrificans. *Biochimica et Biophysica Acta (BBA) – Bioenergetics, 1807*(5), 518–528. https://doi.org/10.1016/j.bbabio.2011.03.004.

Niwa, S., et al. (2014). Structure of the LH1–RC complex from Thermochromatium tepidum at 3.0Å. *Nature, 508*(7495), 228–232. https://doi.org/10.1038/nature13197.

Nogi, T., Hirano, Y., & Miki, K. (2005). Structural and functional studies on the tetraheme cytochrome subunit and its electron donor proteins: The possible docking mechanisms during the electron transfer reaction. *Photosynthesis Research*. https://doi.org/10.1007/s11120-004-2416-5.

Noji, H., et al. (1997). Direct observation of the rotation of F1-ATPase. *Nature, 386*(6622), 299–302. https://doi.org/10.1038/386299a0.

Nottoli, M., et al. (2018). The role of charge-transfer states in the spectral tuning of antenna complexes of purple bacteria. *Photosynthesis Research*. Springer Netherlands, *137*(2), 215–226. https://doi.org/10.1007/s11120-018-0492-1.

Olsen, J. D., et al. (1997). Site-directed modification of the ligands to the bacteriochlorophylls of the light-harvesting LH1 and LH2 complexes of Rhodobacter sphaeroides†. *Biochemistry*. American Chemical Society, *36*, 12625–12632. https://doi.org/10.1021/BI9710481.

Ortega, J. M., Drepper, F., & Mathis, P. (1999). Electron transfer between cytochrome c(2) and the tetraheme cytochrome c in Rhodopseudomonas viridis. *Photosynthesis Research, 59*(2–3), 147–157. https://doi.org/10.1023/A:1006149621029.

Pandit, A. et al. (2003). Investigations of intermediates appearing in the reassociation of the light-harvesting 1 complex of Rhodospirillum rubrum. *Photosynthesis Research, 3*(75), 235–248. Available at: https://www.ncbi.nlm.nih.gov/pubmed/16228604. Accessed 6 Sept 2019.

Pänke, O., et al. (2000). F-ATPase: Specific observation of the rotating c subunit oligomer of EF(o) EF(1). *FEBS Letters, 472*(1), 34–38. https://doi.org/10.1016/s0014-5793(00)01436-8.

Papiz, M. Z., et al. (2003). The structure and thermal motion of the B800-850 LH2 complex from Rps. acidophila at 2.0 Å resolution and 100 K: New structural features and functionally relevant motions. *Journal of Molecular Biology.* https://doi.org/10.1016/S0022-2836(03)00024-X.

Parkes-Loach, P. S., et al. (2000). Articles role of the Core region of the PufX protein in inhibition of reconstitution of the Core light-harvesting complexes of Rhodobacter sphaeroides and Rhodobacter capsulatus †. https://doi.org/10.1021/bi002580i.

Parkes-Loach, P. S., et al. (2004). Interactions stabilizing the structure of the Core light-harvesting complex (LH1) of photosynthetic Bacteria and its subunit (B820) †. *Biochemistry, 43*(22), 7003–7016. https://doi.org/10.1021/bi049798f.

Pogoryelov, D., et al. (2009). High-resolution structure of the rotor ring of a proton-dependent ATP synthase. *Nature Structural & Molecular Biology, 16*(10), 1068–1073. https://doi.org/10.1038/nsmb.1678.

Polívka, T., & Frank, H. A. (2010). Molecular factors controlling photosynthetic light harvesting by carotenoids. *Accounts of Chemical Research*. American Chemical Society, *43*(8), 1125–1134. https://doi.org/10.1021/ar100030m.

Prince, S. M., et al. (1997). Apoprotein structure in the LH2 complex from Rhodopseudomonas acidophila strain 10050: Modular assembly and protein pigment interactions. *Journal of Molecular Biology.* https://doi.org/10.1006/jmbi.1997.0966.

Pugh, R. J. et al. (1998). The LH1–RC core complex of Rhodobacter sphaeroides: Interaction between components, time-dependent assembly, and topology of the PufX protein. *Biochimica et Biophysica Acta (BBA) – Bioenergetics*. Elsevier, *1366*(3), 301–316. https://doi.org/10.1016/S0005-2728(98)00131-5.

Pullerits, T., & Sundström, V. (1996). Photosynthetic light-harvesting pigment–protein complexes: Toward understanding how and why. *American Chemical Society*. https://doi.org/10.1021/AR950110O.

Qian, P. (2017). Structure and function of the reaction centre – Light harvesting 1 Core complexes from purple photosynthetic Bacteria. In *Photosynthesis: Structures, mechanisms, and applications* (pp. 11–31). Cham: Springer International Publishing. https://doi.org/10.1007/978-3-319-48873-8_2.

Qian, P., Hunter, C. N., & Bullough, P. A. (2005). The 8.5 Å projection structure of the core RC–LH1-PufX dimer of Rhodobacter sphaeroides. *Journal of Molecular Biology.* https://doi.org/10.1016/j.jmb.2005.04.032.

Qian, P., Bullough, P. A., & Hunter, C. N. (2008). Three-dimensional reconstruction of a membrane-bending complex: The RC–LH1-PufX core dimer of rhodobacter sphaeroides. *Journal of Biological Chemistry.* https://doi.org/10.1074/jbc.M800625200.

Qian, P., et al. (2013). Three-dimensional structure of the Rhodobacter sphaeroides RC–LH1-PufX complex: Dimerization and quinone channels promoted by PufX. *Biochemistry, 52*(43), 7575–7585. https://doi.org/10.1021/bi4011946.

Qian, P., et al. (2018). Cryo-EM structure of the Blastochloris viridis LH1-RC complex at 2.9 Å. *Nature, 556*(7700), 203–208. https://doi.org/10.1038/s41586-018-0014-5.

Rastogi, V. K., & Girvin, M. E. (1999). Structural changes linked to proton translocation by subunit c of the ATP synthase. *Nature, 402*(6759), 263–268. https://doi.org/10.1038/46224.

Rathgeber, C., Beatty, J. T., & Yurkov, V. (2004). Aerobic phototrophic bacteria: New evidence for the diversity, ecological importance and applied potential of this previously overlooked group. *Photosynthesis Research.* https://doi.org/10.1023/B:PRES.0000035036.49977.bc.

Rieske, J. S., MacLennan, D. H., & Coleman, R. (1964). Isolation and properties of an iron-protein from the (reduced coenzyme Q)-cytochrome C reductase complex of the respiratory chain. *Biochemical and Biophysical Research Communications.* https://doi.org/10.1016/0006-291X(64)90171-8.

Rosen, D., et al. (1983). Interaction of cytochrome c with reaction centers of Rhodopseudomonas sphaeroides R-26: Localization of the binding site by chemical crosslinking and immunochemical studies. *Biochemistry, 22*(2), 335–341. https://doi.org/10.1021/bi00271a016.

Roszak, A. W. et al. (2003). Crystal structure of the RC–LH1 core complex from Rhodopseudomonas palustris. *Science*, 302. Available at: http://science.sciencemag.org/. Accessed 22 July 2019.

Roszak, A. W., et al. (2012). New insights into the structure of the reaction Centre from Blastochloris viridis: Evolution in the laboratory. *Biochemical Journal, 442*, 27–37. https://doi.org/10.1042/BJ20111540.

Sarıbaş, S., et al. (1998). Interactions between the cytochrome b, cytochrome c1, and Fe–S protein subunits at the Ubihydroquinone oxidation site of the bc1 complex of Rhodobacter capsulatus†. *American Chemical Society*. https://doi.org/10.1021/BI973146S.

Scheuring, S., & Sturgis, J. N. (2005). Chromatic adaptation of photosynthetic membranes. *Science, 309*(5733), 484–487. https://doi.org/10.1126/science.1110879.

Scheuring, S., & Sturgis, J. N. (2009). Atomic force microscopy of the bacterial photosynthetic apparatus: Plain pictures of an elaborate machinery. *Photosynthesis Research, 102*(2), 197–211. https://doi.org/10.1007/s11120-009-9413-7.

Scheuring, S., Seguin, J, et al. (2003a). Nanodissection and high-resolution imaging of the Rhodopseudomonas viridis photosynthetic core complex in native membranes by AFM. Atomic force microscopy. *Proceedings of the National Academy of Sciences of the United States of America*. National Academy of Sciences, *100*(4), 1690–1693. https://doi.org/10.1073/pnas.0437992100.

Scheuring, S., Seguin, J., et al. (2003b). AFM characterization of tilt and intrinsic flexibility of Rhodobacter sphaeroides light harvesting complex 2 (LH2). *Journal of Molecular Biology, 325*(3), 569–580. https://doi.org/10.1016/s0022-2836(02)01241-x.

Scheuring, S., et al. (2004a). Watching the photosynthetic apparatus in native membranes. *Sciences-New York, 101*(31), 11293–112967.

Scheuring, S., Rigaud, J. L., & Sturgis, J. N. (2004b). Variable LH2 stoichiometry and core clustering in native membranes of Rhodospirillum photometricum. *EMBO Journal, 23*(21), 4127–4133. https://doi.org/10.1038/sj.emboj.7600429.

Scheuring, S., Busselez, J., & Lévy, D. (2005a). Structure of the dimeric PufX-containing Core complex of *Rhodobacter blasticus* by *in Situ* atomic force microscopy. *Journal of Biological Chemistry, 280*(2), 1426–1431. https://doi.org/10.1074/jbc.M411334200.

Scheuring, S., Lévy, D., & Rigaud, J.-L. (2005b). Watching the components of photosynthetic bacterial membranes and their in situ organisation by atomic force microscopy. *Biochimica et Biophysica Acta (BBA) – Biomembranes*. Elsevier, *1712*(2), 109–127. https://doi.org/10.1016/J.BBAMEM.2005.04.005.

Scheuring, S., et al. (2006). The photosynthetic apparatus of Rhodopseudomonas palustris: Structures and organization. *Journal of Molecular Biology, 358*(1), 83–96. https://doi.org/10.1016/j.jmb.2006.01.085.

Scheuring, S., Boudier, T., & Sturgis, J. N. (2007). From high-resolution AFM topographs to atomic models of supramolecular assemblies. *Journal of Structural Biology, 159*(2 SPEC. ISS), 268–276. https://doi.org/10.1016/j.jsb.2007.01.021.

Scheuring, S., et al. (2014). The architecture of Rhodobacter sphaeroides chromatophores. *BBA – Bioenergetics, 1837*, 1263–1270. https://doi.org/10.1016/j.bbabio.2014.03.011.

Senior, A. E., Nadanaciva, S., & Weber, J. (2002). The molecular mechanism of ATP synthesis by F1F0-ATP synthase. *Biochimica et Biophysica Acta (BBA) – Bioenergetics*. Elsevier, *1553*(3), 188–211. https://doi.org/10.1016/S0005-2728(02)00185-8.

Sheng, Y. U., & Hearst, J. E. (1986). *Regulation of expression of genes for light-harvesting antenna proteins LH-I and LH-II; reaction center polypeptides RC-L, RC-M, and RC-H; and enzymes of bacteriochlorophyll and carotenoid biosynthesis in Rhodobacter capsulatus* by light and oxygen (photosynthetic bacteria/Rhodopseudomonas capsulata/puf operon/polycistronic mRNA/photooxidadve damage), Proceedings of the National Academy of Sciences of the United States of America*. Available at: https://www.pnas.org/content/pnas/83/20/7613.full.pdf. Accessed 29 Aug 2019.

Shopes, R. J., & Wraight, C. A. (1985). The acceptor quinone complex of Rhodopseudomonas viridis reaction centers. *BBA – Bioenergetics*. https://doi.org/10.1016/0005-2728(85)90242-7.

Siebert, C. A., et al. (2004). Molecular architecture of photosynthetic membranes in Rhodobacter sphaeroides: The role of PufX. *The EMBO Journal, 23*, 690–700. https://doi.org/10.1038/sj.emboj.7600092.

Stark, W., et al. (1984). The structure of the photoreceptor unit of *Rhodopseudomonas viridis*; The structure of the photoreceptor unit of *Rhodopseudomonas viridis*. *The EMBO Journal, 3*(4), 777–783. https://doi.org/10.1002/j.1460-2075.1984.tb01884.x.

Sumi, H. (2002). Uphill energy trapping by reaction center in bacterial photosynthesis: Charge separation unistep from antenna excitation, virtually mediated by special-pair excitation. *Journal of Physical Chemistry B*. American Chemical Society, *106*(51), 13370–13383. https://doi.org/10.1021/JP021716E.

Sundström, V. (2004). Ultrafast dynamics of carotenoid excited states–from solution to natural and artificial systems. *American Chemical Society*. https://doi.org/10.1021/CR020674N.

Sundstrom, V., & Van Grondelle, R. (1999). Photosynthetic light-harvesting: Reconciling dynamics and structure of purple bacterial LH2 reveals function of photosynthetic unit. *Journal of Physical Chemistry B, 103*, 2327–2346. https://doi.org/10.1021/jp983722+.

Swingley, W. D., et al. (2007). The complete genome sequence of Roseobacter denitrificans reveals a Mixotrophic rather than photosynthetic metabolism. *Journal of Bacteriology, 189*(3), 683–690. https://doi.org/10.1128/JB.01390-06.

Tabita, R. (2004). *Anoxygenic photosynthetic bacteria*. Available at: https://link.springer.com/content/pdf/10.1007%2F0-306-47954-0.pdf. Accessed 29 Aug 2019.

Takaichi, S. (2000). Characterization of carotenes in a combination of a C18 HPLC column with isocratic elution and absorption spectra with a photodiode-array detector. *Photosynthesis Research*. Kluwer Academic Publishers, *65*(1), 93–99. https://doi.org/10.1023/A:1006445503030.

Tang, K.-H., Tang, Y. J., & Blankenship, R. E. (2011). Carbon metabolic pathways in phototrophic Bacteria and their broader evolutionary implications. *Frontiers in Microbiology*. Frontiers, *2*, 165. https://doi.org/10.3389/fmicb.2011.00165.

Tharia, H. A. et al. (1999). *Characterisation of hydrophobic peptides by RP-HPLC from different spectral forms of LH2 isolated from Rps. palustris, Photosynthesis Research*. Available at: https://link.springer.com/content/pdf/10.1023%2FA%3A1006281532327.pdf. Accessed 17 July 2019.

Tiede, D. M., & Dutton, P. L. (1993). Electron transfer between bacterial reaction centers and mobile c-type cytochromes. In *Photosynthetic reaction center* (pp. 257–288). Academic Press. https://doi.org/10.1016/B978-0-12-208661-8.50014-0.

Todd, J. B., et al. (1998). In vitro reconstitution of the Core and peripheral light-harvesting complexes of *Rhodospirillum m olischianum* from separately isolated components †. *Biochemistry, 37*(50), 17458–17468. https://doi.org/10.1021/bi981114e.

Truper, H. G. T., & Fischer, U. (1982). *Anaerobic oxidation of Sulphur compounds as electron donors for bacterial photosynthesis* (R. Soc. Land. B, Trans.). Available at: https://royalsocietypublishing.org/doi/pdf/10.1098/rstb.1982.0095. Accessed 29 Aug 2019.

Tunnicliffe, R. B., et al. (2006). The solution structure of the PufX polypeptide from Rhodobacter sphaeroides. *FEBS Letters, 580*(30), 6967–6971. https://doi.org/10.1016/j.febslet.2006.11.065.

van Grondelle, R. et al. (1994). Energy transfer and trapping in photosynthesis. *Biochimica et Biophysica Acta (BBA) – Bioenergetics*. Elsevier, *1187*(1), 1–65. https://doi.org/10.1016/0005-2728(94)90166-X.

van Mourik, F., Visschers, R. W., & van Grondelle, R. (1992). Energy transfer and aggregate size effects in the inhomogeneously broadened core light-harvesting complex of Rhodobacter sphaeroides. *Chemical Physics Letters*. North-Holland, *193*(1–3), 1–7. https://doi.org/10.1016/0009-2614(92)85674-Y.

van Oijen, A. M. et al. (1999). Unraveling the electronic structure of individual photosynthetic pigment–protein complexes. *Science (New York, N.Y.)*. American Association for the Advancement of Science, *285*(5426), 400–402. https://doi.org/10.1126/science.285.5426.400.

Vik, S. B., & Antonio, B. J. (1994). A mechanism of proton translocation by F1F0 ATP synthases suggested by double mutants of the a subunit. *The Journal of Biological Chemistry*, *269*(48), 30364–30369. Available at: http://www.ncbi.nlm.nih.gov/pubmed/7982950. Accessed 7 Sept 2019.

Wada, T., et al. (1999). A novel labeling approach supports the five-transmembrane model of subunit *a* of the *Escherichia coli* ATP synthase. *Journal of Biological Chemistry, 274*(24), 17353–17357. https://doi.org/10.1074/jbc.274.24.17353.

Wakao, N. et al. (1996). Discovery of natural photosynthesis using Zn-containing bacteriochlorophyll in an aerobic bacterium Acidiphilium rubrum. *Plant and Cell Physiology*. Narnia, *37*(6), 889–893. https://doi.org/10.1093/oxfordjournals.pcp.a029029.

Weber, J., & Senior, A. E. (2003). ATP synthesis driven by proton transport in F_1 F_0 -ATP synthase. *FEBS Letters*. John Wiley & Sons, Ltd, *545*(1), 61–70. https://doi.org/10.1016/S0014-5793(03)00394-6.

Weber, J., Wilke-Mounts, S., & Senior, A. E. (1994). *Cooperativity and stoichiometry of substrate binding to the catalytic sites of Escherichia coli Fl-ATPase EFFECTS OF MAGNESIUM, INHIBITORS, AND MUTATION*, THE JOURNAL OF BIOIQGICAL CHEMISTRY*. Available at: http://www.jbc.org/content/269/32/20462.full.pdf. Accessed 30 Aug 2019.

Weber, J. et al. (1995). α-aspartate 261 is a key residue in noncatalytic sites of *Escherichia coli* F $_1$ -ATPase. *Journal of Biological Chemistry*. American Society for Biochemistry and Molecular Biology, *270*(36), 21045–21049. https://doi.org/10.1074/jbc.270.36.21045.

Xia, D., et al. (1997). Crystal structure of the cytochrome bc $_1$ complex from bovine heart mitochondria. *Science, 277*(5322), 60–66. https://doi.org/10.1126/science.277.5322.60.

Youvan, D. C. et al. (1984) *Reaction center and light-harvesting I genes from Rhodopseudomonas capsulata (photosynthesis/DNA sequence/enhanced fluorescence mutants/R-prime plasmid/genetic map), Proceedings of the National Academy of Sciences of the United States of America*. Available at: https://www.ncbi.nlm.nih.gov/pmc/articles/PMC344636/pdf/pnas00602-0203.pdf. Accessed 29 Aug 2019.

Yu, L. J., et al. (2018). Structure of photosynthetic LH1-RC supercomplex at 1.9 Å resolution. *Nature, 556*(7700), 209–213. https://doi.org/10.1038/s41586-018-0002-9.

Yurkov, V. V, & Beatty, J. T. (1998). Aerobic anoxygenic phototrophic bacteria. *Microbiology and Molecular Biology Reviews: MMBR*. American Society for Microbiology (ASM), *62*(3), 695–724. Available at: http://www.ncbi.nlm.nih.gov/pubmed/9729607. Accessed 29 Aug 2019.

Zeng, Y., & Koblizek, M. (2017). Phototrophic gemmatimonadetes: A new "purple" branch on the bacterial tree of life. In *Modern topics in the phototrophic prokaryotes: Environmental and applied aspects* (pp. 1–492). https://doi.org/10.1007/978-3-319-46261-5.

Zhang, Z., et al. (1998). Electron transfer by domain movement in cytochrome bc1. *Nature, 392*(6677), 677–684. https://doi.org/10.1038/33612.

Zheng, Q. et al. (2011) Diverse arrangement of photosynthetic gene clusters in aerobic Anoxygenic phototrophic Bacteria. *PLoS One*. Edited by S. Bereswill. Public Library of Science, *6*(9), e25050. https://doi.org/10.1371/journal.pone.0025050.

Zinth, W., & Wachtveitl, J. (2005). The first picoseconds in bacterial photosynthesis? Ultrafast electron transfer for the efficient conversion of light energy. *ChemPhysChem*. John Wiley & Sons, Ltd, *6*(5), 871–880. https://doi.org/10.1002/cphc.200400458.

Zuber, H., & Cogdell, R. (2004) *Advances in photosynthesis. Anoxygenic photosynthetic bacteria*. Available at: https://link.springer.com/content/pdf/10.1007%2F0-306-47954-0.pdf. Accessed 29 Aug 2019.

Redox Potentials of Quinones in Aqueous Solution: Relevance to Redox Potentials in Protein Environments

Hiroshi Ishikita and Keisuke Saito

Abstract Quinones serve as redox-active cofactors in photosynthetic reaction centers. To understand the energetics of electron transfer along the electron transfer pathways in protein environments, the redox potentials (E_m) of the cofactors in water versus normal hydrogen electrode (NHE) are required. However, ubiquinone, menaquinone (phylloquinone), and plastoquinone, which are found in photosynthetic reaction centers, have insoluble hydrophobic isoprene side chains, and thus far only E_m in dimethylformamide (DMF) versus saturated calomel electrode (SCE) had been reported. Recently, E_m in water versus NHE was reported for the quinone species of photosynthetic reaction centers. These results confirmed that $E_m(Q/Q^{\bullet-})$ in water versus NHE was more relevant to $E_m(Q/Q^{\bullet-})$ in protein environments than $E_m(Q/Q^{\bullet-})$ in DMF versus SCE. It has also been demonstrated that E_m for one-electron reduction can also be calculated based on the lowest unoccupied molecular orbital (LUMO) level of the quinone molecules.

Keywords Bacterial photosynthetic reaction centers · *Blastochloris viridis* · Cytochrome b_6f · Cytochrome bc_1 · Photosystem II · *Rhodobacter sphaeroides*

1 Introduction

Quinones are redox-active cofactors in many photosynthetic reaction centers. Ubiquinone serves as an electron acceptor at the Q_A and Q_B binding sites in photosynthetic reaction centers of purple bacteria (PbRC) from *Rhodobacter sphaeroides*

H. Ishikita (✉) · K. Saito
Department of Applied Chemistry, The University of Tokyo,
Tokyo, Japan

Research Center for Advanced Science and Technology, The University of Tokyo,
Tokyo, Japan
e-mail: hiro@appchem.t.u-tokyo.ac.jp

Q. Wang (ed.), *Microbial Photosynthesis*,
https://doi.org/10.1007/978-981-15-3110-1_5

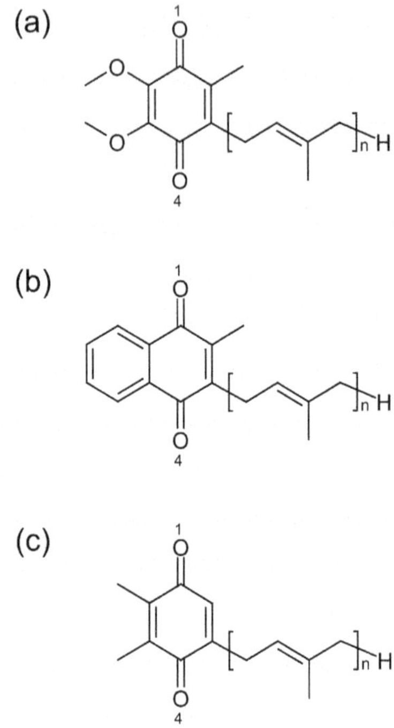

Fig. 1 Molecular structures of (**a**) ubiquinone ($n = 10$), (**b**) menaquinone and phylloquinone ($n = 3$ to 9), and (**c**) plastoquinone ($n = 6$ to 9), where n is the number of isoprene units

and as an electron donor in cytochrome bc_1. Menaquinone (vitamin K_2) is the acceptor at the Q_A site in PbRC from *Blastochloris viridis*, whereas phylloquinone (vitamin K_1) is the active center at the A_{1A} and A_{1B} sites in photosystem I (PSI). Plastoquinone serves as an electron acceptor at the Q_A and Q_B sites in photosystem II (PSII) (Fig. 1) (Robinson and Crofts 1984; Rutherford et al. 1984; Okamura et al. 2000; Brettel and Leibl 2001; Wraight 2004) and as an electron donor in cytochrome b_6f.

Prince et al. measured the redox potential for one-electron reduction, $E_m(Q/Q^{\bullet-})$, of 1,4-quinones, including ubiquinone, menaquinone (phylloquinone), and plastoquinone, in dimethylformamide (DMF) versus saturated calomel electrode (SCE) (Prince et al. 1983). Swallow also measured $E_m(Q/Q^{\bullet-})$ for small 1,4-quinones in water versus normal hydrogen electrode (NHE) (Swallow 1982). As mentioned (Kishi et al. 2017), experimentally measured $E_m(Q/Q^{\bullet-})$ in DMF versus SCE can be practically converted to $E_m(Q/Q^{\bullet-})$ in water versus NHE by adding 480 mV.

2 E_m for Quinones in Water and in Protein Environments

Kishi et al. reported the $E_m(Q/Q^{\bullet-})$ values in water versus NHE as −163 mV for ubiquinone, −260 mV for menaquinone (phylloquinone), and −154 mV for plastoquinone (Table 1) by quantum chemical calculation of the free energy difference

Table 1 Experimentally measured $E_m(Q/Q^{\bullet-})$ (exp.) versus SCE (Prince et al. 1983; Swallow 1982) and calculated $E_m(Q/Q^{\bullet-})$ (calc.) versus NHE (Kishi et al. 2017)

	E_m in DMF (vs. SCE)		E_m in water[a] (vs. NHE)	
	exp. [b]	calc. [c]	exp.	calc. [c]
Ubiquinone-1	−611	−633	*n.d.*	−260
Menaquinone-1	*n.d.*	−738	*n.d.*	−260
Menaquinone-2	−709	−736	*n.d.*	−256
Plastoquinone-1	−640	−626	*n.d.*	−154

n.d. not determined
[a]pH 7
[b]Ref. (Prince et al. 1983)
[c]Ref. (Kishi et al. 2017)

between the neutral state Q and the reduced state $Q^{\bullet-}$ in the aqueous phase (Kishi et al. 2017). Before that study, it was a matter of debate whether $E_m(Q/Q^{\bullet-})$ for quinone in DMF could be relevant to calculate the E_m values in photosynthetic reaction centers when using theoretical approaches, namely, electrostatic calculations. Notably, in electrostatic calculations, only the difference between the E_m of quinone in bulk water and quinone in the protein environment can be computed. Thus, to obtain, for example, $E_m(A_1)$ in PSI, it is necessary to determine the $E_m(Q/Q^{\bullet-})$ for phylloquinone in a reference system (preferentially in water) and add the calculated E_m difference.

Previously, to calculate $E_m(A_1)$ in PSI, Ptushenko et al. used $E_m(Q/Q^{\bullet})$ = −800 mV for phylloquinone in DMF versus NHE, by assuming a liquid junction potential between SCE in DMF and NHE in water (Ptushenko et al. 2008). However, it should be noted that $E_m(Q/Q^{\bullet-})$ for quinones in the two systems differs by 600 mV even in the absence of the liquid junction potential, i.e., the discrepancy between the $E_m(Q/Q^{\bullet-})$ values in the two systems cannot be explained by the liquid junction potential, as previously demonstrated (Kishi et al. 2017).

Although not clearly stated by Ptushenko et al. (2008), it seems likely that in their computational model, the electrostatic interaction between the PSI protein environment and the A_1 phylloquinone molecule was originally underestimated and that they needed the unusually low E_m value of −800 mV for phylloquinone as a reference, mainly to reproduce the reported low $E_m(A_1)$ in PSI (e.g., −810 mV (Vos and van Gorkom 1990), −754 mV (Iwaki and Itoh 1994), and −700 mV (Brettel and Leibl 2001)). Using the unusually low E_m value of −800 mV for phylloquinone in DMF versus NHE allowed them to conveniently match their calculated value to the reported low $E_m(A_1)$ value. However, using the unusually low E_m value would simultaneously cause a problem in reproducing the $E_m(Q_A)$ of −150 mV for the same quinone species (menaquinone) in PbRC (Brettel and Leibl 2001). That is, they must explain how the PbRC protein environment is able to increase $E_m(Q/Q^{\bullet-})$ = −800 mV in DMF versus NHE for phylloquinone to −150 mV at the Q_A site in the PbRC protein environment. Obviously, this is impossible in the PbRC protein electrostatic environment, as demonstrated in numerous electrostatic calculations

(Rabenstein et al. 1998; Ishikita and Knapp 2004; Zhu and Gunner 2005). It seems plausible that the E_m values measured in water versus NHE (−260 mV for menaquinone (phylloquinone)) (Kishi et al. 2017) are more relevant to the E_m values in proteins than the E_m values measured in DMF versus SCE (unless the proteins are solvated in DMF).

This fact would be more understandable when considering E_m of Q_B near the protein bulk surface in PbRC and PSII. $E_m(Q/Q^{\bullet-})$ is −154 mV for plastoquinone in water versus NHE, which is more consistent with $E_m(Q_B) = +90$ mV versus NHE in PSII determined using spectroelectrochemistry (Kato et al. 2016) than $E_m(Q_B) \approx -750$ mV in DMF versus NHE (Kishi et al. 2017). Again, these results confirm that $E_m(Q/Q^{\bullet-})$ in water versus NHE is more relevant to $E_m(Q/Q^{\bullet-})$ in protein environments than $E_m(Q/Q^{\bullet-})$ in DMF versus SCE.

3 Alternative Approach for Calculating E_m of Quinones and Other Cofactors

There are other approaches for calculating E_m of redox-active groups isolated in a solvent, including quinone molecules. As the basis of quantum chemistry, the highest occupied molecular orbital (HOMO) and lowest unoccupied molecular orbital (LUMO) energy levels are associated with E_m for one-electron oxidation and for one-electron reduction, respectively (Watanabe and Kobayashi 1991). Indeed, the experimentally measured E_m for ten 1,4-quinones in dimethylformamide (DMF) versus SCE (Prince et al. 1983) is strongly correlated with the LUMO level of quinone in the neutral state in the aqueous phase (coefficient of determination $R^2 = 0.97$, Fig. 2).

The MO-based approach presented herein requires quantum chemical calculation of the neutral state only, whereas the previous approach reported by Kishi et al. (2017) requires quantum chemical calculation of both the neutral and reduced states. The strong correlation between the calculated E_m values and the LUMO energy levels indicates that the E_m values are in fact determined by the molecular structures in the neutral states (prior to reduction of the quinones) and that structural changes that may be induced in response to reduction of the quinones are negligibly small in terms of E_m. This approach can also be applied to other redox-active cofactors, e.g., chlorophylls (Watanabe and Kobayashi 1991).

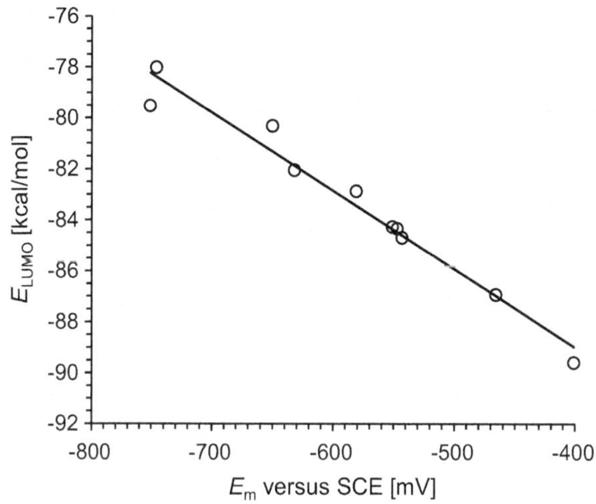

Fig. 2 Correlation between experimentally measured E_m for ten 1,4-quinones in dimethylformamide (DMF) versus saturated calomel electrode (SCE) (Prince et al. 1983) and calculated LUMO energy level (E_{LUMO}) for neutral quinones (coefficient of determination $R^2 = 0.97$). The 1,4-quinones shown are 1,4-benzoquinone, methyl-1,4-benzoquinone, 2,3-dimethyl-1,4-benzoquinone, 2,5-dimethyl-1,4-benzoquinone, 2,6-dimethyl-1,4-benzoquinone, trimethyl-1,4-benzoquinone, tetramethyl-1,4-benzoquinone, 1,4-naphthoquinone, 2-methyl-1,4-naphthoquinone, and 2,3 dimethyl-1,4-naphthoquinone. E_{LUMO} can be calculated using a quantum chemical approach. We employed the restricted density functional theory (DFT) method with the B3LYP functional and 6-31g++** basis sets for Q ($S = 0$), using the Gaussian (Frisch et al. 2004) program code with the PCM method for DMF. Solvent molecules were considered implicitly, using the SCRF = water option and the SCRF = dimethylformamide option with dielectric constants of 78.355 for water and 37.219 for DMF (i.e., default values). (For further details, including the atomic coordinates, see ref. Kishi et al. 2017)

Acknowledgments This research was supported by JST CREST (JPMJCR1656), JSPS KAKENHI (18H01186 to K.S., JP16H06560 to K.S. and H.I., and JP26105012 to H.I.), Japan Agency for Medical Research and Development (AMED), Materials Integration for engineering polymers of Cross-ministerial Strategic Innovation Promotion Program (SIP), and Interdisciplinary Computational Science Program in CCS, University of Tsukuba.

References

Brettel, K., & Leibl, W. (2001). Electron transfer in photosystem I. *Biochimica et Biophysica Acta, 1507*, 100–114.

Frisch, M. J., Trucks, G. W., Schlegel, H. B., Scuseria, G. E., Robb, M. A., Cheeseman, J. R., Montgomery, J., A, J., Vreven, T., Kudin, K. N., Burant, J. C., Millam, J. M., Iyengar, S. S., Tomasi, J., Barone, V., Mennucci, B., Cossi, M., Scalmani, G., Rega, N., Petersson, G. A., Nakatsuji, H., Hada, M., Ehara, M., Toyota, K., Fukuda, R., Hasegawa, J., Ishida, M., Nakajima, T., Honda, Y., Kitao, O., Nakai, H., Klene, M., Li, X., Knox, J. E., Hratchian, H. P., Cross, J. B., Bakken, V., Adamo, C., Jaramillo, J., Gomperts, R., Stratmann, R. E., Yazyev, O., Austin, A. J.,

Cammi, R., Pomelli, C., Ochterski, J. W., Ayala, P. Y., Morokuma, K., Voth, G. A., Salvador, P., Dannenberg, J. J., Zakrzewski, V. G., Dapprich, S., Daniels, A. D., Strain, M. C., Farkas, O., Malick, D. K., Rabuck, A. D., Raghavachari, K., Foresman, J. B., Ortiz, J. V., Cui, Q., Baboul, A. G., Clifford, S., Cioslowski, J., Stefanov, B. B., Liu, G., Liashenko, A., Piskorz, P., Komaromi, I., Martin, R. L., Fox, D. J., Keith, T., Al-Laham, M. A., Peng, C. Y., Nanayakkara, A., Challacombe, M., Gill, P. M. W., Johnson, B., Chen, W., Wong, M. W., Gonzalez, C., & Pople, J. A. (2004). *Gaussian 03*. Wallingford: Gaussian, Inc.

Ishikita, H., & Knapp, E.-W. (2004). Variation of Ser-L223 hydrogen bonding with the Q_B redox state in reaction centers from *Rhodobacter sphaeroides*. *Journal of the American Chemical Society, 126*, 8059–8064.

Iwaki, M., & Itoh, S. (1994). Reaction of reconstituted acceptor quinone and dynamic equilibration of electron transfer in the photosystem I reaction center. *Plant & Cell Physiology, 35*, 983–993.

Kato, Y., Nagao, R., & Noguchi, T. (2016). Redox potential of the terminal quinone electron acceptor Q_B in photosystem II reveals the mechanism of electron transfer regulation. *Proceedings of the National Academy of Sciences of the United States of America, 113*, 620–625.

Kishi, S., Saito, K., Kato, Y., & Ishikita, H. (2017). Redox potentials of ubiquinone, menaquinone, phylloquinone, and plastoquinone in aqueous solution. *Photosynthesis Research, 134*, 193–200.

Okamura, M. Y., Paddock, M. L., Graige, M. S., & Feher, G. (2000). Proton and electron transfer in bacterial reaction centers. *Biochimica et Biophysica Acta, 1458*, 148–163.

Prince, R. C., Dutton, P. L., & Bruce, J. M. (1983). Electrochemistry of ubiquinones: Menaquinones and plastoquinones in aprotic solvents. *FEBS Letters, 160*, 273–276.

Ptushenko, V. V., Cherepanov, D. A., Krishtalik, L. I., & Semenov, A. Y. (2008). Semi-continuum electrostatic calculations of redox potentials in photosystem I. *Photosynthesis Research, 97*, 55–74.

Rabenstein, B., Ullmann, G. M., & Knapp, E.-W. (1998). Energetics of electron-transfer and protonation reactions of the quinones in the photosynthetic reaction center of *Rhodopseudomonas viridis*. *Biochemistry, 37*, 2488–2495.

Robinson, H. H., & Crofts, A. R. (1984). Kinetics of proton uptake and the oxidation-reduction reactions of the quinone acceptor complex of photosystem II from pea chloroplasts. In C. Sybesma (Ed.), *Advances in photosynthesis research* (Vol. 1, pp. 477–480). The Hague: Martinus Nijhoff/Dr. W. Junk Publishers.

Rutherford, A. W., Renger, G., Koike, H., & Inoue, Y. (1984). Thermoluminescence as a probe of photosystem II. The redox and protonation states of the secondary acceptor quinone and the O_2-evolving enzyme. *Biochimica et Biophysica Acta, 767*, 548–556.

Swallow, A. J. (1982). Physical chemistry of semiquinones. In B. L. Trumpower (Ed.), *Function of quinones in energy conserving systems* (pp. 59–72). New York: Academic Press.

Vos, M. H., & Van Gorkom, H. J. (1990). Thermodynamical and structural information on photosynthetic systems obtained from electroluminescence kinetics. *Biophysical Journal, 58*, 1547–1555.

Watanabe, T., & Kobayashi, M. (1991). Electrochemistry of chlorophylls. In H. Scheer (Ed.), *Chlorophylls* (pp. 287–303). Boca Raton: CRC Press.

Wraight, C. A. (2004). Proton and electron transfer in the acceptor quinone complex of photosynthetic reaction centers from *Rhodobacter sphaeroides*. *Frontiers in Bioscience, 9*, 309–337.

Zhu, Z., & Gunner, M. R. (2005). Energetics of quinone-dependent electron and proton transfers in *Rhodobacter sphaeroides* photosynthetic reaction centers. *Biochemistry, 44*, 82–96.

Photosynthesis in *Chlamydomonas reinhardtii*: What We Have Learned So Far?

Hui Lu, Zheng Li, Mengqi Li, and Deqiang Duanmu

Abstract Microalgae play a major role in the global photosynthesis and CO_2 fixation. *Chlamydomonas reinhardtii* represents one of the most extensively studied eukaryotic algal organism, and milestone discoveries related to photosynthesis have been achieved over the years in this alga, including photosynthetic complexes biogenesis, regulation of photosynthetic genes expression, pigments biosynthesis, and the regulation of photosynthetic performance in response to changes of environmental parameters. Comparisons of *Chlamydomonas* photosynthetic mechanisms with other phototrophs, such as cyanobacteria and higher plants, have dramatically expanded our understanding about the robustness and plasticity of photosynthetic machineries during more than 1 billion years of evolution in nature.

Keywords *Chlamydomonas reinhardtii* · Tetrapyrrole · Bilin · LHCSR · Photosystems

1 Introduction

Chlamydomonas reinhardtii (*Chlamydomonas* hereafter) is a single-celled eukaryotic chlorophyte green alga that was first described in 1888 by Dangeard. Major laboratory strains of *Chlamydomonas* were originally derived in 1945 from a single zygospore in a soil sample near Amherst, Massachusetts, USA. Due to the impressive adaptability, quick generation time, the haploid genome, and straightforward mutant construction and phenotypic assays, *Chlamydomonas* has become a classical reference organism for biological research, covering the areas of cilia biogenesis and function, cell cycle and cell division, chloroplast biogenesis and photosynthesis, and environmental acclimation (CO_2, temperature, metal elements, etc.). Over

H. Lu · Z. Li · M. Li · D. Duanmu (✉)
State Key Laboratory of Agricultural Microbiology, College of Life Science and Technology, Huazhong Agricultural University, Wuhan, China

Q. Wang (ed.), *Microbial Photosynthesis*,
https://doi.org/10.1007/978-981-15-3110-1_6

the years, studies of *Chlamydomonas* have provided major research findings in chloroplast-based photosynthesis (Salomé and Merchant 2019).

2 Photosynthetic Complexes Biogenesis and Regulation

Chlamydomonas is among the best characterized unicellular eukaryotic algae that serves as the premier reference organism for investigation of diverse biological processes such as photosynthesis and chloroplast biogenesis, the essential functions derived from more than 1 billion years of evolution since primary endosymbiosis. Over the years, development of mutant resources and availability of multiomics data have enabled the *Chlamydomonas* an excellent resource for understanding photosynthetic complexes biogenesis, environmental regulation, and comparison with higher plants regarding the generality and uniqueness of algal photosynthesis (Salomé and Merchant 2019). This single-celled organism is ideal to compare to higher plants such as *Arabidopsis thaliana* (*Arabidopsis* hereafter) to extend the understanding of photosynthesis regulation in eukaryotic phototrophs.

2.1 Photosynthetic Genes Expression

Light serves both as a source of energy for photosynthesis and provides information about the environment to regulate behavioral and developmental responses of algae, together with many physiological processes such as photosynthetic gene expression. Assessment of light quality and quantity was initially accomplished by a plethora of photoreceptors. Plants mainly use red/far-red photoreceptor phytochrome and blue light receptor cryptochrome to regulate photosynthetic gene expression. Surprisingly, phytochromes are lacking in *Chlamydomonas* (Merchant et al. 2007). Instead, a diverse set of cryptochromes comprising both plant and animal-like cryptochromes were identified in *Chlamydomonas*, and some of those cryptochromes were found to be involved in circadian clock and life cycle regulation (Kottke et al. 2017). Another blue light photoreceptor, phototropin (PHOT), was able to modulate levels of several LHC transcripts accumulation (Im et al. 2006). A systematic analysis of chloroplast genome expression revealed that photosynthetic genes reached peak levels after transcripts of the transcriptional apparatus and translational machinery (Idoine et al. 2014). However, the overall mechanism of transcriptional regulation of photosynthesis-related genes in *Chlamydomonas* remains poorly characterized.

Excess light should be dissipated as chlorophyll (Chl) fluorescence or as heat by nonphotochemical quenching (NPQ). It was proposed that a distinct LHC member called LHCSR3 (light-harvesting complex stress-related protein 3) contributes significantly to NPQ in *Chlamydomonas*, although *Chlamydomonas* also encodes a

PsbS homolog and LHCSR1 in the nucleus genome (Peers et al. 2009). *LHCSR3* shows an acute response during dark-to-light transition in synchronized cultures and was hypothesized to be an NPQ quencher of both photosystems in *Chlamydomonas* (Zones et al. 2015; Strenkert et al. 2019; Girolomoni et al. 2019). The photosystem II (PSII) inhibitor DCMU and a calmodulin inhibitor could suppress transcripts induction of both isoforms of *LHCSR3* under high light, suggesting that photosynthetic electron transfer and calmodulin-mediated signaling pathway is essential for the transcriptional regulation of *LHCSR3* (Maruyama et al. 2014). Blue light photoreceptor PHOT was also required for *LHCSR3* induction, possibly mediated through the E3 ubiquitin ligase CUL4-DDB1^{DET1} pathway (Aihara et al. 2019). A distinct NPQ response in *Chlamydomonas* was recently revealed that is based on UV-B light sensor UVR8 and the downstream E3 ubiquitin ligase complex and a transcription factor CrCO, a homolog of CONSTANS, which controls flowering in *Arabidopsis* (Allorent et al. 2016; Gabilly et al. 2019). These findings provide novel insights into wiring conserved light-signaling pathways for sensing, utilization, and dissipation of light in a concerted process in *Chlamydomonas*.

Regulation of photosynthetic gene expression also occurs at posttranscriptional level. An RNA-binding protein NAB1 exhibited a reversible cysteine modification that controls its sequestration of *LHCBM* mRNA and repressing its translational initiation in the cytosol (Mussgnug et al. 2005; Wobbe et al. 2009). Two OPR (Octotrico Peptide Repeat) proteins were necessary to stabilize small subunits of PSII and cytochrome b_6f (Wang et al. 2015). Several nucleus-encoded chloroplast proteins are required for a coordinated response to iron limitation by stabilizing and/or promoting translation of chloroplast encoded photosystem I (PSI) genes, i.e., PsaA and PsaC (Lefebvre-Legendre et al. 2015; Douchi et al. 2016). These findings have provided new insights into how *Chlamydomonas* is able to employ the elaborate posttranscriptional control mechanisms to fine-tune photosynthetic protein synthesis under various environmental conditions.

Since *Chlamydomonas* chloroplast genome encodes ~100 genes, whereas around 3500–4000 proteins are encoded by nucleus and localized in chloroplast, coordinated expression between nucleus genome and chloroplast genome is essential for chloroplast development, functioning, and acclimation to environmental cues (Maul et al. 2002; Rea et al. 2018). Chloroplast retrograde signals, defined by cues originated in chloroplast and exported to their destination in the nucleus to regulate gene expression, were extensively characterized in higher plants and the unicellular organisms including *Chlamydomonas*. Communication networks between chloroplast and nucleus include photosynthesis and ROS signals, chloroplast gene expression and redox status, tetrapyrrole intermediates, etc. (Rea et al. 2018). In higher plants, bilins are chromophore cofactor for phytochromes, which modulate important developmental processes throughout plants life cycle (Rockwell et al. 2006). *Chlamydomonas* lacks phytochromes but retains the ability of bilin biosynthesis. A retrograde signaling role of bilins was recently discovered in that bilins could regulate a small subset of nuclear genes involved in ROS detoxification during diurnal dark to light transitions (Duanmu et al. 2013; Duanmu et al. 2017).

2.2 Photosynthetic Pigments Biosynthesis

Chlorophylls and carotenoids are major pigments associated with photosynthetic complexes and play essential roles in the light harvesting, electron transport chain, photoprotection, and other cellular processes (Grossman et al. 2004). In plants and algae, chlorophylls are synthesized from glutamate by a tetrapyrrole pathway localized entirely within plastids. Genomic sequencing and genetic and biochemical analysis have clearly characterized genes involved in chlorophyll biosynthesis (Tanaka and Tanaka 2007). Since the misregulation of tetrapyrrole metabolism can lead to formation of damaging reactive oxygen species (ROS), the tetrapyrrole biosynthetic pathway is therefore under strict control at multiple levels to meet the demands of growing cells under various light conditions. Several enzymatic steps of tetrapyrrole biosynthesis are encoded by usually 2~3 homologous proteins (Tanaka and Tanaka 2007). Light-dependent regulation of gene expression was extensively studied in plants but was rarely performed in *Chlamydomonas*. Transcriptional control is important for long-term acclimation to environmental changes. In contrast, posttranslational control of many enzymatic steps, i.e., redox regulation, phosphorylation, complex formation, feedback inhibition by tetrapyrrole molecules or other proteins, can facilitate a rapid response and more accurately balance the flow of metabolites for a tight regulation of tetrapyrrole biosynthesis, particularly for the biosynthesis of the first committed tetrapyrrole precursor 5-aminolevulinic acid (ALA), as well as the branch point of the pathway towards heme and chlorophyll biosynthesis (Czarnecki and Grimm 2012; Brzezowski et al. 2015).

Although *Chlamydomonas* shares a similar chlorophyll biosynthetic pathway as higher plants, several differences in the biochemistry and regulation of key steps of chlorophyll biosynthesis are identified between *Chlamydomonas* and *Arabidopsis*. Chelation of Mg^{2+} into protoporphyrin IX (Proto) is catalyzed by Mg chelatase (MgCh) and GUN4 (GENOMES UNCOUPLED 4) is essential in stimulating MgCh activity by binding of both the enzyme's substrate Proto and the product Mg-Proto IX (Chen et al. 2015). In plants, phosphorylation of GUN4 in the dark has a reduced stimulatory effect on MgCh activity. Absence of this phosphorylation of GUN4 as a control mechanism in reducing MgCh activity at the branch point of the tetrapyrrole pathway in *Chlamydomonas* was proposed to be associated with chlorophyll biosynthesis and less accumulation of photoreactive porphyrin intermediates in darkness (Richter et al. 2016). Indeed, angiosperms depend on the nucleus-encoded light-dependent protochlorophyllide oxidoreductase (LPOR) to reduce protochlorophyllide to chlorophyllide, whereas *Chlamydomonas* harbors an additional chloroplast-encoded light-independent (dark-operative) enzyme complex (DPOR) that is absent from land plants (Reinbothe et al. 2010). The bilin biosynthetic enzyme PCYA1 (phycocyanobilin/ferredoxin oxidoreductase) interacts specifically with LPOR but not DPOR in *Chlamydomonas*, suggesting a possible regulatory role of bilin, a heme degradation product, in the regulation of chlorophyll biosynthesis in *Chlamydomonas* (Zhang et al. 2018).

Carotenoids (Cars) are noncovalently bound to the core and antenna subunits of PSI or PSII and play crucial roles in photosynthesis by quenching of the excited triplet Chl. It has been shown that xanthophylls are also essential for refolding of LHCI and LHCII antenna complexes in vitro. Therefore, inactivation of the Cars biosynthesis or inhibition of key enzymes in the biosynthetic pathway usually leads to plant lethal phenotypes or rapid photobleaching under high light conditions (Trebst and Depka 1997). In *Chlamydomonas*, a Car-less mutant was created that is blocked at phytoene synthesis, the first committed step in Car biosynthetic pathway (McCarthy et al. 2004). This mutant exhibited failure of PSII core and antenna complexes formation and the reduced abundance of Cyt b_6f. However, assembly and the functionality of PSI was not affected (Santabarbara et al. 2013).

Compared to the abundant pigments of Chl and Car present in photosystems, naphthoquinone, or vitamin K_1, is in low abundance and functions as the electron carrier of PSI. *Chlamydomonas menD1* (2-succinyl-6-hydroxy-2,4-cyclohexadiene-1-carboxylate (SHCHC) synthase) mutant was abolished in the naphthoquinone biosynthesis and exhibited the replacement by plastoquinone in PSI as in the case of cyanobacteria. However, distinct from cyanobacteria, accumulation of PSI subunits was not affected, whereas abundance of PSII subunits was dramatically reduced (Lefebvre-Legendre et al. 2007). HPLC analysis of naphthoquinone species in the PSI complex indicates that 5′-monohydroxyphylloquinone and phylloquinone account for ~90% and ~10% of the total naphthoquinone content, respectively (Ozawa et al. 2012). Bioinformatic analysis identified a putative phylloquinone biosynthetic pathway in *Chlamydomonas* that comprises 11 enzymatic steps located in both the chloroplast and the peroxisome (Emonds-Alt et al. 2017). Four mutants affected in different enzymatic steps of the PhQ biosynthesis pathway all primarily impacts PSI activity, further corroborating the essential roles of PhQ in the maintenance of a fully functional PSI in *Chlamydomonas*.

2.3 PSI Biogenesis and Functional Regulation

Photosystem I (PSI) is a large pigment-protein complex that accepts and transfers electrons from plastocyanin to ferredoxin, finally leading to the production of NADPH (Caspy and Nelson 2018). *Chlamydomonas* PSI-LHCI is significantly larger than that of plants, containing five additional LHCAs and red forms of Chl with higher energy, and therefore was able to harvest ~41% more photons than the *Arabidopsis* PSI (Le Quiniou et al. 2015). Cryo-electron microscopy structure of CrPSI-LHCI supercomplex revealed that eight or ten LHCIs are associated with the PSI core and function as peripheral antennae. Each LHCA protein binds 14~17 Chls, compared to 14~15 Chls for plant LHCA proteins, suggesting that the larger antennae and the higher number of Chls in the antennae as an evolutionary consequence of low light adaptation in aquatic environment (Su et al. 2019). LHCA2/LHCA9 heterodimer was found in the side layer and could confer enhanced light-harvesting for PSI (Kubota-Kawai et al. 2019). Dissociation of LHCA2/LHCA9 from CrPSI-LHCI could be involved in the formation of the PSI-LHCI-cyt b_6f supercomplex that is crucial for the cyclic electron flow (Suga et al. 2019). The structural characterization provides deeper insights into antennae protein organization, pigment arrangement, light harvesting, and energy transfer within the supercomplexes and electron transfer with plastocyanin.

Assembly of CrPSI-LHCI is a highly complicated and multistep process. Several protein factors have been identified to be involved in the assembly, stability, and maintenance of PSI, although specific functions of these regulatory factors may differ in cyanobacteria, *Chlamydomonas,* and plants (Yang et al. 2015; Heinnickel et al. 2016). Two modules were found to be major mediators of PSI-LHCI assembly process, consisting of the chloroplast-encoded tetratricopeptide repeat protein Ycf3 and Ycf4, respectively. Ycf3 and its interacting protein CGL59, an ortholog of plant Y3IP1, mainly facilitates the assembly of PSI reaction center subunits, whereas oligomeric Ycf4 is essential for the integration of peripheral subunits and LHCIs into the PSI reaction center (Nellaepalli et al. 2018). Although inactivation of Ycf4 resulted in absence of PSI complex in *Chlamydomonas*, knockout of this gene in cyanobacteria and tobacco still retained a smaller amount of PSI that enabled sufficient photoautotrophic growth of tobacco plants (Boudreau et al. 1997; Wilde et al. 1995; Krech et al. 2012). Ycf3, Y3IP1, and Ycf4 mediate almost the whole assembly process of PSI-LHCI supercomplex, and a five-stage PSI complex assembly model was proposed. Ycf4 and Y3IP1 are reused in the assembly process, whereas Ycf3 turns over rapidly and therefore is replaced with newly synthesized protein (Nellaepalli et al. 2018). Despite these advances, biochemical and genetic approaches should be used to identify novel factors involved in the regulation of PSI function, PSI-LHCI assembly and degradation processes.

During state transitions, LHCII is reversibly associated or disassociated from the two photosystems to balance their light harvesting and trapping efficiency. *Chlamydomonas* PSI-LHCI-LHCII has an increase of the absorption cross section of ~47%. LHCII is loosely connected to PSI-LHCI but still showing a 96% trapping efficiency, indicating that the design of green algal PSI represents an evolutionary path for efficient light harvesting (Le Quiniou et al. 2015). Interestingly, by combining biochemical and spectroscopic approaches, LHCII was found to transfer excitation energy to PSI, not only to PSII, for efficient fluorescence quenching under excess light (Kosuge et al. 2018). Therefore, PSI provides another layer of photoprotection in *Chlamydomonas*.

2.4 PSII Biogenesis and Functional Regulation

Photosystem II (PSII) splits water, initiates electron transfer and reduces plastoquinone in the light reaction of photosynthesis. PSII consists of an oxygen-evolving/electrogenic reaction center (RC) complex and a connected light-harvesting antenna system (LHCII). The PSII core complex is highly conserved in cyanobacteria, algae, and land plants. LHCII plays two general functions in plants and algae, light harvesting to fuel the photosynthetic reactions and energy dissipation or ROS reduction to protect PSII from harmful excess illumination. In *Chlamydomonas*, nine genes (*LhcbM1-M9*) encode LHCII proteins that could be divided into four groups based on their sequence homology. Functional studies of Type 1 LHCII proteins suggest that LHCBM4/6/8 are important for the full amplitude of NPQ activity rather than having an essential function in light harvesting, whereas LHCBM9 serves an essential protective function during stress conditions such as sulfur deficiency (Grewe et al. 2014; Girolomoni et al. 2016).

The isolated *Chlamydomonas* PSII-LHCII supercomplex binds three LHCII trimers on each side of the RC, exhibiting a $C_2S_2M_2L_2$ organization that is larger than that of plants and retains a higher oxygen-evolving capability (Tokutsu et al. 2012). Using an ion-exchange chromatography method, four distinct LHCII trimers were purified showing different compositions of LHCII proteins. However, the four LHCII trimers have similar pigments composition and exhibit similar levels of light-harvesting ability (Kawakami et al. 2019). Structure of the $C_2S_2M_2L_2$-type CrPSII-LHCII supercomplex was recently resolved, consisting of a dimer with three LHCII trimers and two additional peripheral antenna subunits (CP16 and CP29) surrounding individual PSII core (Shen et al. 2019; Burton-Smith et al. 2019). Detailed supramolecular organization and arrangement of pigments and protein subunits in the complete PSII-LHCII complex provided deeper insights into a highly efficient energy transfer network in the *Chlamydomonas* PSII-LHCII compared to plant photosystems, an evolutionary outcome likely due to algal acclimation to low light intensities in aquatic environments (Shen et al. 2019).

Assembly of PSII-LHCII has been extensively investigated in *Chlamydomonas* and *Arabidopsis* and many regulatory factors have been identified that are crucial for the assembly process (Cao et al. 2018). In *Chlamydomonas*, the nuclear gene encoding chloroplast Alb3.1p, a homolog of *Arabidopsis* ALBINO3, functions as a membrane integral chaperone by interacting with the reaction center D1 protein of PSII and facilitating integration of D1 into the thylakoid membrane (Ossenbühl et al. 2004). Another chloroplast-located PSII assembly factor rubredoxin 1 (RBD1) participates together with the cytochrome b_{559} in the protection of PSII assembly intermediates from photodamage during de novo assembly and repair (García-Cerdán et al. 2019). Chloroplast-encoded regulatory proteins were also identified to be essential for PSII biogenesis. The absence of PsbT, a small PSII polypeptide with a single transmembrane helix, resulted in retarded accumulation of PSII complex although the synthesis of PSII proteins were not affected (Ohnishi and Takahashi 2008). The reaction center D1/D2 heterodimer contains pheophytin a (Pheo *a*) besides Chl *a*, indicating that Pheo *a* may also contribute to the formation of the PSII reaction center complex. Indeed, a Mg-dechelatase, STAY-GREEN (SGR) which converts Chl *a* to Pheo *a*, was recently shown to be primarily involved in PSII biogenesis in *Chlamydomonas*. Surprisingly, SGR in *Arabidopsis* was required for Chl degradation, and PSII was synthesized normally in the *Arabidopsis sgr* triple mutant. The distinct physiological functions of SGR in *Arabidopsis* and

Chlamydomonas could be caused by different location of the enzyme or the presence of an unidentified Mg-dechelatase in land plants (Chen et al. 2019).

The D1 subunit of PSII complexes is the major target of photooxidative damage and therefore needs to be degraded and replaced with newly synthesized protein in a process known as PSII repair (Aro et al. 1993). Combining fluorescence in situ hybridization (FISH) and immunofluorescence (IF) staining techniques, de novo PSII assembly was observed in discrete regions surrounding the pyrenoid. Interestingly, synthesis of the D1 subunit for PSII repair occurs throughout regions of the chloroplast with thylakoid membranes, suggesting distinct location of PSII assembly and repair in *Chlamydomonas* (Uniacke and Zerges 2007). REP27, a nucleus-encoded chloroplast-located membrane protein, plays a dual role in the regulation of D1 turnover by permitting cotranslational biosynthesis insertion and activation of the newly synthesized D1 during the PSII repair cycle (Dewez et al. 2009). Photodamaged D1 is degraded by the cooperative action of Deg and FtsH proteases (Kato et al. 2012). A unique GTPases family protein EngA was recently shown to be able to interact with the ATPase domain of FtsH and regulate its stability. Therefore, the rapid turnover of FtsH is also indispensable for maintaining a functional PSII under high light conditions (Kato et al. 2018).

Under high light stress, *Chlamydomonas* PSII was found to be associated with LHCSR3, an active qE effector that could bind Chl *a* and *b* and several Car pigments (Bonente et al. 2011). A PSII-LHCII-LHCSR3 supercomplex was isolated from *Chlamydomonas*, and PSII-LHCII was in an energy-dissipative state instead of a light-harvesting mode in this supercomplex. It was proposed that the induction of qE in *Chlamydomonas* was largely dependent on the protonation of LHCSR3 and the subsequent assembly of PSII-LHCII-LHCSR3 supercomplex to form a quenching center (Tokutsu and Minagawa 2013). PSBR (photosystem II subunit R) is involved in forming a recognition site for LHCSR3 binding to PSII-LHCII under high light (Xue et al. 2015). Fluorescence kinetic analyses further revealed the molecular events of the pH- and LHCSR3-dependent energy quenching process of the PSII-LHCII-LHCSR3 supercomplex (Kim et al. 2017). State transition is another long-term acclimation strategy in plants and algae, involving reversible phosphorylation of LHCII and the relocation between PSII and PSI (Bellafiore et al. 2005). In higher plants, around 10–15% of LHCII moves to PSI in state 2, whereas 80% of the major LHCII dissociates from PSII in *Chlamydomonas* (Delosme et al. 1996). However, it was recently demonstrated that only a fraction of dissociated LHCII attaches to PSI, and the rest LHCII becomes quenched, providing another layer of protection against photodamage (Ünlü et al. 2014).

2.5 Photosynthetic Electron Transport

Photosynthetic light reactions establish a linear electron flow in serial activity of PSII, cyt b_6f, and PSI, leading to the generation of ATP and NADPH at an estimated ratio of 2.7:2. The Calvin-Benson-Bassham (CBB) cycle requires a ratio of 3:2 to sustain CO_2 assimilation (Munekage et al. 2004). Other chloroplast processes and environmental factors should further increase the ATP demands, such as protein synthesis, transport processes, maintenance of an active carbon concentrating mechanism (CCM) under ambient or even lower CO_2 concentrations, etc. (Kono et al. 2014). Cyclic electron flow (CEF) around PSI and the cytochrome b_6f complex was therefore evolved in algae and plants to exclusively generate a proton gradient and ATP without producing reductants (Munekage et al. 2004). In *Chlamydomonas*, two CEF pathways were distinguished, based on the reinjection site of PSI-generated reducing equivalents. NDA2, a type-2 NADPH dehydrogenase, is able to reduce the plastoquinone pool by NADPH (Desplats et al. 2009). In contrast, the PGR pathway, comprising the proton-gradient regulator 5 (PGR5) and PGR5-like1 (PGRL1), represents a major supplier of ATP for the cell (Hertle et al. 2013).

A PSI-LHCI-LHCII-FNR-Cyt b_6f-PGRL1 supercomplex engaged in *Chlamydomonas* CEF was isolated when PSII was preferentially excited. Formation and dissociation of this supercomplex were proposed to control the energy balance of the two photosystems and also maintain chloroplast ATP homeostasis by switching between LEF and CEF (Iwai et al. 2010). Under anaerobic conditions, structural characterization of a CEF-performing PSI-LHCI-Cyt b_6f supercomplex further proposed a model that dissociation of LHCA2 and LHCA9 from PSI could support the CEF complex formation. A constitutively enhanced CEF was indeed observed in a *Δlhca2* null mutant (Steinbeck et al. 2018). Anaerobic Response 1 (ANR1), an algae-specific protein that is inducible under anaerobic conditions, interacts with the chloroplast localized Ca^{2+} sensor (CAS) protein as well as with PGRL1 in vivo, establishing a Ca^{2+}-dependent CEF regulatory module via the combined function of CAS, ANR1, and PGRL1 (Terashima et al. 2012). An ANR1-interacting protein, PETO, a thylakoid membrane protein unique to green algae, also contributes to the stability of CEF supercomplexes and the activity in anoxia (Takahashi et al. 2016). Surprisingly, the maximal rate and the duration of CEF were not significantly changed in the *pgrl1* mutant compared to the wild type, indicating that PGRL1 is not a bona fide ferredoxin-plastoquinone oxidoreductase and therefore is not involved mechanistically in CEF (Nawrocki et al. 2019).

CEF is believed to be mainly controlled by the redox poise of the stroma. In mutants with a disrupted Rubisco or ATPase, the reducing power was not able to be used for CO_2 fixation, and a stimulation of CEF was observed (Alric et al. 2010). CEF plays a crucial role in PSI photoprotection. In the absence of PGR5 and/or PGRL1, PSI was the primary target of photodamage (Mosebach et al. 2017). A complementary role of PGRL1 and LHCSR3 in high light stress adaptation and survival under low oxygen was also revealed by comparing single and double mutant photosynthetic parameters, underscoring an intimate link between CEF and NPQ (qE) in *Chlamydomonas* (Kukuczka et al. 2014). In the *pgrl1* mutant, normal biomass productivity was achieved under steady-state growth conditions, largely due to combined actions of mitochondrial respiration and oxygen photoreduction downstream of PSI in supplying extra ATP for cellular demands. In contrast, the mutant demonstrated reduced growth performance and lower biomass productivity in fluctuating light conditions, indicating that CEF is essential in enabling a rapid response to sudden environmental changes (Dang et al. 2014).

Alternatively, the reducing power can be dissipated by transferring electrons to molecular oxygen as a terminal electron acceptor (Mehler reaction; or flavodiiron proteins (FLV)-mediated enzymatic reaction) or plastid terminal oxidase PTOX (chlororespiration). The Mehler reaction produces superoxide radicals that can be converted to hydrogen peroxide by superoxide dismutase (Mehler 1951; Badger et al. 2000). Specific FLV proteins directly reduce O_2 to H_2O without the release of ROS (Chaux et al. 2017). The action of PTOX generates a transthylakoid membrane proton gradient that is used for ATP production (Nawrocki et al. 2015). *Chlamydomonas* nuclear genome encodes two plastid terminal oxidases, PTOX1 and PTOX2. PTOX2 plays a major role in *Chlamydomonas* chlororespiration. Inactivation of PTOX2 shows constitutively reduced plastoquinone pool under dark-aerobic conditions, and the phototrophic growth of the mutant is also inhibited relative to wild-type cells. PTOX2 therefore serves as a safety valve to pull out electrons from the over-reduced plastoquinone pool under extreme conditions (Houille-Vernes et al. 2011). Overall, alternative outlets allow rerouting of electrons and enable photosynthetic organisms an effective acclimation approach to fluctuating environmental conditions, such as high light, cold, low CO_2, etc.

3 Concluding Remarks

The ~111-megabase nuclear genome of *Chlamydomonas reinhardtii* contains ~19,526 protein-coding transcripts (Merchant et al. 2007). Comparison among the proteins encoded by different organisms identified the GreenCut2 proteins that are unique to the green lineage eukaryotes including *Chlamydomonas* and *Arabidopsis* but absent from the heterotrophs. Half of these ~597 GreenCut2 proteins have an unknown function, serving as a rich source for investigation of molecular details related to photosynthesis and accessory chloroplast processes in these photosynthetic autotrophs (Grossman et al. 2019). A genome-wide mutant library was recently released, covering ~83% of nuclear genes in *Chlamydomonas* (Li et al. 2019). CRISPR-based genome-editing technologies were also successfully established in this alga (Ferenczi et al. 2017; Greiner et al. 2017). Those powerful tools will certainly accelerate the photosynthesis research and enlighten our understanding about photosynthetic regulation and photosystems plasticity in *Chlamydomonas*.

Acknowledgments The research in our laboratory was supported by the National Natural Science Foundation of China (31570233 and 31870220) and the Fundamental Research Funds for the Central Universities (Program No. 2662015PY171).

References

Aihara, Y., Fujimura-Kamada, K., Yamasaki, T., & Minagawa, J. (2019). Algal photoprotection is regulated by the E3 ligase CUL4-DDB1(DET1). *Nat Plants, 5*(1), 34–40.

Allorent, G., Lefebvre-Legendre, L., Chappuis, R., Kuntz, M., Truong, T. B., Niyogi, K. K., Ulm, R., & Goldschmidt-Clermont, M. (2016). UV-B photoreceptor-mediated protection of the photosynthetic machinery in Chlamydomonas reinhardtii. *Proceedings of the National Academy of Sciences of the United States of America, 113*(51), 14864–14869.

Alric, J., Lavergne, J., & Rappaport, F. (2010). Redox and ATP control of photosynthetic cyclic electron flow in Chlamydomonas reinhardtii (I) aerobic conditions. *Biochimica et Biophysica Acta, 1797*(1), 44–51.

Aro, E. M., Mccaffery, S., & Anderson, J. M. (1993). Photoinhibition and D1 protein degradation in peas acclimated to different growth irradiances. *Plant Physiology, 103*(3), 835–843.

Badger, M. R., von Caemmerer, S., Ruuska, S., & Nakano, H. (2000). Electron flow to oxygen in higher plants and algae: Rates and control of direct photoreduction (Mehler reaction) and rubisco oxygenase. *Philosophical Transactions of the Royal Society of London. Series B, Biological Sciences, 355*(1402), 1433–1446.

Bellafiore, S., Barneche, F., Peltier, G., & Rochaix, J. (2005). State transitions and light adaptation require chloroplast thylakoid protein kinase STN7. *Nature, 433*(7028), 892–895.

Bonente, G., Ballottari, M., Truong, T. B., Morosinotto, T., Ahn, T. K., Fleming, G. R., Niyogi, K. K., & Bassi, R. (2011). Analysis of LhcSR3, a protein essential for feedback de-excitation in the green alga *Chlamydomonas reinhardtii*. *PLoS Biology, 9*(1), e1000577.

Boudreau, E., Takahashi, Y., Lemieux, C., Turmel, M., & Rochaix, J. D. (1997). The chloroplast *ycf3* and *ycf4* open reading frames of *Chlamydomonas reinhardtii* are required for the accumulation of the photosystem I complex. *The EMBO Journal, 16*(20), 6095–6104.

Brzezowski, P., Richter, A. S., & Grimm, B. (2015). Regulation and function of tetrapyrrole biosynthesis in plants and algae. *Biochimica et Biophysica Acta, 1847*(9), 968–985.

Burton-Smith, R. N., Watanabe, A., Tokutsu, R., Song, C., Murata, K., & Minagawa, J. (2019). Structural determination of the large photosystem II–light-harvesting complex II supercomplex of *Chlamydomonas reinhardtii* using nonionic amphipol. *The Journal of Biological Chemistry, 294*(41), 15003–15013.

Cao, P., Su, X., Pan, X., Liu, Z., Chang, W., & Li, M. (2018). Structure, assembly and energy transfer of plant photosystem II supercomplex. *Biochimica et Biophysica Acta, 1859*(9), 633–644.

Caspy, I., & Nelson, N. (2018). Structure of the plant photosystem I. *Biochemical Society Transactions, 46*(2), 285–294.

Chaux, F., Burlacot, A., Mekhalfi, M., Auroy, P., Blangy, S., Richaud, P., & Peltier, G. (2017). Flavodiiron proteins promote fast and transient O_2 Photoreduction in Chlamydomonas. *Plant Physiology, 174*(3), 1825–1836.

Chen, X., Pu, H., Wang, X., Long, W., Lin, R., & Liu, L. (2015). Crystal structures of GUN4 in complex with Porphyrins. *Molecular Plant, 8*(7), 1125–1127.

Chen, Y., Shimoda, Y., Yokono, M., Ito, H., & Tanaka, A. (2019). Mg-dechelatase is involved in the formation of photosystem II but not in chlorophyll degradation in *Chlamydomonas reinhardtii*. *The Plant Journal, 97*(6), 1022–1031.

Czarnecki, O., & Grimm, B. (2012). Post-translational control of tetrapyrrole biosynthesis in plants, algae, and cyanobacteria. *Journal of Experimental Botany, 63*(4), 1675–1687.

Dang, K. V., Plet, J., Tolleter, D., Jokel, M., Cuiné, S., Carrier, P., Auroy, P., Richaud, P., et al. (2014). Combined increases in mitochondrial cooperation and oxygen photoreduction compensate for deficiency in cyclic electron flow in Chlamydomonas reinhardtii. *Plant Cell, 26*(7), 3036–3050.

Delosme, R., Olive, J., & Wollman, F. A. (1996). Changes in light energy distribution upon state transitions: An in vivo photoacoustic study of the wide type and photosynthesis mutants from Chlamydomonas reinhardtii. *Biochimica et Biophysica Acta, 1273*(1996), 150–158.

Desplats, C., Mus, F., Cuiné, S., Billon, E., Cournac, L., & Peltier, G. (2009). Characterization of Nda2, a plastoquinone-reducing type II NAD(P)H dehydrogenase in chlamydomonas chloroplasts. *The Journal of Biological Chemistry, 284*(7), 4148–4157.

Dewez, D., Park, S., García-Cerdán, J. G., Lindberg, P., & Melis, A. (2009). Mechanism of REP27 protein action in the D1 protein turnover and photosystem II repair from Photodamage. *Plant Physiology, 151*(1), 88–99.

Douchi, D., Qu, Y., Longoni, P., Legendre-Lefebvre, L., Johnson, X., Schmitz-Linneweber, C., & Goldschmidt-Clermont, M. (2016). A nucleus-encoded chloroplast phosphoprotein governs expression of the photosystem I subunit PsaC in Chlamydomonas reinhardtii. *Plant Cell, 28*(5), 1182–1199.

Duanmu, D., Casero, D., Dent, R. M., Gallaher, S., Yang, W., Rockwell, N. C., Martin, S. S., Pellegrini, M., et al. (2013). Retrograde Bilin signaling enables Chlamydomonas greening and phototrophic survival. *Proceedings of the National Academy of Sciences of the United States of America, 110*(9), 3621–3626.

Duanmu, D., Rockwell, N. C., & Lagarias, J. C. (2017). Algal light sensing and photoacclimation in aquatic environments. *Plant, Cell & Environment, 40*(11), 2558–2570.

Emonds-Alt, B., Coosemans, N., Gerards, T., Remacle, C., & Cardol, P. (2017). Isolation and characterization of mutants corresponding to the MENA, MENB, MENC and MENE enzymatic steps of 5′-monohydroxyphylloquinone biosynthesis in Chlamydomonas reinhardtii. *The Plant Journal, 89*(1), 141–154.

Ferenczi, A., Pyott, D. E., Xipnitou, A., & Molnar, A. (2017). Efficient targeted DNA editing and replacement in Chlamydomonas reinhardtii using Cpf1 ribonucleoproteins and single-stranded DNA. *Proceedings of the National Academy of Sciences of the United States of America, 114*(51), 13567–13572.

Gabilly, S. T., Baker, C. R., Wakao, S., Crisanto, T., Guan, K., Bi, K., Guiet, E., Guadagno, C. R., et al. (2019). Regulation of photoprotection gene expression in Chlamydomonas by a putative E3 ubiquitin ligase complex and a homolog of CONSTANS. *Proceedings of the National Academy of Sciences of the United States of America, 116*(35), 17556–17562.

García-Cerdán, J. G., Furst, A. L., McDonald, K. L., Schünemann, D., Francis, M. B., & Niyogi, K. K. (2019). A thylakoid membrane-bound and redox-active rubredoxin (RBD1) functions in de novo assembly and repair of photosystem II. *Proceedings of the National Academy of Sciences of the United States of America, 116*(33), 16631–16640.

Girolomoni, L., Ferrante, P., Berteotti, S., Giuliano, G., Bassi, R., & Ballottari, M. (2016). The function of LHCBM4/6/8 antenna proteins in *Chlamydomonas reinhardtii. Journal of Experimental Botany, 68*(3), 627–641.

Girolomoni, L., Cazzaniga, S., Pinnola, A., Perozeni, F., Ballottari, M., & Bassi, R. (2019). LHCSR3 is a nonphotochemical quencher of both photosystems in Chlamydomonas reinhardtii. *Proceedings of the National Academy of Sciences of the United States of America, 116*(10), 4212–4217.

Greiner, A., Kelterborn, S., Evers, H., Kreimer, G., Sizova, I., & Hegemann, P. (2017). Targeting of photoreceptor genes in Chlamydomonas reinhardtii via zinc-finger nucleases and CRISPR/Cas9. *Plant Cell, 29*(10), 2498–2518.

Grewe, S., Ballottari, M., Alcocer, M., D'Andrea, C., Blifernez-Klassen, O., Hankamer, B., Mussgnug, J. H., Bassi, R., et al. (2014). Light-harvesting complex protein LHCBM9 is critical for photosystem II activity and hydrogen production in *Chlamydomonas reinhardtii. Plant Cell, 26*(4), 1598–1611.

Grossman, A. R., Lohr, M., & Im, C. S. (2004). Chlamydomonas reinhardtii in the landscape of pigments. *Annual Review of Genetics, 38*, 119–173.

Grossman, A., Sanz-Luque, E., Yi, H., & Yang, W. (2019). Building the GreenCut2 suite of proteins to unmask photosynthetic function and regulation. *Microbiology, 165*(7), 697–718.

Heinnickel, M., Kim, R. G., Wittkopp, T. M., Yang, W., Walters, K. A., Herbert, S. K., & Grossman, A. R. (2016). Tetratricopeptide repeat protein protects photosystem I from oxidative disruption during assembly. *Proceedings of the National Academy of Sciences of the United States of America, 113*(10), 2774–2779.

Hertle, A. P., Blunder, T., Wunder, T., Pesaresi, P., Pribil, M., Armbruster, U., & Leister, D. (2013). PGRL1 is the elusive ferredoxin-plastoquinone reductase in photosynthetic cyclic electron flow. *Molecular Cell, 49*(3), 511–523.

Houille-Vernes, L., Rappaport, F., Wollman, F. A., Alric, J., & Johnson, X. (2011). Plastid terminal oxidase 2 (PTOX2) is the major oxidase involved in chlororespiration in Chlamydomonas. *Proceedings of the National Academy of Sciences of the United States of America, 108*(51), 20820–20825.

Idoine, A. D., Boulouis, A., Rupprecht, J., & Bock, R. (2014). The diurnal logic of the expression of the chloroplast genome in Chlamydomonas reinhardtii. *PLoS One, 9*(10), e108760.

Im, C. S., Eberhard, S., Huang, K., Beck, C. F., & Grossman, A. R. (2006). Phototropin involvement in the expression of genes encoding chlorophyll and carotenoid biosynthesis enzymes and LHC apoproteins in Chlamydomonas reinhardtii. *The Plant Journal, 48*(1), 1–16.

Iwai, M., Takizawa, K., Tokutsu, R., Okamuro, A., Takahashi, Y., & Minagawa, J. (2010). Isolation of the elusive supercomplex that drives cyclic electron flow in photosynthesis. *Nature, 464*(7292), 1210–1213.

Kato, Y., Sun, X., Zhang, L., & Sakamoto, W. (2012). Cooperative D1 degradation in the photosystem II repair mediated by Chloroplastic proteases in Arabidopsis. *Plant Physiology, 159*(4), 1428–1439.

Kato, Y., Hyodo, K., & Sakamoto, W. (2018). The photosystem II repair cycle requires FtsH turnover through the EngA GTPase. *Plant Physiology, 178*(2), 596–611.

Kawakami, K., Tokutsu, R., Kim, E., & Minagawa, J. (2019). Four distinct trimeric forms of light-harvesting complex II isolated from the green alga *Chlamydomonas reinhardtii. Photosynthesis Research.* https://doi.org/10.1007/s11120-019-00669-y.

Kim, E., Akimoto, S., Tokutsu, R., Yokono, M., & Minagawa, J. (2017). Fluorescence lifetime analyses reveal how the high light-responsive protein LHCSR3 transforms PSII light-harvesting complexes into an energy-dissipative state. *The Journal of Biological Chemistry, 292*(46), 18951–18960.

Kono, M., Noguchi, K., & Terashima, I. (2014). Roles of the cyclic electron flow around PSI (CEF-PSI) and O_2-dependent alternative pathways in regulation of the photosynthetic electron flow in short-term fluctuating light in Arabidopsis thaliana. *Plant & Cell Physiology, 55*(5), 990–1004.

Kosuge, K., Tokutsu, R., Kim, E., Akimoto, S., Yokono, M., Ueno, Y., & Minagawa, J. (2018). LHCSR1- dependent fluorescence quenching is mediated by excitation energy transfer from LHCII to photosystem I in *Chlamydomonas reinhardtii. Proceedings of the National Academy of Sciences of the United States of America, 115*(14), 3722–3727.

Kottke, T., Oldemeyer, S., Wenzel, S., Zou, Y., & Mittag, M. (2017). Cryptochrome photoreceptors in green algae: Unexpected versatility of mechanisms and functions. *Journal of Plant Physiology, 217*, 4–14.

Krech, K., Ruf, S., Masduki, F. F., Thiele, W., Bednarczyk, D., Albus, C. A., Tiller, N., Hasse, C., et al. (2012). The plastid genome-encoded Ycf4 protein functions as a nonessential assembly factor for photosystem I in higher plants. *Plant Physiology, 159*(2), 579–591.

Kubota-Kawai, H., Burton-Smith, R. N., Tokutsu, R., Song, C., Akimoto, S., Yokono, M., Ueno, Y., Kim, E., et al. (2019). Ten antenna proteins are associated with the core in the supramolecular organization of the photosystem I supercomplex in *Chlamydomonas reinhardtii. The Journal of Biological Chemistry, 294*(12), 4304–4314.

Kukuczka, B., Magneschi, L., Petroutsos, D., Steinbeck, J., Bald, T., Powikrowska, M., Fufezan, C., Finazzi, G., et al. (2014). Proton gradient Regulation5-Like1-mediated cyclic Electron flow is crucial for acclimation to anoxia and complementary to nonphotochemical quenching in stress adaptation. *Plant Physiology, 165*(4), 1604–1617.

Le Quiniou, C., Tian, L., Drop, B., Wientjes, E., van Stokkum, I. H. M., van Oort, B., & Croce, R. (2015). PSI–LHCI of Chlamydomonas reinhardtii: Increasing the absorption cross section without losing efficiency. *Biochimica et Biophysica Acta, 1847*(4–5), 458–467.

Lefebvre-Legendre, L., Rappaport, F., Finazzi, G., Ceol, M., Grivet, C., Hopfgartner, G., & Rochaix, J. D. (2007). Loss of Phylloquinone in Chlamydomonas affects Plastoquinone Pool size and photosystem II synthesis. *The Journal of Biological Chemistry, 282*(18), 13250–13263.

Lefebvre-Legendre, L., Choquet, Y., Kuras, R., Loubéry, S., Douchi, D., & Goldschmidt-Clermont, M. (2015). A nucleus-encoded chloroplast protein regulated by Iron availability governs expression of the photosystem I subunit PsaA in Chlamydomonas reinhardtii. *Plant Physiology, 167*(4), 1527–1540.

Li, X., Patena, W., Fauser, F., Jinkerson, R. E., Saroussi, S., Meyer, M. T., Ivanova, N., Robertson, J. M., et al. (2019). A genome-wide algal mutant library and functional screen identifies genes required for eukaryotic photosynthesis. *Nature Genetics, 51*(4), 627–635.

Maruyama, S., Tokutsu, R., & Minagawa, J. (2014). Transcriptional regulation of the stress-responsive light harvesting complex genes in Chlamydomonas reinhardtii. *Plant & Cell Physiology, 55*(7), 1304–1310.

Maul, J. E., Lilly, J. W., Cui, L., dePamphilis, C. W., Miller, W., Harris, E. H., & Stern, D. B. (2002). The Chlamydomonas reinhardtii plastid chromosome: Islands of genes in a sea of repeats. *Plant Cell, 14*(11), 2659–2679.

McCarthy, S. S., Kobayashi, M. C., & Niyogi, K. K. (2004). White mutants of Chlamydomonas reinhardtii are defective in phytoene synthase. *Genetics, 168*(3), 1249–1257.

Mehler, A. H. (1951). Studies on reactions of illuminated chloroplasts. II. Stimulation and inhibition of the reaction with molecular oxygen. *Archives of Biochemistry and Biophysics, 34*(2), 339–351.

Merchant, S. S., Prochnik, S. E., Vallon, O., Harris, E. H., Karpowicz, S. J., Witman, G. B., Terry, A., Salamov, A., et al. (2007). The Chlamydomonas genome reveals the evolution of key animal and plant functions. *Science, 318*(5848), 245–250.

Mosebach, L., Heilmann, C., Mutoh, R., Gäbelein, P., Steinbeck, J., Happe, T., Ikegami, T., Hanke, G., et al. (2017). Association of Ferredoxin:NADP(+) oxidoreductase with the photosynthetic apparatus modulates electron transfer in Chlamydomonas reinhardtii. *Photosynthesis Research, 134*(3), 291–306.

Munekage, Y., Hashimoto, M., Miyake, C., Tomizawa, K., Endo, T., Tasaka, M., & Shikanai, T. (2004). Cyclic electron flow around photosystem I is essential for photosynthesis. *Nature, 429*(6991), 579–582.

Mussgnug, J. H., Wobbe, L., Elles, I., Claus, C., Hamilton, M., Fink, A., Kahmann, U., Kapazoglou, A., et al. (2005). NAB1 is an RNA binding protein involved in the light-regulated differential expression of the light-harvesting antenna of Chlamydomonas reinhardtii. *Plant Cell, 17*(12), 3409–3421.

Nawrocki, W. J., Tourasse, N. J., Taly, A., Rappaport, F., & Wollman, F. A. (2015). The plastid terminal oxidase: Its elusive function points to multiple contributions to plastid physiology. *Annual Review of Plant Biology, 66*, 49–74.

Nawrocki, W. J., Bailleul, B., Cardol, P., Rappaport, F., Wollman, F. A., & Joliot, P. (2019). Maximal cyclic electron flow rate is independent of PGRL1 in Chlamydomonas. *Biochimica et Biophysica Acta – Bioenergetics, 1860*(5), 425–432.

Nellaepalli, S., Ozawa, S., Kuroda, H., & Takahashi, Y. (2018). The photosystem I assembly apparatus consisting of Ycf3-Y3IP1 and Ycf4 modules. *Nature Communications, 9*(1), 2439.

Ohnishi, N., & Takahashi, Y. (2008). Chloroplast-encoded PsbT is required for efficient biogenesis of photosystem II complex in the green alga *Chlamydomonas reinhardtii. Photosynthesis Research, 98*(1–3), 315–322.

Ossenbühl, F., Göhre, V., Meurer, J., Krieger-Liszkay, A., Rochaix, J., & Eichacker, L. A. (2004). Efficient assembly of photosystem II in Chlamydomonas reinhardtii requires Alb3.1p, a homolog of Arabidopsis ALBINO3. *Plant Cell, 16*(7), 1790–1800.

Ozawa, S., Kosugi, M., Kashino, Y., Sugimura, T., & Takahashi, Y. (2012). 5′-monohydroxyphylloquinone is the dominant naphthoquinone of PSI in the green alga Chlamydomonas reinhardtii. *Plant & Cell Physiology, 53*(1), 237–243.

Peers, G., Truong, T. B., Ostendorf, E., Busch, A., Elrad, D., Grossman, A. R., Hippler, M., & Niyogi, K. K. (2009). An ancient light-harvesting protein is critical for the regulation of algal photosynthesis. *Nature, 462*(7272), 518–521.

Rea, G., Antonacci, A., Lambreva, M. D., & Mattoo, A. K. (2018). Features of cues and processes during chloroplast-mediated retrograde signaling in the alga Chlamydomonas. *Plant Science, 272*, 193–206.

Reinbothe, C., El Bakkouri, M., Buhr, F., Muraki, N., Nomata, J., Kurisu, G., Fujita, Y., & Reinbothe, S. (2010). Chlorophyll biosynthesis: Spotlight on protochlorophyllide reduction. *Trends in Plant Science, 15*(11), 614–624.

Richter, A. S., Hochheuser, C., Fufezan, C., Heinze, L., Kuhnert, F., & Grimm, B. (2016). Phosphorylation of GENOMES UNCOUPLED 4 alters stimulation of mg Chelatase activity in angiosperms. *Plant Physiology, 172*(3), 1578–1595.

Rockwell, N. C., Su, Y. S., & Lagarias, J. C. (2006). Phytochrome structure and signaling mechanisms. *Annual Review of Plant Biology, 57*, 837–858.

Salomé, P. A., & Merchant, S. S. (2019). A series of fortunate events: Introducing Chlamydomonas as a reference organism. *Plant Cell, 31*(8), 1682–1707.

Santabarbara, S., Casazza, A. P., Ali, K., Economou, C. K., Wannathong, T., Zito, F., Redding, K. E., Rappaport, F., et al. (2013). The requirement for carotenoids in the assembly and function of the photosynthetic complexes in Chlamydomonas reinhardtii. *Plant Physiology, 161*(1), 535–546.

Shen, L., Huang, Z., Chang, S., Wang, W., Wang, J., Kuang, T., Han, G., Shen, J., et al. (2019). Structure of a C2S2M2N2-type PSII–LHCII supercomplex from the green alga *Chlamydomonas reinhardtii. Proceedings of the National Academy of Sciences of the United States of America, 116*(42), 21246–21255.

Steinbeck, J., Ross, I. L., Rothnagel, R., Gäbelein, P., Schulze, S., Giles, N., Ali, R., Drysdale, R., et al. (2018). Structure of a PSI-LHCI-cyt b6f supercomplex in Chlamydomonas reinhardtii promoting cyclic electron flow under anaerobic conditions. *Proceedings of the National Academy of Sciences of the United States of America, 115*(41), 10517–10522.

Strenkert, D., Schmollinger, S., Gallaher, S. D., Salomé, P. A., Purvine, S. O., Nicora, C. D., Mettler-Altmann, T., Soubeyrand, E., et al. (2019). Multiomics resolution of molecular events during a day in the life of Chlamydomonas. *Proceedings of the National Academy of Sciences of the United States of America, 116*(6), 2374–2383.

Su, X., Ma, J., Pan, X., Zhao, X., Chang, W., Liu, Z., Zhang, X., & Li, M. (2019). Antenna arrangement and energy transfer pathways of a green algal photosystem-I-LHCI supercomplex. *Nature Plants, 5*(3), 273–281.

Suga, M., Ozawa, S., Yoshida-Motomura, K., Akita, F., Miyazaki, N., & Takahashi, Y. (2019). Structure of the green algal photosystem I supercomplex with a decameric light-harvesting complex I. *Nature Plants, 5*(6), 626–636.

Takahashi, H., Schmollinger, S., Lee, J. H., Schroda, M., Rappaport, F., Wollman, F. A., & Vallon, O. (2016). PETO interacts with other effectors of cyclic Electron flow in Chlamydomonas. *Molecular Plant, 9*(4), 558–568.

Tanaka, R., & Tanaka, A. (2007). Tetrapyrrole biosynthesis in higher plants. *Annual Review of Plant Biology, 58*, 321–346.

Terashima, M., Petroutsos, D., Hüdig, M., Tolstygina, I., Trompelt, K., Gäbelein, P., Fufezan, C., Kudla, J., et al. (2012). Calcium-dependent regulation of cyclic photosynthetic electron transfer by a CAS, ANR1, and PGRL1 complex. *Proceedings of the National Academy of Sciences of the United States of America, 109*(43), 17717–17722.

Tokutsu, R., & Minagawa, J. (2013). Energy-dissipative supercomplex of photosystem II associated with LHCSR3 in Chlamydomonas reinhardtii. *Proceedings of the National Academy of Sciences of the United States of America, 110*(24), 10016–10021.

Tokutsu, R., Kato, N., Bui, K. H., Ishikawa, T., & Minagawa, J. (2012). Revisiting the supramolecular Organization of Photosystem II in *Chlamydomonas reinhardtii. The Journal of Biological Chemistry, 287*(37), 31574–31581.

Trebst, A., & Depka, B. (1997). Role of carotene in the rapid turnover and assembly of photosystem II in Chlamydomonas reinhardtii. *FEBS Letters, 400*(3), 359–362.

Uniacke, J., & Zerges, W. (2007). Photosystem II assembly and repair are differentially localized in *Chlamydomonas*. *Plant Cell, 19*(11), 3640–3654.

Ünlü, C., Drop, B., Croce, R., & van Amerongen, H. (2014). State transitions in *Chlamydomonas reinhardtii* strongly modulate the functional size of photosystem II but not of photosystem I. *Proceedings of the National Academy of Sciences of the United States of America, 111*(9), 3460–3465.

Wang, F., Johnson, X., Cavaiuolo, M., Bohne, A. V., Nickelsen, J., & Vallon, O. (2015). Two Chlamydomonas OPR proteins stabilize chloroplast mRNAs encoding small subunits of photosystem II and cytochrome b6 f. *The Plant Journal, 82*(5), 861–873.

Wilde, A., Härtel, H., Hübschmann, T., Hoffmann, P., Shestakov, S. V., & Börner, T. (1995). Inactivation of a Synechocystis sp strain PCC 6803 gene with homology to conserved chloroplast open reading frame 184 increases the photosystem II-to-photosystem I ratio. *Plant Cell, 7*(5), 649–658.

Wobbe, L., Blifernez, O., Schwarz, C., Mussgnug, J. H., Nickelsen, J., & Kruse, O. (2009). Cysteine modification of a specific repressor protein controls the translational status of nucleus-encoded LHCII mRNAs in Chlamydomonas. *Proceedings of the National Academy of Sciences of the United States of America, 106*(32), 13290–13295.

Xue, H., Tokutsu, R., Bergner, S. V., Scholz, M., Minagawa, J., & Hippler, M. (2015). PHOTOSYSTEM II SUBUNIT R is required for efficient binding of LIGHT-HARVESTING COMPLEX STRESS-RELATED PROTEIN3 to photosystem II-light-harvesting Supercomplexes in Chlamydomonas reinhardtii. *Plant Physiology, 167*(4), 1566–1578.

Yang, H., Liu, J., Wen, X., & Lu, C. (2015). Molecular mechanism of photosystem I assembly in oxygenic organisms. *Biochimica et Biophysica Acta, 1847*(9), 838–848.

Zhang, W., Zhong, H., Lu, H., Zhang, Y., Deng, X., Huang, K., & Duanmu, D. (2018). Characterization of ferredoxin-dependent Biliverdin reductase PCYA1 reveals the dual function in retrograde Bilin biosynthesis and interaction with light-dependent Protochlorophyllide oxidoreductase LPOR in Chlamydomonas reinhardtii. *Frontiers in Plant Science, 9*, 676.

Zones, J. M., Blaby, I. K., Merchant, S. S., & Umen, J. G. (2015). High-resolution profiling of a synchronized diurnal transcriptome from Chlamydomonas reinhardtii reveals continuous cell and metabolic differentiation. *Plant Cell, 27*(10), 2743–2769.

Part II
Photosynthesis and the Environment

Photosynthetic Performances of Marine Microalgae Under Influences of Rising CO_2 and Solar UV Radiation

Kunshan Gao and Donat-P. Häder

Abstract Marine photosynthesis contributes approximately half of the global primary productivity. Ocean climate changes, such as increasing dissolved CO_2 in seawater and consequently declining pH (known as ocean acidification, OA), may alter marine photosynthetic performance. There are numerous studies on the effects of OA on photosynthetic organisms, but controversial findings indicate positive, neutral, and negative influences. Most of the studies so far have been conducted under controlled conditions that ignored the presence of solar UV radiation. Increased CO_2 availability may play a fertilizing role, while the concurrent pH drop may exert pressure on microalgal cells, especially during the night period. It is known that elevated CO_2 concentrations downregulate CO_2-concentrating mechanisms (CCMs), and intracellular concentrations of dissolved inorganic carbon in diatoms grown under elevated CO_2 levels can be much lower than that in low CO_2-grown ones. Such a reduced CO_2 availability within cells in response to increased CO_2 in the water can lead to enhanced photorespiration due to an increased O_2 to CO_2 ratio around the carboxylating and oxygenating enzyme, RuBisCO. Therefore, negative and positive effects of OA may depend on light levels, since the saved energy due to downregulation of CCMs can benefit growth under light-limited conditions but enhance photoinhibition under light-excessive conditions. OA affects metabolic pathways in phytoplankton. It augments ß-oxidation and the citric acid cycle, which accumulates toxic phenolic compounds. In the upper mixed layer, phytoplankton are exposed to excessive PAR and UV radiation (UVR). The calcareous incrustations of calcified microalgae, known to shield the organisms from UVR, are thinned due to OA, exposing the cells to increased solar UV and further inhibiting their calcification and photosynthesis, reflecting a compounded impact. Such UV and OA interactive effects are expected to reduce primary productivity in oligotrophic pelagic surface waters. In this chapter, we review and analyze recent results on

K. Gao (✉)
State Key Laboratory of Marine Environmental Science/College of Ocean and Earth Sciences, Xiamen University (Xiang-An Campus), Xiamen, China
e-mail: ksgao@xmu.edu.cn

D.-P. Häder
Department of Biology, Friedrich-Alexander University Erlangen-Nürnberg, Erlangen, Germany

Q. Wang (ed.), *Microbial Photosynthesis*,
https://doi.org/10.1007/978-981-15-3110-1_7

effects of OA and UV and their combined effects on marine photosynthesis of microalgae, which falls in the context of marine photosynthesis under changing ocean environments and multiple stressors.

Keywords CO_2 · Ocean acidification · Marine photosynthesis · Phytoplankton · Solar UV radiation

1 Introduction

In marine environments, photosynthesis is the basic driver of the biological CO_2 pump that is thought to absorb 25 million tons of anthropogenic CO_2 per day in the global oceans, mitigating global warming. Therefore, it is essential to understand the relationship between photosynthesis and marine environmental changes in order to assess CO_2 uptake capacities in different regions.

Microalgae are exposed to a number of external factors which control growth and biomass production. In the upper mixed layers (UML) of marine systems, microalgal or phytoplankton cells are exposed to increasing solar UV and visible radiation, mainly due to ocean warming which results in a shallower UML and enhanced stratification (Gao et al. 2012a). Although the release of CFCs (chlorinated fluorocarbons) has been reduced since the Montreal Protocol agreement was signed, these substances are still harmful for the stratospheric ozone because of their long lifetimes (Bais et al. 2015), and several new gases, harmful to the stratospheric ozone layer, are rapidly accumulating in the atmosphere. Consequently, effects of enhanced UV-B radiation (280–315 nm) still receive much attention (Häder and Gao 2015).

Based on a priori insight, both UV and increasing levels of CO_2 can result in negative, neutral, or positive effects, depending on solar radiation levels, nutrient availability, and other chemical and physical conditions and/or species-specific traits. In view of the metabolic responses and associated energetics, changes in chemical and physical environments may interactively (additively, synergistically, or antagonistically) affect physiological processes and ecological functions of marine photosynthetic organisms. Therefore, combined effects of CO_2 and UV on the photosynthetic performances should be examined in a context of marine environmental changing biology.

2 Effects of Increasing CO_2 Concentration and Declining pH

Influx and efflux of gases into or out of water depend on their equilibrium or difference in their partial pressures between the two media; therefore, continuously increasing concentrations of atmospheric CO_2 are taken up by the oceans. The atmospheric CO_2 concentration has increased from 280 to more than 410 ppmv since the late nineteenth century, even though the oceans have absorbed more than a third of the anthropogenically released amount.

CO_2 is an acidic gas, and its dissolution into seawater leads to a decreasing pH:

$$CO_2 + H_2O = H_2CO_3 \quad (1)$$

$$H_2CO_3 = H^+ + HCO_3^- \quad (2)$$

$$HCO_3^- = H^+ + CO_3^{2-} \quad (3)$$

The equilibrium of these reactions depends on temperature and salinity in the water. In the water CO_2 reacts to carbonic acid which dissociates to form HCO_3^- and H^+. As the concentration of H^+ increases with the concentration of HCO_3^- accounting for over 90% of dissolved inorganic carbon, reaction (3) is shifted to the right, causing the CO_3^{2-} concentration to decrease.

Since the Industrial Revolution, the oceans have been taking up anthropogenic CO_2 emissions, which are currently running at an average rate of 25 million tons per day (Sabine et al. 2004). The result is a declining pH of seawater, leading to ocean acidification (OA). Several studies on the effects of OA on phytoplankton had divergent results either showing increased (Riebesell and Tortell 2011) or decreased growth and photosynthesis due to increased mitochondrial respiration and photorespiration (Wu et al. 2010; Gao et al. 2012b). Some studies indicated no effect (Tortell et al. 2000; Kim et al. 2006; Gao and Campbell 2014). Shipboard exposures in the South China Sea indicated that the effect of OA depends on solar irradiance and thus on the depth distribution where the organisms dwell (Gao et al. 2012b). The concentration of diatoms in the surface layers of South China Sea is reduced as acidification increases, which is the primary productivity in these nutrient-poor waters (Gao et al. 2012b). In contrast, at high nutrient diatom growth increases (Wu et al. 2014). In a mesocosm study using coastal nutrient-rich water, the higher CO_2 partial pressure augmented primary productivity of phytoplankton populations with high diatom concentrations (Huang et al. 2018). However, the effect of OA may be different in different diatom species or phenotypes. The coastal diatom *Thalassiosira weissflogii* shows a high growth rate under OA conditions, while the open ocean *Thalassiosira oceanica* has a lower growth rate. One possible explanation is that the large coastal species is better adapted to daily pH changes (Li et al. 2016). Diatoms with a less effective CO_2-concentrating mechanism (CCM) were more affected by OA, resulting in lower growth rates (Shi et al. 2019).

CCMs are known to be downregulated with increased CO_2 availability in the water (Giordano et al. 2005). Even under climate change-induced CO_2 rise, CCMs are significantly downregulated in a diatom (Wu et al. 2010), and its intracellular dissolved inorganic carbon decreases to a level that is much lower than in cells grown under low CO_2, with a CO_2 accumulation factor being reduced about three times (Liu et al. 2017). Since the concentration of CO_2 in the cells decreases with increased external CO_2 levels, the ratio of O_2 to CO_2 increases and consequently brings about higher photorespiration (Gao et al. 2012b; Xu and Gao 2012). This explains that high CO_2-acclimated diatoms or phytoplankton assemblages exhibit higher non-photochemical quenching along with enhanced photorespiration and reduced carbon fixation (Gao et al. 2012b; Li et al. 2018). Nevertheless, enhanced growth of diatoms has been observed under elevated CO_2 concentration (lowered pH) under indoor low light levels (Wu et al. 2010, 2014) and low levels of mean daytime sunlight (Gao et al. 2012b). The saved energy due to downregulated CCMs could be responsible for such a discrepancy (Fig. 1).

It is possible that the impact of pH drop or acidification overwhelms any performance-enhancing effects of increased pCO_2 and energy saving from downregulation of the CCM. When pH in the water declines due to increased pCO_2, cells usually acclimate to maintain the normal operation of redox systems on the plasma membrane (Suffrian et al. 2011). They pump protons out of the cell using a H^+-translocating ATPase which requires additional energy in the form of ATP (Smith and Raven 1979; Madshus 1988; Suffrian et al. 2011). This H^+ efflux locally decreases the pH boundary layer around the cell, which in turn affects the redox systems of the

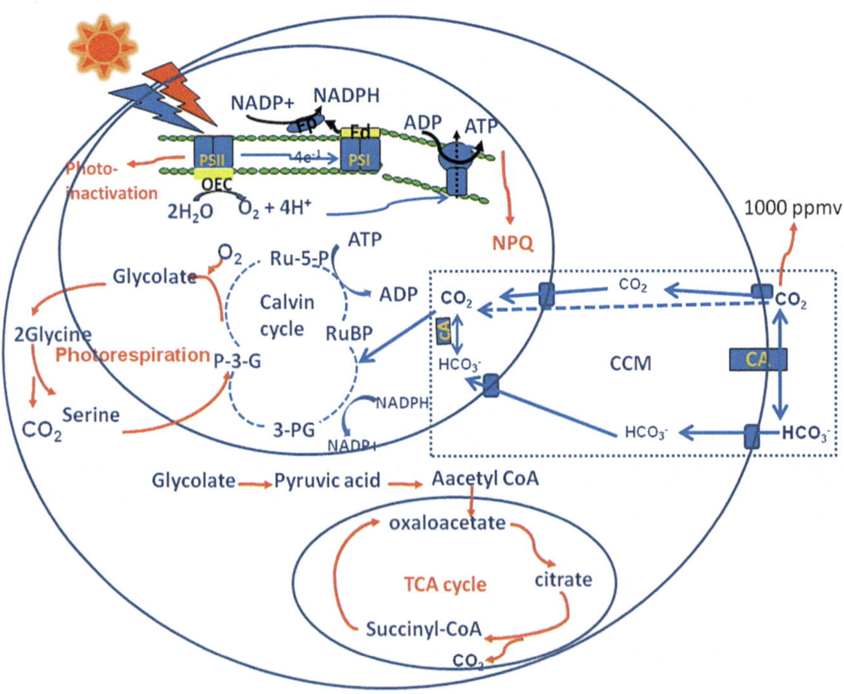

Fig. 1 Biochemical pathways in phytoplankton are upregulated (red and solid) or downregulated (blue and dotted) under OA conditions. CO_2^--concentrating mechanisms (CCMs) are downregulated at enhanced CO_2 levels which save energy expenditure for carbon acquisition. But high light intensities result in additional stress. Meanwhile, it stimulates cellular defenses, referring to increased NPQ, enhancing mitochondrial respiration (TCA cycle) and photorespiration (Gao et al. 2012b)

membrane as well as nutrient (iron, nitrate, etc.) uptake by enzymes located on the apoplastic side of the membrane (Jones and Morel 1988; Moog and Brüggemann 1994). Therefore, nutrient limitation interacts with the CO_2-induced acidification to reduce the photosynthetic rate and enhance mitochondrial respiration in a diatom (Li et al. 2018). While downregulation of the CCM in diatoms saves energy involved in active carbon acquisition, by about 20% with a doubled atmospheric CO_2 level (Hopkinson et al. 2011; Hennon et al. 2014), direct measurements of intracellular pCO_2 levels indeed demonstrated that the inorganic carbon pools within the diatom cells declined with elevated pCO_2 within the range expected from climate change and natural high CO_2 systems in the oceans (Liu et al. 2017). Photorespiration is enhanced by about 20% in three species of diatoms when acclimated to 1000 μatm (Gao et al. 2012b). The above mechanistic responses to increasing CO_2 concentration and declining pH are illustrated in Fig. 1. Since downregulation of CCMs involves energetics, light levels and nutrient availability as well as temperatures are supposed to modulate the effects of increasing CO_2 concentration and declining pH on photosynthesis and primary productivity to different extents in different regions or ecosystems.

OA treatment alters the efficiency of several metabolic pathways in phytoplankton cells such as β-oxidation, the tricarboxylic acid (TCA) cycle, glycolysis, etc., leading to a higher accumulation of toxic phenolics. During degradation by mitochondrial respiration, they obtain additional energy (Fig. 2) (Jin et al. 2015). This implies potential impacts of OA on the food chain via modulated metabolic pathways.

Figure 2. Significantly increased concentrations of phenolics in phytoplankton under influences of ocean acidification. As required for the degradation of phenolics, cells upregulate several metabolic pathways such as β-oxidation, the TCA cycle, and glycolysis (upper panel), so as to gain more energy to resist OA (Jin et al. 2015)

3 UV and Its Effect on Marine Photosynthetic Carbon Fixation

Solar ultraviolet radiation can be divided into UV-A (315–400 nm), UV-B (280–315 nm), and UV-C (<280 nm). The most harmful, UV-C, is quantitatively absorbed by the stratospheric ozone layer; also UV-B is strongly reduced by the ozone but part of it reaches the ground. Although the share of UV-B is only a small fraction of the total solar radiation, it is the most harmful wavelength band and may damage photosynthetic organisms living in the surface water layer.

Solar radiation at different wavelengths is transmitted to different depths. In the Atlantic, UV-B was found to penetrate to 30 m and UV-A to 100 m; blue wavelength was transmitted down to more than 190 m, while longer visible wavelengths reached only 30 m depth (Piazena et al. 2002). The transmittance of visible radiation defines the depth of the euphotic zone. In the open ocean waters of the South China Sea, the

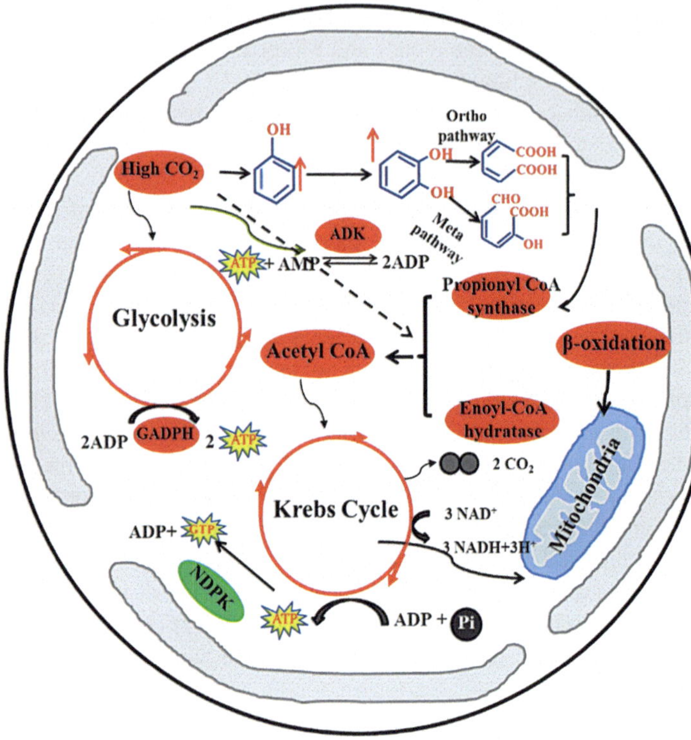

Fig. 2 Significantly increased concentrations of phenolics in phytoplankton under the influence of ocean acidification. Phytoplankton cells upregulated several biochemical pathways including β-oxidation, the tricarboxylic acid cycle, and glycolysis in order to degrade phenolics to obtain more energy to fight against the effects of OA (Jin et al. 2015)

visible radiation was found to transmit to below 80 m, UV-A to 50 m, and UV-B to 38 m (Li et al. 2009).

One of the main targets of UV-B is the photosynthetic electron transport, since it damages the D1 protein in photosystem II, reducing the photosynthetic capacity (García-Gómez et al. 2016). This damage can be repaired by replacing the damaged proteins by newly synthesized molecules, which are augmented at higher temperatures (Gao et al. 2008). Microalgal and cyanobacterial cells are moving actively or passively in the UML; near the surface they are damaged most, while they undergo repair when they are in deeper layers (Helbling et al. 2003). In addition, UV-B damages DNA by inducing cyclobutane pyrimidine dimers (Gao et al. 2008; Rajneesh et al. 2018). If not repaired by a photolyase, these dimers can result in mutations and cell death (García-Gómez et al. 2014).

During a long evolution process, microalgae and cyanobacteria have developed mechanisms to resist UV radiation; for example, they synthesize UV-shielding pigments, such as mycosporine-like amino acids (MAAs), remove active oxygen free radicals, and repair damaged proteins and DNA (Rastogi et al. 2014; Häder et al. 2015). While the cells in surface waters are inevitably affected by UV,

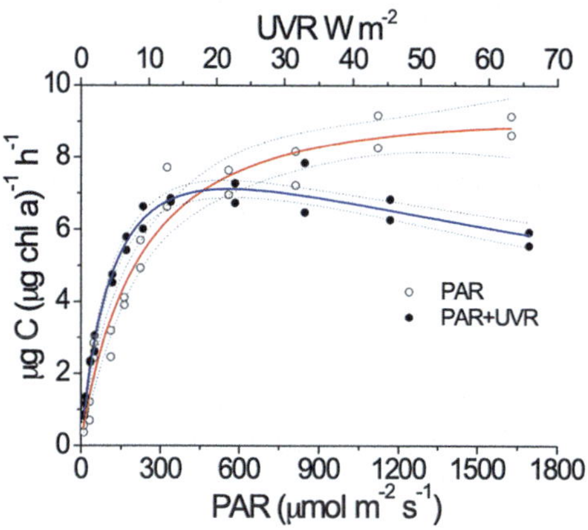

Fig. 3 Photosynthetic carbon fixation rates of coastal phytoplankton assemblages versus irradiance under PAR (open circles) and PAR + UVR (closed circle conditions). Dotted lines represent the 95% confidence limit. The mean solar irradiances (PAR, UV-A, and UV-B) during the incubations (August 6 and 8, 2005) were 321, 50.5, and 2.1 W m^{-2}, respectively (Gao et al. 2007)

resulting in a decrease of photosynthetic carbon fixation (Helbling et al. 2003; Häder and Gao 2015), in deeper layers or under low or moderate levels of solar radiation, UV-A radiation can stimulate photosynthetic carbon fixation (Gao et al. 2007), which was demonstrated in coastal phytoplankton assemblages. After filtering out the visible light, a diatom-dominated phytoplankton community can still carry out photosynthetic carbon fixation, and the fixation rate steadily rises with increased UV-A intensity, therefore leading to enhanced light use efficiency under low levels of solar radiation and reduced photosynthetic rate under high sunlight levels in the presence of UVR compared to PAR alone (Fig. 3) (Gao et al. 2007). Whether or not UV-A has a positive effect depends on the irradiance and the cell size (Li et al. 2011; Li and Gao 2013). Larger phytoplankton species are more capable of using UV-A for carbon fixation than smaller ones. It should be noted that the measured rate of photosynthesis in the presence of the full solar spectrum reflects the balanced effects of UV irradiances, that is, when PAR is limiting, the presence of UV-A increases the light use efficiency; however, when PAR is over-saturating, the presence of UV-A and UV-B can result in stronger inhibition of microalgal photosynthesis (Fig. 3). On the other hand, a meta-analysis on UV-B impacts on microalgae showed that smaller cells are more sensitive to UV-B compared to larger ones and that marine microalgae are more sensitive to it than freshwater species (Jin et al. 2017).

4 The Combined Effects of OA and UV Radiation

There are many reports on the individual effect of UV and OA on microalgae, but only a few look at the combined effects to understand the responses to both stressors (Beardall et al. 2014).

Plants are capable of quickly responding to UV radiation. Under high noon radiation, the photochemical PSII efficiency decreases within minutes. The PSII photochemical efficiency of the diatom *Cylindrotheca closterium f. minutissima* strongly decreases when the cells were cultured at an elevated CO_2 concentration (1000 μatm) and exposed to solar radiation. Even when the organism acclimated the combined effects of OA and UV, they showed a lower electron transport rate (Wu et al. 2012). For *Phaeocystis globosa* grown at 1000 μatm CO_2 (with constant pH) and solar radiation with or without UV radiation, its photochemical efficiency varied with the intensity of solar radiation (a negative correlation with light intensity), with the lowest values at noon under coupling impacts of UV and OA (Chen and Gao 2011). As to the compounded impacts of UV-A and OA, this alga increased its photochemical efficiency (Y′) and decreased its non-photochemical quenching (NPQ) under moderate levels of solar radiation (cloudy days) (Chen and Gao 2011). Regardless of sunny or cloudy days, the combined effects of OA and UV-B lowered its Y′ in this red tide alga and increased its NPQ only on sunny days (Chen and Gao 2011). Under fluctuating solar radiation, OA was shown to antagonize the enhancement of photosynthetic carbon fixation by UV-A while decreasing the photochemical inhibition associated with UV-B (Jin et al. 2013).

The effects of OA and UV have a synergistic impact on calcifying microalgae. OA hampers the calcification in coccolithophorids, and the consequently thinned calcified layer allows more transmission of UV, intensifying the damage by UV and impairing cellular functions (Gao et al. 2009; Xu and Gao 2015). It is obvious that reduced calcification of coccolithophores leads to more damage to the cells, which in turn further lowers the calcification rate and decreases photosynthetic carbon fixation.

5 Perspectives

Microalgae can exhibit evolutionary responses to elevated CO_2 over hundreds to thousands of generations, involving downregulated CCMs in a green alga (Collins et al. 2006), irreversible lost calcifying capability (Tong et al. 2018) and increased cellular N/C ratio (Jin et al. 2013) in a coccolithophorid, and smaller cell size and decreased photosynthesis in a diatom (Li et al. 2017). Nitrogen-fixing cyanobacteria have been grown for hundreds of generations under OA conditions which resulted in an irreversible augmentation of growth and nitrogen fixation which is due to "genetic assimilation of plastic traits into adaptive ones" (Hutchins et al. 2015; Walworth et al. 2016). Consequently, OA-modulated photosynthetic performances must have played a key role for the microalgae to gain evolutionary attributes.

Marine microalgae and cyanobacteria are exposed to degrading environmental conditions, considering the progressive OA and other ocean climate changes, such as ocean warming, stratification, deoxygenation, and enhanced exposures to UV radiation (Gao et al. 2019). Microalgae food quality is supposed to decline in response to OA (Riebesell et al. 2007; Jin et al. 2015), leading to reduced energy supply to grazers per biomass grazed. It is expected that more algal blooms will occur in eutrophicated waters with increasing levels of ocean climate changes. This hypothesis could be tested by laboratory microcosm or mesocosm experiments in combination with in situ investigations.

Understanding photosynthetic performances under multiple drivers, especially the multiple ocean climate change pressures, is gaining increasing attentions. Photosynthetic carbon fixation could be additively, synergistically, antagonistically, or neutrally affected by combinations of different drivers. And, increasing number of environmental drivers may complicate the effects of the key drivers, such as CO_2 and UV radiation (Brennan and Collins 2015). Therefore, effects of rising CO_2 and solar UV radiation on photosynthesis in natural environments should be evaluated against regional environmental conditions and/or future climate changes.

References

Bais, A., McKenzie, R., Bernhard, G., Aucamp, P., Ilyas, M., Madronich, S., & Tourpali, K. (2015). Ozone depletion and climate change: Impacts on UV radiation. *Photochemical & Photobiological Sciences, 14*(1), 19–52.

Beardall, J., Stojkovic, S., & Gao, K. (2014). Interactive effects of nutrient supply and other environmental factors on the sensitivity of marine primary producers to ultraviolet radiation: Implications for the impacts of global change. *Aquatic Biology, 22*, 5–23.

Brennan, G., & Collins, S. (2015). Growth responses of a green alga to multiple environmental drivers. *Nature Climate Change, 5*, 892–897.

Chen, S., & Gao, K. (2011). Solar ultraviolet radiation and CO_2-induced ocean acidification interacts to influence the photosynthetic performance of the red tide alga *Phaeocystis globosa* (Prymnesiophyceae). *Hydrobiologia, 675*, 105–117.

Collins, S., Sültemeyer, D., & Bell, G. (2006). Changes in C uptake in populations of *Chlamydomonas reinhardtii* selected at high CO_2. *Plant Cell Environment, 29*, 1812–1819.

Gao, K., & Campbell, D. (2014). Photophysiological responses of marine diatoms to elevated CO_2 and decreased pH: A review. *Functional Plant Biology, 41*, 449–459.

Gao, K., Wu, Y., Li, G., Wu, H., Villafañe, E. V., & Helbling, E. W. (2007). Solar UV-radiation drives CO2-fixation in marine phytoplankton: A double-edged sword. *Plant Physiology*, 144, 54–59.

Gao, K., Li, P., Watanabe, T., & Helbling, E. W. (2008). Combined effects of ultraviolet radiation and temperature on morphology, photosynthesis, and DNA of *Arthrospira (Spirulina) platensis* (Cyanophyta). *Journal of Phycology, 44*, 777–786.

Gao, K., Ruan, Z., Villafane, V. E., Gattuso, J.-P., & Helbling, E. W. (2009). Ocean acidification exacerbates the effect of UV radiation on the calcifying phytoplankter *Emiliania huxleyi*. *Limnology and Oceanography, 54*(6), 1855–1862.

Gao, K., Helbling, E. W., Häder, D.-P., & Hutchins, D. A. (2012a). Responses of marine primary producers to interactions between ocean acidification, solar radiation, and warming. *Marine Ecology Progress Series, 470*, 167–189.

Gao, K. S., Xu, J. T., Gao, G., Li, Y. H., Hutchins, D. A., Huang, B. Q., Wang, L., Zheng, Y., Jin, P., Cai, X. N., Häder, D. P., Li, W., Xu, K., Liu, N. N., & Riebesell, U. (2012b). Rising CO_2 and increased light exposure synergistically reduce marine primary productivity. *Nature Climate Change, 2*(7), 519–523.

Gao, K., Beardall, J., Häder, D.-P., Hall-Spencer, J. M., Gao, G., & Hutchins, D. A. (2019). Effects of ocean acidification on marine photosynthetic organisms under the concurrent influences of warming, UV radiation and deoxygenation. *Frontiers in Marine Science, 6*, 322.

García-Gómez, C., Gordillo, F. J., Palma, A., Lorenzo, M. R., & Segovia, M. (2014). Elevated CO_2 alleviates high PAR and UV stress in the unicellular chlorophyte *Dunaliella tertiolecta. Photochemical & Photobiological Sciences, 13*(9), 1347–1358.

García-Gómez, C., Mata, M. T., Van Breusegem, F., & Segovia, M. (2016). Low-steady-state metabolism induced by elevated CO_2 increases resilience to UV radiation in the unicellular green-algae *Dunaliella tertiolecta. Environmental and Experimental Botany, 132*, 163–174.

Giordano, M., Beardall, J., & Raven, J. A. (2005). CO_2 concentrating mechanisms in algae: Mechanisms, environmental modulation, and evolution. *Annual Review of Plant Biology, 56*, 99–131.

Häder, D.-P., & Gao, K. (2015). Interactions of anthropogenic stress factors on marine phytoplankton. *Frontiers in Environmental Science, 3*, 14.

Häder, D.-P., Williamson, C. E., Wängberg, S.-A., Rautio, M., Rose, K. C., Gao, K., Helbling, E. W., Sinha, R. P., & Worrest, R. (2015). Effects of UV radiation on aquatic ecosystems and interactions with other environmental factors. *Photochemical & Photobiological Sciences, 14*, 108–126.

Helbling, E. W., Gao, K., Gonçalves, R. J., Wu, H., & Villafañe, V. E. (2003). Utilization of solar UV radiation by coastal phytoplankton assemblages off SE China when exposed to fast mixing. *Marine Ecology Progress Series, 259*, 59–66.

Hennon, G. M., Quay, P., Morales, R. L., Swanson, L. M., & Armbrust, E. V. (2014). Acclimation conditions modify physiological response of the diatom *Thalassiosira pseudonana* to elevated CO_2 concentrations in a nitrate-limited chemostat. *Journal of Phycology, 50*, 243–253.

Hopkinson, B. M., Dupont, C. L., Allen, A. E., & Morel, F. M. M. (2011). Efficiency of the CO_2-concentrating mechanism of diatoms. *Proceedings of the National Academy of Sciences of the United States of America, 108*, 3830–3837.

Huang, Y., Liu, X., Laws, E. A., Bingzhang, C., Li, Y., Xie, Y., Wu, Y., Gao, K., & Huang, B. (2018). Effects of increasing atmospheric CO_2 on the marine phytoplankton and bacterial metabolism during a bloom: A coastal mesocosm study. *Science of the Total Environment, 633*, 618–629.

Hutchins, D. A., Walworth, N. G., Webb, E. A., Saito, M. A., Moran, D., McIlvin, M. R., Gale, J., & Fu, F.-X. (2015). Irreversibly increased nitrogen fixation in *Trichodesmium* experimentally adapted to elevated carbon dioxide. *Nature Communications, 6*(1), 1–7.

Jin, P., Gao, K., Villafañe, V., Campbell, D., & Helbling, W. (2013). Ocean acidification alters the photosynthetic responses of a coccolithophorid to fluctuating ultraviolet and visible radiation. *Plant Physiology, 162*, 2084–2094.

Jin, P., Wang, T., Liu, N., Dupont, S., Beardall, J., Boyd, P. W., Riebesell, U., & Gao, K. (2015). Ocean acidification increases the accumulation of toxic phenolic compounds across trophic levels. *Nature Communications, 6*, 8714.

Jin, P., Ding, J. C., Xing, T., Riebesell, U., & Gao, K. (2017). High levels of solar radiation offset impacts of ocean acidification on calcifying and non-calcifying strains of *Emiliania huxleyi. Marine Ecology Progress Series, 568*, 47–58.

Jones, G. J., & Morel, F. M. (1988). Plasmalemma redox activity in the diatom *Thalassiosira*: A possible role for nitrate reductase. *Plant Physiology, 87*(1), 143–147.

Kim, J.-M., Lee, K., Shin, K., Kang, J.-H., Lee, H.-W., Kim, M., Jang, P.-G., & Jang, M.-C. (2006). The effect of seawater CO_2 concentration on growth of a natural phytoplankton assemblage in a controlled mesocosm experiment. *Limnology and Oceanography, 51*(4), 1629–1636.

Li, G., & Gao, K. (2013). Cell size-dependent effects of solar UV radiation on primary production in coastal waters of the South China Sea. *Estuaries and Coasts, 36*, 728–736.

Li, G., Wu, Y., & Gao, K. (2009). Effects of Typhoon Kaemi on coastal phytoplankton assemblages in the South China Sea, with special reference to the effects of solar UV radiation. *Journal of Geophysical Research-Biogeosciences, 114*, G04029. https://doi.org/10.1029/2008JG000896.

Li, G., Gao, K., & Gao, G. (2011). Differential impacts of solar UV radiation on photosynthetic carbon fixation from the coastal to offshore surface waters in the South China Sea. *Photochemistry and Photobiology, 87*(2), 329–334.

Li, F. T., Wu, Y. P., Hutchins, D. A., Fu, F. X., & Gao, K. S. (2016). Physiological responses of coastal and oceanic diatoms to diurnal fluctuations in seawater carbonate chemistry under two CO_2 concentrations. *Biogeosciences, 13*, 6247–6259.

Li, F., Beardall, J., Collins, S., & Gao, K. (2017). Decreased photosynthesis and growth with reduced respiration in the model diatom *Phaeodactylum tricornutum* grown under elevated CO_2 over 1800 generations. *Global Change Biology, 23*(1), 127–137.

Li, F., Beardall, J., Gao, K., & Sathyendranath, H. e. S. (2018). Diatom performance in a future ocean: Interactions between nitrogen limitation, temperature, and CO_2-induced seawater acidification. *ICES Journal of Marine Science, 75*(4), 1451–1464.

Liu, N., Beardall, J., & Gao, K. (2017). Elevated CO2 and associated seawater chemistry do not benefit a model diatom grown with increased availability of light. *Aquatic Microbial Ecology, 79*, 137–147.

Madshus, I. H. (1988). Regulation of intracellular pH in eukaryotic cells. *Biochemical Journal, 250*(1), 1.

Moog, P. R., & Brüggemann, W. (1994). Iron reductase systems on the plant plasma membrane—A review. *Plant and Soil, 165*(2), 241–260.

Piazena, H., Perez-Rodrigues, E., Häder, D.-P., & Lopez-Figueroa, F. (2002). Penetration of solar radiation into the water column of the central subtropical Atlantic Ocean – optical properties and possible biological consequences. *Deep-Sea Research Part II, 49*, 3513–3528.

Rajneesh, A. C., Pathak, J., Ahmed, H., Singh, V., Singh, D. K., Pandey, A., Singh, S. P., Richa, D.-P. H., & Sinha, R. P. (2018). *Ultraviolet radiation-induced DNA damage and mechanisms of repair in cyanobacteria: An overview. Biotechnology in agriculture, industry and medicine. Trends in life science research. R. P. Sinha and U. P. Shrivastava* (pp. 169–218). New York: Nova Biomedical.

Rastogi, R., Singh, S., Incharoensakdi, A., Häder, D.-P., & Sinha, R. (2014). Ultraviolet radiation-induced generation of reactive oxygen species, DNA damage and induction of UV-absorbing compounds in the cyanobacterium *Rivularia* sp. HKAR-4. *South African Journal of Botany, 90*, 163–169.

Riebesell, U., & Tortell, P. D. (2011). *Effects of ocean acidification on pelagic organisms and ecosystems. Ocean acidification. J. P. Gattuso and L. Hansson* (pp. 99–116). Oxford. Oxford University Press.

Riebesell, U., Schulz, K. G., Bellerby, R., Botros, M., Fritsche, P., Meyerhöfer, M., Neill, C., Nondal, G., Oschlies, A., & Wohlers, J. (2007). Enhanced biological carbon consumption in a high CO_2 ocean. *Nature, 450*(7169), 545.

Sabine, C. L., Feely, R. A., Gruber, N., Key, R. M., Lee, K., Bullister, J. L., Wanninkhof, R., Won, C. S., Wallace, D. W. R., Tilbrook, B., Millero, F. J., Peng, T.-H., Kozyr, A., Ono, T., & Rios, A. F. (2004). The oceanic sink for anthropogenic CO_2. *Science, 305*, 367–371.

Shi, D., Hong, H., Su, X., Liao, L., Chang, S., & Lin, W. (2019). The physiological response of marine diatoms to ocean acidification: Differential roles of seawater pCO_2 and pH. *Journal of Phycology, 55*, 521–533.

Smith, F. A., & Raven, J. A. (1979). Intracellular pH and its regulation. *Annual Review of Plant Physiology, 30*, 289–311.

Suffrian, K., Schulz, K. G., Gutowska, M., Riebesell, U., & Bleich, M. (2011). Cellular pH measurements in *Emiliania huxleyi* reveal pronounced membrane proton permeability. *New Phytologist, 190*(3), 595–608.

Tong, S., Gao, K., & Hutchins, D. A. (2018). Adaptive evolution in the coccolithophore Gephyrocapsa oceanica following 1,000 generations of selection under elevated CO_2. *Global Change Biology, 24*(7), 3055–3064.

Tortell, P. D., Rau, G. H., & Morel, F. M. (2000). Inorganic carbon acquisition in coastal Pacific phytoplankton communities. *Limnology and Oceanography, 45*(7), 1485–1500.

Walworth, N. G., Fu, F.-X., Webb, E. A., Saito, M. A., Moran, D., McIlvin, M. R., Lee, M. D., & Hutchins, D. A. (2016). Mechanisms of increased *Trichodesmium* fitness under iron and phosphorus co-limitation in the present and future ocean. *Nature Communications, 7*, 12081.

Wu, Y., Gao, K., & Riebesell, U. (2010). CO_2-induced seawater acidification affects physiological performance of the marine diatom *Phaeodactylum tricornutum*. *Biogeosciences, 7*, 2915–2923.

Wu, X., Gao, G., Giordano, M., & Gao, K. (2012). Growth and photosynthesis of a diatom grown under elevated CO_2 in the presence of solar UV radiation. *Fundamental and Applied Limnology/Archiv für Hydrobiologie, 180*(4), 279–290.

Wu, Y., Campbell, D. A., Irwin, A. J., Suggett, D. J., & Finkel, Z. V. (2014). Ocean acidification enhances the growth rate of larger diatoms. *Limnology and Oceanography, 59*(3), 1027–1034.

Xu, Z., & Gao, K. (2012). NH_4^+ enrichment and UV radiation interact to affect the photosynthesis and nitrogen uptake of *Gracilaria lemaneiformis* (Rhodophyta). *Marine Pollution Bulletin, 64*(1), 99–105.

Xu, K., & Gao, K. (2015). Solar UV irradiances modulate effects of ocean acidification on the Coccolithophorid *Emiliania huxleyi*. *Photochemistry and Photobiology, 91*(1), 92–101.

Acquisition of Inorganic Carbon by Microalgae and Cyanobacteria

John Beardall and John A. Raven

Abstract Ribulose-1,5-bisphosphate carboxylase/oxygenase (Rubisco) and the Calvin cycle are the dominant features of inorganic carbon assimilation in all cyanobacteria and microalgae. Rubisco carboxylase shows a relatively low affinity for CO_2 and also has an oxygenase activity. These features can lead to inefficiencies in carbon assimilation, involving the process of photorespiration. However, cyanobacteria and algae possess mechanisms that minimise the effects of unfavourable Rubisco kinetics and photorespiration. These involve evolution of Rubiscos with kinetics that are more favourable to carboxylase activity and/or the presence of mechanisms that increase the concentration of CO_2 at the active site of Rubisco (CO_2 concentrating mechanisms, CCMs). CCMs are mostly based on active transport of HCO_3. In one species of marine diatom, there appears to be a biochemical CCM where single cells show a C_3–C_4 intermediate form of C assimilation. In cyanobacteria HCO_3^- accumulation involves active transport of HCO_3^- at the plasmalemma and/or downhill CO_2 entry with energised conversion of CO_2 to HCO_3^- at the thylakoid membrane, with CO_2 accumulated within carboxysomes that contain all of the cellular Rubisco as well as carbonic anhydrase. In eukaryotic microalgae active HCO_3^- transport occurs at either the plasmalemma or chloroplast envelope or both. In those algae that possess them, CO_2 is ultimately concentrated within Rubisco-containing pyrenoids, though it is important to note that the presence of pyrenoids is not an absolute requirement for CCMs. Algae and cyanobacteria can assimilate inorganic carbon in the dark, a process reflecting the need of cells to replenish the supply of C_4 intermediates of the TCA cycle as they are removed for

J. Beardall (✉)
School of Biological Sciences, Monash University, Clayton, VIC, Australia

State Key Laboratory of Marine Environmental Science & College of Ocean and Earth Sciences, Xiamen University, Xiamen, China
e-mail: john.beardall@monash.edu

J. A. Raven
Division of Plant Sciences, University of Dundee at the James Hutton Institute, Invergowrie, Dundee, UK

Climate Change Cluster, University of Technology Sydney, Ultimo, NSW, Australia

School of Biological Sciences, University of Western Australia, Crawley, WA, Australia

Q. Wang (ed.), *Microbial Photosynthesis*,
https://doi.org/10.1007/978-981-15-3110-1_8

biosynthesis. A few can also assimilate organic carbon sources, including in some cases by phagotrophy.

Keywords Algae · CO_2 concentrating mechanisms · Cyanobacteria · Inorganic carbon · Osmomixotrophy · Photorespiration · Rubisco

1 Introduction

Aquatic photoautotrophs are responsible for approximately 50% of the 111–117 Pg C assimilated annually into organic matter (Behrenfeld et al. 2001; Falkowski and Raven 2007). Of this, the majority of inorganic carbon is fixed by the microalgae that comprise the phytoplankton in the open ocean (mostly because there is a lot of ocean), though locally microalgal primary productivity on an areal basis can be high in a range of freshwater water bodies as well as in benthic populations and in biofilms on submerged rocks or plants.

2 Rubisco and the Calvin Cycle Are Central Features of C Acquisition in All Cyanobacteria and Microalgae

In common with terrestrial plants, all cyanobacteria and eukaryotic microalgae ultimately depend on the enzyme ribulose-1,5-bisphosphate carboxylase/oxygenase (Rubisco) and the photosynthetic carbon reduction cycle (PCRC; otherwise known as the Calvin-Benson-Bassham cycle) for net assimilation of inorganic carbon to organic matter. Rubisco is thus the most important carboxylating enzyme on the planet, and it plays a central role in over 99% of primary production (Raven 2009; Beardall and Raven 2016).

Rubisco catalyses the assimilation of CO_2, via a 6C carboxyketone intermediate, into 3-phosphoglycerate (Eq. 1):

$$CO_2 + H_2O + \text{D-ribulose-1,5-bisphosphate} \rightarrow 2 \times \text{3-phosphoglycerate} \quad (1)$$

The 3-phosphoglycerate then undergoes a series of reactions, leading to the net production of one molecule of triose phosphate for every 3 CO_2 assimilated, in the PCRC and the regeneration of one ribulose-1,5-bisphosphate. Each turn of the PCRC uses two NADPH plus three ATP per CO_2 assimilated, involving at least 9 mol photons absorbed per mol CO_2 assimilated (Raven et al. 2014) although the precise energetic costs involved in net incorporation of CO_2 into carbohydrate usually exceed this.

Six possible pathways for the assimilation of CO_2 into organic matter have been identified in autotrophs, including those (C_4 photosynthesis among them) relying on the activity of Rubisco in the PCRC (Fuchs 2011). However, of those pathways found in nature, the only sequence of reactions that is not inhibited by oxygen, and exhibits carboxylase activity with an ecologically relevant affinity for CO_2, is the 3-hydroxypropionate bicycle found in the phototrophic green non-sulphur bacterium *Chloroflexus aurantiacus* and its close relatives in the Chloroflexi (Bar-Even et al. 2010, 2011, 2012; Fuchs 2011; Raven et al. 2012; Raven and Beardall 2016a, b). Of the other pathways, the reductive citric acid cycle (Fuchs 2011) is found in the non-oxygenic phototrophic bacterium *Chlorobium* and a number of non-photosynthetic anaerobic Eubacteria, while none of the other pathways are found in phototrophs (Fuchs 2011). Although there are reports of C_3–C_4 single-cell intermediate photosynthesis in one species of a marine diatom, *Thalassiosira weissflogii* (Reinfelder et al. 2000; Morel et al. 2002; Roberts et al. 2007a, b; Reinfelder 2011; Haimovich-Dayan et al. 2013), this remains the sole microalgal species showing such a pathway, and the remaining alternative pathways for C assimilation described by Fuchs (2011) and Raven et al. (2012) are not represented in cyanobacterial or algal photosynthetic C assimilation. The PCRC is thus the dominant pathway for carbon assimilation in cyanobacteria and the algae.

The basic reactions of the PCRC are similar across the cyanobacteria and algae so far examined, although it has recently been suggested that the chemolithotrophic bacterium *Thermodesulfobium acidiphilum* shows a variant on the PCRC that involves a transaldolase reaction instead of the usual reactions involving sedoheptulose-1,7-bisphosphate aldolase and sedoheptulose-1,7-bisphosphatase (Frolov et al., 2019). However, even in in those algae and cyanobacteria that exhibit a normal set of PCRC reactions, it is becoming apparent that that there are major phylogenetic differences in the way the cycle is regulated (Jensen et al. 2017; Beardall and Raven 2019). Thus, in green algae, the two key enzymes of the PCRC enzymes, phosphoribulokinase (PRK) and glyceraldehyde-3-phosphate dehydrogenase (GAPDH), are subject to redox control, but PRK is not redox-regulated in the marine centric diatom *Odontella sinensis* (Michels et al. 2005) or the freshwater pennate diatom *Asterionella formosa* (Boggetto et al. 2007). In contrast *A. formosa* shows redox- activation of NADPH-dependent GAPDH (Boggetto et al. 2007), although this is lacking in *O. sinensis* GAPDH (Michels et al. 2005). Maberly et al. (2010) investigated the redox regulation of PRK and GAPDH in more detail, including the role of the protein CP12, and have been able to show considerable variation across different algal groups, with the cryptophytes and haptophytes examined in their study showing different regulatory properties to another clade containing chromalveolates and a third with a mix of vascular plants, a diatom, a xanthophycean and an eustigmatophycean.

3 Rubisco Also Has an Oxygenase Activity Which Leads to Inefficiencies in C Assimilation

In addition to its primary role as a carboxylase, Rubisco also catalyses the oxygenation of RuBP (Eq. 2), giving rise to one molecule of phosphoglycerate and one of phosphoglycolate, instead of two molecules of phosphoglycerate:

$$O_2 + \text{D-ribulose-1,5-bisphosphate} \rightarrow \text{3-phosphoglycerate} + \text{2-phosphoglycolate} \quad (2)$$

The carboxylase (Eq. 1) and oxygenase (Eq. 2) activities of Rubisco are competitive and dependent on the ratio of oxygen and CO_2 at the active site of the enzyme, according to Eq. 3, where the selectivity factor S_{rel} defines the ratio of rates of carboxylase to oxygenase reactions:

$$S_{rel} = \frac{K_{0.5}(O_2) \times k_{cat}(CO_2)}{K_{0.5}(CO_2) \times k_{cat}(O_2)} \quad (3)$$

where $k_{cat}(CO_2)$ is the CO_2-saturated specific rate of carboxylase activity of Rubisco (mol CO_2 mol^{-1} active sites s^{-1}), $K_{0.5}(CO_2)$ is the concentration of CO_2 at which the CO_2 fixation rate by Rubisco is half of $k_{cat}(CO_2)$, $k_{cat}(O_2)$ is the O_2-saturated specific rate of oxygenase activity of Rubisco (mol O_2 mol^{-1} active sites s^{-1}) and $K_{0.5}(O_2)$ is the concentration of O_2 at which the O_2 fixation rate by Rubisco is half of $k_{cat}(O_2)$.

Given the kinetic characteristics of cyanobacterial and microalgal Rubisco (discussed below), it is clear that, for autotrophs dependent on diffusive CO_2 entry, the physiology of CO_2 assimilation possesses inherent inefficiencies, such as significant inhibition of CO_2 fixation by oxygen; high CO_2 compensation points; and low affinities for external CO_2. The oxygenase reaction can thus represent a significant source of inefficiency in inorganic carbon acquisition and conversion into organic matter.

The phosphoglycolate formed by the Rubisco oxygenase reaction can be acted upon by 2-phosphoglycolate phosphatase, leading to formation of glycolate. Three of the four carbons in two molecules of glycolate can be recovered and converted to phosphoglycerate by the reactions of the photosynthetic carbon oxidation cycle (PCOC) in the process of photorespiration (Beardall et al. 2003; Raven and Beardall 2005; Raven et al. 2019). However, in this process the other carbon is lost as CO_2 and represents a potentially significant inefficiency in the carbon assimilatory process. In some cases glycolate is excreted, which represents a major loss to the organisms of the C and energy involved in its biosynthesis (Raven et al. 2011; Raven and Beardall 2016a, b). Eisenhut et al. (2008) showed that the cyanobacterium *Synechocystis* has two alternative pathways of glycolate metabolism in addition to the PCOC; in addition to the 'conventional' PCOC involving glycolate dehydrogenase (which in green alga is replaced by glycolate oxidase) and regeneration of glycerate via glyoxylate, glycine, serine and hydroxypyruvate, this cyanobacterium can convert glyoxylate to glycerate via tartronate semialdehyde or, alternatively, via oxalate and formate to CO_2. Deletion of all three of the pathways was lethal, so in this organism glycolate excretion alone is not an adequate sink for glycolate and some form of photorespiratory pathway is obligatory.

4 Cyanobacteria and Microalgae Possess Mechanisms That Minimise the Effects of Unfavourable Rubisco Kinetics and Photorespiration

4.1 Evolution of Rubiscos More Favourable to the Carboxylase Activity

Cyanobacteria and eukaryotic microalgae contain a range of different forms of Rubisco, which show different kinetic properties and selectivity towards oxygen and carbon dioxide. The general trend across all autotrophs is that a low $K_{0.5}(CO_2)$ and a high S_{rel} are correlated with a low $k_{cat}(CO_2)$ and vice versa (Tcherkez et al. 2006; Raven et al. 2012; Bathellier et al. 2018).

The Form IBc Rubiscos, found in most marine and freshwater β-cyanobacteria, have high $K_{0.5}(CO_2)$ values of 105–290 μM (but with most reports giving values in the range of 200–260 μM). S_{rel} of these Rubiscos varies from 38 to 56 mol mol^{-1}, and $k_{cat}(CO_2)$ values range from 2.6 to 11.4 mol CO_2 mol^{-1} active sites s^{-1}. The highest known $K_{0.5}(CO_2)$ of a Form I Rubisco (750 μM) is the Form IAc Rubisco from the marine α-cyanobacterium *Prochlorococcus* MIT9313. This is combined with a moderate $k_{cat}(CO_2)$ of 4.7 mol CO_2 mol^{-1} active sites s^{-1} (Scott et al. 2007). Thus cyanobacterial Rubiscos have relatively low affinities for CO_2 and would perform poorly under present-day CO_2 levels, with air-equilibrium CO_2 concentrations in the region of 10–25 M, depending on factors such as temperature and salinity, if they took up CO_2 by diffusion.

Green algae have, in contrast, Rubiscos with a higher affinity for CO_2, and $K_{0.5}(CO_2)$ values for their Form 1B Rubisco range from 29 to 38 M, with higher S_{rel} values of 61–83 mol mol^{-1}. Values for k_{cat} are, however, lower (Raven and Beardall 2003; Beardall and Raven 2019). Heterokont and haptophyte algae contain Form 1D Rubiscos which have quite variable kinetics, both across and within species. Thus values of $K_{0.5}(CO_2)$ for partially purified Rubisco of the coccolithophore *Emiliania huxleyi* have been reported as 72 μM (Boller et al. 2011) or 200 μM (Shiraiwa et al. 2004). The kinetics of Form ID Rubiscos of diatoms are also variable (Young et al. 2016) with $K_{0.5}(CO_2)$ values from 23 to 68 M, CO_2 selectivity from 57 to 116 mol mol^{-1} and specific reaction rates from 2.1 to 3.7 mol CO_2 mol^{-1} active sites s^{-1}. In the Synurophyceae, relatively high-affinity Rubiscos with $K_{0.5}(CO_2)$ values in vitro of 18.2 μM (*Mallomonas papulosa*), 28.4 M (*Synura petersenii*) and 41.8 M (*Synura uvella*) have been reported (Bhatti and Colman 2008; Raven and Giordano 2017).

Dinoflagellates are unusual as being the only eukaryotes that possess Form II Rubiscos. These enzymes are unstable in vitro and are consequently very poorly characterised. The limited data available suggest, however, that they have very low selectivity factors (with values ~37 mol mol^{-1}). Form II Rubiscos are also found in photosynthetic *Proteobacteria*, from which it is believed the dinoflagellate enzymes originated by lateral gene transfer (Badger et al. 1998; Whitney et al. 1995). These also have very low S_{rel} values (Whitney and Andrews 1998; Leggat et al. 1999; Raven and Beardall 2003), and this suggests that it would be difficult for dinoflagellates to carry out net CO_2 assimilation at air-equilibrium CO_2 levels (Tortell 2000). These considerations apply to basal dinoflagellates; tertiary endosymbiosis has replaced the plastids containing Form II Rubisco with Form IB or Form ID Rubiscos from chlorophyte, cryptophyte, haptophyte or ochrophyte algae (Raven, Suggett and Giordano, in preparation).

Another form of Rubisco, Form III, has been reported in Archaea but to date has been shown to be involved only in metabolism of ribonucleotides and ribonucleosides via the pentose phosphate pathway rather than in CO_2 fixation (Aono et al. 2015). However, Form III Rubisco has recently been shown to participate in CO_2 assimilation via the PCRC in the chemolithotrophic bacterium *Thermodesulfobium acidiphilum* (Frolov et al. 2019), though its kinetics in relation to CO_2 fixation have not yet been characterised.

4.2 *CO_2 Concentrating Mechanisms Increase CO_2:O_2 at the Rubisco Active Site*

Given the relatively low affinities and selectivity factors of Rubiscos from most of the algae and cyanobacteria as discussed above, achievement of significant rates of net photosynthesis necessitates the operation of a CO_2 concentrating mechanisms (CCMs) to elevate CO_2 concentrations and increase CO_2:O_2 ratios at the active site of Rubisco. These mechanisms can take various forms.

4.2.1 Biochemical CCMs

Early work in the late 1970s by Beardall and co-workers proposed the existence of a single-cell, C_4-like, photosynthetic metabolism in diatoms (Beardall et al. 1976) involving an initial fixation of inorganic carbon by the enzyme PEP carboxylase and subsequent decarboxylation of its reaction products (C_4 acids) to concentrate CO_2 at the active site of Rubisco. Later studies (Morris et al. 1978) instead ascribed their labelling patterns and other data to high rates of anaplerotic β-carboxylation through PEPCase, which are necessary to top up the intermediates of the TCA cycle used to support biosynthetic processes such as protein synthesis (Aubry et al. 2011; Chi et al. 2014). Reinfelder et al. (2000) and Morel et al. (2002) revived the notion of C_4

photosynthesis in diatoms, but subsequent work (Reinfelder et al. 2004; Roberts et al. 2007 a, b) has led to the view that there are only two instances in algae of biochemical CCMs based on an initial fixation of inorganic carbon into a 4C acid (C_4 photosynthesis). One is the acellular marine green macroalga *Udotea*, which possesses a single-cell C_4 carbon assimilation pathway (Reiskind et al. 1988; Giordano et al. 2005; Raven et al. 2008), while the only verified instance of C_4 photosynthesis in a microalga is the diatom, *Thalassiosira weissflogii*, which shows C_3–C_4 intermediate C assimilation (Roberts et al. 2007 a, b).

4.2.2 Biophysical CCMs

CO_2 can diffuse across the plasma membrane of microalgae and cyanobacteria, probably with at least some CO_2 uptake effected by aquaporins (Kaldenhoff et al. 2014; Raven and Beardall 2016a, b). However, C acquisition and CO_2 accumulation at the Rubisco active site are driven by biophysical CCMs based on the active transport of bicarbonate or (indirectly) CO_2. These are best characterised in cyanobacteria, in which a number of variants of inorganic carbon transport have been identified. Each of these transport systems has distinct physiological characteristics:

1. *Energised delivery of CO_2*: Either on the cytosolic face of the plasma membrane or on the thylakoid membrane of cyanobacteria, there is an energised conversion of CO_2 to HCO_3^- via a NAD(P)H dehydrogenase, which effectively acts like a Ci-pump, even though direct active transport of CO_2 has not occurred. There are two NAD(P)H dehydrogenase systems: an inducible, high-CO_2 affinity system (NDH-I_3) at the thylakoid membrane and a constitutive, low-affinity one (NDH-I_4, located probably at the plasma membrane) (Price et al. 2008).
2. *Active HCO_3^- transport*: In freshwater β-cyanobacteria, there is a high-affinity low-CO_2-inducible HCO_3^- transporter, BCT1, the genes for which are encoded by the *cmpABCD* operon belonging to the traffic ATPase family (Omata et al. 1999). These genes are absent from the genomes of all marine α- and β-cyanobacteria so far sequenced. In *Synechocystis* 6803 and various other β-cyanobacteria, a high-affinity, inducible, Na^+-dependent, HCO_3^- transporter (SbtA) is present (Shibata et al. 2002). BicA is another Na^+-dependent HCO_3^- transporter, and this is aligned with the SulP family of transporters (Price et al. 2004). Unlike SbtA, this is a low-affinity system. BicA and SbtA may both be forms of Na^+/HCO_3^- symporters, although to date there is no conclusive evidence for this. Whatever the form of inorganic carbon transported, the outcome is HCO_3^- delivery, directly or indirectly, to the cytosol. The HCO_3^- then diffuses into the polyhedral protein-walled bodies termed carboxysomes (Raven and Beardall 2016a, b), which contain all of the Rubisco and show the only carbonic anhydrase (CA) activity in the cell. CA activity within the carboxysomes leads to the buildup to a higher steady-state concentration of CO_2 than in the bulk medium, thereby favouring the carboxylase over the oxygenase activity of Rubisco (Smith and Ferry 2000; Price et al. 2002). Expression of human

carbonic anhydrase in the cytosol of *Synechocystis* PCC7942 creates a high CO_2-requiring phenotype as a result of CO_2 leakage to the medium (Price and Badger 1989).

Biophysical CCMs in eukaryotic microalgae are not as well defined as they are in cyanobacteria and are more complicated because of the additional membranes the DIC needs to traverse. Inorganic carbon needs to be transported across the plasmalemma and then across the membranes of the chloroplast envelope, and CO_2 then needs to be provided at a higher than ambient concentration to the active site of Rubisco, which is within the pyrenoids in those cells that possess them and in the stroma in cells without pyrenoids. It is worth noting that all pyrenoid-containing algae have CCMs (Badger et al. 1998; Raven 1997a, b; Raven and Beardall 2003), but not all algae with CCMs have pyrenoids (Badger et al. 1998; Morita et al. 1998, 1999; Raven 1997b, c; Raven and Beardall 2003; Kevekordes et al. 2006; Raven and Giordano 2017).

Pyrenoids, a microcompartment within the chloroplast, are found in many eukaryotic algae and are analogous structures to the carboxysomes of cyanobacteria. Pyrenoids, where present, contain the majority of the cellular Rubisco, with much smaller amounts in the stroma (McKay and Gibbs 1991). In species lacking pyrenoids, Rubisco is found entirely in the stroma (McKay and Gibbs 1991). The structure, development and composition of pyrenoids in *Chlamydomonas reinhardtii* have been reviewed recently by Meyer et al. (2017), Meyer and Griffiths (2013), Mackinder (2018) and Mackinder et al. (2017) and so are not covered here in detail. In brief, the pyrenoid is commonly surrounded by a starch sheath (although this is not essential for the operation of the CCM [Villarejo et al. 1996], and the starch sheath is limited to algae that store starch) and, critical for CCM activity, is commonly, but not always (Dodge 1973; Badger et al. 1998), traversed by membrane tubules, sometimes termed pyrenoid lamellae, or transpyrenoid thylakoids, which are contiguous with the photosynthetic thylakoid membranes (Engel et al. 2015).

Active transport mechanisms for inorganic carbon could thus be based on the plasma membrane or the inner plastid envelope membrane; the additional one or two membranes round the plastids of algae other than the Chlorophyta and Charophyceae, the pyrenoid tubules or all of these.

Physiological investigations have shown that microalgae can take up CO_2 from the surrounding medium and this, in the absence of hard evidence for an active CO_2 transporter, is assumed to take place by passive diffusion (Patel and Merrett 1986; Colman and Rotatore 1995; Mitchell and Beardall 1996; Johnston and Raven 1996; Korb et al. 1997; Burkhardt et al. 2001; Rost et al. 2003; Trimborn et al. 2008; Kaldenhoff et al. 2014; Raven and Beardall 2016a, b). Diffusion of CO_2 is possibly assisted by aquaporin channels, though a high permeability of cell membranes to CO_2 can create problems associated with leakage (Tchernov et al. 2003; Raven and Beardall 2016a, b). In most cases, therefore, active transport of inorganic carbon and CCMs in microalgae is associated with energised bicarbonate transport.

Although there are some species in which CCM activity appears to be based solely at the plasmalemma (Rotatore and Colman 1990, 1991), in a range of other species, there is also evidence for a role of the plastid envelope. This is largely based on a demonstrable capacity of isolated chloroplasts for active transport of DIC. Thus, photosynthetically active chloroplasts from high- and low-CO_2 grown cells of two species of the Chlorophyceae have been shown to possess low- and high-affinity DIC uptake systems, respectively, as do the corresponding intact cells (Amoroso et al. 1998). The uptake of both CO_2 and HCO_3^-, and CO_2 accumulation, has been demonstrated in isolated chloroplasts of *Chlamydomonas reinhardtii*, *Dunaliella tertiolecta*, *Tetraedron minimum* and *Chlamydomonas noctigama* (Amoroso et al. 1998; van Hunnik et al. 2002). Gee and Niyogi (2017) suggest, on the basis of the location of a carbonic anhydrase essential for photosynthesis, HCO_3^- transport into the epiplastidic endoplasmic reticulum in the eustigmatophycean *Nannochloropsis oceanica*.

Recent developments in molecular biology have allowed us to begin to identify and characterise the various bicarbonate transporters in microalgal cells. Diatom genomes contain sequences encoding a number of solute carrier (SLC)-type transporters, analogous to those found in mammalian systems (Bonar and Casey 2008). A number of these have been implicated in bicarbonate transport. For example, in *Phaeodactylum tricornutum*, the plasmalemma-associated PtSLC4-2 is low-CO_2-inducible and has a high requirement for Na^+ (Nakajima et al. 2013). Other transporters (PtSLC4-1 and PtSLC4-4) are also plasmalemma located in *P. tricornutum* and likewise appear to be involved with HCO_3^- influx from low-CO_2 environments. Nakajima et al. (2013) also showed the existence of orthologous SLC4 genes in another diatom species, *Thalassiosira pseudonana*. Recently Tsuji et al. (2017a) have suggested that in these diatoms plasmalemma HCO_3^- transport is driven, directly or indirectly, by energy generated by linear electron flow from photosystem II to photosystem I. This is in contrast with most previous work showing that CCMs were energised by ATP from cyclic photophosphorylation (Ogawa and Ogren 1985; Ogawa et al. 1985; Palmqvist et al. 1990; Spalding et al. 1984), though some eustigmatophycean algae appear unusual in having a CCM driven by respiratory ATP (Huertas et al. 2002). Plasmalemma-based bicarbonate transporters have also been demonstrated in the green alga *Chlamydomonas reinhardtii* (Ohnishi et al. 2010; Yamano et al. 2015), though the genes encoding these transporters (HLA3 and LCI1) do not appear to be related to the SLC systems in diatoms (Tsuji et al. 2017a, b).

Molecular studies have also identified transporters associated with the chloroplast envelope of microalgae. Thus Nakajima et al. (2013) and Tsuji et al. (2017b) demonstrated the presence of genes for SLC4-type transporters associated with the chloroplast envelope membranes in diatoms, and a number of putative transporters (LCIA, CCP1 and CCP2) have been suggested to be present in the chloroplast envelope of *Chlamydomonas* (Wang et al. 2015; Yamano et al. 2015; Machingura et al. 2017). However, more recent work by Mackinder et al. (2017) suggests that CCP1 and CCP2 may be less important. Diatoms possess four chloroplast envelope membranes, and Matsuda et al. (2017) have postulated the existence of transporters at each of these membranes and on the pyrenoid-penetrating thylakoids, though such transport systems remain uncharacterised and remain speculative (see also Gee and Niyogi 2017).

Within the compartments of cells of cyanobacteria and algae, inorganic carbon species involved in photosynthesis exist in equilibrium as a result of CA activity, aided in the thylakoid lumen by low pH (e.g. Price and Badger 1989; Raven 1997a, b, c; Hopkinson et al. 2011; Mackinder et al. 2017). Operation of CCMs requires the presence of a range of CAs to maintain the equilibrium between CO_2 and bicarbonate in the various cellular compartments (see Di Mario et al. 2017 for a recent review). In many algae utilisation of bicarbonate from the surrounding medium involves an extracellular carbonic anhydrase (CA_{ext}) associated with the cell wall. This speeds up the conversion of bicarbonate to CO_2 at the cell surface, thereby assisting the diffusion of the latter into the cell. Within the cell, internal CAs facilitate the interconversion of bicarbonate and CO_2 with active inorganic transport across the chloroplast envelope then occurring as described above. There are a range of CAs involved, depending on phylogeny; in green algae such as *Chlamydomonas*, the external CA is an α-CA, while in some diatoms, a β-CA or ζ-CA carries out this role (Hopkinson et al. 2013). In dinoflagellates yet another form of external CA (δ-CA_{ext}) has been reported (Lapointe et al. 2008). CA_{ext} activity is inducible by low CO_2 levels and in some cases is only expressed when CO_2 demand from C fixation exceeds the rate of uncatalysed supply from bicarbonate (Smith-Harding et al. 2017), which may explain in part at least the contradictory reports of external CA presence/absence in some species (John-McKay and Colman 1997).

The pyrenoid tubules described above contain α-CA (Cah3), which convert bicarbonate to CO_2 within the lumen of the pyrenoid tubules, the latter then diffusing from the lumen to the Rubisco in the bulk of the pyrenoid. A similar protein, the θ-CA Pt43233, is found in the diatom *Phaeodactylum tricornutum* (Kikutani et al. 2016), suggesting this mechanism could be a widespread mechanism in pyrenoid-containing algae. This mechanism would also require bicarbonate transport into the transpyrenoid thylakoid lumen, driven by the proton-motive force across the thylakoid (Raven 1997a, b, c), but no such transporter has yet been identified, but see Mukherjee et al. (2019). Additional subclasses of carbonic anhydrase in microalgae continue to be discovered (Jensen et al. 2019), but their roles await investigation.

The consequence of CCM operation, irrespective of the mechanism involved, is that photosynthesis and inorganic carbon acquisition are, for most species of cyanobacteria and eukaryotic microalgae, not limited by DIC concentrations under present-day atmospheric CO_2 levels.

4.2.3 The Extent of CCM Activity

The kinetics of Rubisco in most cyanobacteria and algae operating at, or below, air-equilibrium levels of CO_2 require operation of a CO_2 concentrating mechanism (CCM) to improve the supply of CO_2 to the active site, minimise photorespiration and improve net rates of carbon assimilation. There are, however, some exceptions, notably in the Chrysophyceae and Synurophyceae (Maberly et al. 2009; Raven and Giordano 2017), freshwater red macroalgae belonging to the Batrachospermales (Raven et al. 1982, 2005) as well as some marine algae where light levels are low enough to low constrain photosynthesis to the point where CO_2 diffusion satisfies demand by photosynthesis (Kübler and Raven 1994, 1995). Other examples of species lacking CCMs are the coccoid symbiotic trebouxiophycean alga *Coccomyxa* (Raven and Colmer 2016) (though not in the Antarctic species *Coccomyxa subellipsoidea* [Blanc et al. 2012]) and the terrestrial trebouxiophycean species *Stichococcus minor* (Munoz and Merrett 1989). The coccolithophore *Emiliania huxleyi* was originally thought to not have CCM activity, but this is no longer believed to be the case (Rost et al. 2007; Reinfelder 2011; Stojkovic et al. 2013).

Even when present, the expression of CCM activity varies greatly across taxa. Cyanobacteria with low CO_2 affinity Form IA or Form IB Rubisco have highly expressed CCMs, while diatoms and green algae with higher CO_2 affinity show a relatively low capacity for CCM expression. There is a broad inverse relationship between carbon concentration factor and Rubisco specificity factor and a positive relationship between specificity factor and CO_2 levels in the geological past when the given taxa evolved (Tortell 2000; Griffiths et al. 2017). This is complicated however by the fact that environmental factors, including CO_2 levels and light and nutrient availability, can also influence the activity of CCMs (see reviews by Beardall and Giordano 2002; Giordano et al. 2005; Raven et al. 2011; Raven and Beardall 2014).

4.3 Heterotrophic Carbon Assimilation

While not strictly associated with photosynthetic carbon assimilation, it is worth considering the extent to which cyanobacteria and algae can supplement photosynthesis with the use of organic carbon sources. Not all algae are obligate photolithotrophs, i.e. organisms that cannot use exogenous organic carbon for growth, even if organic carbon sources can be taken up by the cells. Many algae, and some cyanobacteria, that are capable of photosynthetic incorporation of inorganic carbon into organic matter are also able to use dissolved organic carbon sources from the environment and, in some cases, use this to stimulate growth in the process of osmomixotrophy (see Beardall and Raven 2016 and references therein). Fewer can grow as heterotrophs in the dark using organic carbon as both energy and carbon source. Some algae however are capable of phagotrophy, with or without also being photosynthetic, the latter type exhibiting phagomixotrophy (Raven et al. 2009; Flynn et al. 2013; Beardall and Raven 2016).

4.3.1 Dark Carbon Fixation

Phototrophs can show assimilation of inorganic carbon in the dark, though rates in the dark are usually much lower than in the light (Morris et al. 1971). This dark fixation reflects the need of cells to replenish the supply of C_4 intermediates of the TCA cycle as they are removed for biosynthesis, particularly of amino acids for protein synthesis. It usually takes the form of anaplerotic fixation of HCO_3^- to produce oxaloacetate using either phosphoenolpyruvate or pyruvate as the acceptor molecule in a $C_3 + C_1$ carboxylation catalysed by phosphoenolpyruvate carboxylase or pyruvate carboxylase, respectively, in what are termed β-carboxylase reactions (Falkowski and Raven 2007). High levels of these enzymes have in the past been misinterpreted as evidence for C_4 photosynthesis (Beardall et al. 1976; Tsuji et al. 2009, 2012).

References

Amoroso, G., Sültemeyer, D. F., Thyssen, C., & Fock, H. P. (1998). Uptake of HCO_3^- and CO_2 in cells and chloroplasts from the microalgae *Chlamydomonas reinhardtii* and *Dunaliella tertiolecta*. *Plant Physiology, 116*, 193–201.

Aono, R., Sato, T., Imanaka, T., & Atomi, H. (2015). A pentose bisphosphate pathway for nucleoside degradation in Archaea. *Nature Chemical Biology, 11*, 355–360.

Aubry, S., Brown, N. J., & Hibberd, J. M. (2011). The role of proteins in C_3 plants prior to their recruitment into the C_4 pathway. *Journal of Experimental Botany, 62*, 3049–3059.

Badger, M. R., Andrews, T. J., Whitney, S. M., Ludwig, M., Yellowlees, D. C., Leggat, W., & Price, G. D. (1998). The diversity and coevolution of Rubiscos, plastids, pyrenoids and chloroplast-based CO_2-concentrating mechanisms in algae. *Canadian Journal of Botany, 76*, 1052–1071.

Bar-Even, A., Noor, E., Lewis, N. E., & Milo, R. (2010). Design and analysis of synthetic carbon fixation pathways. *Proceedings of the National Academy of Sciences of the United States of America, 107*, 8888–8894.

Bar-Even, A., Noor, E., Savir, Y., Liebermeister, W., Davidi, D., Tawfik, D. S., & Milo, R. (2011). The moderately efficient enzyme: Evolutionary and physicochemical trends shaping enzyme parameters. *Biochemistry, 50*, 4402–4404.

Bar-Even, A., Noor, E., & Milo, R. (2012). A survey of carbon fixation pathways through a quantitative lens. *Journal of Experimental Botany, 63*, 2325–2342.

Bathellier, C., Tcherkez, G., Lorimer, G. H., & Farquhar, G. D. (2018). Rubisco is not really so bad. *Plant, Cell & Environment, 41*, 705–716.

Beardall, J., & Giordano, M. (2002). Ecological implications of microalgal and cyanobacterial CCMs and their regulation. *Functional Plant Biology, 29*, 335–347.

Beardall, J., & Raven, J. A. (2016). Carbon acquisition by microalgae. In M. Borowitzka, J. Beardall, & J. A. Raven (Eds.), *The physiology of microalgae* (pp. 89–100). Cham: Springer.

Beardall, J., & Raven, J. A. (2019). Structural and biochemical features of carbon acquisition in algae. In A. W. D. Larkum, A. Grossman, & J. A. Raven (Eds.), *Photosynthesis in the algae* (Vol. 1, 2nd ed.). Dordrecht: Kluwer Academic Publishers. In press.

Beardall, J., Mukerji, D., Glover, H. E., & Morris, I. (1976). The path of carbon in photosynthesis by marine phytoplankton. *Journal of Phycology, 12*, 409–417.

Beardall, J., Quigg, A., & Raven, J. A. (2003). Oxygen consumption: Photorespiration and chlororespiration. In A. W. D. Larkum, S. E. Douglas, & J. A. Raven (Eds.), *Photosynthesis in algae* (pp. 157–181). Dordrecht: Kluwer.

Behrenfeld, M. J., Randerson, J. T., McClain, C. R., Feldman, G. C., Los, S. O., Tucker, C. J., Falkowski, P. G., Field, C. B., Frouin, R., Esaias, W. E., Kolber, D. D., & Pollack, N. H. (2001). Biospheric primary production during an ENSO transition. *Science, 291*, 2594–2597.

Bhatti, S., & Colman, B. (2008). Inorganic carbon acquisition in some synurophyte algae. *Physiologia Plantarum, 133*, 33–40.

Blanc, G., Agarkova, I., Grimwood, J., Kuo, A., Brueggeman, A., Dunigan, D. D., Gurnon, J., Ladunga, I., Lindquist, E., Lucas, S., Pangilinan, J., Pröschold, T., Salamov, A., Schmutz, J.,

Weeks, D., Yamada, T., Lomsadze, A., Borodovsky, M., Claverie, J. M., Grigoriev, I. V., & Van Etten, J. L. (2012). The genome of the polar eukaryotic microalga *Coccomyxa subellipsoidea* reveals traits of cold adaptation. *Genome Biology, 13*, R39. https://doi.org/10.1186/gb-2012-13-5-r39.

Boggetto, N., Gontero, B., & Maberly, S. C. (2007). Regulation of phosphoribulokinase and glyceraldehyde 3-phosphate dehydrogenase in a freshwater diatom, *Asterionella formosa*. *Journal of Phycology, 43*, 1227–1235.

Boller, A. J., Thomas, P. J., Cavenaugh, C. M., & Scott, K. M. (2011). Low stable isotope fractionation by coccolithophore RuBISCO. *Geochimica et Cosmochimica Acta, 75*, 7200–7207.

Bonar, P. T., & Casey, J. R. (2008). Plasma membrane Cl^-/HCO_3^- exchangers: Structure, mechanism and physiology. *Channels, 2*, 337–345.

Burkhardt, S., Amoroso, G., Riebesell, U., & Sültemeyer, D. (2001). CO_2 and HCO_3^- uptake in marine diatoms acclimated to different CO_2 concentrations. *Limnology and Oceanography, 46*, 1378–1391.

Chi, S., Wu, S., Yu, J., Wang, X., Tang, X., & Liu, T. (2014). Phylogeny of C_4-photosynthesis enzymes based on algal transcriptomic and genomic data supports an archaeal/proteobacterial origin and multiple duplication for most C_4-related genes. *PLoS One, 9*, e110154.

Colman, B., & Rotatore, C. (1995). Photosynthetic inorganic carbon uptake and accumulation in two marine diatoms. *Plant, Cell & Environment, 18*, 919–924.

Di Mario, R. J., Machingura, M. C., Waldrop, G. L., & Moroney, J. V. (2017). The many types of carbonic anhydrases in photosynthetic organisms. *Plant Science, 268*, 11–17.

Dodge, J. D. (1973). *The fine structure of algal cells*. London: Academic Press.

Eisenhut, M., Ruth, W., Haimovitch, M., Bauwe, M., Kaplan, A., & Hagemann, M. (2008). The photorespiratory glycolate metabolism is essential for cyanobacteria and may have been conveyed endosymbiotically to plants. *Proceedings of the National Academy of Sciences of the United States of America, 105*, 17199–17204.

Engel, B. D., Schaffer, M., Kuhn Cuellar, L., Villa, E., Plitzko, J. M., & Baumeister, W. (2015). Native architecture of the *Chlamydomonas* chloroplast revealed by in situ cryo-electron tomography. *eLife, 4*, e04889.

Falkowski, P. G., & Raven, J. A. (2007). *Aquatic photosynthesis* (2nd ed.). Princeton: Princeton University Press.

Flynn, K. J., Stoecker, D. K., Mitra, A., Raven, J. A., Glibert, P. M., Hansen, P. J., Granéli, E., & Burkholder, J. M. (2013). A case of mistaken identification: The importance of mixotrophs and the clarification of plankton functional-classification. *Journal of Plankton Research, 35*, 3–11. https://doi.org/10.1093/plankt/fbs062.

Frolov, E. N., Kublanov, I. V., Toshchakov, S. V., Lunev, E. A., Pimenov, N. V., Bonch-Osmolovskaya, E. A., Lebedinsky, A. V., & Chernyh, N. A. (2019). Form III RubisCO-mediated transaldolase variant of the Calvin cycle in a chemolithoautotrophic bacterium. *Proceedings of the National Academy of Sciences of the United States of America, 116*(37), 18638–18646. https://doi.org/10.1073/pnas.1904225116.

Fuchs, G. (2011). Alternative pathways of carbon dioxide fixation: Insights into the early evolution of life? *Annual Review of Microbiology, 65*, 631–658.

Gee, C. W., & Niyogi, K. K. (2017). The carbonic anhydrase CAH1 is an essential component of the carbon-concentrating mechanism of *Nannochloropsis oceanica*. *Proceedings of the National Academy of Sciences of the United States of America, 114*, 4537–4542.

Giordano, M., Beardall, J., & Raven, J. A. (2005). CO_2 concentrating mechanisms in algae: Mechanisms, environmental modulation, and evolution. *Annual Review of Plant Biology, 6*, 99–131.

Griffiths, H., Meyer, M. T., & Rickaby, R. E. M. (2017). Overcoming adversity through diversity: Aquatic carbon concentrating mechanisms. *Journal of Experimental Botany, 68*, 3689–3695.

Haimovich-Dayan, M., Garfinkel, N., Ewe, D., Marcus, Y., Gruber, A., Wagner, H., Kroth, P. G., & Kaplan, A. (2013). The role of C4 metabolism in the marine diatom *Phaeodactylum tricornutum*. *The New Phytologist, 197*, 177–185.

Hopkinson, B. M., Dupont, C. L., Allen, A. E., & Morel, M. M. (2011). Efficiency of the CO_2-concentrating mechanism in diatoms. *Proceedings of the National Academy of Sciences of the United States of America, 108*, 3830–3837.

Hopkinson, B. M., Meile, C., & Shen, C. (2013). Quantification of extracellular carbonic anhydrase in two marine diatoms and investigation of its role. *Plant Physiology, 162*, 1142–1152.

Huertas, I. E., Colman, B., & Espie, G. S. (2002). Inorganic carbon acquisition and its energization in eustigmatophyte algae. *Functional Plant Biology, 29*, 271–277.

Jensen, E., Clement, R., Maberly, S., & Gontero, B. (2017). Regulation of the Calvin–Benson–Bassham cycle in the enigmatic diatoms: Biochemical and evolutionary variations on an original theme. *Philosophical Transactions of the Royal Society of London B, 372*, 20160401.

Jensen, E., Clement, R., Kosta, A., Maberly, S. C., & Gontero, B. (2019). A new widespread subclass of carbonic anhydrase in marine phytoplankton. *The ISME Journal, 13*, 2094–2106.

John-McKay, M., & Colman, B. (1997). Variation in the occurrence of external carbonic anhydrase among strains of the marine diatom *Phaeodactylum tricornutum* (Bacillariophyceae). *Journal of Phycology, 33*, 988–990.

Johnston, A. M., & Raven, J. A. (1996). Inorganic carbon accumulation by the marine diatom *Phaeodactylum tricornutum. European Journal of Phycology, 31*, 285–290.

Kaldenhoff, R., Kai, L., & Uehlein, N. (2014). Aquaporins and membrane diffusion of CO_2 in living organisms. *Biochimica et Biophysica Acta, 1840*, 1592–1595.

Kevekordes, K., Holland, D., Jenkins, S., Koss, R., Roberts, S., Raven, J. A., Scrimgeour, C. M., Shelly, K., Stojkovic, S., & Beardall, J. (2006). Inorganic carbon acquisition by eight species of *Caulerpa. Phycologia, 45*, 442–449.

Kikutani, S., Nakajima, K., Nagasato, C., Tsuji, Y., Miyatake, A., & Matsuda, Y. (2016). Thylakoid luminal θ-carbonic anhydrase critical for growth and photosynthesis in the marine diatom *Phaeodactylum tricornutum. Proceedings of the National Academy of Sciences of the United States of America, 113*, 9828–9833.

Korb, R. E., Saville, P. J., Johnston, A. M., & Raven, J. A. (1997). Sources of inorganic carbon for photosynthesis by three species of marine diatoms. *Journal of Phycology, 33*, 433–440.

Kübler, J. E., & Raven, J. A. (1994). Consequences of light limitation for carbon acquisition in three rhodophytes. *Marine Ecology Progress Series, 110*, 203–209.

Kübler, J. E., & Raven, J. A. (1995). The interaction between inorganic carbon supply and light supply in *Palmaria palmata* (Rhodophyta). *Journal of Phycology, 31*, 369–375.

Lapointe, M., MacKenzie, T. D. B., & Morse, D. (2008). An external δ-carbonic anhydrase in a free-living marine dinoflagellate may circumvent diffusion-limited carbon acquisition. *Plant Physiology, 147*, 1427–1436.

Leggat, W., Badger, M. R., & Yellowlees, D. C. (1999). Evidence for an inorganic carbon-concentrating mechanism in the symbiotic dinoflagellate *Symbiodinium* sp. *Plant Physiology, 121*, 1247–1255.

Maberly, S. C., Ball, L. A., Raven, J. A., & Sültemeyer, D. (2009). Inorganic carbon acquisition by chrysophytes. *Journal of Phycology, 45*, 1057–1061.

Maberly, S. C., Courcelle, C., Grobden, R., & Gontero, B. (2010). Phylogenetically-based variation in the regulation of the Calvin cycle enzymes, phosphoribulokinase and glyceraldehyde-3-phosphate dehydrogenase, in algae. *Journal of Experimental Botany, 61*, 735–745.

Machingura, M. C., Bajsa-Hirschel, J., Laborde, S. M., Schwartzenburg, J. B., Mukherjee, B., Mukherjee, A., Pollock, S. V., Förster, B., Price, G. D., & Moroney, J. V. (2017). Identification and characterization of a solute carrier, CIA8, involved in inorganic carbon acclimation in *Chlamydomonas reinhardtii. Journal of Experimental Botany, 68*, 3879–3890.

Mackinder, L. C. M. (2018). The *Chlamydomonas* CO_2-concentrating mechanism and its potential for engineering photosynthesis in plants. *The New Phytologist, 217*, 54–61.

Mackinder, L., Chen, C., Leib, R., Patena, W., Blum, S. R., Rodman, M., Ramundo, S., Adams, C. M., & Jonikas, M. C. (2017). A spatial interactome reveals the protein organization of the algal CO_2 concentrating mechanism. *Cell, 171*, 133–147.

Matsuda, Y., Hopkinson, B. M., Nakajima, K., Dupont, C. L., & Tsuji, Y. (2017). Mechanisms of carbon dioxide acquisition and CO_2 sensing in marine diatoms: A gateway to carbon metabolism. *Philosophical Transactions of the Royal Society B, 372*, 20160403.
McKay, R. M. L., & Gibbs, S. P. (1991). Composition and function of pyrenoids: Cytochemical and immunocytochemical approaches. *Canadian Journal of Botany, 69*, 1040–1052.
Meyer, M., & Griffiths, H. (2013). Origins and diversity of eukaryotic CO_2-concentrating mechanisms: Lessons for the future. *Journal of Experimental Botany, 64*, 769–786.
Meyer, M. T., Whittaker, C., & Griffiths, H. (2017). The algal pyrenoid: Key unanswered questions. *Journal of Experimental Botany, 68*, 3739–3749.
Michels, A. K., Wedel, N., & Kroth, P. G. (2005). Diatom plastids possess a phosphoribulokinase with an altered regulation and no oxidative pentose phosphate pathway. *Plant Physiology, 137*, 911–920.
Mitchell, C., & Beardall, J. (1996). Inorganic carbon uptake by an Antarctic sea-ice diatom, *Nitzschia frigida*. *Polar Biology, 21*, 310–315.
Morel, F. M. M., Cox, E. H., Kraepiel, A. M. L., Lane, T. W., Milligan, A. J., Schaperdoth, I., Reinfelder, J. R., & Tortell, P. D. (2002). Acquisition of inorganic carbon by the marine diatom *Thalassiosira weissflogii*. *Functional Plant Biology, 29*, 301–308.
Morita, E., Abe, T., Tsuzuki, M., Fujiwara, S., Sato, N., Hirata, A., Sonoike, K., & Nozaki, H. (1998). Presence of the CO_2-concentrating mechanism in some species of the pyrenoid-less free-living algal genus *Chloromonas* (Volvocales, Chlorophyta). *Planta, 204*, 269–276.
Morita, E., Abe, T., Tsuzuki, M., Fujiwara, S., Sato, N., Hirata, A., Sonoike, K., & Nozaki, H. (1999). Role of pyrenoids in the CO_2-concentrating mechanism: Comparative morphology, physiology and molecular phylogenetic analysis of closely related strains of *Chlamydomonas* and *Chloromonas* (Volvocales). *Planta, 208*, 365–372.
Morris, I., Yentsch, C. M., & Yentsch, C. S. (1971). Relationship between light carbon dioxide fixation and dark carbon dioxide fixation by marine algae. *Limnology and Oceanography, 16*, 854–858.
Morris, I., Beardall, J., & Mukerji, D. (1978). The mechanisms of carbon fixation in phytoplankton. *Mitt Internat Verein Limnol, 21*, 174–183.
Mukherjee, A., Lau, C. S., Walker, C. E., Rai, A. K., Prejean, C. I., Yates, G., Emrich-Mills, T., Lemoine, S. G., Vinyard, D. J., Mackinder, L. C. M., & Moroney, J. V. (2019). Thylakoid localized bestrophin-like proteins are essential for the CO_2 concentrating mechanism of *Chlamydomonas reinhardtii*. *Proceedings of the National Academy of Sciences of the United States of America 116*, 16915–16920.
Munoz, J., & Merrett, M. J. (1989). Inorganic carbon transport in some marine eukaryotic microalgae. *Planta, 178*, 450–455.
Nakajima, K., Tanaka, A., & Matsuda, Y. (2013). SLC4 family transporters in a marine diatom directly pump bicarbonate from seawater. *Proceedings of the National Academy of Sciences of the United States of America, 110*, 1767–1772.
Ogawa, T., & Ogren, W. L. (1985). Action spectra for accumulation of inorganic carbon in the cyanobacterium, *Anabaena variabilis*. *Photochemistry and Photobiology, 41*, 583–587.
Ogawa, T., Miyano, A., & Inoue, Y. (1985). Photosystem-I-driven inorganic carbon transport in the cyanobacterium, *Anacystis nidulans*. *Biochimica et Biophysica Acta, 808*, 74–75.
Ohnishi, N., Mukherjee, B., Tsujikawa, T., Yanase, M., Nakano, H., Moroney, J. V., & Fukuzawa, H. (2010). Expression of a low CO_2-inducible protein, LCI1, increases inorganic carbon uptake in the green alga *Chlamydomonas reinhardtii*. *Plant Cell, 22*, 3105–3311.
Omata, T., Price, G. D., Badger, M. R., Okamura, M., Gohta, S., & Ogawa, T. (1999). Identification of an ATP-binding cassette transporter involved in bicarbonate uptake in the cyanobacterium *Synechococcus* sp. strain PCC 7942. *Proceedings of the National Academy of Sciences of the United States of America, 96*, 13571–13576.
Palmqvist, K., Sundblad, L.-G., Wingsle, G., & Samuelsson, G. (1990). Acclimation of photosynthetic light reactions during induction of inorganic carbon accumulation in the green alga *Chlamydomonas reinhardtii*. *Plant Physiology, 94*, 357–366.

Patel, B. N., & Merrett, M. J. (1986). Regulation of carbonic anhydrase activity, inorganic carbon uptake and photosynthetic biomass yield in *Chlamydomonas reinhardtii*. *Planta, 169*, 81–86.
Price, G. D., & Badger, M. R. (1989). Expression of human carbonic anhydrase in the cyanobacterium *Synechocystis* PCC7942 creates a high CO_2-requiring phenotype. *Plant Physiology, 91*, 505–513.
Price, G. D., Maeda, S., Omata, T., & Badger, M. (2002). Modes of active inorganic carbon uptake in the cyanobacterium *Synechococcus* sp. PCC7942. *Functional Plant Biology, 29*, 131–149.
Price, G. D., Woodger, F. J., Badger, M. R., Howitt, S. M., & Tucker, L. (2004). Identification of a SulP-type bicarbonate transporter in marine cyanobacteria. *Proceedings of the National Academy of Sciences of USA, 101*, 18228–18233.
Price, G. D., Badger, M. R., Woodger, F. J., & Long, B. J. (2008). Advances in understanding the cyanobacterial CO_2- concentrating-mechanism (CCM): Functional components, Ci transporters, diversity, genetic regulation and prospects for engineering into plants. *Journal of Experimental Botany, 59*, 1441–1461.
Raven, J. A. (1997a). Inorganic carbon acquisition by marine autotrophs. *Advances in Botanical Research, 27*, 85–209.
Raven, J. A. (1997b). Putting the C in phycology. *European Journal of Phycology, 32*, 319–333.
Raven, J. A. (1997c). CO_2 concentrating mechanisms: A role for thylakoid lumen acidification. *Plant, Cell & Environment, 20*, 147–154.
Raven, J. A. (2009). Contributions of anoxygenic and oxygenic phototrophy and chemolithotrophy to carbon and oxygen fluxes in aquatic environments. *Aquatic Microbial Ecology, 56*, 177–192.
Raven, J. A., & Beardall, J. (2003). CO_2 acquisition mechanisms in algae: Carbon dioxide diffusion and carbon dioxide concentrating mechanisms. In A. W. D. Larkum, S. E. Douglas, & J. A. Raven (Eds.), *Photosynthesis in the algae* (pp. 225–244). Dordrecht: Kluwer Academic Publishers.
Raven, J. A., & Beardall, J. (2005). Respiration in aquatic photolithotrophs. In P. A. del Giorgio & P. J. L. B. Williams (Eds.), *Respiration in aquatic ecosystems* (pp. 36–46). Oxford: Oxford University Press.
Raven, J. A., & Beardall, J. (2014). CO_2 concentrating mechanisms and environmental change. *Aquatic Botany, 118*, 24–37.
Raven, J. A., & Beardall, J. (2016a). Dark respiration and organic carbon loss. In M. Borowitzka, J. Beardall, & J. A. Raven (Eds.), *The physiology of microalgae* (pp. 129–142). Cham: Springer.
Raven, J. A., & Beardall, J. (2016b). The ins and outs of CO_2. *Journal of Experimental Botany, 67*, 1–13.
Raven, J. A., & Colmer, T. D. (2016). Life at the boundary: Photosynthesis at the soil-liquid interface. A synthesis focusing on mosses. *Journal of Experimental Botany, 67*, 1613–1623.
Raven, J. A., & Giordano, M. (2017). Acquisition and metabolism of carbon in the Ochrophyta other than diatoms. *Philosophical Transactions of the Royal Society of London B, 372*, 20160400.
Raven, J. A., Beardall, J., & Griffiths, H. (1982). Inorganic C-sources for *Lemanea*, *Cladophora* and *Ranunculus* in a fast flowing stream: Measurements of gas exchange and of carbon isotope ratio and their ecological significance. *Oecologia, 53*, 68–78.
Raven, J. A., Ball, L., Beardall, J., Giordano, M., & Maberly, S. C. (2005). Algae lacking CCMs. *Canadian Journal of Botany, 83*, 879–890.
Raven, J. A., Cockell, C. S., & De La Rocha, C. L. (2008). The evolution of inorganic carbon concentrating mechanisms in photosynthesis. *Philosophical Transactions of the Royal Society of London B, 363*, 2641–2650.
Raven, J. A., Beardall, J., Flynn, K. J., & Maberly, S. C. (2009). Phagotrophy in the origins of photosynthesis in eukaryotes and as a complementary mode of nutrition in phototrophs: Relation to Darwin's insectivorous plants. *Journal of Experimental Botany, 60*, 3975–3987. https://doi.org/10.1093/jxb/erp282.

Raven, J. A., Giordano, M., Beardall, J., & Maberly, S. (2011). Algal and aquatic plant carbon concentrating mechanisms in relation to environmental change. *Photosynthesis Research, 109*, 281–296.

Raven, J. A., Giordano, M., Beardall, J., & Maberly, S. C. (2012). Algal evolution in relation to atmospheric CO_2: Carboxylases, carbon concentrating mechanisms and carbon oxidation cycles. *Philosophical Transactions of the Royal Society B, 367*, 493–507.

Raven, J. A., Beardall, J., & Giordano, M. (2014). Energy costs of carbon dioxide concentrating mechanisms. *Photosynthesis Research, 121*, 111–124.

Raven, J. A., Beardall, J., & Quigg, A. (2019). Light-driven oxygen consumption in the water-water cycles and photorespiration, and light stimulated mitochondrial respiration. In A. W. D. Larkum, A. R. Grossman, & J. A. Raven (Eds.), *Photosynthesis in Algae* (2nd ed.). Berlin: Springer. In press.

Reinfelder, J. R. (2011). Carbon concentrating mechanisms in eukaryotic marine phytoplankton. *Annual Review of Marine Science, 3*, 291–315.

Reinfelder, J. R., Kraepiel, A. M. L., & Morel, F. M. M. (2000). Unicellular C_4 photosynthesis in a marine diatom. *Nature, 407*, 996–999.

Reinfelder, J. R., Milligan, A. J., & Morel, F. M. M. (2004). The role of C_4 photosynthesis in carbon accumulation and fixation in a marine diatom. *Plant Physiology, 135*, 2106–2111.

Reiskind, J. B., Seaman, P. T., & Bowes, G. (1988). Alternative methods of photosynthetic carbon assimilation in marine macroalgae. *Plant Physiology, 87*, 686–692.

Roberts, K., Granum, E., Leegood, R. C., & Raven, J. A. (2007a). C3 and C4 pathways of photosynthetic carbon assimilation in marine diatoms are under genetic, not environmental, control. *Plant Physiology, 145*, 230–235.

Roberts, K., Granum, E., Leegood, R. C., & Raven, J. A. (2007b). Carbon acquisition by diatoms. *Photosynthesis Research, 93*, 79–88.

Rost, B., Riebesell, U., Burkhardt, S., & Sültemeyer, D. (2003). Carbon acquisition of bloom-forming marine phytoplankton. *Limnology and Oceanography, 48*, 55–67.

Rost, B., Kranz, S. A., Richter, K.-U., & Tortell, P. D. (2007). Isotope disequilibrium and mass spectrometric studies of inorganic carbon acquisition by phytoplankton. *Limnology and Oceanography: Methods, 5*, 328–337.

Rotatore, C., & Colman, B. (1990). Uptake of inorganic carbon by isolated chloroplasts of the unicellular green alga *Chlorella ellipsoidea*. *Plant Physiology, 93*, 1597–1600.

Rotatore, C., & Colman, B. (1991). The localization of active carbon transport at the plasma membrane in *Chlorella ellipsoidea*. *Canadian Journal of Botany, 69*, 1025–1031.

Scott, K. M., Henn-Sax, M., Harmer, T. L., Longo, D. L., Frome, C. H., & Cavenaugh, C. M. (2007). Kinetic isotope effect and biochemical characterisation of form IA Rubisco from the marine cyanobacterium *Prochlorococcus marinus* MIT9313. *Limnology and Oceanography, 55*, 2199–2204.

Shibata, M., Katoh, H., Sonoda, M., Ohkawa, H., Shimoyama, M., Fukuzawa, H., Kaplan, A., & Ogawa, T. (2002). Genes essential to sodium-dependent bicarbonate transport in cyanobacteria: Function and phylogenetic analysis. *The Journal of Biological Chemistry, 277*, 18658–18664.

Shiraiwa, Y., Danbara, A., & Yoke, K. (2004). Characterization of highly oxygen-sensitive photosynthesis in coccolithophorids. *Japanese Journal of Phycology, 52*(Supplement), 87–94.

Smith, K. S., & Ferry, J. G. (2000). Prokaryotic carbonic anhydrases. *FEMS Microbiology Reviews, 24*, 335–366.

Smith-Harding, T. J., Mitchell, J. G., & Beardall, J. (2017). The role of external carbonic anhydrase in photosynthesis during growth of the marine diatom *Chaetoceros muelleri*. *Journal of Phycology, 53*, 1159–1170.

Spalding, M. H., Critchley, C., Govindjee, & Ogren, W. L. (1984). Influence of carbon dioxide concentration during growth on fluorescence induction characteristics of the green alga *Chlamydomonas reinhardtii*. *Photosynthesis Research, 5*, 169–176.

Stojkovic, S., Beardall, J., & Matear, R. (2013). CO_2 concentrating mechanisms in three southern hemisphere strains of *Emiliania huxleyi*. *Journal of Phycology, 49*, 670–679.

Tcherkez, G. G. B., Farqhar, G. D., & Andrews, T. J. (2006). Despite slow catalysis and confused substrate specificity, all ribulose bisphosphate carboxylases may be nearly perfectly optimized. *Proceedings of the National Academy of Sciences of the United States of America, 103*, 7246–7725.

Tchernov, D., Silverman, J., Luz, B., Reinhold, L., & Kaplan, A. (2003). Massive light-dependent cycling of inorganic carbon between oxygenic photosynthetic microorganisms and their surroundings. *Photosynthesis Research, 77*, 95–103.

Tortell, P. (2000). Evolutionary and ecological perspectives on carbon acquisition in phytoplankton. *Limnology and Oceanography, 45*, 744–750.

Trimborn, S., Lundholm, N., Thoms, S., Richter, K. U., Krock, B., Hansen, P. J., & Rost, B. (2008). Inorganic carbon acquisition in potentially toxic and non-toxic diatoms: The effect of pH-induced changes in seawater carbonate chemistry. *Physiologia Plantarum, 133*, 92–105.

Tsuji, Y., Suzuki, I., & Shiraiwa, Y. (2009). Photosynthetic carbon assimilation in the coccolithophorid *Emiliania huxleyi* (Haptophyta): Evidence for the predominant operation of the C_3 cycle and the contribution of β-carboxylases to the active anaplerotic reaction. *Plant & Cell Physiology, 50*, 318–329.

Tsuji, Y., Suzuki, I., & Shiraiwa, Y. (2012). Enzymological evidence for the function of a plastid-located pyruvate carboxylase in the haptophyte alga *Emiliania huxleyi*: A novel pathway for the production of C_4 compounds. *Plant & Cell Physiology, 53*, 1043–1052.

Tsuji, Y., Mahardika, A., & Matsuda, Y. (2017a). Evolutionarily distinct strategies for the acquisition of inorganic carbon from seawater in marine diatoms. *Journal of Experimental Botany, 68*, 3949–3958.

Tsuji, Y., Nakajima, K., & Matsuda, Y. (2017b). Molecular aspects of the biophysical CO_2-concentrating mechanism and its regulation in marine diatoms. *Journal of Experimental Botany, 68*, 3763–3772.

van Hunnik, E., Amoroso, G., & Sültemeyer, D. (2002). Uptake of CO_2 and bicarbonate by intact cells and chloroplasts of *Tetraedon minimum* and *Chlamydomonas noctigama. Planta, 215*, 763–769.

Villarejo, A., Martinez, F., del Pino Plumed, M., & Ramazanov, Z. (1996). The induction of the CO_2 concentrating mechanism in a starch-less mutant of *Chlamydomonas reinhardtii. Physiologia Plantarum, 98*, 798–802.

Wang, Y., Stessman, D. J., & Spalding, M. H. (2015). The CO_2 concentrating mechanism and photosynthetic carbon assimilation in limiting CO_2: How *Chlamydomonas* works against the gradient. *Plant Journal, 82*, 429–448.

Whitney, S. M., & Andrews, T. J. (1998). The CO_2/O_2 specificity of single-subunit ribulose-bisphosphate carboxylase from the dinoflagellate, *Amphidinium carterae. Australian Journal of Plant Physiology, 25*, 131–138.

Whitney, S., Shaw, D., & Yellowlees, D. (1995). Evidence that some dinoflagellates contain a ribulose-1,5-bisphosphate carboxylase/oxygenase related to that of the alpha-proteobacteria. *Proceedings of the Royal Society of London B, 259*, 271–275.

Yamano, T., Sato, E., Iguchi, H., Fukuda, Y., & Fukuzawa, H. (2015). Characterization of cooperative bicarbonate uptake into chloroplast stroma in the green alga *Chlamydomonas reinhardtii. Proceedings of the National Academy of Sciences of the United States of America, 112*, 7315–7320.

Young, J. N., Heureux, A. M., Sharwood, R. E., Rickaby, R. E., Morel, F. M., & Whitney, S. M. (2016). Large variation in the Rubisco kinetics of diatoms reveals diversity among their carbon-concentrating mechanisms. *Journal of Experimental Botany, 67*, 3445–3456.

Circadian Clocks in Cyanobacteria

Susan E. Cohen

Abstract Circadian rhythms, which are oscillations in biological activity that occur with ~24-h periodicity and driven by a circadian clock, function to anticipate daily environmental changes and can be found ubiquitously throughout nature. Cyanobacteria are the only prokaryotes known to have a rigorously tested and robust circadian clock. *Synechococcus elongatus* PCC 7942 is the paradigm for our understanding of circadian rhythms in prokaryotes. The KaiABC posttranslational oscillator drives 24-h rhythmicity in cyanobacteria and has been studied in exquisite detail. Here, the current understanding of the *S. elongatus* circadian oscillator, its synchronization with the environment, and its coordination with cellular activities is highlighted.

Keywords Circadian oscillator · Circadian rhythm · Cyanobacteria · KaiA · KaiB · KaiC · Oscillator

1 Introduction

Over the course of the day, we experience changes in light intensity, temperature, and humidity as a result of the daily rising and setting of the sun. Many organisms have evolved what are known as circadian rhythms that allow them to coordinate their physiology and metabolism over the course of the day. Circadian rhythms are oscillations in biological activity that have a periodicity of ~24 h, meaning that they occur or peak once per day, and can be found ubiquitously throughout nature. Today, the molecular details of circadian rhythms are studied in various model systems that span the evolutionary tree, from microorganisms such as bacteria and fungi to plants, flies, and mammalian systems (Bell-Pedersen et al. 2005).

S. E. Cohen (✉)
Department of Biological Sciences, BS-143, California State University,
Los Angeles, CA, USA
e-mail: scohen8@calstatela.edu

Q. Wang (ed.), *Microbial Photosynthesis*,
https://doi.org/10.1007/978-981-15-3110-1_9

Circadian rhythms are driven by a circadian clock comprised of an oscillator that keeps ~24-h time, as well as pathways to synchronize the oscillator with the external environment and relay temporal information to clock-controlled activities, which we observe as circadian rhythms. While the molecular mechanisms driving circadian rhythms vary among the various model systems, they all share the three fundamental principles of circadian rhythmicity: persistence, entrainment, and temperature compensation. Persistence refers to the fact that circadian rhythms are genetically encoded rhythms that will persist in the absence of an environmental or light: dark (LD) cycle. Although circadian rhythms are not driven by the environment, they can be entrained, meaning that they can be reset to local time. Lastly circadian rhythms are temperature compensated, meaning that the period of the oscillator remains ~24-h intendent of temperature. Thus, the circadian clock runs the same on a hot summer day as it does on a cool winter day.

Although it was originally thought that bacteria were not complex enough to possess a circadian clock, early reports of ~24-h rhythms of photosynthesis and nitrogen fixation hinted at a biological timing mechanism (Chen et al. 1988; Mitsui et al. 1986; Stal and Krumbein 1987). It was the work of T.C. Huang and colleagues that demonstrated rhythms of nitrogen fixation and amino acid uptake that satisfied the three criteria of circadian rhythms (persistence, entrainment, and temperature compensation) and convinced the field that bacteria could indeed have circadian rhythms (Chen et al. 1991; Grobbelaar et al. 1986; Huang et al. 1990). Today cyanobacteria are the only prokaryotes known to have a rigorously tested and robust circadian clock, although reports from archaea (Whitehead et al. 2009), purple photosynthetic bacteria (Min et al. 2005; Ma et al. 2016), as well as enteric bacteria (Paulose et al. 2016) suggest that ~24-h timekeeping mechanisms are not limited to cyanobacteria. Work from the groups of Golden, Johnson, and Kondo established *Synechococcus elongatus* PCC 7942, hereafter *S. elongatus*, as the primary model for understanding the molecular details of the prokaryotic circadian clock. Using luciferase fusions to track the circadian clock, they were able to identify mutants in clock function and track them to mutations in the genes that encode the core oscillator KaiA, KaiB, and KaiC (Ishiura et al. 1998; Kondo et al. 1993). In *S. elongatus*, the *kaiA*, *kaiB*, and *kaiC* genes comprise the core oscillator and regulate global patterns of gene expression (Ishiura et al. 1998; Kondo et al. 1994), the timing of cell division (Dong et al. 2010; Mori et al. 1996), compaction of the chromosome (Smith and Williams 2006; Woelfle et al. 2007), and metabolite partitioning (Diamond et al. 2015).

2 Kai-Based Oscillator

Genetic screens monitoring promoter-luciferase fusions allowed for the identification of mutants with altered circadian rhythms – from 16 h to 60 h periods as well as a few arrhythmic strains (Kondo et al. 1994). These mutants mapped to three neighboring genes, named *kaiA*, *kaiB*, and *kaiC*, where either overexpression or deletion of any of these three genes results in arrhythmicity (Ishiura et al. 1998; Xu et al. 2004). With the exception of KaiC, which has predicted ATP-binding domains, these genes did not display homology to known components of the eukaryotic oscillator nor did they display homology to other proteins that might have elucidated their function (Kondo et al. 1994). Unlike the transcription-translation feedback loops underlying the core oscillator mechanisms in eukaryotic systems, the Kai-based oscillator functions as a posttranslational oscillator. We now know that KaiC undergoes ~24-h rhythm of phosphorylation at two key residues, Ser431 and Thr432, that is governed by rhythmic interactions with KaiA and KaiB (Xu et al. 2004; Nishiwaki et al. 2004). This rhythm of KaiC phosphorylation can be reconstituted in a test tube, by combining purified KaiA, KaiB, and KaiC with ATP, where a temperature-compensated rhythm with ~24-h periodicity is observed, solidifying these three proteins as the core oscillator (Nakajima et al. 2005).

Recent biochemical and structural investigations of the Kai-based oscillator have imparted an unprecedented view of how this three-protein clock keeps time. Each KaiC molecule is comprised of a CI-domain, CII-domain, and A-loop extensions, which protrude from the C-terminus (Pattanayek et al. 2004). KaiC is a hexameric protein that comes together to look like two stacked rings, known as the CI- and CII-rings, and possesses both autokinase and autophosphatase activities (Fig. 1). While the CI- and CII-rings are homologous and are both capable of binding to and hydrolyzing ATP, they play distinct yet important timekeeping roles. In the CI domain, phosphate is transferred from ATP to water, while in the CII-domain, phosphate is transferred from ATP to either Ser431 or Thr432 (Hayashi et al. 2004; Phong et al. 2013). The ATPase activity of the CI-ring is rhythmic and is correlated to the period of the oscillator both in vivo and in vitro (Terauchi et al. 2007; Abe et al. 2015).

At dawn KaiC starts off in a non-phosphorylated state where KaiA can bind to the A-loops, maintaining them in an extended position which promotes KaiC autokinase activity (Kageyama et al. 2006; Iwasaki et al. 2002; Kim et al. 2008). KaiC phosphorylation at Ser431 and T432 occurs in an ordered and sequential manner where T432 is phosphorylated first, followed by phosphorylation on S431 resulting in the doubly or fully phosphorylated form of KaiC (Nishiwaki et al. 2004, 2007; Rust et al. 2007). Once the fully phosphorylated state of KaiC is achieved, there are a series of structural rearrangements in the KaiC hexamer that trigger the nighttime phase of the cycle. In particular, the doubly phosphorylated state of KaiC results in enhanced interactions between the CI-ring and CII-ring exposing a binding site for KaiB on the CII-ring known as the B-loop. Once the B-loop on the CI-ring has been exposed and KaiB is in a KaiC-competent binding state, KaiB will bind to KaiC as well as KaiA, sequestering KaiA in an inactive state (Chang et al. 2015; Tseng et al. 2017). This results in the A-loops being able to retract into the KaiC hexamer, promoting the autophosphatase activity of KaiC. During the nighttime phase of cycle, the dephosphorylation of KaiC occurs in a similarly ordered direction, where Thr432 becomes dephosphorylated first, followed by S431, resulting in the unphosphorylated state at dawn where the cycle can begin anew (Fig. 1).

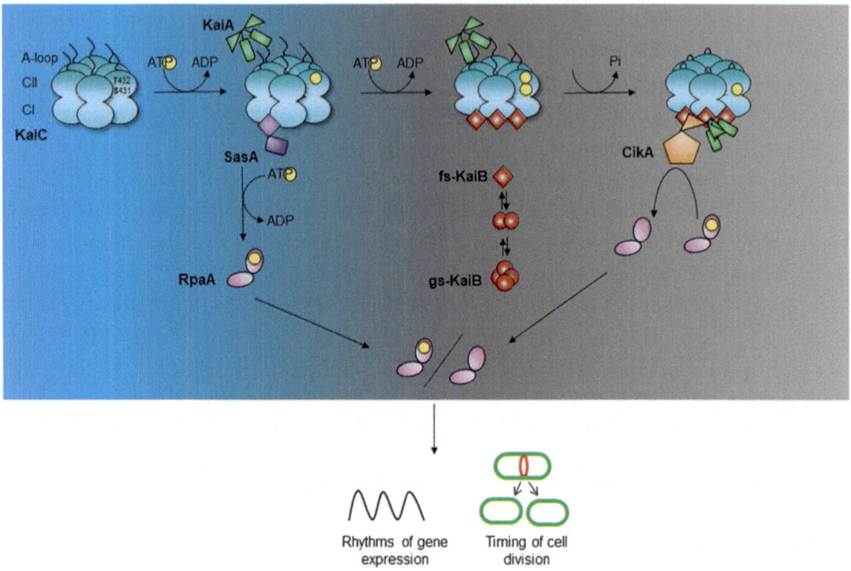

Fig. 1 Mechanism of the cyanobacterial circadian oscillator. KaiC is a hexameric protein comprised of CI- and CII-rings with A-loop tails that extend from the C-terminus of the CII-ring. At dawn (left most part of the figure) KaiC begins in the non-phosphorylated state where KaiA can interact with the A-loop tails, promoting KaiC autophosphorylation during the day. KaiC phosphorylates (yellow circle) in an ordered and sequential manner where T432 becomes phosphorylated first, followed by S431, resulting in the doubly phosphorylated species. SasA can bind to the CI-domain of phosphorylated KaiC, promoting its kinase activity toward RpaA. RpaA phosphorylation increases throughout the day. KaiB reversibly changes from gs-KaiB, which is not capable of binding to KaiC, to fs-KaiB, a conformation that is capable of binding to KaiC. At dusk (blue-grey transition), once KaiC has reached the doubly phosphorylated state, which results in a conformation change exposing the B-loops in the CII-domain, and fs-KaiB are present, fs-KaiB can bind to KaiC, displacing SasA. Fs-KaiB binds to and sequesters an inactivated state of KaiA, allowing the A-loops to retract within the KaiC hexamer and promoting KaiC autophosphatase activity. KaiC dephosphorylation occurs first on T432 following by S431, resulting in a non-phosphorylated state. Additionally, fs-KaiB binds to and activates the phosphatase activity of CikA, resulting in the dephosphorylation if RpaA. RpaA-P peaks at dusk and promotes rhythms of gene expression and the timing of cell division among other rhythmic phenomena in *S. elongatus*

While it was known that KaiB functioned to attenuate KaiA and promote KaiC's autophosphatase activity, for many years it was unclear how this was achieved. Only recently have structural and biochemical studies revealed the role for KaiB in inhibiting KaiA and promoting the nighttime phase of the cycle. Early structural studies found that KaiB, in isolation, is a tetramer that takes on a rare-fold, also referred to as ground-state KaiB (gs-KaiB) (Hitomi et al. 2005; Iwase et al. 2004). However, more recent structural investigations of the KaiB-KaiC complex uncovered that KaiB binds to the CI-domain of KaiC as a monomer (Chang et al. 2015). Moreover, KaiB bound to KaiC no longer adopts a rare-fold but rather takes on a thioredoxin-like fold, known as fold-switched KaiB (fs-KaiB) (Chang et al. 2015). In order for fs-KaiB to adopt this KaiC-binding competent, thioredoxin-like fold, it undergoes a complete rearrangement of its tertiary structure relative to the tetrameric state, classifying KaiB as a metamorphic protein. The conversion of gs-KaiB to fs-KaiB is slow, but once achieved, KaiB binds very quickly to KaiC, and this fold-switching accounts for much of the time delay needed to achieve ~24-h timing (Chang et al. 2015). Moreover, fs-KaiB binds to KaiA and sequesters KaiA in an inactive state at the CI-ring away from the A-loops that protrude from the CII-ring (Tseng et al. 2017). This ensures that KaiA is kept away from the A-loops, allowing the A-loops the retract into the KaiC hexamer, a conformation that promotes KaiC autophosphatase activity. Moreover, KaiA is sequestered in an inactive state, so it is no capable of stimulating the autokinase activity of a neighboring KaiC hexamer (Tseng et al. 2017). In addition, fs-KaiB plays critical roles in determining the activity of several of the output components helping the oscillator to coordinate the timing of clock-controlled behaviors.

3 Synchronization with the Environment

While it was originally thought that a photoreceptor was involved in relaying environmental information to the oscillator, as is the case in eukaryotes, mutation of the photoreceptors in *S. elongatus* demonstrated that they were not involved. Rather, changes in the day-night cycle are perceived indirectly via proteins that monitor changes in metabolites and cellular redox. Clock resetting is monitored in vivo by exposing the cells to a dark pulse when they are not expecting to see it. This dark pulse depending on when administered results in a change in the phase of the oscillation, known as a resetting dark pulse.

Key input factor CikA and core oscillator protein KaiA bind directly to oxidized quinones (Wood et al. 2010; Ivleva et al. 2006). The redox status of the quinone pool changes as a function of photosynthetic activity, where the quinone pool has been shown to be rapidly oxidized upon darkness and signals the beginning of the night-time period to the oscillator (Kim et al. 2012). The addition of oxidized quinone at a time when the cell would usually not expect to see it, during the day in vivo or when KaiC is phosphorylating in vitro, results in phase changes both in vivo and in vitro (Kim et al. 2012). Upon binding to oxidized quinones, KaiA and CikA have been shown to aggregate in vitro, and this aggregation is thought to target them for degradation as the protein level decreases upon exposure to oxidized quinone in vivo (Wood et al. 2010; Ivleva et al. 2006). This aggregation, and potential degradation, potentially changes the stoichiometry of the KaiABC complex resulting in the resetting of the phase of the oscillator to accommodate changes in the day-night cycle.

While oxidized quinone signals the onset of the night period, it is the ATP/ADP ratio in the cell that determines the duration of the night period. KaiC can sense changes in the ATP/ADP ratio [ATP/(ATP + ADP)] which decreases gradually over the course of the night. Transient changes in the ATP/ADP ratio in vitro, when administered when KaiC is phosphorylating, a time that correlates to the day, result in changes to KaiC phosphorylation patterns allowing the oscillator to adapt to changing environments (Rust et al. 2011).

Deletion of *cikA* results in cells that are unable to reset the phase of the oscillation in response to environmental changes, which established CikA as a critical component of the input pathway (Schmitz et al. 2000). In addition, CikA additionally functions a part of the output pathway to regulate circadian outputs. One of these outputs is the rhythmic accumulation of the glucose polymer glycogen during the day which is consumed at night (Diamond et al. 2015; Pattanayak et al. 2014). Mutants lacking *cikA* accumulate more glycogen during the day, compared to the wild type, resulting in increased ATP during the nighttime portion of the cycle, and are less sensitive to resetting dark pulses (Pattanayak et al. 2014). Conversely mutants that accumulate less glycogen during the day are hypersensitive to resetting dark pulses (Pattanayak et al. 2014). Together these data suggest that metabolic rhythms feed back to the oscillator and help the oscillator synchronize with changes in the environment.

The Kai-based oscillator also undergoes a circadian orchestration in its subcellular localization patterns, where KaiA and KaiC are found to be distributed throughout the cell during the day and localized to a single pole of the cell at night (Cohen et al. 2014). KaiA localization is dependent on KaiC and co-localizes with CikA, which is found to be at the pole constitutively throughout the day (Cohen et al. 2014). While the mechanism behind these localization patterns has not been elucidated, these localization patterns would facilitate the binding of CikA and KaiA with the quinone pool in the membrane and may contribute to entrainment and synchronization with the environment.

While protein synthesis and degradation are critical components of the circadian clock in eukaryotic systems, it was not thought to be important in the prokaryotic clock as the oscillator can be reconstituted in vitro. However, this is not the case as mutations in the Clp protease, consisting of ClpX chaperone and ClpP1 and Clp2 protease subunits, display altered circadian rhythms where a long-period rhythm is observed (Imai et al. 2013; Cohen et al. 2018; Holtman et al. 2005). In addition to long-period rhythms, mutants in which the protease activity of the ClpXP1P2 protease is inhibited result in a hyper-resetting phenotype (Cohen et al. 2018). Suggesting that the regulated proteolysis functions not only to determine period length but also to limit the range in which the clock can reset in response to environmental signals. While the targets of the protease that function to limit how much the clock can reset have not yet been identified, some evidence points to KaiC as a substrate of ClpXP1P2 (Cohen et al. 2018).

4 Coordination of Cellular Activities

The output pathway transmits temporal information from the clock-to-clock-controlled behaviors which in *S. elongatus* include genome-wide rhythms of gene expression, the timing of cell division, and metabolite partitioning. While it had been previously thought that input and output from the oscillator were linear and independent pathways, we know now that it is not that simple. Several factors play key roles in either input, oscillator, or output activities, and competition of factors for binding to the Kai oscillator is critical for controlling these output activities.

The key to controlling output hinges on the regulation of the RpaA transcription factor. When RpaA is in a phosphorylated state, RpaA-P, it is active as a transcriptional regulator that can bind to DNA and regulate gene expression (Markson et al. 2013). While RpaA-P only binds directly to ~100 targets in the *S. elongatus* genome, it dictates the rhythmic expression of the entire genome, where it promotes the expression of dusk-peaking genes and inhibits the expression of dusk peaking genes (Markson et al. 2013). Part of the RpaA regulon include other transcription factors and sigma factors (Markson et al. 2013), which are sequentially activated (Fleming and O'Shea 2018), resulting in global rhythms of gene expression. Additionally, the entire chromosome undergoes a circadian rhythm in compaction state, where the chromosome is in its most condensed state at dusk that is controlled by the clock (Smith and Williams 2006; Woelfle et al. 2007). While these compaction rhythms have been proposed to play a role in regulating global rhythms of gene expression, the mechanisms by which they are controlled by or coordinated with RpaA have not been investigated.

RpaA phosphorylation oscillates throughout the day with peak RpaA-P occurring at subjective dusk and is controlled by the opposing effects of the kinase activity of SasA and phosphatase activity of CikA (Gutu and O'Shea 2013) (Fig. 1). The histidine kinase SasA was dually identified as a component of the circadian clock by means of its homology to the N-terminus of KaiB and well its identification as the kinase that transfers the phosphoryl group to cognate response regulator RpaA (Iwasaki et al. 2000). The sensor domain of SasA, which adopts a thioredoxin-like fold, binds to the CI-domain of SpT and pSpT phosphorylated forms of KaiC (Chang et al. 2015; Gutu and O'Shea 2013; Valencia et al. 2012), activating its kinase activity, resulting in enhanced RpaA-P during the day (Gutu and O'Shea 2013; Iwasaki et al. 2000). In line with the homology shared between the N-terminus of KaiB and the sensor domain of SasA, fs-KaiB also adopts a thioredoxin-like fold and is capable of displacing SasA from KaiC, inactivating SasA as a kinase toward RpaA (Chang et al. 2015). In addition, fs-KaiB bound to KaiC binds to CikA activating its phosphatase activity toward RpaA at night (Chang et al. 2015; Tseng et al. 2017). Therefore, the switch from gs-KaiB to the KaiC-binding-competent fs-KaiB serves three critical purposes in the circadian cycle: (1) sequestering KaiA and promoting the dephosphorylation phase of KaiC; (2) displacing SasA from KaiC and inactivating SasA kinase activity; and (3) binding to CikA and promoting its phosphatase activity (Fig. 1).

The clock controls rhythmic expression of nearly every gene in the *S. elongatus* genome (Vijayan et al. 2009; Liu et al. 1995; Ito et al. 2009), which in turn regulates other activities throughout the day including the timing of cell division and metabolic partitioning. While the circadian clock runs independent of the cell division cycle, meaning that the clock runs with ~24-h periodicity whether the cells are growing rapidly or not at all (Kondo et al. 1997), the clock disallows or specifically inhibits cell division for a 4–6 h period of time, known as circadian gating of cell division (Dong et al. 2010; Mori et al. 1996). Cell division is specifically inhibited by the clock in the early subjective night, a time in the cycle that correlates to when the lights should be off, but the cells are growing in conditions of constant light. While the mechanism of how the clock controls the timing of cell division has not been fully elucidated, the current model proposes that RpaA-P, either directly or indirectly, is responsible for the expression of a factor(s) that function to inhibit FtsZ, the bacterial tubulin homolog, from forming the cytokinetic ring (Dong et al. 2010). Moreover, the model predicts that this inhibition imposed by the clock is then relieved by the ClpX chaperone ~4–6 h later (Cohen et al. 2018). Presumably ClpX unfolds or inactivates this factor, allowing for the cytokinetic ring to form and cell division to presume.

The first indications that cyanobacteria might have a circadian clock occurred in diazotrophic organisms, where it was observed that oxygenic photosynthesis occurred during the day while the oxygen-sensitive process of nitrogen fixation occurred at night. However, the clock is more than a mechanism to separate two incompatible processes. During the day the cells are performing photosynthesis and cell division and storing photosynthate in the form of a glucose polymer glycogen. Glycogen is consumed during the night; thus, it is of critical importance that enough glycogen is accumulated during the day. Indeed, several clock mutants cannot survive when grown in light: dark conditions due to perturbations in glycogen accumulation (Diamond et al. 2015). The circadian clock coordinates metabolism and physiology to best allow the cells to survive in the face of the changes that occur in response to the day-night cycle. Competition experiments have shown that circadian clock confers a competitive fitness advantage (Ouyang et al. 1998). While individual clock mutants don't show a growth defect when grown in isolation, a circadian clock that resonates with the external environment allows the cells to outcompete cells that do not, meaning that a wild-type strain will outcompete a mutant that lacks a clock when grown in 24-h light dark conditions of 12-h light and 12-h darkness. Likewise a mutant with a long period will outcompete the wild-type strain when grown in long-day conditions, and a short-period mutant will outcompete the wild-type strain when grown under short-day conditions (Ouyang et al. 1998).

5 Conclusions

Since the discovery that bacteria do indeed process circadian timekeeping mechanisms, work in *S. elongatus* has provided an exquisite understanding of the molecular mechanism underlying these rhythms. While circadian rhythmicity among bacteria is likely not limited to cyanobacteria, as examples of rhythmic phenomena have been reported in purple photosynthetic bacteria, eubacteria, and archaea (Min et al. 2005; Paulose et al. 2016; Ottesen et al. 2014; Edgar et al. 2012; Maniscalco et al. 2014), *S. elongatus* remains the only bacterial organisms where a rigorously tested and robust clock has been investigated. Additionally, hourglass timers, which require environmental cues in order to be sustained, are also important for coordinating activity over the day. Continued research in these areas is required to determine their mechanisms and potential roles for *kai* homologs.

Acknowledgments Research on circadian rhythms in the Cohen laboratory is supported by a grant from the National Science Foundation (MCB1845953).

References

Abe, J., et al. (2015). Circadian rhythms. Atomic-scale origins of slowness in the cyanobacterial circadian clock. *Science, 349*, 312–316. https://doi.org/10.1126/science.1261040.

Bell-Pedersen, D., et al. (2005). Circadian rhythms from multiple oscillators: Lessons from diverse organisms. *Nature Reviews. Genetics, 6*, 544–556.

Chang, Y. G., et al. (2015). A protein fold switch joins the circadian oscillator to clock output in cyanobacteria. *Science, 349*, 324–328. https://doi.org/10.1126/science.1260031. science.1260031 [pii].

Chen, T. H., Huang, T. C., & Chow, T. J. (1988). Calcium requirement in nitrogen fixation in the cyanobacterium *Synechococcus* RF-1. *Planta, 173*, 253–256. https://doi.org/10.1007/BF00403017.

Chen, T. H., Chen, T. L., Hung, L. M., & Huang, T. C. (1991). Circadian rhythm in amino acid uptake by *Synechococcus* RF-1. *Plant Physiology, 97*, 55–59.

Cohen, S. E., et al. (2014). Dynamic localization of the cyanobacterial circadian clock proteins. *Current Biology, 24*, 1836–1844. https://doi.org/10.1016/j.cub.2014.07.036. S0960-9822(14)00899-9 [pii].

Cohen, S. E., McKnight, B. M., & Golden, S. S. (2018). Roles for ClpXP in regulating the circadian clock in Synechococcus elongatus. *Proceedings of the National Academy of Sciences of the United States of America, 115*, E7805–E7813. https://doi.org/10.1073/pnas.1800828115.

Diamond, S., Jun, D., Rubin, B. E., & Golden, S. S. (2015). The circadian oscillator in Synechococcus elongatus controls metabolite partitioning during diurnal growth. *Proceedings of the National Academy of Sciences of the United States of America, 112*, E1916–E1925. https://doi.org/10.1073/pnas.1504576112.

Dong, G., et al. (2010). Elevated ATPase activity of KaiC constitutes a circadian checkpoint of cell division in *Synechococcus elongatus*. *Cell, 140*, 529–539.

Edgar, R. S., et al. (2012). Peroxiredoxins are conserved markers of circadian rhythms. *Nature, 485*, 459–464. https://doi.org/10.1038/nature11088. nature11088 [pii].

Fleming, K. E., & O'Shea, E. K. (2018). An RpaA-dependent sigma factor Cascade sets the timing of circadian transcriptional rhythms in Synechococcus elongatus. *Cell Reports, 25*, 2937–2945 e2933. https://doi.org/10.1016/j.celrep.2018.11.049.

Grobbelaar, N., Huang, T. C., Lin, H. Y., & Chow, T. J. (1986). Dinitrogen-fixing endogenous rhythm in *Synechococcus* RF-1. *FEMS Microbiology Letters, 37*, 173–177.

Gutu, A., & O'Shea, E. K. (2013). Two antagonistic clock-regulated histidine kinases time the activation of circadian gene expression. *Molecular Cell, 50*, 288–294. https://doi.org/10.1016/j.molcel.2013.02.022. S1097-2765(13)00178-0 [pii].

Hayashi, F., et al. (2004). Roles of two ATPase-motif-containing domains in cyanobacterial circadian clock protein KaiC. *The Journal of Biological Chemistry, 279*, 52331–52337. https://doi.org/10.1074/jbc.M406604200. M406604200 [pii].

Hitomi, K., Oyama, T., Han, S., Arvai, A. S., & Getzoff, E. D. (2005). Tetrameric architecture of the circadian clock protein KaiB. A novel interface for intermolecular interactions and its impact on the circadian rhythm. *The Journal of Biological Chemistry, 280*, 19127–19135. https://doi.org/10.1074/jbc.M411284200. M411284200 [pii].

Holtman, C. K., et al. (2005). High-throughput functional analysis of the *Synechococcus elongatus* PCC 7942 genome. *DNA Research, 12*, 103–115. https://doi.org/10.1093/dnares/12.2.103. 12/2/103 [pii].

Huang, T. C., Tu, J., Chow, T. J., & Chen, T. H. (1990). Circadian rhythm of the prokaryote *Synechococcus* sp. RF-1. *Plant Physiology, 92*, 531–533.

Imai, K., Kitayama, Y., & Kondo, T. (2013). Elucidation of the role of clp protease components in circadian rhythm by genetic deletion and overexpression in cyanobacteria. *Journal of Bacteriology, 195*, 4517–4526. https://doi.org/10.1128/JB.00300-13. JB.00300-13 [pii].

Ishiura, M., et al. (1998). Expression of a gene cluster *kaiABC* as a circadian feedback process in cyanobacteria. *Science, 281*, 1519–1523.

Ito, H., et al. (2009). Cyanobacterial daily life with Kai-based circadian and diurnal genome-wide transcriptional control in *Synechococcus elongatus. Proceedings of the National Academy of Sciences of the United States of America, 106*, 14168–14173.

Ivleva, N. B., Gao, T., LiWang, A. C., & Golden, S. S. (2006). Quinone sensing by the circadian input kinase of the cyanobacterial circadian clock. *Proceedings of the National Academy of Sciences of the United States of America, 103*, 17468–17473. https://doi.org/10.1073/pnas.0606639103. 0606639103 [pii].

Iwasaki, H., et al. (2000). A KaiC-interacting sensory histidine kinase, SasA, necessary to sustain robust circadian oscillation in cyanobacteria. *Cell, 101*, 223–233. https://doi.org/10.1016/S0092-8674(00)80832-6. S0092-8674(00)80832-6 [pii].

Iwasaki, H., Nishiwaki, T., Kitayama, Y., Nakajima, M., & Kondo, T. (2002). KaiA-stimulated KaiC phosphorylation in circadian timing loops in cyanobacteria. *Proceedings of the National Academy of Sciences of the United States of America, 99*, 15788–15793. https://doi.org/10.1073/pnas.222467299. 222467299 [pii].

Iwase, R., et al. (2004). Crystallization and preliminary crystallographic analysis of the circadian clock protein KaiB from the thermophilic cyanobacterium *Thermosynechococcus elongatus* BP-1. *Acta Crystallographica. Section D, Biological Crystallography, 60*, 727–729. https://doi.org/10.1107/S0907444904002112. S0907444904002112 [pii].

Kageyama, H., et al. (2006). Cyanobacterial circadian pacemaker: Kai protein complex dynamics in the KaiC phosphorylation cycle in vitro. *Molecular Cell, 23*, 161–171. https://doi.org/10.1016/j.molcel.2006.05.039. S1097-2765(06)00381-9 [pii].

Kim, Y. I., Dong, G., Carruthers, C. W., Jr., Golden, S. S., & LiWang, A. (2008). The day/night switch in KaiC, a central oscillator component of the circadian clock of cyanobacteria. *Proceedings of the National Academy of Sciences of the United States of America, 105*, 12825–12830. https://doi.org/10.1073/pnas.0800526105. 0800526105 [pii].

Kim, Y. I., Vinyard, D. J., Ananyev, G. M., Dismukes, G. C., & Golden, S. S. (2012). Oxidized quinones signal onset of darkness directly to the cyanobacterial circadian oscillator. *Proceedings of the National Academy of Sciences of the United States of America, 109*, 17765–17769. https://doi.org/10.1073/pnas.1216401109. 1216401109 [pii].

Kondo, T., et al. (1993). Circadian rhythms in prokaryotes: Luciferase as a reporter of circadian gene expression in cyanobacteria. *Proceedings of the National Academy of Sciences of the United States of America, 90*, 5672–5676.

Kondo, T., et al. (1994). Circadian clock mutants of cyanobacteria. *Science, 266*, 1233–1236.

Kondo, T., et al. (1997). Circadian rhythms in rapidly dividing cyanobacteria. *Science, 275*, 224–227.

Liu, Y., et al. (1995). Circadian orchestration of gene expression in cyanobacteria. *Genes & Development, 9*, 1469–1478.

Ma, P., Mori, T., Zhao, C., Thiel, T., & Johnson, C. H. (2016). Evolution of KaiC-dependent timekeepers: A proto-circadian timing mechanism confers adaptive fitness in the purple bacterium rhodopseudomonas palustris. *PLoS Genetics, 12*, e1005922. https://doi.org/10.1371/journal.pgen.1005922.

Maniscalco, M., et al. (2014). Light-dependent expression of four cryptic archaeal circadian gene homologs. *Frontiers in Microbiology, 5*, 79. https://doi.org/10.3389/fmicb.2014.00079.

Markson, J. S., Piechura, J. R., Puszynska, A. M., & O'Shea, E. K. (2013). Circadian control of global gene expression by the cyanobacterial master regulator RpaA. *Cell, 155*, 1396–1408. https://doi.org/10.1016/j.cell.2013.11.005. S0092-8674(13)01418-9 [pii].

Min, H., Guo, H., & Xiong, J. (2005). Rhythmic gene expression in a purple photosynthetic bacterium, *Rhodobacter sphaeroides*. *FEBS Letters, 579*, 808–812. https://doi.org/10.1016/j.febslet.2005.01.003. S0014-5793(05)00029-3 [pii].

Mitsui, A., et al. (1986). Strategy by which nitrogen-fixing unicellular cyanobacteria grow photoautotrophically. *Nature, 323*, 720–722.

Mori, T., Binder, B., & Johnson, C. H. (1996). Circadian gating of cell division in cyanobacteria growing with average doubling times of less than 24 hours. *Proceedings of the National Academy of Sciences of the United States of America, 93*, 10183–10188.

Nakajima, M., et al. (2005). Reconstitution of circadian oscillation of cyanobacterial KaiC phosphorylation in vitro. *Science, 308*, 414–415.

Nishiwaki, T., et al. (2004). Role of KaiC phosphorylation in the circadian clock system of *Synechococcus elongatus* PCC 7942. *Proceedings of the National Academy of Sciences of the United States of America, 101*, 13927–13932. https://doi.org/10.1073/pnas.0403906101. 0403906101 [pii].

Nishiwaki, T., et al. (2007). A sequential program of dual phosphorylation of KaiC as a basis for circadian rhythm in cyanobacteria. *The EMBO Journal, 26*, 4029–4037. https://doi.org/10.1038/sj.emboj.7601832. 7601832 [pii].

Ottesen, E. A., et al. (2014). Ocean microbes. Multispecies diel transcriptional oscillations in open ocean heterotrophic bacterial assemblages. *Science, 345*, 207–212. https://doi.org/10.1126/science.1252476. 345/6193/207 [pii].

Ouyang, Y., Andersson, C. R., Kondo, T., Golden, S. S., & Johnson, C. H. (1998). Resonating circadian clocks enhance fitness in cyanobacteria. *Proceedings of the National Academy of Sciences of the United States of America, 95*, 8660–8664.

Pattanayak, G. K., Phong, C., & Rust, M. J. (2014). Rhythms in energy storage control the ability of the cyanobacterial circadian clock to reset. *Current Biology, 24*, 1934–1938. https://doi.org/10.1016/j.cub.2014.07.022. S0960-9822(14)00848-3 [pii].

Pattanayek, R., et al. (2004). Visualizing a circadian clock protein: Crystal structure of KaiC and functional insights. *Molecular Cell, 15*, 375–388. https://doi.org/10.1016/j.molcel.2004.07.013. S1097276504004356 [pii].

Paulose, J. K., Wright, J. M., Patel, A. G., & Cassone, V. M. (2016). Human gut bacteria are sensitive to melatonin and express endogenous circadian rhythmicity. *PLoS One, 11*, e0146643. https://doi.org/10.1371/journal.pone.0146643.

Phong, C., Markson, J. S., Wilhoite, C. M., & Rust, M. J. (2013). Robust and tunable circadian rhythms from differentially sensitive catalytic domains. *Proceedings of the National Academy of Sciences of the United States of America, 110*, 1124–1129. https://doi.org/10.1073/pnas.1212113110.

Rust, M. J., Markson, J. S., Lane, W. S., Fisher, D. S., & O'Shea, E. K. (2007). Ordered phosphorylation governs oscillation of a three-protein circadian clock. *Science, 318*, 809–812. https://doi.org/10.1126/science.1148596. 1148596 [pii].

Rust, M. J., Golden, S. S., & O'Shea, E. K. (2011). Light-driven changes in energy metabolism directly entrain the cyanobacterial circadian oscillator. *Science, 331*, 220–223. https://doi.org/10.1126/science.1197243. 331/6014/220 [pii].

Schmitz, O., Katayama, M., Williams, S. B., Kondo, T., & Golden, S. S. (2000). CikA, a bacteriophytochrome that resets the cyanobacterial circadian clock. *Science, 289*, 765–768. 8723 [pii].

Smith, R. M., & Williams, S. B. (2006). Circadian rhythms in gene transcription imparted by chromosome compaction in the cyanobacterium *Synechococcus elongatus*. *Proceedings of the National Academy of Sciences of the United States of America, 103*, 8564–8569.

Stal, L. J., & Krumbein, W. E. (1987). Temporal separation of nitrogen fixation and photosynthesis in the filamentous, non-heterocystous cyanobacterium *Oscillatoria* sp. *Archives of Microbiology, 149*, 76–80.

Terauchi, K., et al. (2007). ATPase activity of KaiC determines the basic timing for circadian clock of cyanobacteria. *Proceedings of the National Academy of Sciences of the United States of America, 104*, 16377–16381. https://doi.org/10.1073/pnas.0706292104. 0706292104 [pii].

Tseng, R., et al. (2017). Structural basis of the day-night transition in a bacterial circadian clock. *Science, 355*, 1174–1180. https://doi.org/10.1126/science.aag2516.

Valencia, S. J., et al. (2012). Phase-dependent generation and transmission of time information by the KaiABC circadian clock oscillator through SasA-KaiC interaction in cyanobacteria. *Genes to Cells, 17*, 398–419. https://doi.org/10.1111/j.1365-2443.2012.01597.x.

Vijayan, V., Zuzow, R., & O'Shea, E. K. (2009). Oscillations in supercoiling drive circadian gene expression in cyanobacteria. *Proceedings of the National Academy of Sciences of the United States of America, 106*, 2564–22568.

Whitehead, K., Pan, M., Masumura, K., Bonneau, R., & Baliga, N. S. (2009). Diurnally entrained anticipatory behavior in archaea. *PLoS One, 4*, e5485. https://doi.org/10.1371/journal.pone.0005485.

Woelfle, M. A., Xu, Y., Qin, X., & Johnson, C. H. (2007). Circadian rhythms of superhelical status of DNA in cyanobacteria. *Proceedings of the National Academy of Sciences of the United States of America, 104*, 18819–18824.

Wood, T. L., et al. (2010). The KaiA protein of the cyanobacterial circadian oscillator is modulated by a redox-active cofactor. *Proceedings of the National Academy of Sciences of the United States of America, 107*, 5804–5809. https://doi.org/10.1073/pnas.0910141107. 0910141107 [pii].

Xu, Y., et al. (2004). Identification of key phosphorylation sites in the circadian clock protein KaiC by crystallographic and mutagenetic analyses. *Proceedings of the National Academy of Sciences of the United States of America, 101*, 13933–13938. https://doi.org/10.1073/pnas.0404768101.

Iron Deficiency in Cyanobacteria

Dan Cheng and Qingfang He

Abstract Iron is very important for photosynthetic microorganisms, including cyanobacteria, because the electron transport system of photosynthesis is highly iron-dependent. However, iron limitation is frequently occurred in natural habitats, limiting the photosynthetic activity and biomass production of cyanobacteria and other photosynthetic microorganisms. Cyanobacteria have evolved various strategies to adapt the conditions of iron deficiency. In this chapter, we will review the iron stress responses of cyanobacteria, talking about both the physiological changes and the molecular mechanisms.

Keywords Iron deficiency · Cyanobacteria · Photosynthesis

1 The Challenges of Iron Deficiency in Cyanobacteria

Iron is an essential element required for growth and development of most living organisms, including microorganisms (Hantke 2001). Iron plays a critical role in a wide variety of biochemical processes such as photosynthesis, respiration, nitrogen fixation, DNA synthesis, and many other redox reactions (Boyer et al. 1987; Straus 1994; Ferreira and Straus 1994; Behrenfeld and Kolber 1999; Raven et al. 1999; Richardson 2000; Andrews et al. 2003; Shcolnick and Keren 2006; Bellenger et al. 2011). Oxygenic photosynthesis is strikingly dependent upon iron because the photosynthetic apparatus all contains iron in heme, non-heme, or iron-sulfur clusters (Martin and Fitzwater 1988; Wilhelm 1995; Webb et al. 2001; Eldridge et al. 2004).

Cyanobacteria, also known as blue-green algae, are photosynthesizing microorganisms that are abundant in most marine and freshwater habitats. A significant part of global primary productivity is provided by cyanobacteria. In cyanobacteria, the functional photosynthetic apparatus requires 12 iron atoms for photosystem I (PS I),

D. Cheng
NCI, NIH, Bethesda, MD, USA

Q. He (✉)
Department of Biology, University of Arkansas at Little Rock, Little Rock, AR, USA
e-mail: qfhe@ualr.edu

Q. Wang (ed.), *Microbial Photosynthesis*,
https://doi.org/10.1007/978-981-15-3110-1_10

3 iron atoms for photosystem II (PS II), and 7 iron atoms for the cytochrome b6/f complex and ferredoxin (Fig. 1) (Ferreira and Straus 1994; Baniulis et al. 2008). Therefore, cyanobacterial iron requirements far exceed those of non-photosynthetic prokaryotes and are exceptionally high even among other photosynthetic organisms (Finney and O'Halloran 2003; Keren et al. 2004; Kranzler et al. 2013). The cyanobacterium *Synechocystis* sp. PCC 6803 (hereafter *Synechocystis* 6803) contains approximately 4×10^6 iron atoms per cell, which is about tenfold higher than the iron quota of *Escherichia coli* (Shcolnick and Keren 2006; Keren et al. 2004).

Although iron is the fourth most plentiful element in the Earth's crust, its bioavailability is very limited (Neilands 1982; Bagg and Neilands 1987; Braun et al. 1990). In aqueous solutions, iron exists in two oxidation states: reduced ferrous iron and oxidized ferric iron (Fraústo da Silva and Williams 2001). The ferrous iron is relatively soluble at physiological pH range and, therefore, considered readily bioavailable. Upon the evolution of oxygenic photosynthesis, the Earth's aerobic environment leads to rapid oxidation of the ferrous iron to the ferric iron (Bekker et al. 2004; Allen and Vermaas 2010). At physiological pH, the ferric iron is poorly soluble and severely reducing the biological availability of iron (Kranzler et al. 2013; Brand et al. 1983; Rich and Morel 1990; Ratledge and Dover 2000; Morel et al. 2008). The concentration of free environmental ferric iron ranges from 10^{-9} to 10^{-18} M, whereas virtually living microorganisms require a minimum concentration of 10^{-8} M to live and at least 10^{-7}–10^{-5} M to achieve optimal growth (Andrews et al. 2003). Therefore, iron deficiency occurs frequently in natural habitats and severely

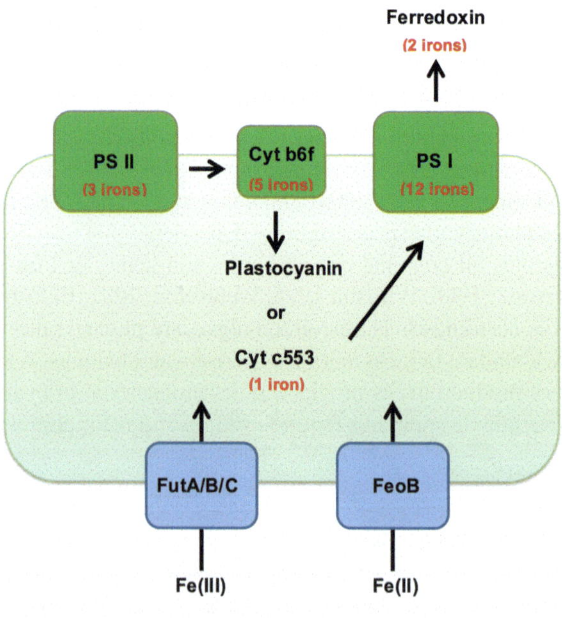

Fig. 1 A diagrammatic representation of photosynthetic iron and iron transport system in cyanobacteria. Fe(III): ferric iron, Fe(II): ferrous iron

limits photosynthetic activity and overall biomass production of photosynthetic microorganisms (Behrenfeld and Kolber 1999; Martin and Fitzwater 1988; Martin et al. 1994; Behrenfeld et al. 1996). It has been observed that iron limitation restricts primary productivity in as much as 40% of the world's ocean, supporting the iron hypothesis, which suggested that iron is a factor limiting phytoplankton growth (Martin et al. 1994; Achilles et al. 2003; Falkowski and Raven 2007; Boyd and Ellwood 2010).

Since cyanobacteria have a significantly high demand for iron, they are highly vulnerable to iron deficiency (Straus 1994; Wilhelm 1995; Behrenfeld et al. 1996; Brand 1991). On the other hand, excess iron can catalyze the formation of reactive oxygen species (ROS), which are damaging to the cells (Shcolnick and Keren 2006; Guerinot and Yi 1994; Latifi et al. 2009; Gonzalez et al. 2012). Indeed, cyanobacteria have evolved several strategies to adapt to iron-deficient environments and maintain iron homeostasis (Straus 1994).

2 The Strategies for Adaptation to Iron Deficiency in Cyanobacteria

As Straus stated, microorganisms, including cyanobacteria, developed several strategies to cope with iron limitation: retrenchment (reduction in cellular structures and physiological activities), compensation (replacement of iron-containing enzymes by functionally similar iron-free counterparts), and acquisition (enhancement of iron acquisition and iron transport ability) (Straus 1994; Ferreira and Straus 1994; Carr and Mann 1994).

2.1 Retrenchment

In cyanobacteria the most obvious alterations under iron limitation conditions are the significant decrease in cellular pigmentation (Straus 1994; Guikema and Sherman 1983, 1984; Odom et al. 1993; Riethman and Sherman 1988; Sherman and Sherman 1983). The synthesis of major light harvesting phycobilisome complexes relies upon iron-dependent metabolic pathways, thus phycobilisome content declines under iron limitation (Grossman et al. 1993; Sandström et al. 2002; Schrader et al. 2011). Iron availability also affects the chlorophyll synthesis and low iron concentrations result in loss of chlorophyll, which accompanied by the reduction of photosynthetic activities (Ivanov et al. 2000).

Because PS I has the highest iron content among photosynthetic apparatus (12–14 iron atoms per PS I unit), it is therefore no surprise that PS I is the primary target of iron limitation (Keren et al. 2004; Jordan et al. 2001). During the early stages of iron deprivation, PS I trimers become to monomers and reduce the capacity for state transition (Ivanov et al. 2006). The content of PS I decreases substantially more than that of PS II upon iron starvation, leading to a large drop in PS I/PS II ratio (Straus 1994; Falk et al. 1995; Sandström et al. 2001). It has been reported that cyanobacteria reduce the PS I/PS II ratio from a 4:1 to 1:1 under iron limitation conditions (Straus 1994). In the cyanobacterium *Synechocystis* sp. PCC 6714, moderate iron stress decreases the electron transport capacity by 83% for PS I and 46% for PS II (Sandmann 1985).

As response to iron starvation, the cell size reduction and/or morphological changes have also been observed (Sherman and Sherman 1983; Mann and Chisholm 2000; Walworth et al. 2016). For example, a reduction in the number of thylakoid membranes and carboxysomes has been shown when the cyanobacterium *Synechococcus elongatus* sp. PCC 7942 (hereafter *Synechococcus* 7942) was grown in low-iron conditions (Sherman and Sherman 1983).

Additionally, the nitrogen-fixing cyanobacteria like *Trichodesmium* reduce the amount of the iron-rich nitrogenase enzyme (28–36 iron atoms per enzyme complex) under conditions of iron deficiency before downregulating their photosynthetic systems (Richier et al. 2012; Rueter et al. 1990; Shi et al. 2007).

2.2 Compensation

Another strategy the cyanobacteria evolved to overcome iron limitation is the replacement of iron-containing proteins with functional analogs that demand no iron (Ferreira and Straus 1994). For instance, the iron-dependent Cyt c553 is replaced by its copper-dependent counterpart, plastocyanin, and iron-requiring ferredoxin is replaced by iron-free flavodoxin (Straus 1994; Hutber et al. 1977; Sandmann and Malkin 1983). Flavodoxin upregulation has been traditionally used as a molecular indicative of iron stress in field studies (Geiss et al. 2001).

Flavodoxin is encoded by *isiB* gene and can replace ferredoxin as an electronic carrier in a number of reactions (Sandström et al. 2002; Fillat et al. 1990; Lodeyro et al. 2012; Pierella Karlusich et al. 2015; Razquin et al. 1995; Vigara et al. 1998). In photosynthesis, flavodoxin serves as an alternative intermediate for the photosynthetic electron transfer chain as ferredoxin does and functions as the main distributor of the reducing power (Lodeyro et al. 2012; Tognetti et al. 2007). Although flavodoxin and ferredoxin are isofunctional, they don't show any significant similarity in primary, secondary, or tertiary structures, yet the efficiency of flavodoxin function is comparable to that of ferredoxin, allowing the distribution of light energy as reducing power remains unchanged even under iron-deficient conditions (Fillat et al. 1990, 1988; Vigara et al. 1998). However, it has been shown that flavodoxin is not able to functionally replace heterocyst ferredoxin (Razquin et al. 1995).

2.3 Acquisition

Iron deprivation triggers induction of the iron acquisition systems (Andrews et al. 2003; Katoh et al. 2001a, b; Kranzler et al. 2014). Like most microorganisms, cyanobacteria also improve their ability to scavenge from the environment through production of siderophores (Leonhardt and Straus 1992). Siderophores are strong ferric iron chelators whose production and secretion occur especially under iron starvation conditions (Kranzler et al. 2013). Once binding to the ferric iron, the formed ferri-siderophore complexes are actively transported across the outer membrane into the periplasmic space via a TonB-dependent transport system (Miethke and Marahiel 2007; Nicolaisen et al. 2008; Nicolaisen et al. 2010; Noinaj et al. 2010; Rudolf et al. 2015). Siderophores are structurally diverse and can be divided into three main classes, the catecholate-type siderophore, the hydroxamate-type siderophore, and the mixed-type siderophore (Boyer et al. 1987; Straus 1994; Miethke and Marahiel 2007; Drechsel and Winkelmann 1997; Neilands 1995; Trick and Kerry 1992). The cyanobacterium *Anabaena* sp. PCC 7120 (hereafter *Anabaena* 7120) is known to produce the mixed-type (α-carboxylate-hydroxamate) siderophore (Simpson and Nielands 1976). *Synechococcus* sp. PCC 7002 synthesizes a hydroxamate-type siderophore (Arnistroug and Van Baalen 1979). However, many cyanobacteria species do not possess genes for the biosynthesis and/or transport of siderophores (Hopkinson and Morel 2009; Mirus et al. 2009). For example, bioinformatics analysis indicated that *Synechocystis* 6803 lacks siderophore biosynthesis genes (Hopkinson and Morel 2009). It has been reported that *Synechocystis* 6803 is capable of acquiring exogenous ferri-siderophores secreted by the filamentous cyanobacterium *Anabaena variabilis* ATCC 29413 (Babykin et al. 2018; Obando et al. 2018).

In addition, iron depletion also induces the expression of proteins involved in iron transport (Shcolnick et al. 2009). Once into the periplasmic space, the ferric iron is recognized and transported by an ATP-binding cassette (ABC) transporter system. The major ABC-type ferric iron transporter system of *Synechocystis* 6803 is encoded by four *fut* genes: *futA1* (*slr1295*), *futA2* (*slr0513*), *futB* (*slr0327*), and *futC* (*sll1878*) (Katoh et al. 2001a; Koropatkin et al. 2007; Waldron et al. 2007; Badarau et al. 2008; Brandt et al. 2009). The high affinity ferric-binding proteins FutA1 and FutA2 bind ferric ions and shuttle them into ferric permeases FutB and FutC in inner membrane (Fig. 1) (Katoh et al. 2001a; Rocap et al. 2003). Alternatively, the ferric ion is reduced to the ferrous iron and cross the inner membrane through ferrous iron transporters like FeoB (Fig. 1) (Kranzler et al. 2014). FutABC and FeoB transport systems are responsible for most of the iron accumulation in cyanobacteria (Katoh et al. 2001a, b; Badarau et al. 2008). Shcolnick and co-workers observed an increase in the transcript levels of *futA2*, *futC*, and *feoB* genes under iron limitation conditions (Shcolnick et al. 2009).

3 Important Iron-Deficiency Proteins in Cyanobacteria

3.1 IsiA

The induction of *isiA* (iron stress-induced protein A) is one of the most noticeable responses to iron deficiency (Guikema and Sherman 1983; Geiss et al. 2001; Kouřil et al. 2003; Latifi et al. 2005). IsiA is initially called CP43' due to its homology to the core light-harvesting antenna protein CP43 of photosystem II (PSII) (Falk et al. 1995; Burnap et al. 1993). Similar to CP43, IsiA also has six transmembrane helices and binds to chlorophyll a (Leonhardt and Straus 1992; Bricker and Frankel 2002). IsiA expression causes a blue-shift of 5–8 nm in chlorophyll a absorbance at room temperature and the presence of a prominent peak at around 685 nm in the 77K fluorescence emission spectrum (Guikema and Sherman 1983; Falk et al. 1995; Burnap et al. 1993; Öquist 1971; Pakrasi et al. 1985).

It has been shown that IsiA can form large ring structures around PSI in cyanobacteria when grown under iron-deficient conditions (Bibby et al. 2001a, b; Boekema et al. 2001; Yeremenko et al. 2004). During the early stage of iron depletion, IsiA associates with PS I to form a complex that consists of a trimeric PSI core, surrounded by a ring of 18 IsiA proteins (Bibby et al. 2001a, b; Boekema et al. 2001). Under prolonged iron deficiency, IsiA is found to form ring structures around PS I monomers and sometimes form large aggregates without PS I (Yeremenko et al. 2004; Aspinwall et al. 2004; Chauhan et al. 2011; Kouřil et al. 2005). For example, Yeremenko et al. observed a large complex with 12–14 IsiA in an inner ring and 19–21 IsiA in an outer ring around a PS I monomer (Yeremenko et al. 2004). In addition, Wang et al. observed a novel IsiA-PS I-PS II supercomplex and an IsiA-PS I supercomplex (other than IsiA-PS I trimers) in *Synechocystis* PCC 6803 under extensive iron limitation and high light conditions (Wang et al. 2010). It also has been shown that IsiA is a component of high light-inducible carotenoid-binding protein complex (HLCC), which protects thylakoid membranes from prolonged photooxidative damage (Daddy et al. 2015).

There are several possible functions of IsiA that have been proposed (Kouřil et al. 2005; Sun and Golbeck 2015). (1) IsiA serves as an alternative antenna complex for both PS I and PS II (Riethman and Sherman 1988; Pakrasi et al. 1985; Andrizhiyevskaya et al. 2002; Melkozernov et al. 2003). In the 18 IsiA ring structure, the PS I antenna size is increased by about 70%, and the light harvesting capacity of trimer is approximately doubled compared to that of the PS I core complex (Bibby et al. 2001b; Melkozernov et al. 2003; Andrizhiyevskaya et al. 2004). (2) IsiA has been suggested to protect PS I and PS II from photoinhibition through excitation energy dissipation (Sandström et al. 2001; Latifi et al. 2005; Park et al. 1999). (3) IsiA can act as a chlorophyll a storage protein to facilitate its rapid recovery after iron stress (Riethman and Sherman 1988; Burnap et al. 1993). Moreover, IsiA is induced under many other stress conditions, such as high light stress, oxidative stress, salt stress, and heat stress, indicating multiple roles of IsiA in stress responses (Fulda and Hagemann 1995; Hagemann et al. 1999; Havaux et al. 2005; Jeanjean et al. 2003; Li et al. 2004; Vinnemeier and Hagemann 1999; Yousef et al. 2003).

IsiA expression is regulated at both the transcriptional level and the posttranscriptional level. There is a Fur-binding site located in promoter region of *isiA*; thus it has been assumed that Fur controls the transcription of *isiA* (Ghassemian and Straus 1996; Kunert et al. 2003). In cyanobacterium *Synechococcus* 7942, it has been shown that the expression of IsiA is regulated by Fur (Ghassemian and Straus 1996). In addition, Pkn22 and PfsR are also reported to control IsiA expression at the transcription level (Cheng and He 2014; Xu et al. 2003). The posttranscriptional regulation of IsiA expression is mediated by the antisense RNA IsrR (iron stress-repressed RNA) (Dühring et al. 2006). IsrR is transcribed from the noncoding strand of *isiA* and forms heteroduplexes with *isiA* mRNA, and the IsrR-isiA mRNA heteroduplexes are then targeted for selective degradation (Dühring et al. 2006).

3.2 Fur

Fur (ferric-uptake regulator) was first identified in the Gram-negative bacteria *Salmonella typhimurium* and *Escherichia coli* as a Crp/Fnr family transcriptional regulator, which regulates many genes involved in iron metabolism (Ernst et al. 1978; Hantke 1981). Fur acts as a transcription repressor only in the presence of its corepressor, ferrous iron. Under iron-sufficient conditions, Fur dimer binds to specific DNA sequences, termed Fur boxes, located in the promoter region. Fur binding prevents the RNA polymerase from binding to the promoter of target genes and thus represses the transcription of those genes (Escolar et al. 1999; Faraldo-Gomez and Sansom 2003; Hantke and Braun 1998; Yu and Genco 2012). At low iron concentrations, ferrous irons are released from Fur and reduce Fur's affinity for the Fur boxes, thus allowing the RNA polymerase to access the promoters and initiate transcription of target genes (Bagg and Neilands 1987; Hernández et al. 2006). Therefore, activation of iron-regulated genes is induced by dissociation of Fur.

As in most Gram-negative and several Gram-positive bacteria, Fur is also a global transcriptional regulator of iron acquisition in cyanobacteria (Andrews et al. 2003). Fur has been identified in a number of cyanobacteria species, such as *Synechococcus* 7942, *Synechocystis* 6803, *Anabaena* 7120, and *Microcystis aeruginosa* PCC 7806 (Ghassemian and Straus 1996; Bes et al. 2001; Hernández et al. 2002; Kaneko et al. 1996; Martin-Luna et al. 2006). Actually, the promoter regions of *isiAB*, *irpA*, and *mapA* genes all contain sequences resembling Fur boxes (Straus 1994). It has been shown that *isiA* and *isiB* genes are transcriptionally repressed by Fur in *Synechococcus* 7942 and *Anabaena* 7120 (Gonzalez et al. 2010, 2011, 2012, 2014; Ghassemian and Straus 1996; Bes et al. 2001; Leonhardt and Straus 1994). There are three fur-type genes in *Anabaena* 7120, but only one of them, *furA*, appears directly involved in regulation of iron-response genes (Gonzalez et al. 2012; Hernández et al. 2004; Ludwig et al. 2015). FurA is a master regulator of iron homeostasis in *Anabaena* 7120 and presumably in other cyanobacteria species (Gonzalez et al. 2010, 2011, 2012, 2014, 2016).

3.3 IdiA

Another prominent iron-regulated protein of cyanobacteria is IdiA (iron-deficiency-induced protein A) (Michel and Pistorius 1992; Michel et al. 1996). IdiA is only weakly expressed under iron-sufficient conditions, but its expression is highly elevated under iron deficiency in *Synechococcus elongatus* PCC 6301 and *Synechococcus* 7942 (Michel and Pistorius 1992). Early cell-fractionation experiments indicated that IdiA was mainly associated with the cytoplasmic side of thylakoid membranes (Michel et al. 1996, 1998).

The blue-native gel electrophoresis and ion exchange chromatography showed that part of the cellular IdiA associates with PS II (Exss-Sonne et al. 2000). Electron microscopic analysis showed that IdiA directly interacts with CP43 and D1 of PSII (Lax et al. 2007). Additionally, when grown under iron-deficient conditions, the IdiA depletion mutant showed reduced PS II activity, whereas the PS I activity was not affected (Exss-Sonne et al. 2000). All these evidences support that IdiA plays an important role in protecting the acceptor side of PS II, which is more exposed due to phycobilisome degradation during iron limitation (Shcolnick et al. 2009; Nodop et al. 2008). Moreover, IdiA shows considerable sequence similarity to some members of ABC transporter systems (such as FutA), and the association between IdiA and outer membrane has been detected in some freshwater cyanobacteria strains, implying a potential role of IdiA in iron transport (Fulda et al. 2000; Tolle et al. 2002).

The expression of IdiA is regulated by the transcriptional activator IdiB (Michel et al. 2001). IdiB is a helix-turn-helix transcriptional regulator of Crp/Fnr family (Michel and Pistorius 2004). IdiB is induced upon iron deprivation and also regulates other genes involved in iron stress responses (Yousef et al. 2003).

3.4 PfsR

Another protein is of much importance for iron stress response in cyanobacteria is PfsR (photosynthesis, Fe homeostasis, and stress response regulator), although it has not been studied extensively. PfsR is encoded by the gene *sll1392* in *Synechocystis* 6803 and is identified in a genetic screen for the suppressor that alleviates the lethality of high light to the high light-sensitive strain (Jantaro et al. 2006). The transcription of PfsR is induced by iron deficiency (Cheng and He 2014). It has been shown that in *Synechocystis* 6803, the *pfsR* deletion mutant has higher iron quota, displays enhanced viability, accumulates more photosynthetic pigments, and maintains high-level photosynthetic apparatus compared to the wild type during iron limitation (Cheng and He 2014). The *pfsR* deletion mutant also exhibits higher photosynthetic activity and efficiency than the wild type under iron-limiting conditions (Cheng and He 2014).

PfsR resembles the tetracycline repressor (TetR) family of transcriptional regulators, implying its role in regulating gene expression (Jantaro et al. 2006). Cheng and

He have shown that PfsR autoregulates its own transcription by binding to its own promoter, and PfsR also regulates the transcription of a set of genes that involved in iron homeostasis, including *isiA*, *furA*, and *fut* genes, *feoB* and *bfr* genes, and *ho* genes (Cheng and He 2014).

4 Conclusion

Iron deficiency is a particular challenge for cyanobacteria, because cyanobacteria require large amount of iron to maintain various biological activities, especially photosynthesis. To survive under iron-limiting conditions, cyanobacteria have developed many smart strategies to optimize iron acquisition and iron utilization while adjusting their photosynthesis activities. At the molecular level, the expression or activity of diverse proteins (such as IsiA, FurA, IdiA, and PfsR) is regulated by iron availability, and these proteins are further involved in the adaptation responses to iron stress. The cyanobacterial studies pave the way for understanding the mechanisms to cope with iron stress in photosynthetic microorganisms.

References

Achilles, K. M., Church, T. M., Wilhelm, S. W., et al. (2003). Bioavailability of iron to Trichodesmium colonies in the western subtropical Atlantic Ocean. *Limnology and Oceanography, 48*, 2250–2255.

Allen, J. F., & Vermaas, W. F. J. (2010). Evolution of photosynthesis. In *eLS*. Chichester: Wiley.

Andrews, S. C., Robinson, A. K., & Rodríguez-Quiñones, F. (2003). Bacterial iron homeostasis. *FEMS Microbiology Reviews, 27*, 215–237.

Andrizhiyevskaya, E. G., Schwabe, T. M., Germano, M., et al. (2002). Spectroscopic properties of PSI-IsiA supercomplexes from the cyanobacterium Synechococcus PCC 7942. *Biochimica et Biophysica Acta-Bioenergetics, 1556*, 265–272.

Andrizhiyevskaya, E. G., Frolov, D., van Grondelle, R., et al. (2004). Energy transfer and trapping in the photosystem I complex of Synechococcus PCC 7942 and in its supercomplex with IsiA. *Biochimica et Biophysica Acta, 1656*, 104–113.

Arnistroug, J. E., & Van Baalen, C. (1979). Iron transport in microalgae: The isolation and biochemical'activity of a hydroxamate siderophore from the blue-green' alga, Agmenellum quadruplicatum. *Journal of General Microbiology, 111*, 253–262.

Aspinwall, C. L., Duncan, J., Bibby, T., et al. (2004). The trimeric organisation of photosystem I is not necessary for the iron-stress induced CP43' protein to functionally associate with this reaction centre. *FEBS Letters, 574*, 126–130.

Babykin, M. M., Obando, S. T. A., & Zinchenko, V. V. (2018). TonB-dependent utilization of dihydroxamate xenosiderophores in Synechocystis sp. PCC 6803. *Current Microbiology, 75*, 117–123.

Badarau, A., Firbank, S. J., Waldron, K. J., et al. (2008). FutA2 is a ferric binding protein from Synechocystis PCC 6803. *The Journal of Biological Chemistry, 283*, 12520–12527.

Bagg, A., & Neilands, J. B. (1987). Molecular mechanism of regulation of siderophore-mediated iron assimilation. *Microbiological Reviews, 51*, 509–518.

Baniulis, D., Yamashita, E., Zhang, H., et al. (2008). Structure–function of the cytochrome b6f complex. *Photochemistry and Photobiology, 84*, 1349–1358.

Behrenfeld, M. J., & Kolber, Z. S. (1999). Widespread iron limitation of phytoplankton in the South Pacific Ocean. *Science, 283*, 840–843.

Behrenfeld, M. J., Bale, A. J., Kolber, Z. S., et al. (1996). Confirmation of iron limitation of phytoplankton photosynthesis in the equatorial pacific ocean. *Nature, 383*, 508–511.

Bekker, A., Holland, H. D., Wang, P. L., et al. (2004). Dating the rise of atmospheric oxygen. *Nature, 427*, 117–120.

Bellenger, J. P., Wichard, T., Xu, Y., & Kraepiel, A. M. L. (2011). Essential metals for nitrogen fixation in a free-living N2-fixing bacterium: Chelation, homeostasis and high use efficiency. *Environmental Microbiology, 13*, 1395–1411.

Bes, M. T., Herna'ndez, J. A., Peleato, M. L., et al. (2001). Cloning, overexpression and interaction of recombinant Fur from the cyanobacterium Anabaena PCC 7119 with isiB and its own promoter. *FEMS Microbiology Letters, 194*(2), 187–192.

Bibby, T. S., Nield, J., & Barber, J. (2001a). Iron deficiency induces the formation of an antenna ring around trimeric photosystem I in cyanobacteria. *Nature, 412*, 743–745.

Bibby, T. S., Nield, J., & Barber, J. (2001b). Three-dimensional model and characterization of the iron stress-induced CP43'-photosystem I supercomplex isolated from the cyanobacterium Synechocystis PCC 6803. *The Journal of Biological Chemistry, 276*, 43246–43252.

Boekema, E. J., Hifney, A., Yakushevska, A. E., et al. (2001). A giant chlorophyll-protein complex induced by iron deficiency in cyanobacteria. *Nature, 412*, 745–748.

Boyd, P. W., & Ellwood, M. J. (2010). The biogeochemical cycle of iron in the ocean. *Nature Geoscience, 3*, 675–682.

Boyer, G. L., Gillam, A. H., & Trick, C. (1987). Iron chelation and uptake. In P. Fay & C. Van Baalen (Eds.), *The cyanobacteria* (pp. 415–436). Amsterdam: Elsevier Scientific Publishers.

Brand, L. E. (1991). Minimum iron requirements of marine phytoplankton and the implications for the biogeochemical control of new production. *Limnology and Oceanography, 36*, 1756–1771.

Brand, L. E., Sunda, W. G., & Guillard, R. R. L. (1983). Limitation of marine phytoplankton reproductive rates by zinc, manganese, and iron. *Limnology and Oceanography, 28*, 1182–1198.

Brandt, A. M., Raksajit, W., Mulo, P., et al. (2009). Transcriptional regulation and structural modeling of the FutC subunit of an ABC-type iron transporter in Synechocystis sp. strain PCC 6803. *Archives of Microbiology, 191*, 561–570.

Braun, V., Schaffer, S., Hantke, K., et al. (1990). Regulation of gene expression by iron. In G. Hauska & R. Thauer (Eds.), *The molecular basis of bacterial metabolism* (pp. 35–51). Berlin: VCH-Verlagsgesellschaft.

Bricker, T. M., & Frankel, L. K. (2002). The structure and function of CP47 and CP43 in photosystem II. *Photosynthesis Research, 72*, 131–146.

Burnap, R. L., Troyan, T., & Sherman, L. A. (1993). The highly abundant chlorophyll-protein complex of iron-deficient Synechococcus sp. PCC7942 (CP43') is encoded by the isiA gene. *Plant Physiology, 103*(3), 893–902.

Carr, N. G., & Mann, N. H. (1994). The oceanic cyanobacterial picoplankton. In D. A. Bryant (Ed.), *The molecular biology of cyanobacteria* (Advances in photosynthesis) (Vol. 1, pp. 27–48). Dordrecht: Springer.

Chauhan, D., Folea, I. M., Jolley, C. C., et al. (2011). A novel photosynthetic strategy for adaptation to low-iron aquatic environments. *Biochemistry, 50*, 686–692.

Cheng, D., & He, Q. (2014). PfsR is a key regulator of iron homeostasis in Synechocystis PCC 6803. *PLoS One, 9*(7), e101743.

Fraústo da Silva, J. J. R., & Williams, R. J. P. (2001). *The biological chemistry of the elements: The inorganic chemistry of life*. Oxford: Oxford University Press.

Daddy, S., Zhan, J., Jantaro, S., et al. (2015). A novel high light-inducible carotenoid-binding protein complex in the thylakoid membranes of Synechocystis PCC 6803. *Scientific Reports, 5*, 9480.

Drechsel, H., & Winkelmann, G. (1997). Iron chelation and siderophores. In G. Winkelmann & C. J. Carrano (Eds.), *Transition metals in microbial metabolism* (pp. 1–49). Amsterdam: Harwood Academic Publishers.

Dühring, U., Axmann, I. M., Hess, W. R., et al. (2006). An internal antisense RNA regulates expression of the photosynthesis gene isiA. *Proceedings of the National Academy of Sciences of the United States of America, 103*, 7054–7058.

Eldridge, M., Trick, C., Alm, M., et al. (2004). Phytoplankton community response to a manipulation of bioavailable iron in HNLC waters of the subtropical Pacific Ocean. *Aquatic Microbial Ecology, 35*, 79–91.

Ernst, J. F., Bennett, R. L., & Rothfield, L. I. (1978). Constitutive expression of the iron-enterochelin and ferrichrome uptake systems in a mutant strain of Salmonella typhimurium. *Journal of Bacteriology, 135*, 928–934.

Escolar, L., Perez-Martin, J., & de Lorenzo, V. (1999). Opening the iron box: Transcriptional metalloregulation by the Fur protein. *Journal of Bacteriology, 181*, 6223–6229.

Exss-Sonne, P. J., Toelle, K. P., Bader, E. K., et al. (2000). The IdiA protein of Synechococcus sp. PCC 7942 functions in protecting photosystem II under oxidative stress. *Photosynthesis Research, 63*, 145–157.

Falk, S., Samson, G., Bruce, D., et al. (1995). Functional analysis of the iron-stress induced CP43P polypeptide of PSII in the cyanobacterium Synechococcus sp. PCC 7942. *Photosynthesis Research, 45*, 51–60.

Falkowski, P. G., & Raven, J. A. (2007). *Aquatic photosynthesis* (2nd ed.). Princeton: Princeton University Press.

Faraldo-Gomez, J. D., & Sansom, M. S. P. (2003). Acquisition of siderophores in gram-negative bacteria. *Nature Reviews. Molecular Cell Biology, 4*, 105–116.

Ferreira, F., & Straus, N. A. (1994). Iron deprivation in cyanobacteria. *Journal of Applied Phycology, 6*, 199–210.

Fillat, M. F., Sandmann, G., & Gómez-Moreno, C. (1988). Flavodoxin from the nitrogen-fixing cyanobacterium Anabaena PCC 7119. *Archives of Microbiology, 150*, 160–164.

Fillat, M. F., Edmondson, D. E., & Gomez-Moreno, C. (1990). Structural and chemical properties of a flavodoxin from Anabaena PCC 7119. *Biochimica et Biophysica Acta, 1040*(2), 301–307.

Finney, L. A., & O'Halloran, T. V. (2003). Transition metal speciation in the cell: Insights fromthechemistryofmetal ionreceptors. *Science, 300*, 931–936.

Fulda, S., & Hagemann, M. (1995). Salt treatment induces accumulation of flavodoxin in the cyanobacterium Synechocystis sp. PCC 6803. *Journal of Plant Physiology, 146*(4), 520–526.

Fulda, S., Huang, F., Nilsson, F., et al. (2000). Proteomics of Synechocystis sp. strain PCC 6803: Identification of periplasmic proteins in cells grown at low and high salt concentrations. *European Journal of Biochemistry, 267*, 5900–5907.

Geiss, U., Vinnemeier, J., Kunert, A., et al. (2001). Detection of the isiA gene across cyanobacterial strains: Potential for probing iron deficiency. *Applied and Environmental Microbiology, 67*(11), 5247–5253.

Ghassemian, M., & Straus, N. A. (1996). Fur regulates the expression of iron-stress genes in the cyanobacterium Synechococcus sp. strain PCC 7942. *Microbiology, 142*, 1469–1476.

Gonzalez, A., Bes, M. T., Barja, F., et al. (2010). Overexpression of FurA in Anabaena sp. PCC 7120 reveals new targets for this regulator involved in photosynthesis, iron uptake and cellular morphology. *Plant & Cell Physiology, 51*, 1900–1914.

Gonzalez, A., Bes, M. T., Peleato, M. L., et al. (2011). Unraveling the regulatory function of FurA in Anabaena sp. PCC 7120 through 2-D DIGE proteomic analysis. *Journal of Proteomics, 74*, 660–671.

Gonzalez, A., Bes, M. T., Valladares, A., et al. (2012). FurA is the master regulator of iron homeostasis and modulates the expression of tetrapyrrole biosynthesis genes in Anabaena sp. PCC 7120. *Environmental Microbiology, 14*(12), 3175–3187.

Gonzalez, A., Angarica, V. E., Sancho, J., et al. (2014). The FurA regulon in Anabaena sp. PCC 7120: In silico prediction and experimental validation of novel target genes. *Nucleic Acids Research, 42*, 4833–4846.

Gonzalez, A., Bes, M. T., Peleato, M. L., et al. (2016). Expanding the role of FurA as essential global regulator in cyanobacteria. *PLoS One, 11*(3), e0151384.

Grossman, A. R., Schaefer, M. R., Chiang, G. G., & Collier, J. L. (1993). The phycobilisome, a light-harvesting complex responsive to environmental conditions. *Microbiological Reviews, 57*, 725–749.

Guerinot, M. L., & Yi, Y. (1994). Iron: Nutritious, noxious, and not readily available. *Plant Physiology, 104*, 815–820.

Guikema, J. A., & Sherman, L. A. (1983). Organization and function of chlorophyll in membranes of cyanobacteria during iron-starvation. *Plant Physiology, 73*, 250–256.

Guikema, J. A., & Sherman, L. A. (1984). Influence of iron deprivation on the membrane composition of Anacystis nidulans. *Plant Physiology, 74*, 90–95.

Hagemann, M., Jeanjean, R., Fulda, S., et al. (1999). Flavodoxin accumulation contributes to enhanced cyclic electron flow around photosystem I in salt-stressed cells of Synechocystis sp strain PCC 6803. *Physiologia Plantarum, 105*, 670–678.

Hantke, K. (1981). Regulation of ferric iron transport in Escherichia coli K12: Isolation of a constitutive mutant. *Molecular & General Genetics, 182*, 288–292.

Hantke, K. (2001). Iron and metal regulation in bacteria. *Current Opinion in Microbiology, 4*, 172–177.

Hantke, K., & Braun, V. (1998). Control of bacterial iron transport by regulatory proteins. In S. Silver & W. Walden (Eds.), *Metal ions in gene regulation* (pp. 11–45). New York: International Thomson Publishing.

Havaux, M., Guedeney, G., Hagemann, M., et al. (2005). The chlorophyllbinding protein IsiA is inducible by high light and protects the cyanobacterium Synechocystis PCC6803 from photo-oxidative stress. *FEBS Letters, 579*, 2289–2293.

Hernández, J. A., Artieda, M., Peleato, M. L., et al. (2002). Iron stress and genetic response in cyanobacteria: Fur genes from Synechococcus PCC 7942 and Anabaena PCC 7120. *Annales de Limnologie, 38*(1), 3–11.

Hernández, J. A., Lopez-Gomollon, S., Bes, M. T., et al. (2004). Three fur homologues from Anabaena sp. PCC7120: Exploring reciprocal protein-promoter recognition. *FEMS Microbiology Letters, 236*(2), 275–282.

Hernández, J. A., López-Gomollón, S., Muro-Pastor, A., et al. (2006). Interaction of FurA from Anabaena sp PCC 7120 with DNA: A reducing environment and the presence of Mn2+ are positive effectors in the binding to isiB and furA promoters. *Biometals, 19*, 259–268.

Hopkinson, B. M., & Morel, F. M. (2009). The role of siderophores in iron acquisition by photosynthetic marine microorganisms. *Biometals, 22*, 659–669.

Hutber, G. N., Hutson, K. G., & Rogers, L. J. (1977). Effect of iron deficiency on levels of two ferredoxins and flavodoxin in a cyanobacterium. *FEMS Microbiology Letters, 1*, 193–196.

Ivanov, A. G., Park, Y. I., Miskiewicz, E., et al. (2000). Iron stress restricts photosynthetic intersystem electron transport in Synechococcus sp. PCC 7942. *FEBS Letters, 485*, 173–177.

Ivanov, A. G., Krol, M., Sveshnikov, D., et al. (2006). Iron deficiency in cyanobacteria causes monomerization of photosystem I trimers and reduces the capacity for state transitions and the effective absorption cross section of photosystem I in vivo. *Plant Physiology, 141*, 1436–1445.

Jantaro, S., Ali, Q., Lone, S., et al. (2006). Suppression of the lethality of high light to a quadruple HLI mutant by the inactivation of the regulatory protein PfsR in Synechocystis PCC 6803. *The Journal of Biological Chemistry, 281*(41), 30865–30874.

Jeanjean, R., Zuther, E., Yeremenko, N., et al. (2003). A photosystem I psaFJ-null mutant of the cyanobacterium Synechocystis PCC 6803 expresses the isiAB operon under iron replete conditions. *FEBS Letters, 549*, 52–56.

Jordan, P., Fromme, P., Witt, H. T., et al. (2001). Three-dimensional structure of cyanobacterial photosystem I at 2.5 A ° resolution. *Nature, 411*, 909–917.

Kaneko, T., Sato, S., Kotani, H., et al. (1996). Sequence analysis of the genome of the unicellular cyanobacterium Synechocystis sp. strain PCC6803. II. Sequence determination of the entire genome and assignment of potential protein-coding regions. *DNA Research, 3*(3), 109–136.

Katoh, H., Hagino, N., & Grossman, A. R. (2001a). Genes essential to iron transport in the cyanobacterium Synechocystis sp. strain PCC 6803. *Journal of Bacteriology, 183*, 2779–2784.

Katoh, H., Hagino, N., & Ogawa, T. (2001b). Iron-binding of FutA1 subunit of an ABC-type iron transporter in the cyanobacterium Synechocystis sp. strain PCC 6803. *Plant & Cell Physiology, 42*, 823–827.

Keren, N., Aurora, R., & Pakrasi, H. B. (2004). Critical roles of bacterioferritins in iron storage and proliferation of cyanobacteria. *Plant Physiology, 135*, 1666–1673.

Koropatkin, N., Randich, A. M., Bhattacharyya-Pakrasi, M., et al. (2007). The structure of the iron-binding protein, FutA1, from Synechocystis 6803. *The Journal of Biological Chemistry, 282*, 27468–27477.

Kouřil, R., Yeremenko D'Haene, N. S., et al. (2003). Photosystem I trimers from Synechocystis PCC 6803 lacking the PsaF and PsaJ subunits bind an IsiA ring of 17 units. *Biochimica et Biophysica Acta, 1607*, 1–4.

Kouřil, R., Arteni, A. A., Lax, J., et al. (2005). Structure and functional role of supercomplexes of IsiA and photosystem I in cyanobacterial photosynthesis. *FEBS Letters, 579*, 3253–3257.

Kranzler, C., Rudolf, M., Keren, N., et al. (2013). Iron in cyanobacteria. *Advances in Botanical Research, 65*, 57–105.

Kranzler, C., Lis, H., Finkel, O. M., et al. (2014). Coordinated transporter activity shapes high-affinity iron acquisition in Cyanobacteria. *ISMEJ, 8*, 409–417.

Kunert, A., Vinnemeier, J., Erdmann, N., et al. (2003). Repression by Fur is not the main mechanism controlling the iron-inducible isiAB operon in the cyanobacterium Synechocystis sp. PCC 6803. *FEMS Microbiology Letters, 227*, 255–262.

Latifi, A., Jeanjean, R., Lemeille, S., et al. (2005). Iron starvation leads to oxidative stress in Anabaena sp. strain PCC 7120. *Journal of Bacteriology, 187*, 6596–6598.

Latifi, A., Ruiz, M., & Zhang, C. C. (2009). Oxidative stress in cyanobacteria. *FEMS Microbiology Reviews, 33*(2), 258–278.

Lax, J. E. M., Arteni, A. A., Boekema, E., et al. (2007). Structural response of photosystem II to iron deficiency: Characterization of a new photosystem II-IdiA complex from the cyanobacterium Thermosynechococcus elongatus BP-1. *Biochimica et Biophysica Acta, 1767*, 528–534.

Leonhardt, K., & Straus, N. A. (1992). An iron stress operon involved in photosynthetic electron transport in the marine cyanobacterium Synechococcus sp. PCC 7002. *Journal of General Microbiology, 138*, 1613–1621.

Leonhardt, K., & Straus, N. A. (1994). Photosystem II genes isiA, psbDI and psbC in Anabaena sp. PCC 7120: Cloning, sequencing and the transcriptional regulation in iron-stressed and iron-repleted cells. *Plant Molecular Biology, 24*(1), 63–73.

Li, H., Singh, A. K., McIntyre, L. M., et al. (2004). Differential gene expression in response to hydrogen peroxide and the putative PerR regulon of Synechocystis sp. strain PCC 6803. *Journal of Bacteriology, 186*, 3331–3345.

Lodeyro, A. F., Ceccoli, R. D., Pierella Karlusich, J. J., et al. (2012). The importance of flavodoxin for environmental stress tolerance in photosynthetic microorganisms and transgenic plants. Mechanism, evolution and biotechnological potential. *FEBS Letters, 586*(18), 2917–2924.

Ludwig, M., Chua, T. T., Chew, C. Y., et al. (2015). Fur-type transcriptional repressors and metal homeostasis in the cyanobacterium Synechococcus sp. PCC 7002. *Frontiers in Microbiology, 6*, 1217.

Mann, E. L., & Chisholm, S. W. (2000). Iron limits the cell division rate of Prochlorococcus in the eastern equatorial Pacific. *Limnology and Oceanography, 45*(5), 1067–1076.

Martin, J. H., & Fitzwater, S. E. (1988). Iron deficiency limits phytoplankton growth in the north-East Pacific subarctic. *Nature, 331*, 341–343.

Martin, J. H., Coale, K. H., Johnson, K. S., et al. (1994). Testing the iron hypothesis in ecosystems of the equatorial pacific ocean. *Nature, 371*, 123–129.

Martin-Luna, B., Hernandez, J. A., Bes, M. T., et al. (2006). Identification of a ferric uptake regulator from Microcystis aeruginosa PCC7806. *FEMS Microbiology Letters, 254*(1), 63–70.

Melkozernov, A. N., Bibby, T. S., Lin, S., et al. (2003). Time-resolved absorption and emission show that the CP43' antenna ring of iron-stressed Synechocystis sp. PCC6803 is efficiently coupled to the photosystem I reaction center core. *Biochemistry, 42*, 3893–3903.

Michel, K. P., & Pistorius, E. K. (1992). Isolation of a photosystem II associated 36 kDa polypeptide and an iron stress 34 kDa polypeptide from thylakoid membranes of the cyanobacterium Synechococcus PCC 6301 grown under mild iron deficiency. *Z Naturforsch Teil C Biochem Biophys Biol Virol, 47*, 867–874.

Michel, K. P., & Pistorius, E. K. (2004). Adaptation of the photosynthetic electron transport chain in cyanobacteria to iron deficiency: The function of IdiA and IsiA. *Physiologia Plantarum, 119*, 1–15.

Michel, K. P., Thole, H. H., & Pistorius, E. K. (1996). IdiA, a 34 kDa protein in the cyanobacteria Synechococcus sp. strains PCC 6301 and PCC 7942, is required for growth under iron and manganese limitations. *Microbiology, 142*(Pt9), 1635–1645.

Michel, K. P. P., Exss-Sonne, G., Scholten-Beck, U., et al. (1998). Immunocytochemical localization of IdiA, a protein expressed under iron or manganese limitation in the mesophilic cyanobacterium Synechococcus PCC 6301 and the thermophilic cyanobacterium Synechococcus elongatus. *Planta, 205*, 73–81.

Michel, K. P., Pistorius, E. K., & Golden, S. S. (2001). Unusual regulatory elements for iron deficiency induction of the idiA gene of Synechococcus elongatus PCC 7942. *Journal of Bacteriology, 183*, 5015–5024.

Miethke, M., & Marahiel, M. A. (2007). Siderophore-based iron acquisition and pathogen control. *Microbiology and Molecular Biology Reviews, 71*, 413.

Mirus, O., Strauss, S., Nicolaisen, K., et al. (2009). TonB-dependent transporters and their occurrence in cyanobacteria. *BMC Biology, 7*, 68.

Morel, F. M. M., Kustka, A. B., & Shaked, Y. (2008). The role of unchelated Fe in the iron nutrition of phytoplankton. *Limnology and Oceanography, 53*, 400–404.

Neilands, J. B. (1982). Microbial envelope proteins related to iron. *Annual Review of Microbiology, 36*, 285–309.

Neilands, J. B. (1995). Siderophores: Structure and function of microbial iron transport compounds. *The Journal of Biological Chemistry, 270*, 26723–26726.

Nicolaisen, K., Moslavac, S., Samborski, A., et al. (2008). Alr0397 is an outer membrane transporter for the siderophore schizokinen in Anabaena sp strain PCC 7120. *Journal of Bacteriology, 190*, 7500–7507.

Nicolaisen, K., Hahn, A., Valdebenito, M., et al. (2010). The interplay between siderophore secretion and coupled iron and copper transport in the heterocyst-forming cyanobacterium Anabaena sp. PCC 7120. *Biochimica et Biophysica Acta, 1798*, 2131–2140.

Nodop, A., Pietsch, D., Höcker, R., et al. (2008). Transcript profiling reveals new insights into the acclimation of the emesophilic fresh-water cyanobacterium Synechococcus elongatus PCC 7942 to iron starvation. *Plant Physiology, 147*, 747–763.

Noinaj, N., Guillier, M., Barnard, T. J., et al. (2010). TonB-dependent transporters: Regulation, structure, and function. *Annual Review of Microbiology, 64*, 43–60.

Obando, S. T. A., Babykin, M. M., & Zinchenko, V. V. (2018). A cluster of five genes essential for the utilization of dihydroxamate xenosiderophores in Synechocystis sp. PCC 6803. *Current Microbiology, 75*, 1165–1173.

Odom, W. R., Hodges, R., Chitnis, P. R., et al. (1993). Characterization of Synechocystis sp. PCC 6803 in iron-supplied and iron-deficient media. *Plant Molecular Biology, 23*, 1255–1264.

Öquist, G. (1971). Changes in pigment composition and photosynthesis induced by iron-deficiency in blue-green-alga Anacystis nidulans. *Physiologia Plantarum, 25*, 188–191.

Pakrasi, H. B., Goldenberg, A., & Sherman, L. A. (1985). Membrane development in the cyanobacterium, Anacystis nidulans, during recovery from Iron starvation. *Plant Physiology, 79*(1), 290–295.

Park, Y. I., Sandstrom, S., Gustafsson, P., et al. (1999). Expression of the isiA gene is essential for the survival of the cyanobacterium Synechococcus sp. PCC 7942 by protecting photosystem II from excess light under iron limitation. *Molecular Microbiology, 32*(1), 123–129.

Pierella Karlusich, J. J., Ceccoli, R. D., Grana, M., et al. (2015). Environmental selection pressures related to iron utilization are involved in the loss of the flavodoxin gene from the plant genome. *Genome Biology and Evolution, 7*(3), 750–767.

Ratledge, C., & Dover, L. G. (2000). Iron metabolism in pathogenic bacteria. *Annual Review of Microbiology, 54*, 881–941.

Raven, J. A., Evans, M. C. W., & Korb, R. E. (1999). The role of trace metals in photosynthetic electron transport in O2-evolving organisms. *Photosynthesis Research, 60*, 111–149.

Razquin, P., Schmitz, S., & Peleato, M. L. (1995). Differential activities of heterocyst ferredoxin, vegetative cell ferredoxin, and flavodoxin as electron carriers in nitrogen-fixation and photosynthesis in Anabaena sp. *Photosynthesis Research, 43*(1), 35–40.

Rich, H. W., & Morel, F. M. M. (1990). Availability of well-defined iron colloids to the marine diatom Thalassiosira weissflogii. *Limnology and Oceanography, 35*(3), 652–662.

Richardson, D. J. (2000). Bacterial respiration: A flexible process for a changing environment. *Microbiology, 146*(Pt 3), 551–571.

Richier, S., Macey, A. I., Pratt, N. J., et al. (2012). Abundances of iron-binding photosynthetic and nitrogen-fixing proteins of trichodesmium both in culture and in situ from the North Atlantic. *PLoS One, 7*, e35571.

Riethman, H. C., & Sherman, L. A. (1988). Purification and characterization of an iron stress-induced chlorophyll-protein from the cyanobacterium Anacystis nidulans R2. *Biochimica et Biophysica Acta, 935*, 141–151.

Rocap, G., Larimer, F. W., Lamerdin, J., et al. (2003). Genome divergence in two Prochlorococcus ecotypes reflects oceanic niche differentiation. *Nature, 424*(6952), 1042–1047.

Rudolf, M., Kranzler, C., Lis, H., et al. (2015). Multiple modes of iron uptake by the filamentous, siderophore-producing cyanobacterium, Anabaena sp. PCC 7120. *Molecular Microbiology, 97*, 577–588.

Rueter, J. G., Ohki, K., & Fujita, Y. (1990). The effect of iron nutrition on photosynthesis and nitrogen fixation in cultures of Trichodesmium (cyanophyceae) 1. *Journal of Phycology, 26*, 30–35.

Sandmann, G. (1985). Consequences of iron deficiency on photosynthetic and respiratory electron transport in blue-green algae. *Photosynthesis Research, 6*, 261–271.

Sandmann, G., & Malkin, R. (1983). Iron-sulfur centers and activities of the photosynthetic electron transport chain in iron-deficient cultures of the blue-green alga Aphanocapsa. *Plant Physiology, 73*, 724–728.

Sandström, S., Park, Y. I., Oquist, G., et al. (2001). CP43', the isiA gene product, functions as an excitation energy dissipator in the cyanobacterium Synechococcus sp. PCC 7942. *Photochemistry and Photobiology, 74*, 431–437.

Sandström, S., Ivanov, A. G., Park, Y. I., et al. (2002). Iron stress responses in the cyanobacterium Synechococcus sp. PCC 7942. *Physiologia Plantarum, 116*, 255–263.

Schrader, P. S., Milligan, A. J., & Behrenfeld, M. J. (2011). Surplus photosynthetic antennae complexes underlie diagnostics of iron limitation in a cyanobacterium. *PLoS One, 6*, e18753.

Shcolnick, S., & Keren, N. (2006). Metal homeostasis in cyanobacteria and chloroplasts. Balancing benefits and risks to the photosynthetic apparatus. *Plant Physiology, 141*, 805–810.

Shcolnick, S., Summerfield, T. C., Reytman, L., et al. (2009). The mechanism of iron homeostasis in the unicellular cyanobacterium synechocystis sp. PCC 6803 and its relationship to oxidative stress. *Plant Physiology, 150*, 2045–2056.

Sherman, D. M., & Sherman, L. A. (1983). Effect of iron deficiency and iron restoration on ultrastructure of Anacystis nidulans. *Journal of Bacteriology, 156*, 393–401.

Shi, T., Sun, Y., & Falkowski, P. G. (2007). Effects of iron limitation on the expression of metabolic genes in the marine cyanobacterium Trichodesmium erythraeum IMS101. *Environmental Microbiology, 9*, 2945–2956.

Simpson, F. B., & Nielands, J. B. (1976). Siderochromes in cyanophyceae: Isolation and characterization of schizokinen from Anabaena sp. *Journal of Phycology, 12*(44), 48.

Straus, N. A. (1994). Iron deprivation: Physiology and gene regulation. In D. A. Bryant (Ed.), *The molecular biology of cyanobacteria* (Advances in photosynthesis) (Vol. 1, pp. 731–750). Dordrecht: Springer.

Sun, J., & Golbeck, J. H. (2015). The presence of the IsiA-PSI supercomplex leads to enhanced photosystem I electron throughput in iron-starved cells of Synechococcus sp. PCC 7002. *The Journal of Physical Chemistry. B, 119*(43), 13549–13559.

Tognetti, V. B., Zurbriggen, M. D., Morandi, E. N., et al. (2007). Enhanced plant tolerance to iron starvation by functional substitution of chloroplast ferredoxin with a bacterial flavodoxin. *Proceedings of the National Academy of Sciences of the United States of America, 104*(27), 11495–11500.

Tolle, J., Michel, K. P., Kruip, J., et al. (2002). Localization and function of the IdiA homologue Slr1295 in the cyanobacterium Synechocystis sp. strain PCC 6803. *Microbiology, 148*, 3293–3305.

Trick, C. G., & Kerry, A. (1992). Isolation and purification of siderophores produced by cyanobacteria, Synechococcus sp. PCC 7942 and Anabaena variabilis ATCC 29413. *Current Microbiology, 24*, 241–245.

Vigara, A. J., Inda, L. A., & Vega, J. M. (1998). Flavodoxin as an electronic donor in photosynthetic inorganic nitrogen assimilation by iron-deficient Chlorella fusca cells. *Photochemistry and Photobiology, 67*(4), 446–449.

Vinnemeier, J., & Hagemann, M. (1999). Identification of salt-regulated genes in the genome of the cyanobacterium Synechocystis sp. strain PCC 6803 by subtractive RNA hybridization. *Archives of Microbiology, 172*, 377–386.

Waldron, K. J., Tottey, S., Yanagisawa, S., et al. (2007). A periplasmic iron-binding protein contributes toward inward copper supply. *The Journal of Biological Chemistry, 282*, 3837–3846.

Walworth, N. G., Fu, F. X., Webb, E. A., et al. (2016). Mechanisms of increased Trichodesmium fitness under iron and phosphorus co-limitation in the present and future ocean. *Nature Communications, 7*, 12081.

Wang, Q., Hall, C. L., Al-Adami, M. Z., et al. (2010). IsiA is required for the formation of photosystem I supercomplexes and for efficient state transition in Synechocystis PCC 6803. *PLoS One, 5*, e10432.

Webb, E. A., Moffett, J. W., & Waterbury, J. B. (2001). Iron stress in open-ocean cyanobacteria (Synechococcus, Trichodesmium, and Crocosphaera spp.): Identification of the IdiA protein. *Applied and Environmental Microbiology, 67*, 5444–5452.

Wilhelm, S. W. (1995). Ecology of iron-limited cyanobacteria: A review of physiological responses and implications for aquatic systems. *Aquatic Microbial Ecology, 9*, 295–303.

Xu, W., Jeanjean, R., Liu, Y., et al. (2003). pkn22 (alr2502) encoding a putative Ser/Thr kinase in the cyanobacterium Anabaena sp. PCC 7120 is induced by both iron starvation and oxidative stress and regulates the expression of isiA. *FEBS Letters, 553*(1–2), 179–182.

Yeremenko, N., Kouril, R., Ihalainen, J. A., et al. (2004). Supramolecular organization and dual function of the IsiA chlorophyll-binding protein in cyanobacteria. *Biochemistry, 43*, 10308–10313.

Yousef, N., Pistorius, E. K., & Michel, K. P. (2003). Comparative analysis of idiA and isiA transcription under iron starvation and oxidative stress in Synechococcus elongatus PCC 7942 wild-type and selected mutants. *Archives of Microbiology, 180*, 471–483.

Yu, C., & Genco, C. A. (2012). Fur-mediated global regulatory circuits in pathogenic Neisseria species. *Journal of Bacteriology, 194*, 6372–6381.

Adaptive Mechanisms of the Model Photosynthetic Organisms, Cyanobacteria, to Iron Deficiency

Hai-Bo Jiang, Xiao-Hui Lu, Bin Deng, Ling-Mei Liu, and Bao-Sheng Qiu

Abstract Cyanobacteria are the oldest *oxygen*-evolving photosynthetic organisms on the Earth. They are widely distributed in marine, freshwater, and terrestrial environments and contribute about 25% of global primary productivity. They are thought to be responsible for the conversion of the Earth's atmosphere from anaerobic to aerobic about 2.4 billion years ago. This development permitted the evolution of aerobic bacteria, algae, plants, and animals. However, due to the emergence of oxidative environments on the Earth's surface, soluble ferrous iron (Fe^{2+}) was almost completely oxidized to hardly soluble ferric iron (Fe^{3+}) in aquatic environments. The extremely low bioavailability of iron in the ocean has been considered as an important factor that is limiting global primary productivity. As photosynthetic organisms, cyanobacteria have higher iron demand than other non-photosynthetic organisms to meet the needs of photosynthetic electron transport and chlorophyll synthesis. The nitrogen-fixing cyanobacteria need even more iron to fix the inert dinitrogen gas. The contradiction between the high iron demand of cyanobacteria and their iron-limiting habitats has forced them to evolve special strategies to overcome iron deficiency during the long-period evolution. In this review, we summarized the recent perspectives on the physiological responses and special strategies of cyanobacteria to overcome the changing iron bioavailability in freshwater, coastal, and open-ocean environments.

Keywords Cyanobacteria · Primary productivity · Iron deficiency · Adaptive mechanisms

---

The original version of this chapter was revised. A correction to this chapter can be found at https://doi.org/10.1007/978-981-15-3110-1_16

H.-B. Jiang (✉)
School of Life Sciences, Hubei Key Laboratory of Genetic Regulation and Integrative Biology, Central China Normal University, Wuhan, Hubei, People's Republic of China

Southern Marine Science and Engineering Guangdong Laboratory (Zhuhai), Zhuhai, Guangdong, People's Republic of China
e-mail: haibojiang@mail.ccnu.edu.cn

X.-H. Lu · B. Deng · L.-M. Liu · B.-S. Qiu
School of Life Sciences, Hubei Key Laboratory of Genetic Regulation and Integrative Biology, Central China Normal University, Wuhan, Hubei, People's Republic of China

Q. Wang (ed.), *Microbial Photosynthesis*,
https://doi.org/10.1007/978-981-15-3110-1_11

1 The Feature of Cyanobacterial Cell Wall

Cyanobacteria, also known as blue-green algae, are the most ancient and simple organisms capable of *oxygen*-evolving photosynthesis. Cyanobacteria are considered as a special group of Gram-negative bacteria. Like other Gram-negative bacteria, they are enclosed with two membranes including plasma membrane and outer membrane, and a peptidoglycan layer between them (Fig. 1) (Liberton et al. 2006). Both of the membranes are composed of lipid bilayers which protect cells but also set up two barriers for nutrient import. Generally, the cell wall of Gram-positive bacteria has a thick (20–80 nm) and dense peptidoglycan layer, which accounts for 60–90% of the cell wall components, while Gram-negative bacteria usually have relatively thin cell walls (15–20 nm) and loose peptidoglycan layer (Liberton et al. 2006). Although the structure of cyanobacterial cell wall is similar to Gram-negative bacteria, its peptidoglycan layer is much thicker and denser than most of Gram-negative bacteria. The components of the outer membrane of cyanobacteria are also special, as their phospholipid bilayer has fewer phosphates and deoxyoctanoic acid (Mikheyskaya et al. 1977; Schmidt et al. 1980) and has some special components, such as beta-hydroxyhexadecanoic acid (Schrader et al. 1981). In addition, the cross-linking tightness of cyanobacterial peptidoglycan was more similar to those of Gram-positive bacteria. Of cause, the most special feature of cyanobacteria among the Gram-negative bacteria is that they have an independent membrane system for photosynthetic oxygen production, that is, thylakoid membrane (Fig. 1). Cyanobacteria are the evolutionary ancestors of chloroplasts in higher plants. It has

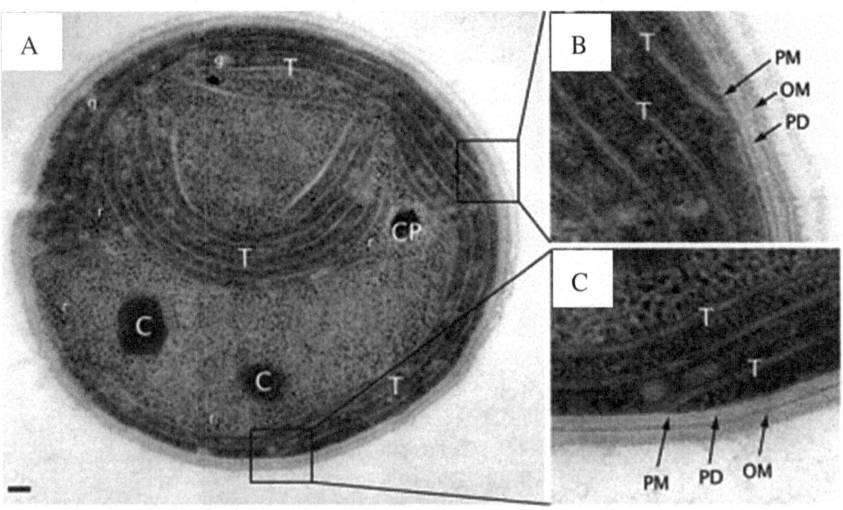

Fig. 1 The subcellular structure of cyanobacteria. (**a**) The electron micrograph through a *Synechocystis* sp. PCC 6803 cell. (**b**) and (**c**) are enlargements of the boxed areas, showing close proximity of thylakoid membrane and plasma membrane. *T* thylakoid membrane, *PM* plasma membrane, *OM* outer membrane, *PD* peptidoglycan layer. Bar: 100 nm (Liberton et al. 2006)

been known that cyanobacteria and the chloroplasts from higher plants share similar genome and structure. For example, the similarities between the genome of the model cyanobacterium *Synechocystis* sp. PCC 6803 and that of *Arabidopsis thaliana* chloroplasts are as high as 35% (Abdallah et al. 2000). Therefore, cyanobacteria are commonly considered as a good model for photosynthesis-related research.

2 The Distribution of Cyanobacteria and Its Significance in Global Primary Productivity

Cyanobacteria are widely distributed in oceans, freshwater, wetlands, deserts, reefs, soils, Antarctica, and most of the stromatolite ecosystems (de los Ríos et al. 2007). Cyanobacteria are the main phytoplankton in the ocean and the main contributors to global primary productivity (Field et al. 1998). About half of the global photosynthesis on Earth is accomplished by the ocean ecosystem, of which cyanobacteria contribute half of the marine photosynthesis. Thus, cyanobacteria have been estimated to contribute about 25% of the global primary productivity per year. Among them, *Prochlorococcus* and *Synechococcus* are the most dominant two species distributed almost anywhere of tropical and subtropical oceans (Fig. 2) (Flombaum et al. 2013). Cyanobacteria absorb CO_2 through photosynthesis and convert it into organic carbon, which forms the base of the marine food chain obtaining material and energy. One third of the organic carbon fixed by cyanobacteria precipitates into the deep ocean, isolating this carbon source from the atmosphere for hundreds to thousands of years (Volk and Hoffert 1985). This process, called "bio-carbon pump," plays an important role in mitigating the greenhouse effect caused by elevated atmospheric CO_2 concentration. Some cyanobacteria perform nitrogen fixation and provide new available nitrogen source supporting the ocean's food web. The concentration of nitrogen sources in the ocean directly affects the biomass of other species, thus affecting the utilization of many other nutrients by other organisms. Therefore, cyanobacteria play an important role in biogeochemistry cycle of carbon, nitrogen, sulfur, iron, and other chemical elements in the ocean ecosystem (Falkowski et al. 1998).

3 The Indissoluble Bond Between Cyanobacteria and Iron

Cyanobacteria have been living in the Earth for more than 2.7 billion years and played a crucial role in the transformation of the Earth's surface from anaerobic environment to the present aerobic environment (Schopf 2012). Over the next 300 million years after the emergence of cyanobacteria, the Earth's Great Oxygenation Event killed more than 99% of the early living organisms and led to the transformation of dominant life forms from anaerobe to aerobe, which is of great significance to the biodiversity of the Earth (Lane 2010). Before the Earth's

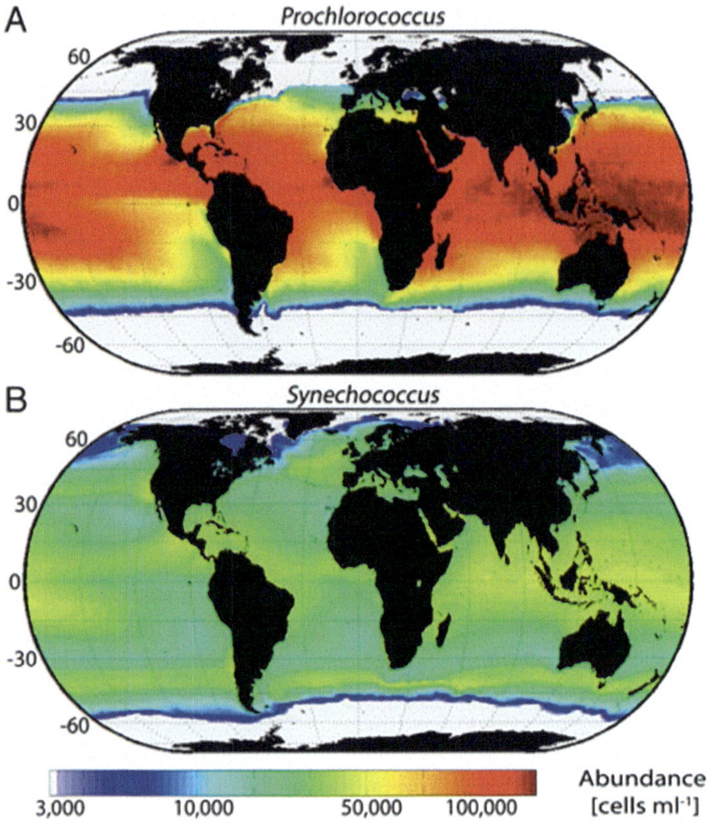

Fig. 2 Present global distribution of the most abundant cyanobacterial species *Prochlorococcus* and *Synechococcus*. (**a**) *Prochlorococcus* and (**b**) *Synechococcus* mean annual abundances at the sea surface (Flombaum et al. 2013)

Great Oxygenation, the Earth's surface was an anaerobic atmosphere, and iron in the crust mostly existed in high soluble ferrous form; therefore the bioavailable iron concentration was very high. Along with the Earth's surface which was changed from anaerobic environments to aerobic environments, most of the ferrous iron was oxidized into trivalent state and precipitated as $Fe(OH)_3$, which resulted in a dramatically low bioavailability of iron in the ocean (Fig. 3) (Lodeyro et al. 2012). The concentration of iron in the sea surface of the aquatic environment ranges from 10^{-9} M to 10^{-12} M. Especially in open oceans, the concentration of bioavailable iron was extremely low since there were no new sources of iron from the dusts and rivers. Thus, after the Earth's Great Oxygenation Event, almost all the living organisms on the Earth were facing the challenge of oxidative stress and iron deficiency stress. It has been shown with strong evidences that the oxidative stress and iron deficiency usually resulted to similar physiological and gene expression responses among different organisms.

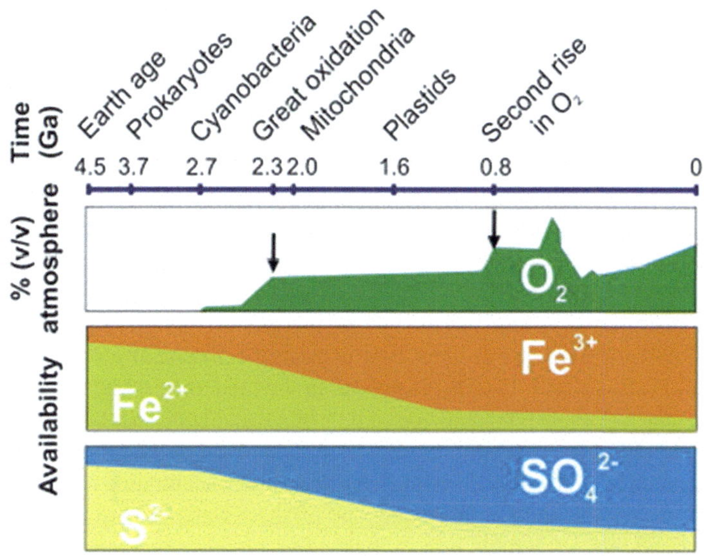

Fig. 3 Temporal relationships between atmospheric O_2 concentrations, availability of iron and sulfur, and major evolutionary events along the Earth's life. (Adapted from Lodeyro et al. 2012)

Cyanobacteria were probably the initiator of evil to transform the Earth into an iron-deficient environment. Ironically, cyanobacteria themselves have higher demand of iron to meet their requirements for photosynthetic electron transport, respiration electron transport, nitrogen fixation, chlorophyll synthesis, etc. Iron plays an important role in the redox reaction of cyanobacteria. In photosynthetic systems, iron mainly participates in the form of iron-sulfur center and heme. Cytochrome b_6f complex has one iron-sulfur center and 4 hemes, and photosynthetic system I contains 3 iron-sulfur centers and 12 iron atoms (Kurisu et al. 2003; Stroebel et al. 2003). Compared to higher plants, cyanobacteria have relatively high ratio of PSI/PSII and thus have higher demand of iron since that PSI needs more iron atoms than PSII. Besides photosynthetic apparatus, respiration electron transfer chain also needs lots of iron atoms to form the cofactor of the electron transport carriers. Some cyanobacteria can also perform nitrogen fixation, and the nitrogenase needs more iron and molybdenum to form catalytic functions. Therefore, it is ironical that cyanobacteria have high demand of iron, but they put themselves into a challenge limiting environment. The contradictory conditions forced them to acclimate to the iron deficiency during the long-term evolution history.

4 Existence Form and Availability of Iron

The forms of iron in natural water systems can be particle iron (insoluble iron, greater than 0.4 μm), colloidal iron (chelated with organic matter), and inorganic free iron (Fe′) (Bruland et al. 1979; Gordon et al. 1982). There are also studies that the iron-containing substances larger than 0.2 μm are considered as granular iron, and those less than 0.2 μm are dissolved iron (FeL) (Barbeau et al. 2001; Croot et al. 2008). According to different valence states, iron is referred into Fe^{2+} and Fe^{3+}. These two valence states of iron can bind with particles or organic compounds to form Fe-complex in natural water. Fe^{2+} is much more soluble than Fe^{3+} (Sunda 2001). The main forms of divalent iron in water environment are Fe^{2+}, $Fe(OH)^{+}$, $Fe(OH)_2^{0}$, Fe(II) $_{coll}$, and Fe(II)L. However, in the modern aerobic water where pH is higher than 5, Fe^{2+} will be oxidized to Fe^{3+} by O_2 and H_2O_2 in a few minutes. Fe^{3+} is very stable but its solubility is very low. If it exceeds Fe^{3+} dissolution limit, $Fe(OH)_3$ precipitation will be formed. Therefore, the concentration of inorganic free iron (Fe′) is very low even in freshwater and offshore waters where iron sources are abundant, and 99% of soluble iron is chelated by organic compounds (Shaked and Lis 2012). The main forms of Fe(III)′ in water include $Fe(OH)_n(3\text{-}n)^{+})$, Fe(III)L, and $Fe(III)_{coll}$, which only account for less than 1% of the total soluble iron in the surface of the natural water (Shaked and Lis 2012).

In natural water environments, various forms of iron can be transformed into each other (Fig. 4) (Shaked and Lis 2012). Particulate iron is a form of iron that cannot be utilized by cyanobacteria, but it is an important iron pool in the water environments and plays an important role in the primary productivity of cyanobacteria (Kuma et al. 1998; Nishioka et al. 2001; Wu et al. 2001). Particulate iron can be transferred to Fe^{2+} through photochemistry and biochemistry, which is then oxidized to Fe^{3+} and forms Fe(III)L (Stumm and Sulzberger 1992; Yoshida et al. 2006). Fe^{2+}, Fe^{3+}, and Fe(III)L are the main dissolved iron forms and are biologically available forms of iron in natural waters. Studies have shown that blue light, respiration, O_2^- and reductase can catalyze the transformation of trivalent iron into bivalent iron and increase bioavailable iron (Fujii et al. 2011; Kranzler et al. 2011; Rose et al. 2005). In water

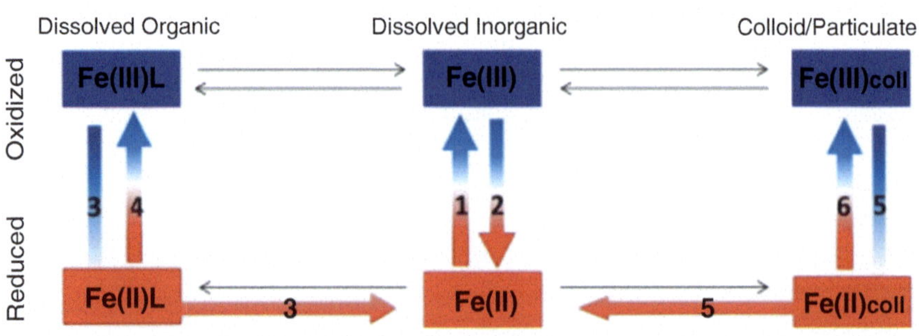

Fig. 4 Redox reactions of different Fe species in aquatic environments (Shaked and Lis 2012)

environments, Fe^{2+} is oxidized to form Fe^{3+} not only by biological factors but also by abiotic factors (pH, salinity and temperature, etc.) (Croot et al. 2001, 2008; Roy et al. 2008). When pH < 4, Fe^{2+} is the main form of divalent iron in water. When 4 < pH < 5, $Fe(OH)_2^0$ is the main form of the existence of divalent iron. Compared to Fe^{2+}, $Fe(OH)^+$ and $Fe(OH)_2^0$ are more easily oxidized into trivalent iron. When 5 < pH < 8, the relative concentration of $Fe(OH)_2^0$ increases with the increase of pH, and the residence time of bivalent state decreases. When pH > 8, the relative concentration of $Fe(OH)_2^0$ is not affected by the change of pH value, and the oxidation rate of divalent iron is not affected by the change of pH value either (Gledhill and Buck 2012). Studies have shown that the release and utilization of oxygen by biologically released redox substances, photosynthesis, and respiration all affect the oxidation of iron dioxide. In high, windless, and stable water bodies, biological factors play an important role in the oxidation of divalent iron, while in low biomass waters, biological factors play a secondary role. At the same time, Fe^{2+} can also combine with organic matter to increase the retention time of Fe^{2+}. Organic substances with low temperature, low pH, and close binding to Fe^{2+} are beneficial to prolonging the existence time of Fe^{2+} (Roy et al. 2008).

Free inorganic Fe^{2+} and Fe^{3+} are the forms of iron that cyanobacteria can use directly (Morel et al. 2008). Due to the thermodynamic instability, Fe^{2+} is generally considered to be an intermediate product that increases bioavailable iron in various reduction pathways (Sunda and Huntsman 1997; Yoshida et al. 2006). The bioavailability of Fe(II)L formed in the environment is not clear (Croot et al. 2008; Roy et al. 2008). Fe^{3+} is the main form of iron used by phytoplankton in natrural seawater. Despite its low concentration, the availability of Fe′ is much higher than that of most Fe(III)L (Morel et al. 2008) in both laboratory and natural environments. As shown in Fig. 5, Fe availability is vary according to iron forms in different cyanobacterial species or from different natural assemblages (Shaked and Lis 2012).

There are many kinds of organic chelators of iron, such as siderophore, cell degradation products, carbohydrate polymers, and so on. According to the dissociation constants of organic chelators and iron elements, iron organic chelates can be divided into two categories: one is siderophores substance (L1) with strong chelation with iron, and the other is substance (L2) with weak chelation with iron (Gledhill and van den Berg 1994; Ibisanmi et al. 2011). Siderophores are a group of small molecule substance secreted by organisms under iron restriction, which can chelate iron elements. According to chemical properties, siderophores can be classified into four categories: phenols, hydroxamic acids, hydroxy acids, and complexes of the first three substances (Miethke and Marahiel 2007). Biological uptake rates of iron and iron chelate complexes are much lower than that of inorganic iron (Shaked and Lis 2012). Under certain conditions, some organisms can efficiently absorb their own secreted iron, such as *Anabaena* sp. PCC 7120 (Rudolf et al. 2015; Goldman et al. 1983). The availability of chelated iron depends on whether organisms have efficient iron absorption systems. Polysaccharides (EPS) secreted by microorganisms and their derivatives (TE) can chelate iron and promote iron uptake by organisms (Decho 1990; Passow 2002; Wotton 2004; Croot et al. 2007). Whether

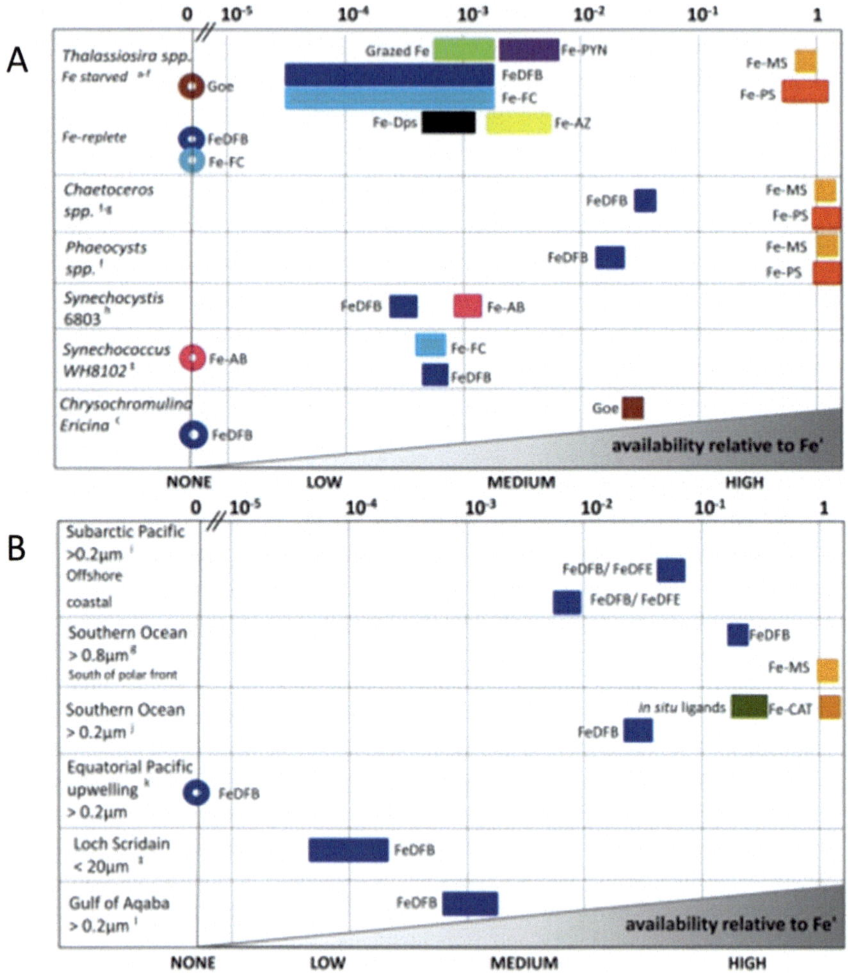

Fig. 5 Relative scale of Fe availability established from phytoplankton uptake rates obtained for (**a**) cultures (**b**) and natural assemblages. Different siderophores are highlighted as follows: catecholates are in red, phenolates are in orange, hydroxamates are in pale yellow, α-hydroxy-carboxylates (deriving from citrate units) are in light green, and α-keto-carboxylates (deriving from 2-oxo-glutarate units) are in blue-green (Shaked and Lis 2012)

EPS and TEP can be directly utilized by organisms or whether they can promote the adaptation of organisms to iron-limited environment by increasing soluble iron content needs further study.

5 The Physiological Functions of Iron in Cyanobacteria

Iron is an essential trace element for photosynthetic organisms (Raven et al. 1999). Cyanobacteria have higher iron demand than eukaryotic photosynthetic organisms. For example, the model unicellular cyanobacterium *Synechocystis* sp. PCC 6803 contains 10^6 atoms/cell (Kustka et al. 2002; Keren et al. 2004). Iron plays an important role in the growth and reproduction of cyanobacteria, photosynthesis, respiration, synthesis of photosynthetic pigments, nitrogen fixation, removal of reactive oxygen species, and synthesis of amino acids and fatty acids (Hantke 2001; Klausner et al. 1993). Because a large amount of iron is involved in photosynthetic electron transport, the intracellular iron content of photosynthetic organisms is ten times higher than that of non-photosynthetic cells of the same size (Terry and Low 1982; Keren et al. 2004).

Iron is an important component of photosynthetic electron transport chain. Photosystem II (PSII) contains two or three iron atoms: one or two heme-binding cytochrome $Cytb_{559}$ and one non-heme iron atom (Zouni et al. 2001; Kamiya and Shen 2003) that binds to D_1 protein and participates in the electron transfer from Q_A to Q_B. Cytochrome b_6/f complex ($Cytb_6$/f) contains six iron atoms: two cytochromes bind to one heme; cytochrome f binds to one heme, one atypical heme and one non-heme 2Fe-2S protein (Kurisu et al. 2003; Stroebel et al. 2003). At the same time, the $Cytb_6$/f complex also has some movable electron carriers such as PETC (Schneider et al. 2004). Plastocyanin (PC) and cytochrome c_6 ($Cytc_6$) are electron carriers between $Cytb_6$/f and PSI. $Cytc_6$ contains one heme and is replaced by copper-containing plastocyanin (PC) under iron restriction. In some algae strains, plastocyanin (PC) is the only electronic carrier.

Photosystem I (PSI) contains 3 4Fe-4S iron-sulfur clusters and 12 iron atoms (Jordan et al. 2001). Ferredoxin contains a 2Fe-2S center. In the linear electron transfer chain, according to PSII (1): $Cytb_6$/f (1): $Cytc_6$ (1): PSI (1), there are about 24 iron atoms in the linear electron transfer chain. The proportion and composition of electron transporters in the linear electron transfer chain vary with different algae strains and environments. Ferredoxin was replaced by flavodoxin (Entsch and Smillie 1972), which did not contain iron. At present, the research on the cyclic electron transport chain is not very clear, and many functional proteins have not yet been identified. NAD (P) oxidoreductase, PsaE, and FesM play an important role in this process. NAD (P) oxidoreductase contains 8–18 iron atoms, and FesM contains 4 iron atoms (Xu et al. 2005; Myers 1987; Yu et al. 1993). Photosynthetic electron transfer chain includes linear electron transfer chain (H_2O → PSII → PQ → $Cytb_6f$ → PC/$Cytc_6$ → PSI → Fd → FNR → $NADP^+$), cyclic electron transfer chain around PSI (PSI → Fd → $Cytb_6f$ → PC → PSI), and pseudo-cyclic electron transfer chain around water-water cycle (H_2O → PSII → PQ → $Cytb_6f$ → PC → PS I → Fd → O_2).

Iron also plays crucial roles in the respiration process. The known components of the respiratory electron transport chain in cyanobacteria include six enzyme complexes: NADH oxidoreductase, succinate dehydrogenase, Cytc oxidoreductase, Cytc oxidase and $Cytb_6f$, PC or $Cytc_6$ involved in photosynthetic electron transport chain. NADH oxidoreductase has 2–4 Fe-S centers, succinate dehydrogenase contains 3 Fe-S proteins, Cytc-reductase contains 5 Fe atoms (2 Cytb, 1 Cytc1, and 1 Fe-S protein), Cytc-cytochrome oxidase has 2 Fe atoms (1 Cyta and 1 Cyta3). Each cytochrome contains one iron atom, and each Fe-S protein contains two iron atoms (Binder 1982; Vermaas 2001; Hart et al. 2005). Iron-containing superoxide dismutase (Fe-SOD, 1 iron atom), catalase (CAT, 4 Fe atoms), and anti-cyclic acid oxidase (APX, 1 Fe atom) all contain iron atoms, which play an important role in scavenging reactive oxygen species and avoiding oxidative damage to cells (Canini et al. 1992; Shcolnick and Keren 2006; Raven et al. 1999). If the concentration of free iron in the cell is too high, iron will react to produce reactive oxygen species and cause oxidative damage to cells. Cyanobacterial cells have two types of ferritin, bacterial ferritin and ferritin. Excess free iron in cyanobacterial cells is saved to maintain the dynamic balance of intracellular iron. At the same time, under iron restriction, cyanobacterial cells are the source of iron for normal metabolic activities (Keren et al. 2004; Shcolnick et al. 2009).

Cyanobacteria can not only utilize nitrate but also fix N_2 (Raven 1988). Nitrate reductase, nitrite reductase, and nitrogenase all use iron as coenzyme factors (Geider and La Roche 1994). Nitrogenase consisting of heterodimers of ferritin and molybdenum ferritin contains 30 iron atoms (Bishop and Premakumar 1992). Iron atom is not only a component of nitrogen-fixing enzymes but also a coenzyme factor of superoxide dismutase, which has a great influence on the activity of nitrogen-fixing enzymes. In some algae which have been restricted by iron for a long time, iron atom can be replaced by other metal elements, such as Ni (Eitinger 2004). Nitrogenase is easily inactivated in an aerobic environment. Studies have shown that Fe-SOD is the most abundant and active superoxide dismutase in cyanobacteria cells, which scavenges reactive oxygen species and promotes nitrogen fixation (Regelsberger et al. 2004). Fe-SOD also plays an important role in the formation of nitrogen-fixing heterocysts. In the process of heterosexual cell differentiation, a large amount of energy is needed to form more reactive oxygen species, and Fe-SOD is expressed in this process (Li et al. 2002). Iron atoms play an important role in the activity of nitrogenase and the formation of heteromorphic cells. The iron requirement of assimilated NH_4^+ cyanobacteria was 1.6 times that of non-nitrogen-fixing cyanobacteria (Raven 1988, 1990). The iron requirement of fixed N_2 cyanobacteria was 2.5 ~ 5.2 times that of non-nitrogen-fixing cyanobacteria (Sañudo-Wilhelmy et al. 2001).

6 Iron Limitation Hypothesis

About one third of the oceans is high-nutrient low-chlorophyll (HNLC) area, which is rich in N and P nutrients and has a very low chlorophyll content, such as the Antarctic Ocean (Pacific, the Atlantic, and Southern Indian Ocean), the North Pacific, the equatorial Pacific, and so on (Boyd et al. 2007). In 1988, John Martin first used trace-metal-clean technology to test the iron in the northeast of the subarctic Pacific Ocean. In 1990, the iron addition experiment at 64 km^2 and 450 kg in the equatorial Pacific showed that iron supply is the main limiting factor of marine primary productivity. In 1990, Prof. Martin proposed that iron is the main limiting factor for marine primary productivity. Iron influences future geochemical cycles, climate, and life evolution through the use of inorganic C and inorganic N by photosynthetic organisms (Martin and Fitzwater 1988; Martin 1990).

Martin's iron hypothesis has attracted worldwide attention. Since then, bio-oceanographers have conducted many large-scale field experiments on iron addition. From 1993 to 2005, scientists conducted 12 mesoscale iron tests in the eastern equatorial Pacific, northeast and northwest Pacific, Antarctic and Atlantic Oceans, and southern polar regions of New Zealand (Fig. 6) (Boyd et al. 2007). The iron hypothesis was confirmed by the increase of chlorophyll concentration, photosynthesis, nutrient concentration, CO_2 concentration in the upper air, and the increase of organic C deposited into the ocean after the addition of iron in these HNLC sea areas (Landry et al. 2000; Bakker et al. 2001; Croot et al. 2001).

Over the past 250 years, the concentration of atmospheric CO_2 has increased by at least 40% due to the massive burning of fossil fuels and deforestation. The concentration of atmospheric CO_2 has increased from 280 ppmv (parts per million volume) to 384 ppmv in 2007 (Solomon et al. 2007) and now exceeds 400 ppmv. The

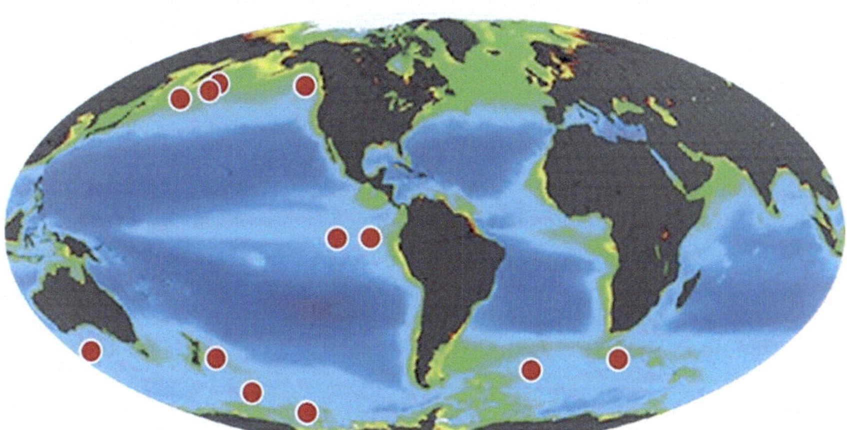

Fig. 6 Twelve field iron fertilization experiments were conducted, and the results confirmed Martin's iron hypothesis (Adapted from Boyd et al. 2007 and Trujillo 2011)

rising concentration of CO_2 in the atmosphere will lead to greenhouse effect and ocean acidification. At present, the average concentration of dissolved iron in marine surface water is 0.07 nM (Rivers 2009). Marine acidification will further reduce the concentration of dissolved iron in the sea and reduce the availability of iron. Primary productivity in the ocean will be further limited by iron, and the role of regulating the greenhouse effect of the Earth will be reduced (Feely et al. 2009; Shi et al. 2010). The absorption and utilization of iron in ocean by cyanobacteria and the availability of iron in ocean may be of great significance to the future climate change and life evolution of the Earth.

7 Physiological Response of Cyanobacteria to Iron Limitation

Iron is involved in photosynthesis, respiration, nitrogen fixation, and other important metabolic processes in cyanobacteria. When cyanobacteria face iron limitation, they showed a series of physiological and biochemical responses, including changes in utilization of iron stored in cells, reduction of absorption and utilization of nitrogen fixation, degradation of phycobilisome protein, changes of photosynthetic electron transport chain and membrane structure, respiration and cell division rate, etc. (Sandmann 1985; Fraser et al. 2013).

7.1 Photosynthesis

Iron is needed for chlorophyll synthesis. The synthesis of aminolevulinic acid (ALA), an intermediate product of chlorophyll synthesis, requires NADPH and organic acids of the tricarboxylic acid cycles (Chereskin and Castelfranco 1982; Miller et al. 1982). The tricarboxylic acid cycle is also regulated by the cis-aconitase Fe-S protein and directly catalyzed by fecal porphyrinogen oxidase containing iron to form chlorophyll lipids (Miller et al. 1982). The content of iron-binding protein and photosynthetic pigments in iron-deficient cells decreased, which led to the structural changes of cell membrane. In iron-deficient *Synechococcus cedrorum* UTEX cells, the membrane structure was reduced by nearly 50% and recovered rapidly after iron addition (James et al. 1983). Under sufficient iron conditions, *Aphanocapsa* cells contained at least 800 chlorophyll molecules per photosynthetic electron transport chain. But under iron deficiency conditions, chlorophyll molecules decreased to 39% (Sandmann 1985). Ferredoxin (Fd) is the most susceptible component in photosynthetic electron transport chain. Under moderate iron-deficient conditions (10% of the normal medium of iron concentration), Fd is replaced by flavodoxin (Fld) that requires no iron and cytochrome c-553 was replaced by plastocyanin (PC) which requires no iron either. The photosynthetic pigments were affected at lower iron concentration (Sandmann and Malkin 1983). PSI and Cytb$_6$f have relatively higher iron requirement, accounting for 60% of intracellular iron content. PSI contained 12 iron atoms. Cytb$_6$f complex contained six iron atoms. PSI was most susceptible to iron deficiency (Berman-frank et al. 2001). Under iron restriction, intracellular PSI and Cytb$_6$f were more susceptible than PSII (Bailey et al. 2008; Fraser et al. 2013). Under iron-sufficient conditions, the PSI/PSII in cyanobacteria cells is higher to ensure that PSI can receive the electrons transferred from PSII to the plastid quinone bank at the maximum speed, such as the *Synechococcus* PSI/PSII ratio is 2:3, and the *Synechocystis* PSI/PSII ratio is 2:5 (Vermaas 2001). However, under iron-deficient conditions, *Synechococcus* PSI/PSII ratio decreased to 0:4, and *Synechocystis* PSI/PSII ratio decreased to 1:1 (Fraser et al. 2013; Ryan-Keogh et al. 2012). Under iron-replete conditions, Cytb$_6$f/PSII ratio in *Synechocystis* was usually 1:2 which decreased to 0:3 under iron-deficient conditions (Fraser et al. 2013). The Fe-S center of PSI, cytochrome b-559, and cytochrome b-563/f-557 of PSII in iron-deficient *Aphanocapsa* were 1/10, 1/2, and 1/3 of that in iron-sufficient cells, respectively (Sandmann 1985). Immunoblotting showed that the levels of D1, CP43, and CP47 in PSII reaction center decreased to 60% of that in iron-limiting cells (Vassiliev et al. 1995). Iron deficiency can induce the synthesis of CP43 (IsiA) protein. Studies have shown that the content of IsiA in iron-deficient cells is eight times higher than that in iron-sufficient cells. IsiA can bind to chlorophyll or PSI to form supercomplex and protect the photosystems (Kouřil et al. 2005).

7.2 Respiration

Compared with the effects of iron deficiency on photosynthesis, the influence of iron deficiency on respiratory process was moderate. It has been shown that the respiration rate of *Trichodesmium* decreased by 30%, and the photosynthetic rate decreased by 50% under iron deficiency (Sandmann and Malkin 1983). This physiological response to iron deficiency indicates a redistribution of iron in algae cells. It may also due to that the iron content in photosynthetic electron transport chain is much higher than that in respiratory electron transport chain, so photosynthesis is more susceptible to iron deficiency. On the other hand, the photosynthetic and respiratory electron transport chains of cyanobacteria share common components; thus these two parts are interacting.

7.3 Nitrogen Fixation

Nitrogen uptake and utilization by cyanobacteria is one of the metabolic processes requiring the most iron and is very sensitive to iron limitation (Raven and Richard 1988). Biological nitrogen fixation is limited by available iron content in 75% of the world's oceans (Berman-frank et al. 2001; Kustka et al. 2003). Biological nitrogen fixation is the main source of available nitrogen in the ocean. *Trichodesmium* is the most important nitrogen-fixing organism in the ocean, mainly distributed in tropical and subtropical waters (Capone et al. 1997, 2005; Westberry and Siegel 2006). Under iron limitation, the biomass of *Trichodesmium* decreased, the expression of intracellular nitrogenase decreased by 60%, and the activity of nitrogenase is almost lost (Küpper et al. 2008). What's more, the number of nitrogen-fixing heterocysts decreased significantly (Regelsberger et al. 2004; Guikema and Sherman 1983; Xing et al. 2007; Küpper et al. 2008). Iron fertilization assays in both laboratory and fields showed that enhanced iron concentration can increase biological nitrogen fixation (Falkowski et al. 1998; Hutchins et al. 2007).

7.4 Oxidative Stress

Since the Great Oxygenation Events on the Earth, all the living organisms faced both iron limitation and oxidative stress, which are tightly related. Reactive oxygen species (ROS) in algae cells mainly include superoxide anions (O_2^-), singlet oxygen, hydrogen peroxide, hydroxyl radicals, and so on. Photosynthetic apparatus is the main metabolic process that produces ROS. The chlorophyll molecule absorbs the energy of light and is excited and reacts with the oxygen in the cell to form singlet oxygen ($^1O_2$). Under photoinhibition, PSII is the main target site of reactive oxygen species. When photosynthetic electron transport is inhibited, hydrogen peroxide (H_2O_2) is produced during water photolysis in PSII (Krieger and Rutherford 1998). At the same time, PSII transfers some electrons to oxygen molecules to form superoxide anion (O_2^-), which reacts with non-heme ferritin of PSI to form hydroxyl radicals (Pospisil et al. 2004). Under oxidative stress, reactive oxygen species (ROS) react with the Fe-S centers of the photosystem to degrade the photosystem and form hydroxyl radicals (OH•). Therefore, under iron-limited conditions, photodegradation produces a large number of free chlorophylls, decreases in iron superoxide dismutase activity, and increases in the proportion of PSII/PSI which results in the production and accumulation of ROS in cells and causes severe oxidative stresses (Vermaas 2001). It has been shown that under iron-limited conditions, the accumulation of intracellular ROS in *Anabaena* sp. PCC 7120 and *Synechocystis* sp. PCC 6803 was ten times higher than that of iron-replaced cells (Latifi et al. 2005).

8 Adaptative Strategies of Cyanobacteria to Iron Limitation

Under low iron concentrations, cyanobacteria have complicated and perfect adaptation mechanisms. These include the following:

(1) Biosynthesis and secretion of iron chelators such as siderophores to assist in iron acquisition

(2) Induction of protective proteins such as isiA to enhance the tolerance of cells to low-iron environments

(3) Reduction of iron demand and maintenance of a lower metabolic level

(4) Induction of iron uptake transporters and regulation of the balance between active transport and passive transport

(5) Optimization of the ferrous and ferric iron transport

(6) Development of special cell surface structure to facilitate iron adhesion and uptake

(7) Reduction of the cell size and increase of specific surface area/volume ratio to facilitate passive diffusion of iron, especially in oligotrophic ocean species

8.1 Biosynthesis and Secretion of Iron Chelators

Siderophores are widely distributed in bacteria and fungi, which are found in marine, freshwater, and terrestrial environments (Årstøl and Hohmann-Marriott 2019). Siderophores are biosynthesized and excreted under iron starvation, when the intracellular iron concentration drops under a certain threshold required for functionality (Řezanka et al. 2018). Studies have shown that many cyanobacteria secrete iron chelators (siderophores) with high affinity for trivalent iron ions (Fe^{3+}). Iron chelators are a kind of low molecular weight substance (about 1000 Da) (Moeck and Coulton 1998). Its synthesis and secretion are induced by iron deficiency. Under extremely iron-deficient environment, the amount of iron chelator secreted by some bacteria can reach a high concentration of 200 mg/L (Croot et al. 2008). At present, nearly 500 kinds of iron chelators have been identified and can be classified into many kinds according to their different structures (Fig. 7). Ferric chelate complexes bound with Fe^{3+} cannot enter cyanobacteria cell membranes by free diffusion because their molecular weight is larger than the porin channel size (600 Da). Compared with Gram-positive bacteria, cyanobacteria have a layer of outer membrane. The process of uptake by iron chelator is as follows: firstly, iron chelators bind to receptor protein on outer membrane (TBDT, a TonB-dependent transporter protein). The energy of this process comes from the inner membrane, mediated by TonB-ExbB-ExbD system. The complex can be transported inside the cell. The reduction of Fe^{3+} facilitates its release, as siderophores do not typically chelate ferrous iron (Fe^{2+}) (Hider and Kong 2010).

8.1.1 Types of Siderophores

Typically, siderophores form a strong hexadentate octahedral complex with ferric iron (Fe^{3+}). Based on the primary oxygen-donating ligands that bind the iron, siderophores are divided into four different types (Miethke and Marahiel 2007; Khan et al. 2018). They are the hydroxamates, catecholates, carboxylates, and mixed types of these functional groups (Fig. 7).

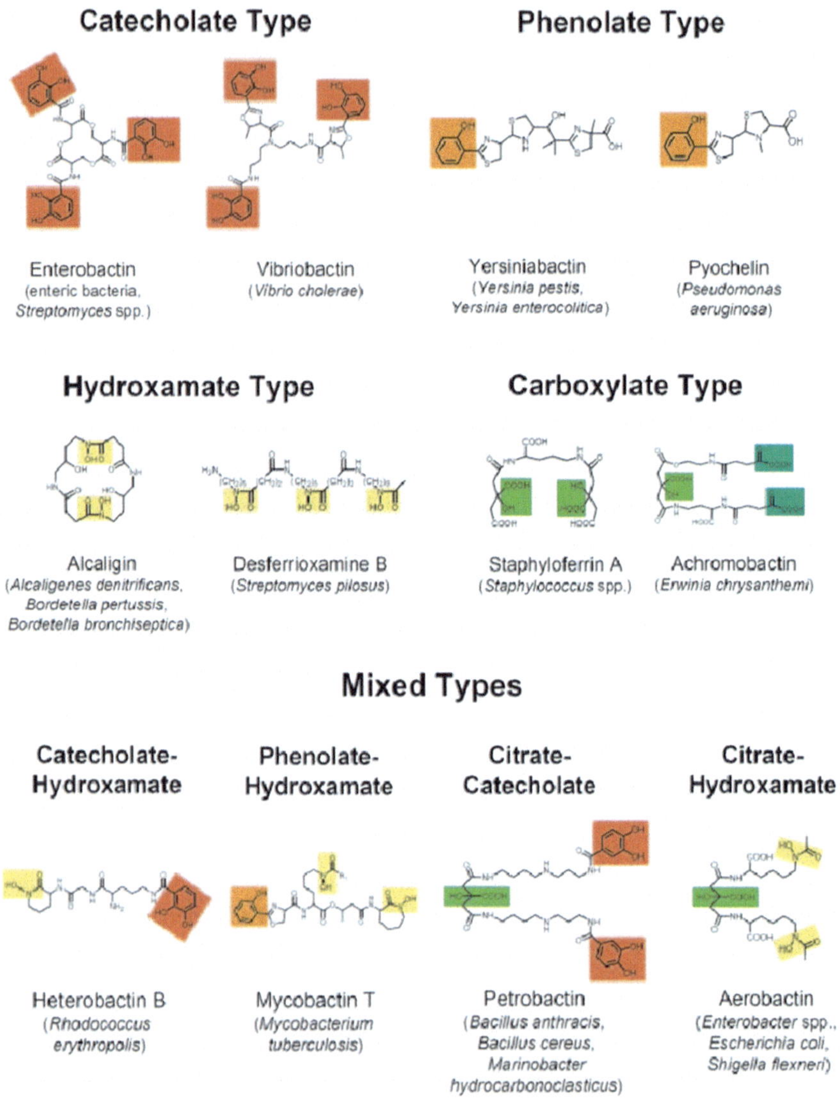

Fig. 7 Different types of siderophores. Catecholates are in red, phenolates are in orange, hydroxamates are in pale yellow, α-hydroxy-carboxylates (deriving from citrate units) are in light green, and α-keto-carboxylates (deriving from 2-oxo-glutarate units) are in blue-green (Miethke and Marahiel 2007)

(1) Hydroxamate type

Hydroxamate-type siderophores appear to be the most common type, being widely distributed in both bacteria (Saha et al. 2016) and fungi (Garnerin et al. 2017). Many cyanobacteria produce hydroxamates, but only two hydroxamate

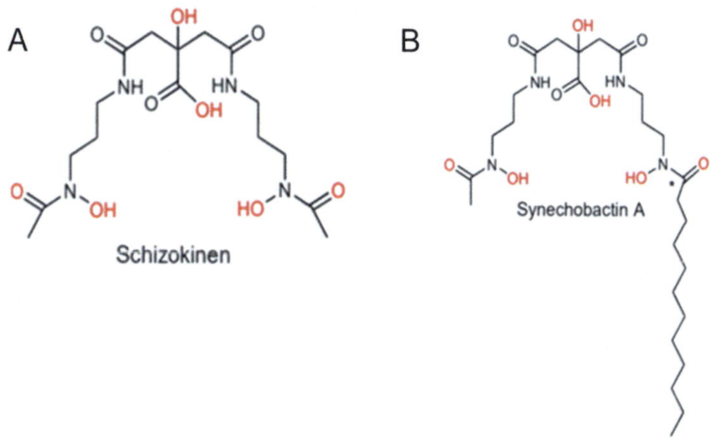

Fig. 8 The structures of schizokinen (**a**), produced by some *Anabaena* strains, and synechobactin, (**b**) produced by *Synechococcus* sp. PCC 7002. Iron-binding oxygen atoms are marked in red. The synechobactins are a suite of siderophore with different fatty acid chain lengths in the (*) mark marked position (Årstøl and Hohmann-Marriott 2019)

siderophore structures have been determined, schizokinen and synechobactin (Fig. 8) (Årstøl and Hohmann-Marriott 2019). Schizokinen is produced by some strains of the freshwater species, such as *Anabaena* sp. strain PCC 6411 and 7120 (Simpson and Neilands 1976; Goldman et al. 1983). Synechobactin is secreted by the coastal algae *Synechococcus* sp. PCC 7002. The length of the fatty acid tail varies, with the most common C12, C10, and C8 varieties being named synechobactins A, B, and C (Ito and Butler 2005; Årstøl and Hohmann-Marriott 2019). In terms of global marine primary productivity, the siderophore exhibits a high concentration of amphiphilic structures in an open-ocean low-iron environment that helps anchor the siderophore in the cyanobacterial membrane (Řezanka et al. 2018).

(2) Catecholate type

Each catecholate group contains two oxygen atoms for iron chelation, which form a hexagonal octahedric complex (Fig. 7) where the ferric ion (Fe^{3+}) has a very high stability constant $K = 10^{52}$ mol/L. Such a high constant signifies that cyanobacteria can live in environments with very low iron concentrations (Řezanka et al. 2018). The structure is characterized by a C3 symmetrical trilactone ring from which the exocyclic nitrogen atoms are acylated by the dihydroxybenzoic acid group.

(3) Carboxylate type

Siderophore containing the carboxylate group which are often produced by cyanobacteria. This type of siderophore binds to iron through carboxyl and hydroxyl groups. As a model compound, the basic building unit is citric acid (Řezanka et al. 2018).

8.1.2 Siderophore Biosynthesis and Phylogenetical Distribution in Cyanobacteria

All cyanobacterial siderophores that have been identified so far are either hydroxamates or catecholates. The phylogenetic and environmental distribution of known siderophore-producing cyanobacteria is shown in Fig. 9, including unicellular, filamentous, and heterocyst-forming cyanobacteria in terrestrial, freshwater, and marine environments (Årstøl and Hohmann-Marriott 2019). Phylogenetically, siderophores are widely distributed in cyanobacteria, but so far have not been identified in the earliest-branching cyanobacterial clades, which also contain fewer NRPS and PKS coding genes than later branching clades (Shih et al. 2013).

The biosynthesis of siderophore involves a combination of either non-ribosomal peptide synthases (NRPS) and polyketide synthase (PKS) or NRPS-independent synthases (NIS). Citrate-based siderophores such as schizokinen and synechobactin are produced by the latter system (Khan et al. 2017). Typically, NIS biosynthesis

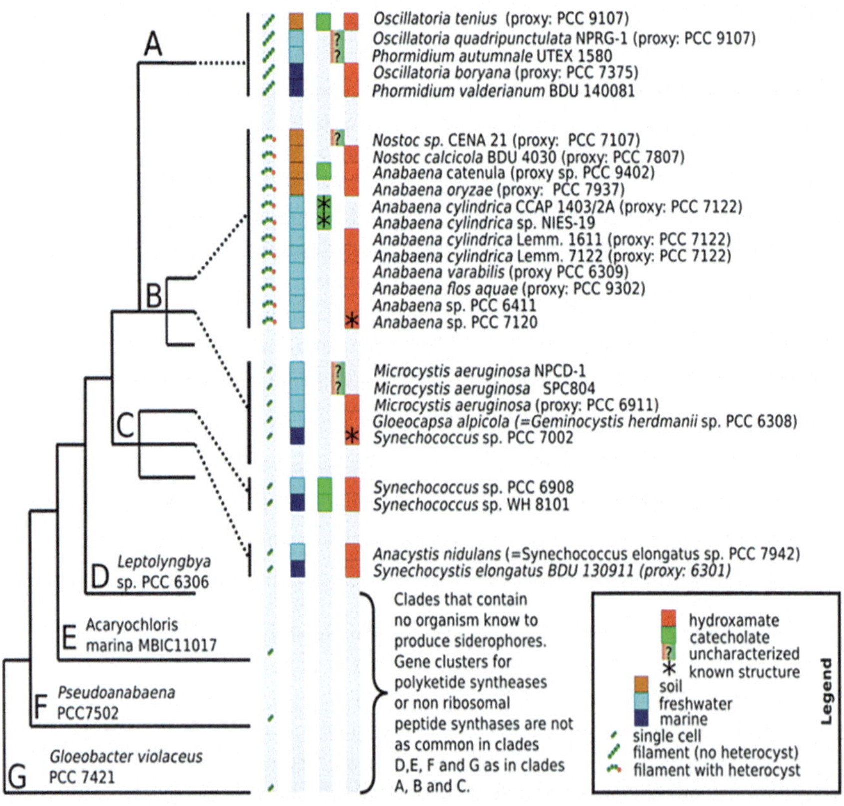

Fig. 9 Siderophore biosynthesis occurs via two pathways: the non-ribosomal peptide synthetase (NRPS) pathway and the NRPS-independent siderophore (NIS) synthetase pathway. (Quoted from Årstøl and Hohmann-Marriott 2019)

pathways of siderophore contain at least one enzyme with conserved N-terminal iron uptake chelate (IucA/IucC) domains, with a C-terminal domain used for iron metabolism or transport. NIS synthases can be categorized by the formation of a peptide bond between a hydroxamate-amine with a carboxylic acid substrate—citric acid (type A and A' NIS) and α-ketoglutaric acid (type B NIS)—or a succinic and citric acid derivative (type C and C′ NIS) (Ovescostales et al. 2009; Carroll and Moore 2018).

8.1.3 Siderophore Secretion and Uptake in Cyanobacteria

The secretory mechanism of siderophore is well studied in most pathogenic and nonpathogenic bacteria, while research in cyanobacteria is limited, mainly involving three types of proteins: the major facilitator superfamily (MFS), resistance nodulation and cell division (RND) superfamily, and the ATP-binding cassette (ABC) superfamily. The only cyanobacterial siderophore secretion pathway studied in detail, so far, is in *Anabaena* sp. PCC 7120 (Nicolaisen et al. 2008; Nicolaisen and Schleiff 2010).

Siderophores are large and cannot across the extracellular space through passive diffusion through the outer membrane (OM) and cytoplasmic membrane (CM), thus requiring active transport mechanisms. The import of iron-loaded siderophore is broadly similar across many bacteria and is illustrated in Fig. 10 (Årstøl and Hohmann-Marriott 2019). The transport of incoming siderophore-iron complexes into the periplasmic space is enabled by receptor proteins, TonB-dependent transporters (TBDT) located in the outer membrane. TBDTs consist of a β-barrel domain and a "plug" domain in the barrel interior that acts in concert with a TonB box on the periplasmic side. Energy for transport is derived from the proton-motive force and is mediated through the association of TonB with inner membrane proteins ExbB and ExbD, forming a TonB-ExbB-ExbD complex. Once in the periplasm, Fe^{3+}-siderophores are transported across the inner membrane, often by FhuBCD and FecABC transporters (Noinaj et al. 2010). Once inside the cell, iron has to be removed from the siderophore. This can happen either through the reduction of iron from Fe^{3+} to Fe^{2+} by ferric-siderophore reductases or by ferric-siderophore hydrolases (Miethke and Marahiel 2007) or by the photolysis of iron-bound siderophore, as described for synechobactin (Årstøl and Hohmann-Marriott 2019).

8.2 Induction of Protective Proteins Such as IsiA to Avoid Photooxidation of Photosystem I

Under iron limitation, some protective proteins were induced, such as IsiA (iron stress-induced protein A), DpsA (DNA-protected protein under starved conditions) and FutA/IdiA (iron de enantioselective protein A). Except few cyanobacterial strains such as *Synechococcus* WH8102, most of the strains contain IsiA protein,

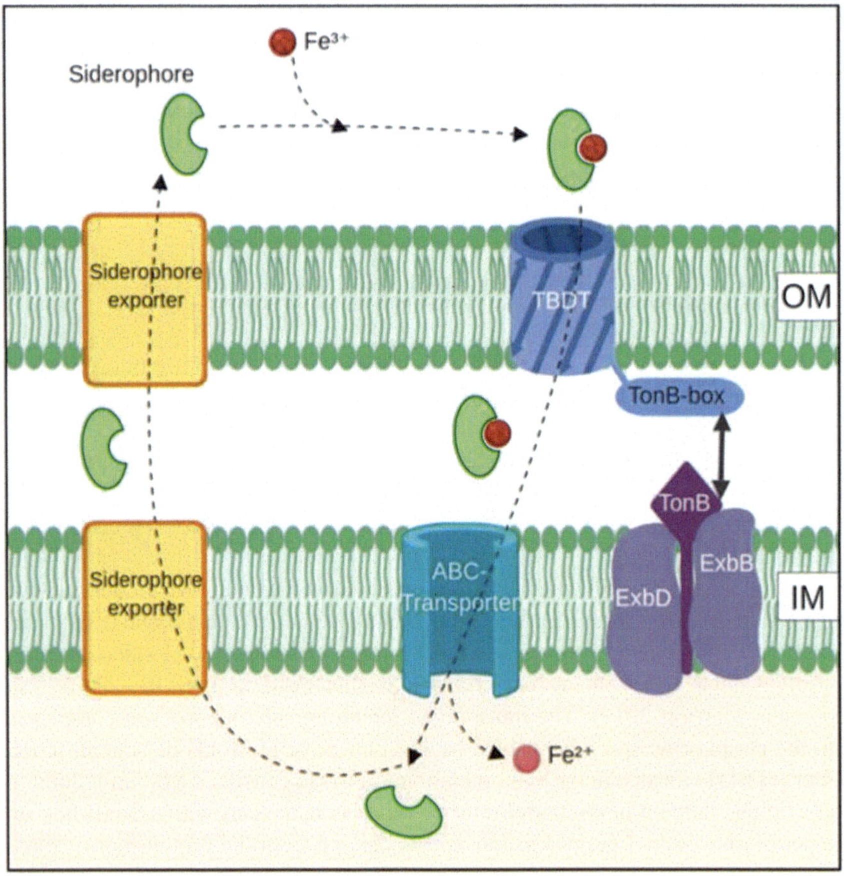

Fig. 10 An overview of siderophore cycling in a generic organism. Typically, iron-loaded siderophores are imported through TonB-dependent transporters (TBDTs) in the outer membrane, with power supplied by a TonB-ExbB-ExbD complex in the inner membrane. Further transport is done by ABC-transporters. Export of siderophore can happen by a variety of transporter types (Årstøl and Hohmann-Marriott 2019)

which is upregulated significantly under iron limitation. In *Synechocystis* sp. PCC 6803 *isiA* is upregulated 22 times by iron deficiency (Singh et al. 2003, 2004). Because IsiA protein and PSI structural protein CP43 are homologous, they are also called CP43′ (Falk et al. 1995; Bricker and Frankel 2002). IsiA protein can bind to PSI trimer, PSI monomer, and chlorophyll molecule (Bibby et al. 2001, 2003; Boekema et al. 2001; Yeremenko et al. 2004). There are two forms of IsiA-PSI_3 supercomplex: one is that 18 IsiA form a ring around the PSI trimer to form $IsiA_{18}$-PSI_3 supercomplex (Fig. 11) (Bibby et al. 2001); the other is that 35 IsiA form two ring structures and PSI trimer to form $IsiA_{35}$-PSI_3 supercomplex (Nield et al. 2003; Andrizhiyevskaya et al. 2002). The first supercomplex increased the light capture ability of PSI by 70% (Nield et al. 2003), and the second supercomplex enhanced the light capture ability of PSI by 7 times (Andrizhiyevskaya et al. 2002). To some extent, it can

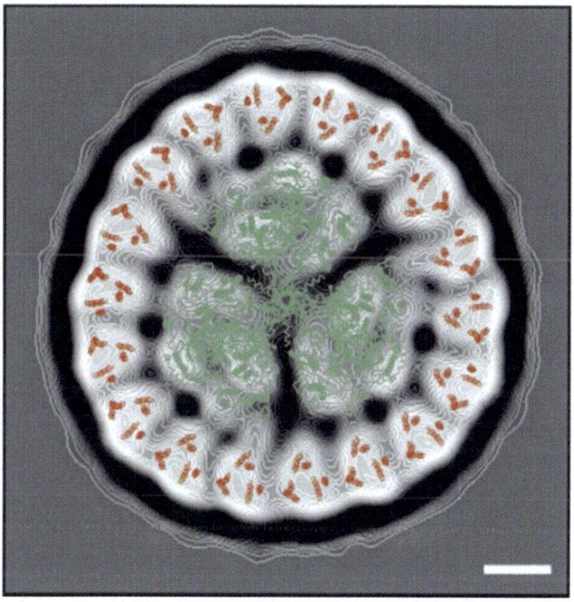

Fig. 11 The PSI trimer structure and CP43 helix organization derived from X-ray crystallography overlaid onto the projection map of the CP43′-PSI complex. Scale bar, 5 nm. Good correlation between the X ray and electron microscopy data confirms the presence of a PSI trimer within the center of an 18-member ring of CP43′ (Bibby et al. 2001)

compensate for the decrease of PSI/PSII ratio under iron limitation, which facilitates photosynthetic electron transport and thus reduce the production of reactive oxygen species in photosynthetic electron transport chain. At the same time, IsiA proteins convert excessive light energy absorbed by phycobilisomes into heat energy and reduce the production of reactive oxygen species (Wilson et al. 2007). PSI-free IsiA proteins can alleviate photooxidation damage to PSII (Burnap et al. 1993). IsiA and chlorophyll bind as a storage of chlorophyll to prevent chlorophyll degradation, enabling cells to rapidly restore the assembled light system under sufficient iron conditions (Burnap et al. 1993). When IsiA protein was inactivated, algae cells could not grow under iron-limiting and high light conditions (Burnap et al. 1993; Singh and Sherman 2006).

DpsA is a DNA-binding protein containing heme, which protects genomic DNA from oxidative stress (Dwivedi et al. 1997). DpsA-bound DNA has almost no sequence specificity between each other (Martinez and Kolter 1997; Wolf et al. 1999). DpsA is widely distributed in cyanobacteria, including *Synechococcus* (except *Synechococcus* WH8102), freshwater algae, and a few *Prochlorococcus* (PMT 2218 and MIT 9313) (Palenik et al. 2006). Gene *dpsA* is upregulated in iron limitation and encoded DpsA to bind DNA (Michel et al. 2003; Rivers 2009). In freshwater cyanobacteria *Synechococcus elongates*, the knockout of *dpsA* is lethal under iron limitation (Sen et al. 2000). In addition to its protective effect,

DpsA also affects the transcription of intracellular iron signaling genes. The transcription level of *isiA/B* gene in *Synechococcus elongatus* PCC 7942 DpsA-inactivated mutant strain was less than twice that of wild type under iron limitation. In iron-deficient mutant cells, it has been observed the following physiological changes such as the degradation of phycobilisomes, the decrease of PSI and PSI activity, the decrease of IsiA-PSI supercomplex, the enhancement of respiration, the redistribution of iron deficiency regulatory system and intracellular iron balance, etc. (Michel et al. 2003; Yousef et al. 2003).

IdiA protein was upregulated in iron-deficient cells and was located in plasma membrane and thylakoid membrane to protect PSII and promote iron uptake (Fulda et al. 2000; Tölle et al. 2002). It has been well studied in *Synechococcus* sp. PCC 6301, *Synechococcus* sp. PCC 7942, and *Synechocystis* sp. PCC 6803. In freshwater cyanobacteria, IdiA participates in iron uptake (Michel et al. 1996, 1998; Katoh et al. 2001; Michel and Pistorius 2004). However, there is no direct evidence that IdiA participates in iron uptake by marine algae, but multiple copies of *idiA* gene were found in the genomes of several marine *Synechococcus* strain, suggesting that IdiA has protective effects and is likely to participate in iron uptake. There are 1.5 copies of *idiA* gene in the genome of each marine *Synechococcus* strain and 0.6 copies in the genome of *Prochlorococcus* (Rivers 2009).

8.3 Decrease Iron Demand and Maintain a Lower Metabolic Level

In addition to the secretion of siderophores and photo-protection mechanisms, cyanobacteria also developed various interesting low-iron-adaptative mechanisms during evolution. Iron-free proteins in freshwater algae can replace similar iron-containing proteins to reduce iron demand under iron-limited conditions, such as flavodoxin replacing ferredoxin, plastocyanin replacing cytochrome c, decrease of PSI (12 iron atoms)/PSII (3 iron atoms) ratio, and reduction of phycobilisomes requiring iron in synthesis minimize genome of marine cyanobacteria (Barber et al. 2006). It has been reported that PSI/PSII ratio varies from 3:1 under iron-replete conditions to less than or close to 1:1 under iron-limited conditions (Behrenfeld and Milligan 2013). Under the condition of iron limitation, the content and activity of superoxide dismutase (SOD) containing manganese can significantly increase and replace the function of Fe-SOD. It has also reported that Fe-SOD can also be replaced by Ni-SOD in some cyanobacterial species (Dufresne et al. 2003; Palenik et al. 2003; Rocap 2003; Eitinger 2004). Rusch et al. (2010) analyzed 73 genomes from the cyanobacterial strains with low iron concentration and found that compared to other ecotypes, two ecotypes of *Prochlorococcus* decreased their ferritin by 10% and lost many genes encoding iron-containing proteins including nitrogenase, PTOX (plastoquinol oxidase), ferredoxin, cytochrome cM, etc. (Rusch et al. 2010). This genotype may imply a strategy for cyanobacteria to adapt to low-iron environment by reducing photosynthetic efficiency. In the field experiments with iron addition, it was shown that the two ecotypes of cyanobacteria did not respond to the increase of iron concentration.

In addition to the reduction of intracellular iron-containing proteins described above, marine cyanobacterial cells can also reduce iron demand through the biological rhythm. It was found that the photosynthetic and nitrogen-fixing proteins of nitrogen-fixing cyanobacteria *Crocosphaera* can alternate according to day and night. Photosynthesis-related proteins are expressed during the day and degraded at night. Nitrogen-fixing related proteins are expressed at night and degraded during the day. At the same time, bacterial ferritin changes with the change of intracellular iron demand and plays a balancing role in these two systems (Tuit et al. 2004; Stockel et al. 2008). The intracellular iron content of *Crocosphaera* may be half of that of *Trichodesmium*, a cyanobacteria that fixes nitrogen during the day (Kustka et al. 2003; Saito et al. 2011).

There are two kinds of iron storage proteins in cyanobacteria, one is ferritin and the other is bacterial ferritin. Each ferritin and bacterial ferritin consists of 21 monomers that bind 2000–3000 iron atoms (Andrews et al. 1993; Keren et al. 2004). Ferritin and bacterial ferritin are widely distributed in marine cyanobacteria such as *Synechococcus*, *Prochlorococcus*, *Trichodesmium*, *Crocosphaera*, and freshwater cyanobacteria (Morrissey and Bowler 2012), which is a common strategy for cyanobacteria to adapt to iron-limited environment. When the iron concentration is high, ferritin and bacterial ferritin are upregulated at the gene level and protein level, storing free iron elements in the cells, avoiding excessive reactions of free iron ions in the cells to produce reactive oxygen species (Kim et al. 2007; Hernández-Prieto et al. 2012; Toulzal et al. 2012; Shcolnick et al. 2009).

8.4 *Increase of Iron Uptake Capacity and Balance Active and Passive Transport*

In natural environments, iron exists in the form of free state and chelated state. The cyanobacteria can obtain iron from environments by active transport and passive diffusion. The active transport process requires energy in the form of proton-motive force and a complex of three inner membrane proteins, TonB-ExbB-ExbD, to transduce this energy to the outer membrane (Shaked and Lis 2012; Noinaj et al. 2010). It has been well studied that TonB-dependent active transport process plays role in the uptake of both siderophore-bound iron and inorganic unchelated iron. At the same time, porin-mediated passive diffusion can facilitate the import of inorganic free iron (Fe′).

8.4.1 Active Transport of Siderophore-Chelated Iron and Unchelated, Inorganic Iron (Fe′)

In natural water, the concentration of unchelated free iron is extremely low, and most of the dissolved iron is chelated by organic matters (Shaked and Lis 2012). Siderophores are produced by microorganisms to enable iron acquisition under iron-limited conditions. Siderophores can combine iron in the environment to form a siderophore-iron complex and then enter the cell through a specific transport system, increasing the availability of iron (Müller et al. 1985; Romheld and Marschner 1983; Miethke and Marahiel 2007; Årstøl and Hohmann-Marriott 2019).

Gram-negative bacteria possess an outer membrane (OM) as well as a cytoplasmic membrane (CM), which presents an additional barrier to the exchange of solutes. As ferric-siderophores are too large to passively diffuse through the OM porins, they must be actively transported across the membrane by specific receptor proteins (Miethke and Marahiel 2007; Noinaj et al. 2010). The uptake of ferric-siderophore in cyanobacteria depends on TonB-dependent transporters (also OM receptors). OM receptors are the "first gate" of ferric-siderophore recognition. All of them consist of a 22-β-strand barrel formed by C-terminal residues, while 150 N-terminal residues fold inside the barrel to form a "plug" or "cork" domain. The OM receptors/transporters bind the ferric-siderophore complexes and directly interact with the energizing TonB-ExbB-ExbD complex in the inner membrane to allow the iron complex to be transported into the periplasmic space. This transport process involves three components: (i) OM-localized transporters, (ii) a CM-localized TonB-ExbB-ExbD complex, and (iii) ion electrochemical potential (Noinaj et al. 2010). Energy for the transport is supplied by the TonB-ExbB-ExbD complex. It is thought that the TonB protein "transduces" the proton motive force energy to the receptor to allow substrate translocation. TonB makes contacts with both ExbB and ExbD via its N-terminal domain (Koedding et al. 2004). Additionally, TonB interacts via its C-terminal domain with the N-terminal domain of the OM receptors (TBDTs).

In the genomes of freshwater cyanobacteria and marine strains, genes encoding TBDTs frequently appear, while ExbB-ExbD is highly conserved in amino acid sequence. Table 1 shows the distribution of TBDTs in the genomes of freshwater cyanobacteria and marine strains, indicating TonB-dependent iron transport systems are likely to be ubiquitous in freshwater and coastal cyanobacteria (Hopkinson and Morel 2009).

Previous studies have shown that cyanobacteria can uptake unchelated inorganic iron very efficiently (Fujii et al. 2010; Kranzler et al. 2011), but the mechanism is not clear. Our laboratory has proved that TonB-dependent transporter is also involved in the absorption of inorganic iron and ExbB-ExbD is involved in inorganic iron uptake and is an essential part of the iron acquisition pathway in cyanobacteria. The involvement of an ExbB-ExbD system for inorganic iron uptake may allow cyanobacteria to more tightly maintain iron homeostasis, particularly in variable environments where iron concentrations range from limiting to sufficient (Jiang et al. 2015).

Table 1 Putative siderophore biosynthesis and transport proteins in cyanobacterial genomes (Hopkinson and Morel 2009)

Organisms	IucA/IucC	IucD	NRPS	TBDT
Freshwater cyanobacteria			14	23
Anabaena sp. PCC 7120	2	1	21	10
Anabaena variabilis ATCC 29413	2	1	13	1
Cyanothece sp. ATCC 51142	0	0	14	1
Cyanothece sp. CCY0110	0	0	0	32
Gloeobacter violaceus PCC 7421	0	0	19	0
Microcystis aeruginosa NIES-843	0	0	51	2
Nostoc punctiforme PCC 73102	0	0	0	0
Synechococcus elongatus PCC 6301	0	0	0	0
Synechococcus elongatus PCC 7942	0	0	0	2
Synechococcus sp. JA-2-3B'a(2–13)	0	0	0	2
Synechococcus sp. JA-3-3Ab	0	0	0	4
Synechocystis sp. PCC 6803	0	0	0	0
Thermosynechococcus elongatus BP-1	0	0		
Marine cyanobacteria				
Acaryochloris marina MBIC11017	0	0	10	12
Crocosphaera watsonii WH 8501	0	0	19	0
Lyngbya aestuarii CCY9616	0	0	0	0
Nodularia spumigena CCY9414	0	0	30	3
Synechococcus sp. PCC 7002	2	1	0	7
Trichodesmium erythraeum IMS101	0	0	2	0
Marine picocyanobacteria				
Synechococcus spp.	0/10	0/10	0/10	0/10
Prochlorococcus spp.	0/12	0/12	0/12	0/12

8.4.2 Passive Diffusion: Uptake of Inorganic Free Iron

A class of porins on the outer membrane of Gram-negative bacteria consists of closely linked homotrimers. Each subunit consists of 16 or 18 antiparallel beta sheets, and the interaction between the layers of beta sheets forms a transmembrane channel allowing free passage of solutes with molecular weights less than 1000 Da. Inorganic free iron can be utilized by cells by passive diffusion of porins into the periplasmic space (Qiu et al. 2018).

Inorganic free iron can also diffuse into cyanobacteria cells at a lower rate through passive diffusion mediated by outer membrane porins. When iron is sufficient, the passive diffusion can meet the physiological needs of cyanobacteria cells, which process without energy consumption. Under iron-limited conditions, cyanobacteria may express active transport to acquire iron to maintain the physiological activities of cells (Fig. 12) (Qiu et al. 2018).

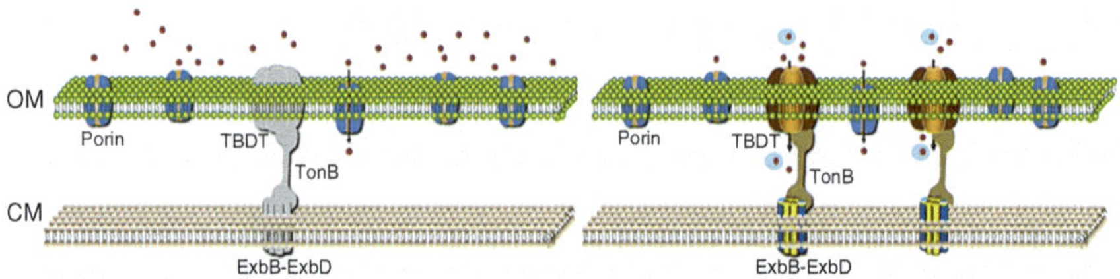

Fig. 12 Iron uptake pathway model for *Synechocystis* sp. PCC 6803. Fe-replete (left) and Fe-deplete (right) (Qiu et al. 2018)

8.5 Optimize Ferrous and Ferric Iron Transport

The Great Oxygenation Event made the bioavailability of iron greatly decreased, and biological activity in many aquatic and terrestrial environments is thus limited by iron (Andrews et al. 2003). In many aquatic environments, iron is predominantly complexed by a pool of strong organic chelators (Kranzler et al. 2011). Reductive iron uptake is a transport strategy in which iron substrates undergo reduction from Fe^{3+} to Fe^{2+} before transport through innermenbrane. The reduction-based iron uptake strategy is well suited for acquiring iron from a range of substrates including multiple Fe-siderophore complexes and free inorganic iron in aquatic environments (Kranzler et al. 2011). Given its broad range of substrates, reductive iron uptake may be applied to a variety of Fe sources (Kranzler et al. 2011). However, the location of the iron reduction process is still unknown. It was suggested that it may happen in the periplasmic space. Fe^{2+}-specific chelator ferrozine (FZ) inhibited iron uptake by 85–90% in wild-type *Synechocystis*, indicating that Fe^{3+} reduction is important and Fe^{2+} may be the predominant form for iron uptake (Kranzler et al. 2014). It is likely that uptake-associated iron reduction occurs inside the periplasmic space, (Kranzler et al. 2014). The *Synechocystis* sp. PCC 6803 alternating respiratory terminal oxidase (ARTO) is located exclusively in the plasma membrane (Pils and Schmetterer 2001; Berry et al. 2002), a location that could and may facilitate the supply of electrons to reduce periplasmic Fe^{3+} (Kranzler et al. 2014).

However, recent studies showed that most of the iron needs to undergo another re-oxidation process after reduction process (Fig. 13) (Xu et al. 2016). The whole process is as follows: (1) extracellular inorganic free iron and chelated iron can be transported across the outer membrane (OM) via the TonB-ExbB-ExbD-dependent transport system (TBDT) or porins (Jiang et al. 2015; Qiu et al. 2018); (2) after entering the periplasm space, some Fe^{3+}-siderophores then directly transport through the plasma membrane with the help of FhuB. Most of the inorganic free iron and Fe^{3+} chelated by siderophore could be reduced to Fe^{2+} and released from siderophores. ATRO is thought to play roles in the iron reduction process; (3) Fe^{2+} released from siderophores must be quickly transported through the plasma membrane via an efficient Fe^{2+} transporter, FeoB (Kranzler et al. 2011, 2014); (4) If Fe^{2+} cannot be transported into the cytoplasm in short time, the remaining Fe^{2+} will be re-oxidized. Fe^{3+} transporters such as cation diffusion facilitators (CDF) (Jiang et al. 2012),

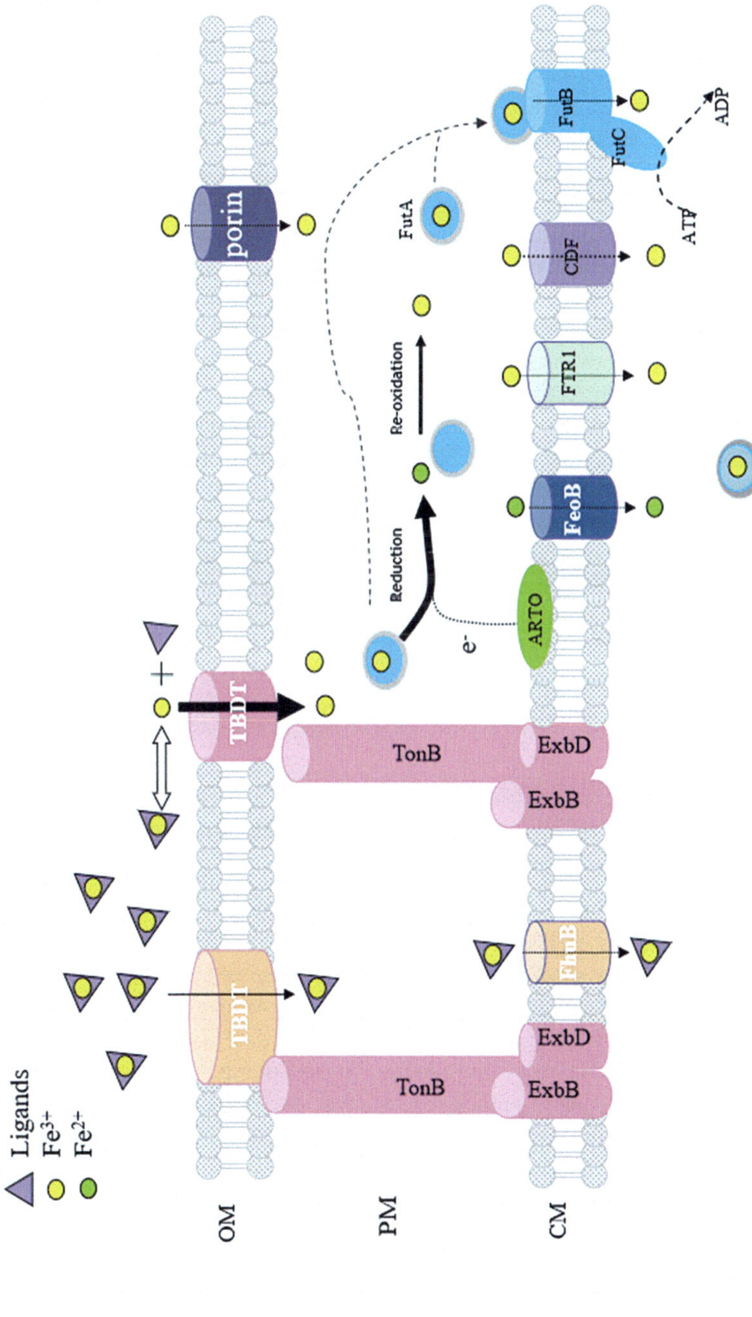

Fig. 13 Iron reductive/re-oxidation pathway of *Synechocystis* sp. PCC 6803. (Adapted from Xu et al. 2016)

cFTR1 (Xu et al. 2016) and FutB (Katoh et al. 2001) can further transport Fe^{3+} into the cytoplasm. The majority of Fe^{3+} is complexed by FutA2 and transported by the FutBC complex transporter. It is very interesting that most of reductive Fe^{2+} has been reported to be re-oxidized to Fe^{3+} in aerobic environment.

The reductive/re-oxidation pathway is an important strategy for cyanobacteria to adapt to the aerobic environment. The ferrous iron transporter FeoB is widely distributed in many archaea as a "living fossil" (Marlovits et al. 2002; Hantkel 2003) and usually plays an important role in iron uptake under anaerobic conditions (Velayudhan et al. 2000; Kim et al. 2015). When the Earth changed from anaerobic to aerobic environment, it is probably that cyanobacteria have gradually evolved ferric iron transport pathways including the reductive/re-oxidation process. Iron reduction step in the reductive/re-oxidation pathway is significantly important for cyanobacteria. Firstly, studies have shown that about 90% of FutA2-Fe^{3+} can be reduced by the reduction way (Kranzler et al. 2014); Secondly, under aerobic conditions, most of the reductive iron will be rapidly oxidize to ferric iron and transported into cytoplasm through ferric iron transport system. Thus, the reduction of iron seems more important under anaerobic conditions. Thirdly, with the reduction process, cyanobacteria can utilize different kinds of substrates to acquire iron including iron chelated by siderophores. Under diverse conditions, cyanobacteria can optimize the transformation and assignation of Fe^{3+} and Fe^{2+} through the re-oxidation step, which seems to be in line with the evolution of the Earth from anaerobic to aerobic. As the oldest oxygen-producing autotrophic organism on the Earth, cyanobacteria play an important role in the event of the Great Oxygenation Event of the Earth. It not only created a big aerobic environment on Earth but also provided an aerobic microenvironment in cells. So it was suggested that O_2 they product may be important for the re-oxidation of Fe^{2+} (Xu et al. 2016).

8.6 Development of Special Cell Surface Structure to Facilitate Iron Adhesion and Uptake

Various iron transport systems are involved in the absorption and utilization of iron in cyanobacteria. Some special structures on the surface of cyanobacteria or substances secreted by cells might also play an important role in promoting the absorption and utilization of iron. It is interesting that formation of colonies sometimes also make sense in survival of cyanobacteria. Extracellular polysaccharide (EPS) is one of the main substances synthesized by cyanobacteria and secreted into the periphery or around cells. It plays a role in the adaptation of cyanobacteria to complex and variable environments, such as the process of resisting drought and ultraviolet radiation (Ren et al. 2013). It has been shown that the gelatinous sheath on the surface of *Microcystis aeruginosa* cells can efficiently adsorb certain metals through the EPS in the gelatinous sheath (Parker et al. 2000). It has been reported that the EPS of the *Thiobacillus thiooxidans* have an adsorption effect on both Fe^{2+} and Fe^{3+}. Most cyanobacterial extracellular polysaccharides are anionic (Ren et al. 2013). The adsorption relationship between EPS and extracellular metal ions is through mutual attraction between charged ions. Biofilm refers to a group of organized bacteria attached to the surface of a living or inanimate object that is surrounded by extracellular macromolecules of bacteria. There are various major biological macromolecules such as proteins, polysaccharides, DNA, RNA, peptidoglycan, lipids, and phospholipids in biofilms. It is useful for adhesion of cells.

It was reported that biofilm can transfer electron in other species. This means that insoluble iron can be utilized by reduction. It is possible that biofilm or other structures in cyanobacteria may have the same function in iron utilization. The S-layer protein is synthesized by cyanobacteria and secreted to the outside of cells through a type I secretion system, which forms clathrin outside surface the cell (Oliveira et al. 2016). It is still uncertain whether this dense and regular clathrin formed on the cell surface may be involved in the adsorption of extracellular iron. In addition, some cyanobacteria have some filaments that are thinner and longer than flagella (Schuergers and Wilde 2015). They are called pili, which are widely located on the cell surface like hair strands. They can shrink and prolong, not only involved in cell adhesion, exercise but also in DNA absorption. It was reported that pilus may also play an important role in the utilization of iron in the environment (Bhaya et al. 2000; Yoshihara et al. 2001; Lamb et al. 2014).

In oligotrophic oceans, iron becomes one of the limiting factors that restrict growth of phytoplankton (Tagliabue et al. 2017). Sources of iron in the ocean include the input of external iron and the reuse of iron in seawater. The dusts caused by atmospheric deposition are an important source of iron for marine life (Mahowald et al. 2015; Boyd and Ellwood 2010). Some cyanobacteria form colonies and use the interaction between cells to collect iron from the dusts. For example, *Trichodesmium*, a filamentous dinitrogen-fixing cyanobacteria forms a community under natural conditions. The community-formed *Trichodesmium* have been reported to be able to effectively capture dusts and actively transfer them to the center of the colony (Fig. 14) (Rubin et al. 2011). It has been suggested that it is an important strategy to acquire iron from dusts. Recent studies have shown that *Trichodesmium* prefers to collect iron-rich dust particles compared to iron-free particles (Fig. 15) (Kessler et al. 2019). These cyanobacterial communities are very effective in enhancing the dissolution and utilization of iron in iron-limiting areas. Interestingly, *Trichodesmium* forms colonies under natural conditions but not in laboratory culture conditions. It is still unclear of the relationship between the formation of the formation of *Trichodesmium* colonies and iron acquisition.

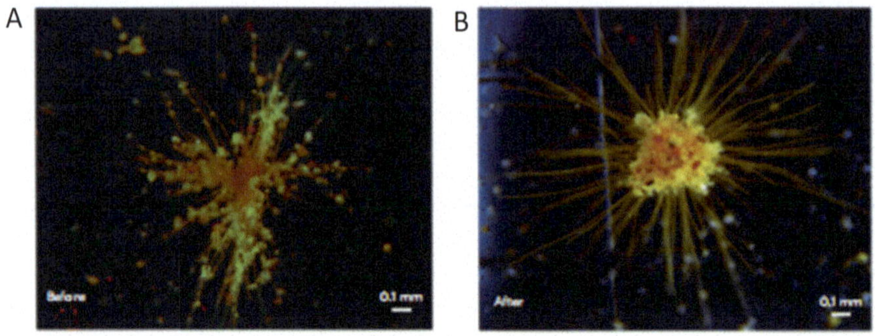

Fig. 14 Micrographs of natural *Trichodesmium* colonies mixed with dust showing efficient dust retention and active centering of the dust in the colony core. (**a**) Natural puffs were highly efficient in retaining desert dust. These pictures were taken subsequent to dust addition and before its shuttling to the colony core. (**b**) Desert dust that was actively centered in the core of natural puff colonies (Rubin et al. 2011)

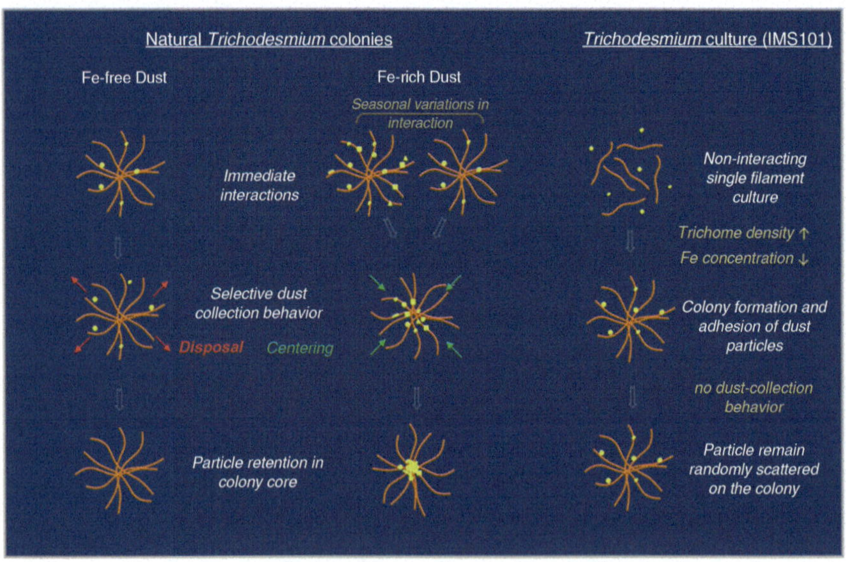

Fig. 15 Schematic representation of *Trichodesmium*-dust interactions observed in this study and the role of Fe in these interactions (Kessler et al. 2019)

8.7 Reduce the Cell Size and Increase Specific Surface Area to Facilitate Passive Diffusion of Iron

One of the more prominent features of cells cultured under iron-limiting conditions is a decrease in cell diameter and an increase in specific surface area (Li et al. 2016). Firstly, the smaller cells have lower nutrient requirements. Secondly, the greater the ratio of cell surface area to volume (specific surface area) under iron restriction, the

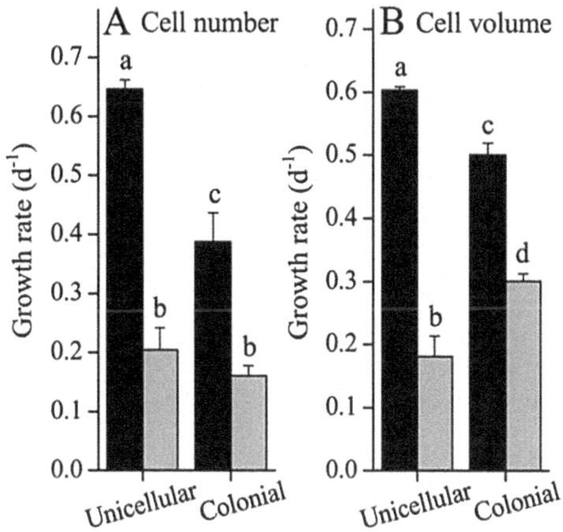

Fig. 16 The growth of the unicellular and colonial forms of *Microcystis* sp. under iron-replete (Fe-R) and iron-deficient (Fe-D) conditions. (**a**) Specific growth rates normalized to cell numbers (d^{-1}); (**b**) Specific growth rates expressed as total cell volume (d^{-1}) (Li et al. 2016)

better it is to absorb external nutrients (Li et al. 2016). Under iron-deficient conditions, cyanobacteria will reduce the size of the cells and adapt to the iron-deficient environment. A typical freshwater cyanobacteria *Microcystis* strain grows as small colonies under bloom conditions (Cao et al. 2005). Both unicellular and colony forms of *Microcystis* showed reduced cell size under iron-deficient conditions, although the cell size of the colonial form decreased much less than that of the unicellular culture (Fig. 16). With decreasing iron concentrations, the diameter of cyanobacteria is reduced to increase the specific surface area of the cells in order to adapt to the iron-deficient environment. The rate of iron absorption varies with the change of specific surface area. The smaller the volume of cells, the larger the relative surface area of the cells. The higher the efficiency of material transport, the more favorable it is for cells with small diameters to grow under iron-limited conditions (Li et al. 2016). Thus, increased surface-to-volume ratios with decreasing cell size will increase nutrient uptake capabilities (Hudson and Morel 1993; Sunda and Huntsman 1997; Raven 1999; Leynaert et al. 2004).

The iron source in the ocean is mostly from the flowing dust. The open ocean is far away from the land, so the iron content in the water will be much less than freshwater. Studies have compared different coastal, freshwater, and open-ocean cyanobacteria and concluded that open-ocean cyanobacteria have a smaller cell size and therefore have a larger ratio of surface area to volume, thereby increasing the ability of iron to uptake (Fig. 17) (Lis et al. 2015).

Crocosphaera watsonii is a single-cell nitrogen-fixing cyanobacteria that lives in the ocean. In order to study the changes of cells under iron deficiency conditions, the *C. watsonii* was cultured under conditions of sufficient iron (dFe = 403.3 nM) and iron limitation (dFe = 3.3 nM). Scanning electron microscopy of cyanobacteria cells (Fig. 18) concluded that the volume of cyanobacteria cells with iron restriction was significantly reduced and the cell volume was linearly correlated with iron concentration (Violaine et al. 2014).

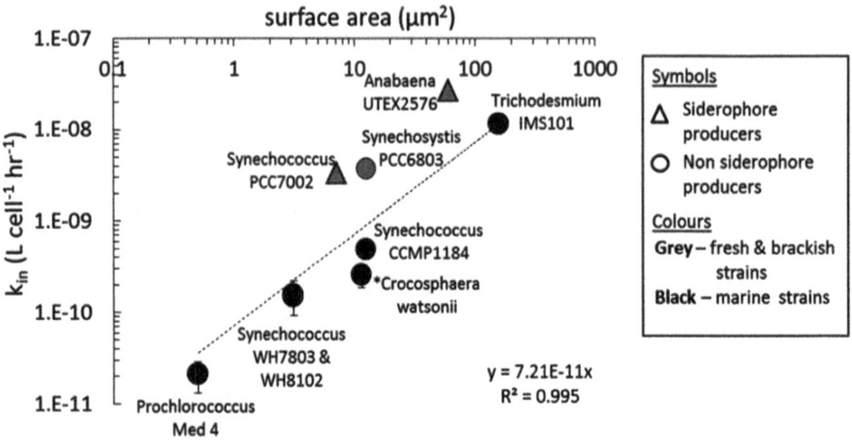

Fig. 17 Dissolved inorganic iron (Fe′) uptake rate constants (k_{in} = uptake rate/[Fe′]) of Fe-limited cyanobacteria as a function of cell surface area (μm^2) on a log-log plot. Each data point represents averaged rate constants from a single study for a single organism (Lis et al. 2015)

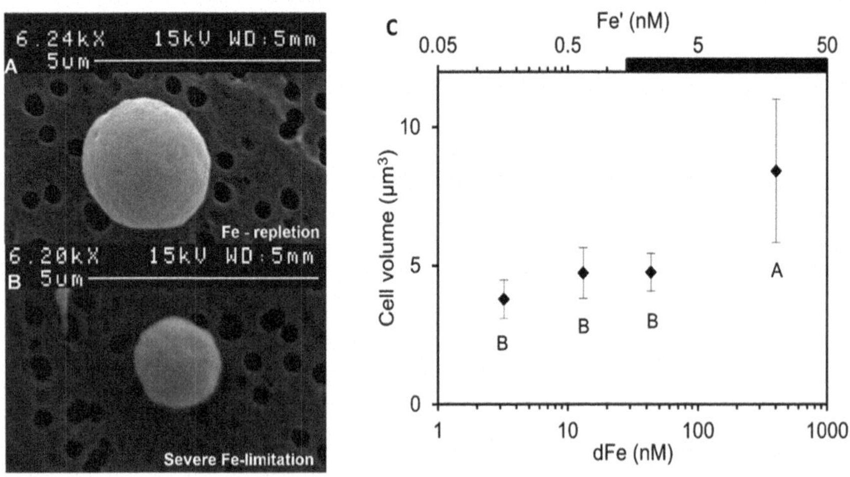

Fig. 18 Influence of Fe availability on *C. watsonii* biovolume. (**a**) Fe-replete condition; (**b**) severe Fe-limited condition. (**c**) Mean biovolume of *C. watsonii* related to dFe and Fe′ concentrations (Violaine et al. 2014)

Reducing cell size in response to iron restriction appears to be a general trend, and cell volume reduction can be seen as a mechanism by which cyanobacteria cells adapt to low iron. The decrease in cell volume can reduce the energy required to maintain the metabolisms of the cell and can increase the specific surface area of the cell. It has been widely proved that the increase of the specific surface area of the cell is very beneficial to passive diffusion of ions. Thus, reducing cell size appears to be an important strategy for cyanobacteria to adapt to iron limitation (Hudson and Morel 1993; Sunda and Huntsman 1997; Raven 1999).

9 Signal Transduction of Iron Deficiency in Cyanobacteria

Intracellular iron concentration must be strictly controlled in the cyanobacteria. The above text has described a series of iron-deficient responses of cyanobacteria, while high iron is also toxic to the cells. In order to regulate this contradiction, the transcriptional regulation to iron deficiency signals is particularly important. Scientists have identified three most common transcriptional regulators in different bacterial species: Fur, DtxR, and RirA (Kakhlon and Cabantchik 2002; Delany et al. 2001). In cyanobacteria, however, much little is known about the transcriptional regulators of iron deficiency response. The global transcriptional regulator Fur has been reported in some cyanobacterial species, such as All1691 in *Anabaena* sp. PCC 7120, Sll0567 in *Synechocystis* sp. PCC 6803, etc. Some researchers have found that PfsR, a transcription regulator of TetR family, may be also involved in iron deficiency signal transduction. In addition, some small RNA plays an important role in iron deficiency signal transduction (González et al. 2014).

9.1 The Global Regulator Fur

Fur protein (Ferric uptake regulator) is a major regulator of intracellular iron balance in phytoplankton. It regulates the transcription of a large number of genes related to intracellular iron homeostasis and oxidative stress (Escolar et al. 1999; Hantke 2001; Baichoo et al. 2002; Martin-Luna et al. 2006). At the same time, intracellular reactive oxygen species can also regulate some iron-related genes through PerR-like transcriptional regulators (Horsburgh et al. 2001; Yousef et al. 2003). Under the condition of sufficient intracellular iron, Fur protein binds to Fe^{2+}, which changes the conformation of Fur protein and forms active dimer. Activated Fur proteins bind to promoters containing Fur box to promote or inhibit the transcription of these genes. Metabolic processes requiring large amounts of iron were inhibited under iron restriction, while some proteins that increased iron restriction tolerance in cyanobacteria were upregulated (de Lorenzo et al. 1988; Coy and Neilands 1991). In addition to Fur protein, other regulatory factors such as IdiB and IsrR play an important role in iron balance (Yousef et al. 2003; Dühring et al. 2006). Under iron restriction, the related transcription factors were upregulated. For example, in iron-deficient cyanobacteria cells, cAMP protein kinase was downregulated 6.25 times, photochrome-related transcription factors were upregulated (Singh et al. 2003), and fur was upregulated 2.5 times (López-Gomollón et al. 2007) and downregulated after iron addition. Under iron-limiting conditions, the genes encoding iron-containing proteins that involved in photosystem components, photosynthetic apparatus assembly and chlorophyll biosynthesis are downregulated. Meanwhile, the enzymes involved in nitrogen assimilation, such as nitrogen fixation and nitrogen uptake are downregulated by Fur transcriptional factor. The expression of NADH dehydrogenase (NDH) and ATPase related to the metabolic processes such as CO_2 immobilization, respiration, and PSI-centered circular electron transport decreased. Maintaining low metabolic levels under iron restriction may be an important strategy for cyanobacterial cells to adapt to low-iron environment (Singh et al. 2003; Hernández-Prieto et al. 2012). The global transcriptional factor Fur can also actively up-regulate some genes that encoding proteins to replace iron-containing proteins. It is well known that flavodoxin can replace ferritin and iron-free catalase AphC replacing heme-containing catalase KatG (Tichy and Vermaas 1999; Singh et al. 2003; Hernández-Prieto et al. 2012). The upregulation of genes related to basic cellular metabolic processes, such as Fe-S protein assembly system, protease, and molecular chaperone, is likely to be related to iron redistribution in cells (Singh et al. 2003; Hernández-Prieto et al. 2012). Iron uptake-related genes such as FutABC family genes, protective genes such as *isiA* (involved in the protection of PSI, etc.), *idiA*, and *dpsA* genes are upregulated. Some noncoding small RNAs also play an important role in the process of iron deficiency transcription regulation of cyanobacteria (Hernández et al. 2002; Yousef et al. 2003; Cao et al. 2005). Cyanobacteria cells adapt to iron-limited environment by altering cellular metabolism and protein components through transcription factors to reduce iron demand and oxidative damage caused by iron deficiency. Iron regulation system plays an important role in cyanobacteria adapting to different concentrations of iron. The iron regulatory system of marine algae is simple. *Synechococcus* WH 8102 has only five histidine kinase sensors and nine response regulatory factors. However, *Synechococcus* CC 9311 has 11 histidine kinase sensors and 17 response regulators, twice as many as *Synechococcus* WH 8102 (Palenik et al. 2006). The iron regulation system of cyanobacteria in the ocean is much simpler than that of freshwater and coastal algae, probably because the iron concentration in the ocean is relatively stable while that in freshwater and coastal waters is variable.

For most bacteria and cyanobacteria, iron has a dual role. On the one hand, transition metal iron acts as a coenzyme factor and participates in many metabolic reactions, such as photosynthesis, respiration, DNA synthesis, nitrogen fixation, and so on. Therefore, cells form many uptake systems to supply iron (Wandersman and Delepelaire 2004). On the other hand, it is also because of the relatively active chemical properties of iron itself, which makes it easy to react in cells. It forms a lot of reactive oxygen species and causes damage to cells. Therefore, intracellular iron needs to be strictly regulated (Kakhlon and Cabantchik 2002). The regulation of intracellular iron homeostasis is mainly mediated by transcription factor Fur (Lee and Helmann 2007). Current studies have shown that Fur is a global iron regulatory factor, which can inhibit the expression of iron uptake genes in cells without iron deficiency and can also relieve the inhibition of iron uptake genes in cells without iron deficiency (Delany et al. 2001). Studies have shown that in *Escherichia coli*, Fur regulates more than 90 genes related to intracellular iron homeostasis (Escolar et al. 2000). Fur is structurally composed of N-terminal binding to DNA and C-terminal rich in histidine. Fur depends on dimers formed by N-terminal binding to metal ions at C-terminal. The experimental results show that the binding ability of Fur with metal ions increases 1000 times (Tronnet et al. 2017).

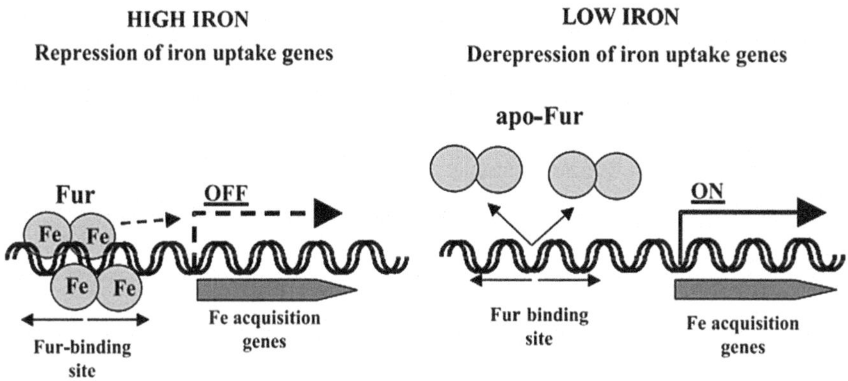

Fig. 19 The expression regulate pattern of iron uptake-related genes by Fur (Andrews et al. 2003)

Fur-Fe^{2+}dimer is usually bound to a 19 bp Fur box region that is usually located at about 35 bp to 10 bp upstream of the transcription initiation site of the Fur-regulated genes. Fur-Fe^{2+} in dimer form binds to Fur box upstream of the gene and regulates its target gene. In *Escherichia coli*, genes related to the formation and secretion of iron carrier (enterobactin), iron absorption, and iron deficiency are inhibited by Fur-Fe^{2+}, while genes related to iron storage and utilization are activated by Fur-Fe^{2+} (Andrews et al. 2003). Its working model is shown in Fig. 19.

Fur is also widely distributed in cyanobacteria, including *Synechococcus* sp. PCC 7942, *Anabaena* sp. PCC 7119, *Synechocystis* sp. PCC 6803, *Microcystis aeruginosa*, and other cyanobacteria (Teresa et al. 2001; González et al. 2014). Researchers also found that the upstream of Fur-regulated genes in cyanobacteria also existed with Fur box in *Escherichia coli*. These genes with Fur box include many genes related to iron uptake, including outer membrane receptor protein (TBDT), trivalent iron uptake Fut system genes, and so on. At present, it is believed that the regulation pattern of Fur in cyanobacteria is similar to that of *Escherichia coli*. Fur of cyanobacteria can be divided into three main categories: FurA, FurB, and FurC. It is generally believed that FurA is mainly related to the regulation of iron uptake, while FurB and FurC are mainly responsible for the functions of cyanobacteria under oxidative stress. In addition, some researchers have found that the Fur family also has Zur related to zinc regulation and PerR involved in oxidative stress caused by iron excess (Li et al. 2004; Parent et al. 2013; Ludwig et al. 2015).

9.2 PfsR

PfsR, a transcription factor of TetR family, has been reported to play a role in iron deficiency signal transduction in *Synechocystis* PCC 6803. It has been found that *pfsR* deletion mutant may be more adapted to iron deficiency than wild-type strain.

The main manifestation was that under iron deficient conditions, the *pfsR* mutant accumulated more chlorophyll a, carotenoids, and phycocyanin than the wild type. After iron deficiency, the complex of photosynthetic system I and photosynthetic system II of the mutant was more than that of the wild type. Detailed mechanisms of PfsR involved in the transcriptional regulation in the cyanobacteria still need to be clarified. (Dan et al. 2014).

9.3 Noncoding RNA

Noncoding RNAs refer to a number of RNA molecules that are not capable of translating proteins. Some of these noncoding RNAs can participate in the response to environmental factors and regulate gene expression. In cyanobacteria, IsiA (iron stress-induced protein A), which is one of the most sensitive genes to iron deficiency, is regulated by a noncoding RNA named IsrR (iron stress-repressed RNA). When IsrR was overexpressed, the formation of IsiA-PSI complex in cyanobacteria was significantly reduced. If IsrR in cyanobacteria was knocked out, the presence of IsiA protein could be detected in cyanobacteria without iron deficiency (Dühring et al. 2006). Recently, researchers have also found that another noncoding RNA-IsaR (iron stress activated RNA) plays an important role in the iron deficiency response of cyanobacteria. IsaR regulates iron response genes by interaction with the transcription factor Fur (Georg et al. 2017). Further studies are needed to deeply demonstrate the complex signal transduction process cyanobacteria to iron of iron deficiency.

10 Outlooks on Cyanobacterial Adaptive Strategies to Marine Iron Limitation

Cyanobacteria in modern oceans have a variety of strategies to adapt to iron limitation, which reflect the long-term evolutionary choices of them for environment. As one of the most important limiting factors to ocean primary productivity, iron limitation is also tightly connected to other global environment changes, such as elevated CO_2 concentration, global warming, ocean acidification, etc. How climate changes will affect iron limitation and other key limiting factors of the global primary productivity still need to be deeply investigated. There are some interesting issues to be addressed, including the interaction between warming and iron availability and iron use efficiency, co-limitation of iron and phosphorus or nitrogen, and the cross-links of multiple limiting factors and global climate changes. Recently, it has been reported that global warming could alleviate ocean iron limitation and increase the iron use efficiency of nitrogen-fixing cyanobacteria (Jiang et al. 2018), which let us to predict that the future warmer ocean possibly will increase new input of nitrogen into the ocean. Such kinds of laboratory and field researches still need to be more extensive and deeper. Since the adaptive strategies of cyanobacteria to iron limitation will undoubtedly be influenced by physical, chemical, and other abiotic factors, global climate change will have far-reaching impact on ocean iron limitation. Thus comprehensive knowledge on the principles of ocean primary production is still a big challenge. Indeed, it will be very difficult to scientists to accurately predict the primary productivity in the future ocean, until big data are built, more extensive field studies are made, and stronger modeling are set up with multiple environment factors.

References

Abdallah, F., Salamini, F., & Leister, D. (2000). A prediction of the size and evolutionary origin of the proteome of chloroplasts of *Arabidopsis*. *Trends in Plant Science, 5*, 141–142.

Andrews, S. C., Smith, J. M. A., Hawkins, C., Williams, J. M., Harrison, P. M., & Guest, J. R. (1993). Overproduction, purification and characterization of the bacterioferritin of *Escherichia coli* and a C-terminally extended variant. *European Journal of Biochemistry, 213*, 329–338.

Andrews, S. C., Robinson, A. K., & Rodrıguez-Quinones, F. (2003). Bacterial iron homeostasis. *FEMS Microbiology Reviews, 27*, 215–237.

Andrizhiyevskaya, E. G., Schwabe, T. M. E., Germano, M. D., Haene, S., Kruip, J., van Grondelle, R., & Dekker, J. P. (2002). Spectroscopic properties of PSI–IsiA supercomplexes from the cyanobacterium *Synechococcus* PCC 7942. *Biochimica et Biophysica Acta, 1556*, 265–272.

Årstøl, E., & Hohmann-Marriott, M. F. (2019). Cyanobacterial siderophores-physiology, structure, biosynthesis, and applications. *Marine Drugs, 17*(5), 281.

Baichoo, N., Wang, T., Ye, R., & Helmann, J. D. (2002). Global analysis of the Bacillus subtilis Fur regulon and the iron starvation stimulon. *Molecular Microbiology, 45*, 1613–1629.

Bailey, S., Melis, A., Mackey, K. R. M., Cardol, P., Finazzi, G., et al. (2008). Alternative photosynthetic electron flow to oxygen in marine *Synechococcus*. *Biochimica et Biophysica Acta, 1777*, 269–276.

Bakker, D. C. E., Watson, A. J., & Law, C. S. (2001). Southern Ocean iron enrichment promotes inorganic carbon drawdown. *Deep Sea Research Part II, 48*, 2483–2507.

Barber, J., Nield, J., Duncan, J., & Bibby, T. S. (2006). Accessory chlorophyll proteins in cyanobacterial photosystem I. In J. H. Golbeck (Ed.), *Photosystem I* (pp. 99–117). Dordrecht, Springer.

Barbeau, K., Rue, E. L., Bruland, K. W., & Butler, A. (2001). Photochemical cycling of iron in the surface ocean mediated by microbial iron(III)-binding ligands. *Nature, 413*, 409–413.

Behrenfeld, M. J., & Milligan, A. J. (2013). Photophysiological expressions of iron stress in phytoplankton. *Annual Review of Marine Science, 5*, 217–246.

Berman-frank, I., Cullen, J. T., Shaked, Y., Sherrell, R., & Falkowski, P. G. (2001). Iron availability, cellular quotas, and nitrogen fixation in *Trichodesmium*. *Limnology and Oceanography, 46*, 1249–1260.

Berry, S., Schneider, D., Vermaas, W. F. J., & Rogner, M. (2002). Electron transport routes in whole cells of *Synechocystis* sp strain PCC 6803: The role of the cytochrome bd-type oxidase. *Biochemistry, 41*, 3422–3429.

Bhaya, D., Bianco, N. R., Bryant, D., & Grossman, A. (2000). Type iv pilus biogenesis and motility in the cyanobacterium *synechocystis* sp. PCC 6803. *Molecular Microbiology, 37*, 941–951.

Bibby, T. S., Nield, J., & Barber, J. (2001). Iron deficiency induces the formation of an antenna ring around trimeric photosystem I in cyanobacteria. *Nature, 412*, 743–745.

Bibby, T. S., Mary, I., Nield, J., Partensky, F., & Barber, J. (2003). Low-light-adapted *Prochlorococcus* species possess specific antennae for each photosystem. *Nature, 424*, 1051–1054.
Binder, A. (1982). Respiration and photosynthesis in energy-transducing membranes of cyanobacteria. *Journal of Bioenergetics and Biomembranes, 14*, 271–286.
Bishop, P. E., & Premakumar, R. (1992). Alternative nitrogen fixation systems. In G. Stacey, R. H. Burris, & D. J. Evans (Eds.), *Biological nitrogen fixation* (pp. 736–762). New York: Chapman & Hall.
Boekema, E. J., Hifney, A., Yakushevska, A. E., Piotrowski, M., Keegstra, W., Berry, S., Michel, K. P., Pistorius, E. K., & Kruip, J. (2001). A giant chlorophyll-protein complex induced by iron deficiency in cyanobacteria. *Nature, 412*, 745–748.
Boyd, P. W., & Ellwood, M. J. (2010). The biogeochemical cycle of iron in the ocean. *Nature Geoscience, 3*, 675–682.
Boyd, P., Jickells, T., Law, C. S., Blain, S., Boyle, E. A., Buesseler, K. O., et al. (2007). Mesoscale iron enrichment experiments 1993–2005: Synthesis and future directions. *Science, 315*, 612–617.
Bricker, T. M., & Frankel, L. K. (2002). The structure and function of CP47 and CP43 in photosystem II. *Photosynthesis Research, 72*, 131–146.
Bruland, K. W., Franks, R. P., Kanuer, G., & Martin, J. H. (1979). Sampling and analytical methods for the determination of copper, cadmium, zinc, and nickel in seawater. *Analytica Chimica Acta, 105*, 233–245.
Burnap, R. L., Troyan, T., & Sherman, L. A. (1993). The highly abundant chlorophyll–protein complex of iron-deficient *Synechococcus* sp. PCC 7942 (CP43′) is encoded by the isiA gene. *Plant Physiology, 103*, 893–902.
Canini, A., Civitareale, P., Marini, S., Grilli, C. A., & Rotilio, G. (1992). Purification of iron superoxide dismutase from the cyanobacterium Anabaena cylindrica Lemm and localization of the enzyme in heterocysts by immunogold labeling. *Planta, 187*, 438–444.
Cao, H. S., Kong, F. X., Tan, J. K., Zhang, X. F., Tao, Y., & Yang, Z. (2005). Recruitment of total phytoplankton, chlorophytes and cyanobacteria from lake sediments recorded by photosynthetic pigments in a large, shallow lake (Lake Taihu, China). *International Review of Hydrobiology, 90*, 347–357.
Capone, D. G., Zehr, J. P., Paerl, H. W., Bergman, B., & Carpenter, E. J. (1997). *Trichodesmium*, a globally significant marine cyanobacterium. *Science, 276*, 1221–1229.
Capone, D. G., Burns, J. A., Montoya, J. P., Subramaniam, A., Mahaffey, C., Gunderson, T., Michaels, A. F., & Carpenter, E. J. (2005). Nitrogen fixation by *Trichodesmium* spp: An important source of new nitrogen to the tropical and subtropical North Atlantic Ocean. *Global Biogeochemical Cycles, 19*.
Carroll, C. S., & Moore, M. M. (2018). Ironing out siderophore biosynthesis: A review of non-ribosomal peptide synthetase (NRPS)-independent siderophore synthetases. *Critical Reviews in Biochemistry and Molecular Biology, 53*(4), 356–381.
Chereskin, B., & Castelfranco, P. (1982). Effects of iron and oxygen on the biosynthetic pathway in etiochloroplasts II observations on isolated etiochloroplasts. *Plant Physiology, 69*, 112–116.
Coy, M., & Neilands, J. B. (1991). Structural dynamics and functional domains of the Fur protein. *Biochemistry, 30*, 8201–8210.
Croot, P. L., Bowie, A. R., Frew, R. D., Maldonado, M. T., Hall, J. A., Safi, K. A., La Roche, J., Boyd, P. W., & Law, C. S. (2001). Retention of dissolved iron and Fe-II in an iron induced Southern Ocean phytoplankton bloom. *Geophysical Research Letters, 28*, 3425–3428.
Croot, P. L., Passow, U., Assmy, P., Jansen, S., & Strass, V. H. (2007). Surface active substances in the upper water column during a Southern Ocean Iron fertilization experiment (EIFEX). *Geophysical Research Letters, 34*, C06015.
Croot, P. L., Bluhm, K., Schlosser, C., Streu, P., Breitbarth, E., Frew, R., & Van Ardelan, M. (2008). Regeneration of Fe(II) during EIFeX and SOFeX. *Geophysical Research Letters, 35*, L19606.
Dan, C., Qingfang, H., & Araujo, W. L. (2014). PfsR is a key regulator of iron homeostasis in *Synechocystis* PCC 6803. *PLoS One, 9*, e101743.

de Lorenzo, V., Giovannini, F., Herrero, M., & Neilands, J. B. (1988). Metal ion regulation of gene expression. Fur repressor-operator interaction at the promoter region of the aerobactin system of pColV-K30. *Journal of Molecular Biology, 203*, 875–884.

de los Ríos, A., Grube, M., Sancho, L. G., & Ascaso, C. (2007). Ultrastructural and genetic characteristics of endolithic cyanobacterial biofilms colonizing Antarctic granite rocks. *FEMS Microbial Ecology, 2*, 386–395.

Decho, A. (1990). Microbial exopolymer secretions in ocean environments: Their role(s) in food webs and marine processes. *Oceanography and Marine Biology. Annual Review, 28*, 73–153.

Delany, I., Spohn, G., Rappuoli, R., et al. (2001). The fur repressor controls transcription of iron-activated and-repressed gene in *Helicobacter pylori. Molecular Microbiology, 42*, 1297–1309.

Dufresne, A., Salanoubat, M., Partensky, F., et al. (2003). Genome sequence of the cyanobacterium Prochlorococcus marinus SS120, a nearly minimal oxyphototrophic genome. *Proceedings of the National Academy of Sciences of the United States of America, 100*, 10020–10025.

Dühring, U., Axmann, I. M., Hess, W. R., & Wilde, A. (2006). An internal antisense RNA regulates expression of the photosynthesis gene isiA. *Proceedings of the National Academy of Sciences of the United States of America, 103*, 7054–7058.

Dwivedi, K., Sen, A., & Bullerjahn, G. S. (1997). Expression and mutagenesis of the dpsA gene of *Synechococcus* sp. PCC 7942, encoding a DNA-binding protein involved in oxidative stress protection. *FEMS Microbiology Letters, 155*, 85–91.

Eitinger, T. (2004). In vivo production of active nickel superoxide dismutase from *Prochlorococcus marinus* MIT9313 is dependent on its cognate peptidase. *Journal of Bacteriology, 186*, 7821.

Entsch, B., & Smillie, R. M. (1972). Oxidation-reduction properties of phytoflavin, a flavoprotein from blue-green algae. *Archives of Biochemistry and Biophysics, 151*, 378–386.

Escolar, L., Perez-Martin, J., & de Lorenzo, V. (1999). Opening the iron box: Transcriptional metalloregulation by the Fur protein. *Journal of Bacteriology, 181*, 6223–6229.

Escolar, L., Perez-Martin, J., & De Lorenzo, V. (2000). Evidence of an unusually long operator for the Fur repressor in the aerobactin promoter of Escherichia coli[J]. *Journal of Biological Chemistry, 275*, 24709–24714.

Falk, S., Samson, G., Bruce, D., Huner, N. P. A., & Laudenbach, D. E. (1995). Functional analysis of the iron-stress induced CP43' polypeptide of PSII in the cyanobacterium *Synechococcus* sp.PCC 7942. *Photosynthesis Research, 45*, 51–60.

Falkowski, P. G., Barber, R. T., & Smetacek, V. (1998). Biogeochemical controls and feedbacks on ocean primary production. *Science, 281*, 200–207.

Feely, R. A., Doney, S. C., & Cooley, S. R. (2009). Ocean acidification: Present conditions and future changes in a high-CO^2 world. *Oceanography, 22*, 36–47.

Field, C. B., Behrenfeld, M. J., Randerson, J. T., & Falkowski, P. (1998). Primary production of the biosphere: Integrating terrestrial and oceanic components. *Science, 281*, 237–240.

Flombaum, P., Gallegos, J. L., Gordillo, R. A., Rincón, J., Zabala, L. L., Jiao, N., et al. (2013). Present and future global distributions of the marine cyanobacteria Prochlorococcus and Synechococcus. *Proceedings of the National Academy of Sciences of the United States of America, 110*(24), 9824–9829.

Fraser, J. M., Tulk, S. E., Jeans, J. A., Campbell, D. A., Bibby, T. S., & Cockshutt, A. M. (2013). Photophysiological and photosynthetic complex changes during iron starvation in *Synechocystis* sp. PCC 6803 and *Synechococcus elongatus* PCC 7942. *PLoS ONE, 8*(3), e59861.

Fujii, M., Rose, A. L., Omura, T., & Waite, T. D. (2010). Effect of Fe(II) and Fe(III) transformation kinetics on iron acquisition by a toxic strain of *Microcystis aeruginosa. Environmental Science & Technology, 44*, 1980–1986.

Fujii, M., Dang, T. C., Rose, A. L., Omura, T., & Waite, T. D. (2011). Effect of light on iron uptake by the freshwater cyanobacterium microcystis *aeruginosa. Environmental Science & Technology, 45*, 1391–1398.

Fulda, S., Huang, F., Nilsson, F., Hagemann, M., & Norling, B. (2000). Proteomics of *Synechocystis* sp. strain PCC 6803: Identification of periplasmic proteins in cells grown at low and high salt concentrations. *European Journal of Biochemistry, 267*, 5900–5907.

Garnerin, T., Dassonville-Klimpt, A., & Sonnet, P. (2017). (2017) Fungal hydroxamate siderophores: Biosynthesis, chemical synthesis and potential medical applications. In A. Méndez-Vilas (Ed.), *Antimicrobial Research: Novel Bioknowledge and Educational Programs*. Badajoz, Spain: Formatex Research Center.

Geider, R. J., & La Roche, J. (1994). The role of iron in phytoplankton photosynthesis, and the potential for iron-limitation of primary productivity in the sea. *Photosynthesis Research, 39*, 275–301.

Georg, J., Kostova, G., Vuorijoki, L., et al. (2017). Acclimation of oxygenic photosynthesis to iron starvation is controlled by the sRNA IsaR1[J]. *Current Biology, 27*, 1425–1436. e7.

Gledhill, M., & Buck, K. N. (2012). The organic complexation of iron in the marine environment: A review. *Frontiers in Microbiology, 3*, 128–144.

Gledhill, M., & van den Berg, C. M. G. (1994). Determination of complexation of iron(III) with natural organic complexing ligands in seawater using cathodic stripping voltammetry. *Marine Chemistry, 47*, 41–54.

Goldman, S. J., Lammers, P. J., Berman, M. S., & Sanders-Loehr, J. (1983). Siderophore-mediated iron uptake in different strains of *Anabaena* sp. *Journal of Bacteriology, 156*, 1144–1150.

González, A., Angarica, V. E., Sancho, J., et al. (2014). The FurA regulon in *Anabaena* sp. PCC 7120: In silico prediction and experimental validation of novel target genes. *Nucleic Acids Research, 42*, 4833–4846.

Gordon, R. M., Marin, J. H., & Knauer, G. A. (1982). Iron in north-east Pacific waters. *Nature, 299*, 611–612.

Guikema, J. A., & Sherman, L. A. (1983). Organization and function of chlorophyll in membranes of cyanobacteria during iron starvation. *Plant Physiology, 73*, 250–256.

Hantke, K. (2001). Iron and metal regulation in bacteria. *Current Opinion in Microbiology, 4*, 172–177.

Hantkel, K. (2003). Is the bacterial ferrous iron transporter FeoB a living fossil? *Trends in Microbiology, 11*, 192–195.

Hart, S. E., Schlarb-Ridley, B. G., Bendall, D. S., & Howe, C. J. (2005). Terminal oxidases of cyanobacteria. *Biochemical Society Transactions, 33*, 832–835.

Hernández, J. A., Artieda, M., Peleato, M. L., Fillat, M. F., & Bes, M. T. (2002). Iron stress and genetic response in cyanobacteria: Fur genes from *Synechococcus* PCC 7942 and *Anabaena* PCC 7120. *Annales de Limnologie, 38*, 3–11.

Hernández-Prieto, M. A., Schön, V., Georg, J., Barreira, L., Varela, J., Hess, W. R., & Futschik, M. E. (2012). Iron deprivation in *Synechocystis*: Inference of pathways, non-coding RNAs, and regulatory elements from comprehensive expression profiling. *G3, 2*, 175–195.

Hider, R. C., & Kong, X. (2010). Chemistry and biology of siderophores. *Natural Product Reports, 27*, 637–657.

Hopkinson, B. M., & Morel, F. M. (2009). The role of siderophores in iron acquisition by photosynthetic marine microorganisms. *Biometals, 22*, 659–669.

Horsburgh, M. J., Clements, M. O., Crossley, H., Ingham, E., & Foster, S. J. (2001). PerR controls oxidative stress resistance and iron storage proteins and is required for virulence in *Staphylococcus aureus*. *Infection and Immunity, 69*, 3744–3754.

Hudson, R. J. M., & Morel, F. M. M. (1993). Trace metal transport by marine microorganisms: Implications of metal coordination kinetics. *Deep Sea Research, 140*, 129–150.

Hutchins, D. A., Fe, F.-X., Zhang, Y., Warner, M. E., Feng, Y., et al. (2007). CO2 control of *Trichodesmium* N2 fixation, photosynthesis, growth rates, and elemental ratios: Implications for past, present, and future ocean biogeochemistry. *Limnology and Oceanography, 52*, 1293–1304.

Ibisanmi, E., Sander, S. G., Boyd, P. W., Bowie, A. R., & Hunter, K. A. (2011). Vertical distributions of iron-(III) complexing ligands in the Southern Ocean. *Deep Sea Research Part II: Topical Studies Oceanography, 58*, 2113–2125.

Ito, Y., & Butler, A. (2005). Structure of synechobactins, new siderophores of the marine cyanobacterium *Synechococcus* sp. PCC 7002. *Limnology and Oceanography, 50*, 1918–1923.

James, A., Guikema, S., & L. A. (1983). Organization and function of chlorophyll in membranes of cyanobacteria during iron starvation. *Plant Physiology, 73*, 250–225.

Jiang, H. B., Lou, W. J., Du, H. Y., Price, N. M., & Qiu, B. S. (2012). Sll1263, a unique cation diffusion facilitator protein that promotes iron uptake in the cyanobacterium *Synechocystis* sp. strain PCC 6803. *Plant and Cell Physiology, 53*, 1404–1417.

Jiang, H. B., Lou, W. J., Ke, W. T., Song, W. Y., Price, N. M., & Qiu, B. S. (2015). New insights into iron acquisition by cyanobacteria: An essential role for ExbB-ExbD complex in inorganic iron uptake. *The ISME Journal, 9*, 297–309.

Jiang, H. B., Fu, F. X., et al. (2018). Ocean warming alleviates iron limitation of marine nitrogen fixation. *Nature Climate Change, 8*(8), 709–712.

Jordan, P., Fromme, P., Witt, H. T., Klukas, O., Saenger, W., & Krauss, N. (2001). Three-dimensional structure of cyanobacterial photosystem I at 2.5 A resolution. *Nature, 411*, 909–917.

Kakhlon, O., & Cabantchik, Z. I. (2002). The labile iron pool: Characterization, measurement, and participation in cellular processes[J]. *Free Radical Biology and Medicine, 33*, 1037–1046.

Kamiya, N., & Shen, J. R. (2003). Crystal structure of oxygen-evolving photosystem II from Thermosynechococcus vulcanus at 3.7-A resolution. *Proceedings of the National Academy of Sciences of the United States of America, 100*, 98–103.

Katoh, H., Hagino, N., Grossman, A. R., & Ogawa, T. (2001). Genes essential to iron transport in the cyanobacterium *Synechocystis* sp. strain PCC 6803. *Journal of Bacteriology, 183*(9), 2779–2784.

Keren, N., Aurora, R., & Pakrasi, H. B. (2004). Critical roles of bacterioferritins in iron storage and proliferation of cyanobacteria. *Plant Physiology, 135*, 1666–1673.

Kessler, N., Armoza-Zvuloni, R., Wang, S., Basu, S., Weber, P., Stuart, R., & Shaked, Y. (2019). Selective collection of iron-rich dust particles by natural *Trichodesmium* colonies. *The ISME Journal, 14*, 1–13.

Khan, A., Singh, P., & Srivastava, A. (2017). Synthesis, nature and utility of universal iron chelator – Siderophore: A review. *Microbiological Research*, S0944501317306730.

Khan, A., Singh, P., & Srivastava, A. (2018). Synthesis, nature and utility of universal iron chelator—Siderophore: A review. *Microbiological Research, 212–213*, 103–111.

Kim, K. S., Chang, Y. J., Chung, Y. J., Park, C. U., & Seo, H. Y. (2007). Enhanced expression of high-affinity iron transporters via H-ferritin production in yeast. *Journal of Biochemistry and Molecular Biology, 40*, 82–87.

Kim, H., Lee, H., & Shin, D. (2015). Lon-mediated proteolysis of the FeoC protein prevents Salmonella enterica from accumulating the Fe(II) transporter FeoB under highoxygen conditions. *Journal of Bacteriology, 197*, 92–98.

Klausner, R. D., Rouault, T. A., & Harford, J. B. (1993). Regulating the fate of mRNA: The control of cellular iron metabolism. *Cell, 72*, 19–28.

Koedding, J., Polzer, P., Killig, F., Howard, S. P., Gerber, K., Seige, P., Diederichsand, K., & Weltea, W. (2004). Crystallization and preliminary X-ray analysis of a C-terminal TonB fragment from *Escherichia coli. Acta Crystallographica Section D*, 1281–1283.

Kouřil, R., Arteni, A. A., Lax, J., Yeremenko, N., Haene, S. D., Rögner, M., Matthijs, H. C. P., Dekker, J. P., & Boekema, E. J. (2005). Structure and functional role of supercomplexes of IsiA and photosystem I in cyanobacterial photosynthesis. *FEBS Letters, 579*, 3253–3257.

Kranzler, C., Lis, H., Shaked, Y., & Keren, N. (2011). The role of reduction in iron uptake processes in a unicellular, planktonic cyanobacterium. *Environmental Microbiology, 13*, 2990–2999.

Kranzler, C., Lis, H., Finkel, O. M., Schmetterer, G., Shaked, Y., & Keren, N. (2014). Coordinated transporter activity shapes high-affinity iron acquisition in cyanobacteria. *The ISME Journal, 8*, 409–417.

Krieger, A., & Rutherford, A. W. (1998). The involvement of H2O2 produced by photosystem II in photoinhibition. In G. Garab (Ed.), *Photosynthesis: Mechanisms and effects* (Vol. 3, pp. 2135–2213). Dordrecht: Kluwer Academic Publishers.

Kuma, K., Katsumoto, A., Kawakami, H., Takatori, F., & Matsunaga, K. (1998). Spatial variability of Fe(III) hydroxide solubility in the water column of the northern North Pacific Ocean. *Deep Sea Research Part I: Oceanographic Research Papers, 45*, 91–113.

Küpper, H., Šetlík, I., Seibert, S., Prášil, O., Šetlikova, E., Strittmatter, M., Levitan, O., Lohscheider, J., Adamska, I., & Berman-Frank, I. (2008). Iron limitation in the marine cyanobacterium *Trichodesmium* reveals new insights into regulation of photosynthesis and nitrogen fixation. *The New Phytologist, 179*, 784–798.

Kurisu, G., Zhang, H.-m., Smith, J. L., & Cramer, W. A. (2003). Structure of the cytochrome b6f complex of oxygenic photosynthesis: Tuning the cavity. *Science, 302*, 1009–1114.

Kustka, A., Carpenter, E. J., & Sanudo-Wilhelmy, S. A. (2002). Iron and marine nitrogen fxation: Progress and future directions. *Research in Microbiology, 153*, 255–262.

Kustka, A. B., Sañudo-Wilhelmy, S. A., & Carpenter, E. J. (2003). Iron requirement for dinitrogen and ammonium-supported growth in cultures of *Trichodesmium* (IMI 101): Comparison with nitrogen fixation rate and iron: Carbon ratios of field populations. *Limnology and Oceanography, 48*, 1869–1884.

Lamb, J. J., Hill, R. E., Eaton-Rye, J. J., & Hohmann-Marriott, M. F. (2014). Functional role of pila in iron acquisition in the cyanobacterium *Synechocystis* sp. PCC 6803. *PLOS ONE, 9*.

Landry, M. R., Constantinou, J., Latasa, M., Brown, S. L., Bidigare, R. R., & Ondrusek, M. E. (2000). Biological response to iron fertilization in the eastern equatorial Pacific (IronEx II). III.Dynamics of phytoplankton growth and microzooplankton grazing. *Marine Ecology Progress Series, 201*, 57–72.

Lane, N. (2010). First breath: Earth's billion-year struggle for oxygen. *New Science*, 36–39.

Latifi, A., Jeanjean, R., Lemeille, S., Havaux, M., & Zhang, C. C. (2005). Iron starvation leads to oxidative stress in *Anabaena* sp. Strain PCC 7120. *Journal of Bacteriology, 187*, 6596–6598.

Lee, J. W., & Helmann, J. D. (2007). Functional specialization within the Fur family of metalloregulators[J]. *Biometals, 20*, 485.

Leynaert, A., Bucciarelli, E., Claquin, P., Dugdale, R. C., Martin-Jézéquel, V., Pondaven, P., & Ragueneau, O. (2004). Effect of iron deficiency on diatom cell size and silicic acid uptake kinetics. *Limnology and Oceanography, 49*, 1134–1143.

Li, T., Huang, X., Zhou, R., Liu, Y., Li, B., Nomura, C., & Zhao, J. (2002). Differential expression and localization of Mn and Fe superoxide dismutases in the heterocystous cyanobacterium *Anabaena* sp. strain PCC 7120. *Journal of Bacteriology, 184*, 5096–5103.

Li, H., Singh, A. K., Mcintyre, L. M., et al. (2004). Differential gene expression in response to hydrogen peroxide and the putative PerR regulon of *Synechocystis* sp. strain PCC 6803[J]. *Journal of Bacteriology, 186*, 3331–3345.

Li, Z. K., Dai, G. Z., Juneau, P., Qiu, B. S., & Post, A. (2016). Capsular polysaccharides facilitate enhanced iron acquisition by the colonial cyanobacterium *Microcystis* sp. isolated from a freshwater lake. *Journal of Phycology, 52*(1), 105–115.

Liberton, M., Berg, R. H., Heuser, J., Roth, R., & Pakrasi, H. B. (2006). Ultrastructure of the membrane systems in the unicellular cyanobacterium *Synechocystis* sp. strain PCC 6803. *Protoplasma, 227*, 129–138.

Lis, H., Kranzler, C., Keren, N., & Shaked, Y. (2015). A comparative study of iron uptake rates and mechanisms amongst marine and fresh water cyanobacteria: Prevalence of reductive iron uptake. *Life, 5*(1), 841–860.

Lodeyro, A. F., Ceccoli, R. D., Karlusich, J. J. P., & Carrillo, N. (2012). The importance of flavodoxin for environmental stress tolerance in photosynthetic microorganisms and transgenic plants. Mechanism, evolution and biotechnological potential. *FEBS Letters, 586*(18), 2917–2924.

López-Gomollón, S., Hernández, J. A., Wolk, C. P., Peleato, M. L., & Fillat, M. F. (2007). Expression of furA is modulated by NtcA and strongly enhanced in heterocysts of *Anabaena* sp. PCC 7120. *Microbiology, 153*, 42–50.

Ludwig, M., Chua, T. T., Chew, C. Y., et al. (2015). Fur-type transcriptional repressors and metal homeostasis in the cyanobacterium *Synechococcus* sp. PCC 7002[J]. *Frontiers in Microbiology, 6*, 1217.

Mahowald, N. M., Baker, A. R., Bergametti, G., Brooks, N., Duce, R. A., Jickells, T. D., et al. (2015). Atmospheric global dust cycle and iron inputs to the ocean. *Global Biogeochemical Cycles, 19*, 1–15.

Marlovits, T. C., Haase, W., Herrmann, C., Aller, S. G., & Unger, V. M. (2002). The membrane protein FeoB contains an intramolecular G protein essential for Fe(II) uptake in bacteria. *Proceedings of the National Academy of Sciences of the United States of America, 99*, 16243–16248.

Martin, J. H. (1990). Glacial interglacial CO_2 change: The iron hypothesis. *Paleoceanography, 5*, 12–13.

Martin, J. H., & Fitzwater, S. E. (1988). Iron deficiency limits phytoplankton growth in the north-east Pacific subarctic. *Nature, 331*, 341–343.

Martinez, A., & Kolter, R. (1997). Protection of DNA during oxidative stress by the non-specific DNA- binding protein Dps. *Journal of Bacteriology, 179*, 5188–5194.

Martin-Luna, B., Hernandez, J. A., Bes, M. T., Fillat, M. F., & Peleato, M. L. (2006). Identification of a Ferric uptake regulator from *Microcystis aeruginosa* PCC 7806. *FEMS Microbiology Letters, 254*, 63–70.

Michel, K. P., & Pistorius, E. K. (2004). Adaptation of the photosynthetic electron transport chain in cyanobacteria to iron deficiency: The function of IdiA and IsiA. *Physiologia Plantarum, 120*, 36–50.

Michel, K. P., Thole, H. H., & Pistorius, E. K. (1996). IdiA, a 34 kDa protein in the cyanobacteria *Synechococcus* sp. strains PCC 6301 and PCC 7942, is required for growth under iron and manganese limitations. *Microbiology, 142*, 2635–2645.

Michel, K. P., Exss-Sonne, P., Scholten-Beck, G., Kahmann, U., Ruppel, H. G., & Pistorius, E. K. (1998). Immunocytochemical localization of IdiA, a protein expressed under iron or manganese limitation in the mesophilic cyanobacterium *Synechococcus* PCC 6301and the thermophilic cyanobacterium *Synechococcus elongatus*. *Planta, 205*, 73–81.

Michel, K.-P., Berry, S., Hifney, A., & Kruip, J. (2003). Adaptation to iron deficiency: A comparison between the cyanobacterium *Synechococcus elongatus* PCC 7942 wild-type and a DpsA-free mutant. *Photosynthesis Research, 75*, 71–84.

Miethke, M., & Marahiel, M. A. (2007). Siderophore-based iron acquisition and pathogen control. *Microbiology and Molecular Biology Reviews, 71*, 413–451.

Mikheyskaya, L. V., Ovodova, R. G., & Ovodova, Y. S. (1977). Isolation and characterization of lipopolysaccharides from cell walls of blue-green algae of the genus Phormidium. *Journal of Bacteriology, 130*, 1–3.

Miller, G. W., Denney, A., Pushnik, J., & Yu, M.-H. (1982). The formation of delta-aminolevulinate, a precursor of chlorophyll, in barley and the role of iron. *Journal of Plant Nutrition, 5*, 289–300.

Moeck, G. S., & Coulton, J. W. (1998). TonB-dependent iron acquisition: mechanisms of siderophore-mediated active transport. *Molecular Microbiology, 28*(4), 675–81

Morel, F. M. M., Kustka, A. B., & Shaked, Y. (2008). The role of unchelated Fe in the iron nutrition of phytoplankton. *Limnology and Oceanography, 53*, 400–404.

Morrissey, J., & Bowler, C. (2012). Iron utilization in marine cyanobacteria and eukaryotic algae. *Frontiers in Microbiology, 3*, 43.

Müller, G., Isowa, Y., & Raymond, K. N. (1985). Stereospecificity of siderophore-mediated iron uptake in Rhodotorula pilimanae as probed by Enantiorhodotoruli acid and isomers of chromic Rhodotorulate. *The Journal of Biological Chemistry, 260*, 13921–13926.

Myers, J. (1987). Is there significant cyclic electron flow around photosystem I in cyanobacteria. *Photosynthesis Research, 14*, 55–69.

Nicolaisen, K., & Schleiff, E. (2010). Iron dependency of and transport by cyanobacteria. In S. Andrews & P. Cornelis (Eds.), *Iron uptake in microorganisms* (pp. 203–229). Norfolk: Caister Academic Press.

Nicolaisen, K., Moslavac, S., Samborski, A., Valdebenito, M., Hantke, K., Maldener, I., et al. (2008). Alr0397 is an outer membrane transporter for the siderophore schizokinen in *Anabaena* sp. strain PCC 7120. *Journal of Bacteriology, 190*(22), 7500–7507.

Nield, J., Morris, E. P., Bibby, T. S., & Barber, J. (2003). Structural analysis of the photosystem I supercomplex of cyanobacteria induced by iron deficiency. *Biochemistry, 42*, 3180–3188.

Nishioka, J., Takeda, S., Wong, C. S., & Johnson, W. K. (2001). Size-fractionated iron concentrations in the Northeast Pacific Ocean: Distribution of soluble and small colloidal iron. *Marine Chemistry, 74*, 157–179.

Noinaj, N., Guillier, M., Barnard, T. J., & Buchanan, S. K. (2010). TonB-dependent transporters: Regulation, structure, and function. *Annual Review of Microbiology, 64*, 43–60.

Ovescostales, D., Kadi, N., & Challis, G. L. (2009). The long-overlooked enzymology of a nonribosomal peptide synthetase-independent pathway for virulence-conferring siderophore biosynthesis. *ChemInform, 41*, 6530–6541.

Oliveira, P., Martins, N. M., Santos, M., Pinto, F., Büttel, Z., Couto, N. A. S., et al. (2016). The versatile tolc-like slr1270 in the cyanobacterium *Synechocystis* sp. PCC 6803. *Environmental Microbiology, 18*.

Palenik, B., Brahamsha, B., Larimer, F. W., Land, M., Hauser, L., Chain, P., Lamerdin, J., Regala, W., Allen, E. E., McCarren, J., Paulsen, I., Dufresne, A., Partensky, F., Webb, E. A., & Waterbury, J. (2003). The genome of a motile marine *Synechococcus*. *Nature, 424*, 1037–1042.

Palenik, B., Ren, Q., Dupont, C. L., Myers, G. S., Heidelberg, J. F., Badger, J. H., et al. (2006). Genome sequence of *Synechococcus* PCC 9311: Insights into adaptation to a coastal environment. *Proceedings of the National Academy of Sciences of the United States of America, 103*, 13555–13559.

Parent, A., Caux-Thang, C., Signor, L., et al. (2013). Single glutamate to aspartate mutation makes ferric uptake regulator (Fur) as sensitive to H_2O_2 as peroxide resistance regulator (PerR)[J]. *Angewandte Chemie, 125*, 10529–10533.

Parker, D. L., Mihalick, J. E., Plude, J. L., Plude, M. J., Clark, T. P., Egan, L., et al. (2000). Sorption of metals by extracellular polymers from the cyanobacterium *Microcystis aeruginosa* fo. flos-aquae strain c3-40. *Journal of Applied Phycology, 12*, 219–224.

Passow, U. (2002). Transparent exopolymer particles (TEP) in aquatic environments. *Progress in Oceanography, 55*, 287–333.

Pils, D., & Schmetterer, G. (2001). Characterization of three bioenergetically active respiratory terminal oxidases in the cyanobacterium *Synechocystis* sp strain PCC 6803. *FEMS Microbiology Letters, 203*, 217–222.

Pospisil, P., Arato, A., & Krieger-Liszkay, A. (2004). Hydroxyl radical generation by photosystem II. *Biochemistry, 43*, 6783–6792.

Qiu, G. W., Lou, W. J., Sun, C. Y., Yang, N., & Qiu, B. S. (2018). Characterization of outer membrane iron uptake pathways in the model cyanobacterium *Synechocystis* sp. PCC 6803. *Applied and Environmental Microbiology, 84*(19), AEM.01512-18.

Raven, J. A. (1988). The iron and molybdenum use efficiencies of plant growth with different energy, carbon and nitrogen sources. *The New Phytologist, 109*, 279–287.

Raven, J. A. (1990). Predictions of Mn and Fe use efficiencies of phototrophic growth as a function of light availability for growth and of C assimilation pathway. *The New Phytologist, 116*, 1–18.

Raven, J. A. (1999). The size of cells and organisms in relation to the evolution of embryophytes. *Plant Biology, 1*, 2–12.

Raven, J. A., & Richard, J. G. (1988). Temperature and Algal Growth. *The New Phytologist, 110*, 411–461.

Raven, J. A., Evans, M. C., & Korb, R. E. (1999). The role of trace metals in photosynthetic electron transport in O^2-evolving organism. *Photosynthesis Research, 60*, 111–150.

Regelsberger, G., Laaha, U., Dietmann, D., Rüker, F., Grilli-Caiola, A. C. M., Furtmuüller, P. G., Jakopitsch, C., Peschek, G. A., & Obinger, C. (2004). The iron superoxide dismutase from the filamentous cyanobacterium nostoc PCC 7120. *The Journal of Biological Chemistry, 279*, 44384–44393.

Ren, X. X., Jiang, H., Leng, X., & An, S. Q. (2013). Ecological significance and industrial application of extracellular polysaccharides from cyanobacteria: A review. *Chinese Journal of Ecology, 291*, 428–435.
Řezanka, T., Palyzová, A., & Sigler, K. (2018). Isolation and identification of siderophores produced by cyanobacteria. *Folia Microbiologica, 63*(5), 569–579.
Rivers, A. R. (2009). Iron limitation and the role of siderophores in marine *Synechococcus*. New College of Florida. Papper for the degree of Doctor. pp. 3.
Rocap, G. (2003). Genome divergence in two *Prochlorococcus* ecotypes reflects oceanic niche differentiation. *Nature, 424*, 1042–1047.
Romheld, V., & Marschner, H. (1983). Mechanism of iron uptake by peanut plants: I. Fe reduction, chelate splitting, and release of phenolics. *Plant Physiology, 71*, 949–954.
Rose, A. L., Salmon, T. P., Lukondeh, T., Neilan, B., & Waite, T. D. (2005). Use of superoxide as an electron shuttle for iron acquisition by the marine cyanobacterium *Lyngbya majuscule*. *Environmental Science & Technology, 39*, 3708–3715.
Roy, E. G., Wells, M. L., & King, D. W. (2008). Persistence of iron(II) in surface waters of the western subarctic Pacific. *Limnology and Oceanography, 53*, 89–98.
Rusch, D. B., Martiny, A. C., Dupont, C. L., & Venter, A. L. H. C. (2010). Characterization of prochlorococcus clades from iron-depleted oceanic regions. *Proceedings of the National Academy of Sciences of the United States of America, 107*, 16184–16189.
Rubin, M., Bermanfrank, I., & Shaked, Y. (2011). Dust- and mineral-iron utilization by the marine dinitrogen-fixer *Trichodesmium*. *Nature Geoscience, 4*, 529–534.
Rudolf, M., Kranzler, C., Lis, H., Margulis, K., Stevanovic, M., Keren, N., & Schleiff, E. (2015). Multiple modes of iron uptake by the filamentous, siderophore producing cyanobacterium, *Anabaena* sp. PCC 7120. *Molecular Microbiology*.
Ryan-Keogh, T. J., Macey, A. I., Cockshutt, A. M., Moore, C. M., & Bibby, T. S. (2012). The cyanobacterial chlorophyll-binding-protein IsiA acts to increase the in vivo effective absorption cross-section of PSI under iron limitation. *Journal of Phycology, 48*, 145–154.
Saha, M., Sarkar, S., Sarkar, B., Sharma, B. K., Bhattacharjee, S., & Tribedi, P. (2016). Microbial siderophores and their potential applications: A review. *Environmental Science and Pollution Research, 23*, 3984–3999.
Saito, M. A., Bertrand, E. M., Dutkiewicz, S., Bulygin, V. V., Moran, D. M., Monteiro, F. M., Follows, M. J., Valois, F. W., & Waterbury, J. B. (2011). Iron conservation by reduction of metal-loenzyme inventories in the marine diazotroph *Crocosphaera watsonii*. *Proceedings of the National Academy of Sciences of the United States of America, 108*, 2184–2189.
Sandmann, G. (1985). Consequences of iron deficiency on photosynthetic and respiratory electron transport in blue-green algae. *Photosynthesis Research, 6*, 261–271.
Sandmann, G., & Malkin, R. (1983). Iron-sulfur centers and activities of the photosynthetic electron transport chain in iron-deficient cultures of the blue-green alga, Aphanocapsa. *Plant Physiology, 73*, 724–728.
Sañudo-Wilhelmy, S. A., Kustka, A. B., Gobler, C. J., Hutchins, D. A., Yang, M., Lwiza, K., Burns, J., Capone, D. G., Ravenk, J. A., & Carpenter, E. J. (2001). Phosphorus limitation of nitrogen fixation by *Trichodesmium* in the central Atlantic Ocean. *Nature, 411*, 66–69.
Schmidt, W., Drews, G., Weckesser, J., Fromme, J., & Borowiak, D. (1980). Characterization of the lipopolysaccharides from eight *Synechococcus* strains. *Archives of Microbiology, 127*, 217–222.
Schneider, D., Berry, S., Volkmer, T., Seidler, A., & Rogner, M. (2004). PetC1 is the major rieske iron-sulfur protein in the cytochrome b6f complex of *Synechocystis* sp. PCC 6803. *The Journal of Biological Chemistry, 279*, 39383–39388.
Schopf, J. W. (2012). Springer Netherlands the fossil record of cyanobacteria. In B. A. Whitton & B. A. Whitton (Eds.), *Ecology of cyanobacteria II* (pp. 15–36). Dordrecht: Springer.
Schrader, M., Drews, G., & Weckesser, J. (1981). Chemical analyses on cell wall constituents of the thermophilic cyanobacterium *Synechococcus* PCC 6716. *FEMS Microbiology Letters, 11*, 37–40.

Schuergers, N., & Wilde, A. (2015). Appendages of the cyanobacterial cell. *Life, 5*, 700–715.
Sen, A., Dwivedi, K., Rice, K. A., & Bullerjahn, G. S. (2000). Growth phase and metal-dependent regulation of the dpsa gene in *Synechococcus* sp. strain PCC 7942. *Archives of Microbiology, 173*, 352–357.
Shaked, Y., & Lis, H. (2012). Disassembling iron availability to phytoplankton. *Frontiers in Microbiology, 3*, 1–25.
Shcolnick, S., & Keren, N. (2006). Metal homeostasis in cyanobacteria and chloroplasts. Balancing benefits and risks to the photosynthetic apparatus. *Plant Physiology, 141*, 805–810.
Shcolnick, S., Summerfield, T. C., Reytman, L., Sherman, L. A., & Keren, N. (2009). The mechanism of Iron homeostasis in the unicellular cyanobacterium *Synechocystis* sp. PCC 6803 and its relationship to oxidative stress. *Plant Physiology, 150*, 2045–2056.
Shi, D., Xu, Y., Hopkinson, B. M., & Morel, F. M. M. (2010). Effect of ocean acidification on iron availability to marine phytoplankton. *Science, 327*, 676–679.
Shih, P. M., Wu, D., Latifi, A., Axen, S. D., Fewer, D. P., Talla, E., et al. (2013). Improving the coverage of the cyanobacterial phylum using diversity-driven genome sequencing. *Proceedings of the National Academy of Sciences, 110*, 1053–1058.
Simpson, F. B., & Neilands, J. B. (1976). Siderochromes in cyanophyceae: isolation and characterization of schizokinen from Anabaena sp. 1. *Journal of Phycology, 12*, 44–48.
Singh, A. K., & Sherman, L. A. (2006). Iron-independent dynamics of IsiA production during the transition to stationary phase in the cyanobacterium *Synechocystis* sp. PCC 6803. *EMS Microbiol Letters, 256*, 159–164.
Singh, A. K., McIntyre, L. M., & Sherman, L. A. (2003). Microarray analysis of the genome-wide response to iron deficiency and iron reconstitution in the cyanobacterium *Synechocystis* sp. PCC 6803. *Plant Physiology, 132*, 1825–1839.
Singh, A. K., Li, H., & Sherman, L. A. (2004). Microarray analysis and redox control of gene expression in the cyanobacterium *Synechocystis* sp. PCC 6803. *Physiologia Plantarum, 120*, 27–35.
Solomon, S., Qin, D., Manning, M., Chen, Z., Marquis, M., et al. (2007). *Climate change 2007: The physical science basis: Contribution of working group I to the fourth assessment report of the intergovernmental panel on climate change*. New York: Cambridge University Press.
Stockel, J., Welsh, E. A., Liberton, M., Kunnvakkam, R., Aurora, R., & Pakrasi, H. B. (2008). Global transcriptomic analysis of Cyanothece 51142 reveals robust diurnal oscillation of central metabolic processes. *Proceedings of the National Academy of Sciences of the United States of America, 105*, 6156–6161.
Stroebel, D., Choquet, Y., Popot, J.-L., & Picot, D. (2003). An atypical haem in the cytochrome b6f complex. *Nature, 426*, 413–418.
Stumm, W., & Sulzberger, B. (1992). The cycling of iron in natural environments: Considerations based on laboratory studies of heterogeneous redox processes. *Geochimica et Cosmochimica Acta, 56*, 3233–3257.
Sunda, W. G. (2001). Bioavailability and bioaccumulation of iron in the sea. In K. Hunter & D. Turner (Eds.), *Biogeochemistry of Fe in seawater* (pp. 41–84). West Sussex: Wiley.
Sunda, W. G., & Huntsman, S. A. (1997). Interrelated influence of iron, light and cell size on marine phytoplankton growth. *Nature, 390*, 389–392.
Tagliabue, A., Bowie, A. R., Boyd, P. W., Buck, K. N., Johnson, K. S., & Saito, M. A. (2017). The integral role of iron in ocean biogeochemistry. *Nature, 543*, 51–59.
Teresa, B. M., Hernández, J. A., Luisa, P. M., et al. (2001). Cloning, overexpression and interaction of recombinant Fur from the cyanobacterium *Anabaena* PCC 7119 with isiB and its own promoter. *FEMS Microbiology Letters, 194*(2), 187–192.
Terry, N., & Low, G. (1982). Leaf chlorophyll content and its relation to the intracellular location of iron. *Journal of Plant Nutrition, 5*, 301–310.
Tichy, M., & Vermaas, W. (1999). In vivo role of catalase-peroxidase in *Synechocystis* sp. strain PCC 6803. *Journal of Bacteriology, 181*, 1875–1882.

Tölle, J., Michel, K. P., Kruip, J., Kahmann, U., Preisfeld, A., & Pistorius, E. K. (2002). Localization and function of the IdiA homologue Slr1295 in the cyanobacterium *Synechocystis* sp. strain PCC 6803. *Microbiology, 148*, 3293–3305.

Toulzal, E., Tagliabue, A., Blain, S., & Piganeau, G. (2012). Analysis of the Global Ocean Sampling (GOS) project for trends in iron uptake by surface ocean microbes. *PLoS One, 7*(2), e30931.

Tronnet, S., Garcie, C., Brachmann, A. O., et al. (2017). High iron supply inhibits the synthesis of the genotoxin colibactin by pathogenic Escherichia coli through a non-canonical Fur/RyhB-mediated pathway[J]. *Pathogens and Disease, 75*, ftx066.74.

Trujillo, A. (2011). *The iron hypothesis*. http://www.homepages.ed.ac.uk/shs/Climatechange/Carbon%20sequestration/Martin%20iron.htm

Tuit, C., Waterbury, J., & Ravizza, G. (2004). Diel variation of molybdenum and iron in marine diazotrophic cyanobacteria. *Limnology and Oceanography, 49*, 978–990.

Vassiliev, L. R., Kolber, Z., Wyman, K. D., Mauzerall, D., Shukla, V. K., & Falkowsk, P. C. (1995). Effects of iron limitation on photosystem II composition and light utilization in Dunaliella tertiolecta. *Plant Physiology, 109*, 963–972.

Velayudhan, J., Hughes, N. J., Mccolm, A. A., Bagshaw, J., Clayton, C. L., Andrews, S. C., et al. (2000). Iron acquisition and virulence in Helicobacter pylori: A major role for FeoB, a high-affinity ferrous iron transporter. *Molecular Microbiology, 37*, 274–286.

Vermaas, W. F. (2001). *Photosynthesis an respiration in cyanobacteria* (Encyclopedia of Life Sciences, pp. 1–7). London: Nature Publishing Group.

Violaine, J., Céline, R., Stéphane, L.'. H., Fanny, K., Alain, S., & Andrew, C. D. (2014). Response of the unicellular diazotrophic cyanobacterium crocosphaera watsonii to iron limitation. *PLoS ONE, 9*(1), e86749.

Volk, T., & Hoffert, M. I. (1985). In the carbon cycle and atmospheric CO_2: Natural variations Archean to present. *American Geophysical Union, 32*, 99–110.

Wandersman, C., & Delepelaire, P. (2004). Bacterial iron sources: From siderophores to hemophores[J]. *Annual Review of Microbiology, 58*, 611.

Westberry, T. K., & Siegel, D. A. (2006). Spatial and temporal distribution of *Trichodesmium* blooms in the worlds oceans. *Global Biogeochemical Cycles, 20*.

Wilson, A., Boulay, C., Wilde, A., Kerfeld, C. A., & Kirilovsky, D. (2007). Light-induced energy dissipation in Iron-starved cyanobacteria: Roles of OCP and IsiA proteins. *Plant Cell, 19*, 656–672.

Wolf, S. G., Frenkiel, D., Arad, T., Finkel, S. E., Kolter, R., & Minsky, A. (1999). DNA protection by stress-induced biocrystallization. *Nature, 400*, 83–85.

Wotton, R. S. (2004). The ubiquity and many roles of exopolymers (EPS) in aquatic systems. *Scientia Marina, 68*, 13–21.

Wu, J. F., Boyle, E., Sunda, W., & Wen, L. S. (2001). Soluble and colloidal iron in the oligotrophic North Atlantic and North Pacific. *Science, 293*, 847–849.

Xing, W., Huang, W. M., Li, D. H., & Liu, Y. D. (2007). Effects of iron on growth, pigment content, photosystem II efficiency, and siderophores production of *Microcystis aeruginosa* and Microcystis wesenbergii. *Current Microbiology, 55*, 94–98.

Xu, D., Liu, X., Zhao, J., & Zhao, J. (2005). FesM, a membrane iron-sulfur protein, is required for cyclic electron flow around Photosystem I and photoheterotrophic growth of the cyanobacterium *Synechococcus* sp. PCC 7002. *Plant Physiology, 138*, 1586–1595.

Xu, N., Qiu, G. W., Lou, W. J., Li, Z. K., Jiang, H. B., Price, N. M., et al. (2016). Identification of an iron permease, cftr1, in cyanobacteria involved in the iron reduction/re-oxidation uptake pathway. *Environmental Microbiology., 18*, 5005–5017.

Yeremenko, N., Kouril, R., Ihalainen, J. A., et al. (2004). Supramolecular organization and dual function of the IsiA chlorophyll-binding protein in cyanobacteria. *Biochemistry, 43*, 10308–10313.

Yoshida, M., Kuma, K., Iwade, S., Isoda, Y., Takata, H., & Yamada, M. (2006). Effect of aging time on the availability of freshly precipitated ferric hydroxide to coastal marine diatoms. *Marine Biology, 149*, 379–392.

Yoshihara, S., Geng, X. X., Okamoto, S., Yura, K., Murata, T., Go, M., Ohmori, M., & Ikeuchi, M. (2001). Mutational analysis of genes involved in pilus structure, motility and transformation competency in the unicellular motile cyanobacterium *Synechocystis* sp. PCC 6803. *Plant and Cell Physiology, 42*, 63–73.

Yousef, N., Pistorius, E. K., & Michel, K.-P. (2003). Comparative analysis of idiA and isiA transcription under iron starvation and oxidative stress in *Synechococcus elongatus* PCC 7942 wild-type and selected mutants. *Archives of Microbiology, 180*, 471–483.

Yu, L., Zhao, J., Muhlenhoff, U., Bryant, D. A., & Golbeck, J. H. (1993). PsaE is required for in vivo cyclic electron flow around photosystem I in the cyanobacterium *Synechococcus* sp. PCC 7002. *Plant Physiology, 103*, 171–180.

Zouni, A., Witt, H. T., Kern, J., Fromme, P., Krauss, N., Saenger, W., & Orth, P. (2001). Crystal structure of photosystem II from *Synechococcus elongatus* at 3.8 A resolution. *Nature, 409*, 739–743.

The Roles of sRNAs in Regulating Stress Responses in Cyanobacteria

Jinlu Hu and Qiang Wang

Abstract Small RNAs (sRNAs) are transcriptional and posttranscriptional regulators of gene expression that play important roles in virtually every aspect of the life cycle of an organism, such as metabolic reactions, plasmid control, pathogenesis, and quorum sensing. Emerging evidence indicates that sRNAs also function in various stress responses. The large number of sRNAs present in organisms reflects the diverse roles of these molecules in adaptation to various environmental conditions. During the long period of evolution, sRNA regulatory networks have provided an economical and efficient way for organisms to adapt to various environmental stresses, as the regulation of gene expression by sRNAs likely requires fewer resources than protein-based gene regulation. Cyanobacteria, one of the most ancient life forms, include all types of photoautotrophic bacteria; these organisms are present in almost all environments on Earth. Recent investigations of sRNA-mediated gene regulation in cyanobacteria have uncovered numerous novel mechanisms that display far more mechanistic complexity and regulatory features than those identified in other model microorganisms, such as *Escherichia coli* and yeast. These complex RNA-based regulatory mechanisms are thought to help cyanobacteria adapt to changes in their diverse environments. In this chapter, we critically review recent efforts to identify the roles of cyanobacterial sRNAs in stress responses and their potential implications for biological evolution. This chapter focuses on the biogenesis, conservation, functions, and traits of sRNAs, setting the course for future research on sRNAs in photosynthetic cyanobacteria.

Keyword Biological evolution · Cyanobacteria · Photoautotrophic bacteria · sRNAs · Stress responses

J. Hu
School of Life Sciences, Northwestern Polytechnical University, Xi'an, China

Q. Wang (✉)
State Key Laboratory of Crop Stress Adaptation and Improvement, School of Life Sciences, Henan University, Kaifeng, China
e-mail: wangqiang@henu.edu.cn

Q. Wang (ed.), *Microbial Photosynthesis*,
https://doi.org/10.1007/978-981-15-3110-1_12

1 Introduction

RNAs that do not function as messenger RNAs (mRNAs), ribosomal RNAs (rRNAs), or transfer RNAs (tRNAs) include a large class of molecules usually referred to as small RNAs (sRNAs) or noncoding RNAs (ncRNAs) (Waters and Storz 2009; Storz 2002). ncRNAs involved in epigenetic regulation include microRNAs (miRNAs), small interfering RNAs (siRNAs), Piwi-interacting RNAs (piRNAs), small nucleolar RNAs (snoRNAs), and long noncoding RNAs (lncRNAs). RNA molecules that function as regulators were discovered in bacteria years before the first miRNAs and siRNAs were discovered in eukaryotes. The first sRNA discovered, RNAI, blocks replication of the ColE1 plasmid via base pairing in *Escherichia coli* (*E. coli*) (Tomizawa et al. 1981). Subsequently, the 174-nt MicF RNA was identified as the first chromosomally encoded sRNA in *E. coli*. This sRNA plays a role in inhibiting translation of the mRNA encoding OmpF, a major outer membrane porin (Mizuno et al. 1984). These sRNAs were identified based on multicopy phenotypes, by gel analysis due to their abundance, or serendipitously (Wassarman et al. 1999; Waters and Storz 2009). Due to recent technical advancements, e.g., deep sequencing, computational searches, and tiling microarrays with full-genome coverage (Sittka et al. 2008; Weinberg et al. 2007; Landt et al. 2008; Hu et al. 2017), many new candidate sRNAs have been predicted in *E. coli* and other bacteria. Over the past two decades, the importance of this remarkable group of RNAs in various organisms has been widely recognized (Ravasi et al. 2006; Huttenhofer et al. 2002; Storz 2002; Raghavan et al. 2011; Kopf et al. 2014).

In bacteria, sRNAs usually range from 50 to 500 nt. These molecules achieve diverse outcomes via multiple mechanisms, such as altering protein binding, base pairing with other RNAs, and interacting with DNA (Fig. 1). A small class of sRNAs must interact with specific target proteins in order to function (Fig. 1a), as exemplified by 6S RNA (Watanabe et al. 1997; Barrick et al. 2005), CsrB (Babitzke and Romeo 2007), RNase P RNA (Tous et al. 2001), and tmRNA (Watanabe et al. 1998). The largest and most extensively studied class of sRNAs acts through base pairing with their target mRNAs to regulate gene expression, usually by modulating the translation and stability of mRNAs (Fig. 1b). Cis-encoded antisense RNAs (asRNAs) are transcribed from the opposite strand of the target gene and share complete complementarity with their target mRNAs. By contrast, for trans-encoded sRNAs, the chromosomal location of the sRNA gene has little correlation with the location of the gene encoding the target mRNA (Fig. 1c). In contrast to cis-encoded asRNAs, trans-encoded sRNAs modulate the expression of more than one target by forming partial RNA-RNA duplexes (Georg et al. 2014). Since sRNAs function as both transcriptional and posttranscriptional regulators of gene expression, they play important roles in every aspect of an organism's life cycle, including maintaining the structural integrity of chromosomes (Volpe et al. 2002), regulating the stability and translation of mRNAs (Storz et al. 2004) and the translocation and stability of proteins (Hüttenhofer and Vogel 2006), and mediating metabolic reactions (Sun et al. 2018), stress responses (Hu et al. 2017; Klahn et al. 2015b), and pathogenesis (Lee and Groisman 2010).

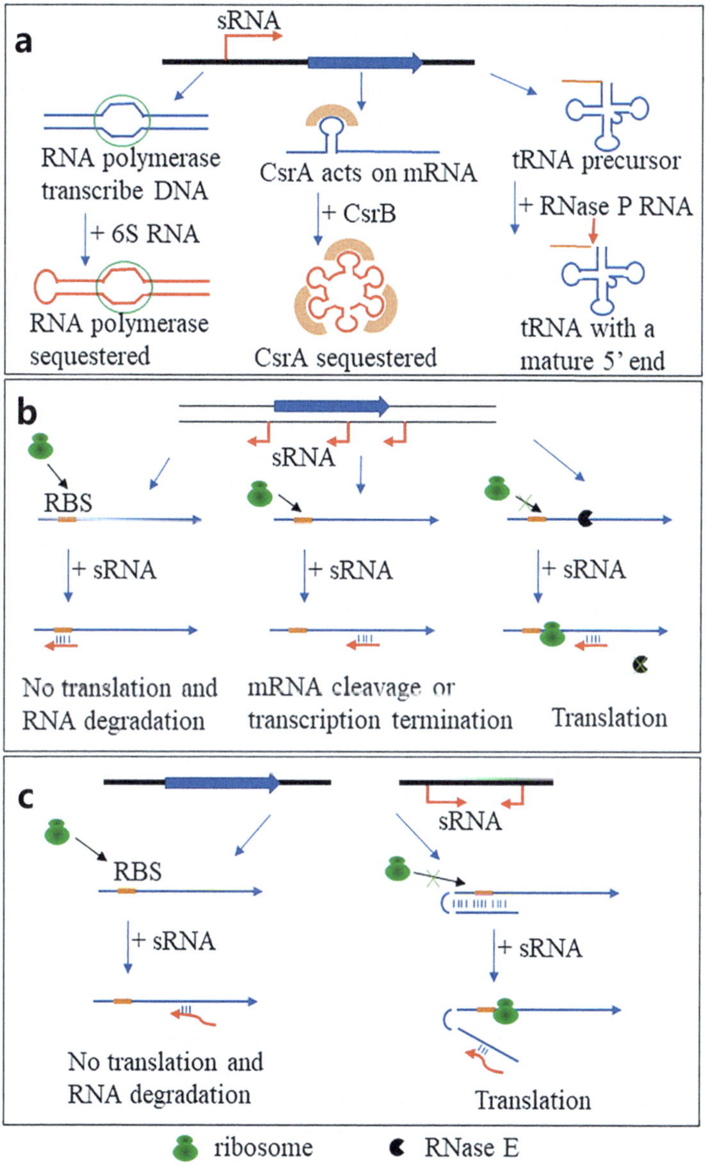

Fig. 1 Categories of regulatory RNA action. (**a**) Regulatory RNA (red) as regulatory protein. (Left panel) σ^{70} RNA polymerase (green oval) binds promoter DNA (blue). When 6S levels increase, the sRNA sequesters σ^{70} RNA polymerase away from some promoters, reducing transcription of certain genes. (Middle panel) This is modeled on the CsrA/CsrB system, in which CsrA protein (orange arc) is a translational inhibitor and inhibition is relieved by binding of multiple molecules of CsrA by CsrB. (Right panels) Ribonuclease P (RNase P) is a universal enzyme necessary for tRNA maturation. The RNA subunit (RNase P RNA) of the enzyme was the catalytic component and can precisely perform endonucleolytic cleavage of tRNA precursors (blue) to produce a mature 5′ end. In all of the figures, the regulatory RNAs are shown in red. (**b**) Regulatory RNA as

In their natural habitats, bacteria are constantly exposed to stressful and changing environmental conditions. Bacteria have evolved intricate mechanisms that sense the surrounding environment and adequately respond by altering their gene expression patterns and thus phenotypes, allowing them to withstand stress and proliferate under these conditions (Hong et al. 2012; Gierga et al. 2012; Hengge-Aronis 2002). Bacterial sRNAs are important regulators that often transmit environmental signals when cells encounter suboptimal or stressful growth conditions. These sRNAs help modulate cellular metabolism, optimizing the utilization of available nutrients and improving the chances for survival (Holmqvist and Wagner 2017; Kopf and Hess 2015).

Cyanobacteria and other photosynthetic bacteria use sunlight as their major source of energy. Therefore, they are exposed to a particular set of additional regulatory challenges distinct from those of other bacteria. Cyanobacteria are the focus of numerous scientific studies investigating oxygenic photosynthesis, the ecological role of these important primary producers, and their possible use in sustainable biofuel production (Soo et al. 2017; Puente-Sanchez et al. 2018; Savakis and Hellingwerf 2015). The connection between stress and sRNAs was initially proposed when the expression patterns of the first well-studied sRNA in *Synechocystis* sp. PCC 6803 were characterized (Dühring et al. 2006). IsrR is induced by three environmental stress conditions: iron stress, high-light stress, and oxidative stress. Other recent examples include NsiR4 (nitrogen stress; (Klahn et al. 2015b, 2018)), RblR (carbon deficiency; (Hu et al. 2017)), CoaR (butanol tolerance; (Sun et al. 2017)), PsrR1 (high-light stress; (Georg et al. 2014)), and As1_flv4 (inorganic carbon; (Eisenhut et al. 2012)).

2 Methods for Studying the Noncoding Transcriptomes of Cyanobacteria and Identifying Stress-Responsive sRNAs

In the past 20 years, many sRNAs have been predicted, identified, and characterized in cyanobacteria using a variety of approaches. Several studies have focused on identifying and functionally characterizing stress-responsive sRNAs in cyanobacteria,

Fig. 1 (continued) cis-encoded antisense RNA (red). Many sRNAs share complete complementarity with their target mRNAs (blue), to change its behavior. Cis-encoded asRNA can act negatively by blocking ribosome binding (left panel) in the 5′ untranslated region (5′ UTR) of its target mRNA and/or targeting the sRNA-mRNA duplex for mRNA cleavage or transcriptional termination (middle panel). (Right panels) An asRNA functions positively in stabilizing its target mRNA against RNase E Attack. (**c**) Genes encoding trans-encoded sRNAs (red) are located separately from the genes encoding their target RNAs (blue) and only have limited complementarity. (Left panel) An sRNA can act negatively by targeting the sRNA-mRNA duplex for degradation by RNases. (Right panel) An sRNA acts positively by preventing the formation of an inhibitory structure, which sequesters the ribosome-binding site (RBS)

leading to the identification of various sRNAs related to the regulation of stress responses. Several different approaches have been used to identify the noncoding transcriptomes of photosynthetic bacteria. Biocomputational prediction, tiling microarrays, and RNA deep sequencing have revealed many stress-responsive candidate sRNAs in cyanobacterial model organisms.

Computational predictions in conjunction with experimental verification have been used to investigate stress-responsive sRNAs in model cyanobacteria including *Synechocystis* PCC6803, marine *Synechococcus*, and *Prochlorococcus*. Several cyanobacterial functional RNAs (Yfr1-7) were computationally predicted and their presence biochemically verified in cyanobacteria of the *Prochlorococcus-Synechococcus* lineage (Axmann et al. 2005). The transcript accumulation of *Yfr1-7* from *Prochlorococcus* MED4 is affected to various degrees by the following conditions: nutrient depletion (nitrogen, phosphate, or iron), blue light or dark stress, oxidative stress mediated by 2 μM 3-(3,4-dichlorophenyl)-1, 1-N-N′-dimethylurea (DCMU), high and low temperatures (30 °C and 15 °C, respectively), and high-light stress (50 μE). Sequence signatures, predicted secondary structures, and experimental data suggest that the Yfr2 family of ncRNAs is present in nearly all cyanobacteria sequenced to date (Gierga et al. 2009). *Yfr2a* expression increased in cultures exposed to elevated light intensities in *Synechocystis* PCC6803 (Voss et al. 2009). In addition, two Yfr2 sRNA homologs containing CGRE1 motifs were strongly induced in *Prochlorococcus* MED4 in response to high-light stress and nitrogen starvation (Lambrecht et al. 2018).

Tiling or high-resolution microarrays have been used to analyze *Prochlorococcus* MED4, *Synechocystis* PCC6803, and *Synechococcus* sp. WH7803. Analysis of Affymetrix microarray expression data from intergenic regions of *Prochlorococcus* MED4 cells subjected to various stress conditions (changes in light quality and quantity, phage infection, or phosphorus starvation) revealed 276 novel transcriptional units (Steglich et al. 2008). Large-scale analysis using a tiling microarray revealed 73 asRNAs and 60 ncRNAs under 9 different conditions in *Synechocystis* PCC6803 (Georg et al. 2009). In *Synechococcus* sp. WH7803, 15 transcripts differentially regulated under environmentally relevant conditions were identified, some of which may be involved in responses to distinct types of nutrient limitation, redox stress, or phage infection (Gierga et al. 2012).

High-throughput RNA sequencing was used to analyze the transcriptomes of *Synechococcus elongatus* UTEX 2973, *Synechocystis* PCC6803, *Synechococcus* PCC7942, and *Prochlorococcus*, leading to the genome-wide identification of transcriptional start sites (TSS) and noncoding transcripts. Differential RNA-sequencing (dRNA-Seq) enabled the genome-wide identification of 4808 TSSs in *S. elongatus* UTEX 2973 under standard, high-light, high-light coupled with high-temperature, and dark conditions using a background reduction algorithm (Tan et al. 2018). High light promoted the transcription of genes associated with central metabolic pathways, and darkness induced transcriptome remodeling, with a decline in the expression of genes for carbon fixation and other major metabolic pathways and increased expression of genes for glycogen catabolism and the Calvin cycle inhibitor CP12. Genome-wide sRNA sequencing combined with systematic analysis of transcriptomic and proteomic data for *Synechocystis* PCC6803 under exogenous biofuel treatment or environmental perturbation revealed 133 trans-encoded sRNA transcripts (Pei et al. 2017). This analysis allowed the authors to identify and functionally confirm sRNAs involved in tolerance to biofuel stress.

High-light stress has a substantial impact on the photosynthetic apparatus (Demmig-Adams and Adams III 1992). Cyanobacteria have developed a complex RNA-based regulatory mechanism that allows them to withstand this stress (Xu et al. 2014; Kopf and Hess 2015). Several studies using dRNA-Seq have revealed sRNA candidates that participate in the gene expression regulatory network under high-light stress in *Synechocystis* PCC6803 (Hu et al. 2017; Mitschke et al. 2011a). Voigt et al. performed a comprehensive comparison of the structures and compositions of the transcriptomes of two *Prochlorococcus* strains representing the two most diverse clades within the two major ecotypes adapted to high- and low-light conditions (Voigt et al. 2014). A high degree of antisense transcription and novel ncRNAs were identified. The TSSs of asRNAs and protein-coding genes were not conserved between the strains (except for photosynthesis-related genes), suggesting a high degree of variability in their transcriptional architecture (Voigt et al. 2014). Kopf et al. analyzed the response of the primary transcriptome of *Synechocystis* PCC6803 to ten environmentally relevant stimuli (Kopf et al. 2014). Bills et al. provided insight into mechanisms of stress acclimation by performing comparative transcriptomics between *Synechococcus* PCC7942 and *Synechocystis* PCC6803 under five different stress conditions (Billis et al. 2014). Furthermore, the construction of a genome-wide map of >10,000 TSSs in *Anabaena* sp. PCC7120 provided insight into the dynamic changes in transcriptional organization at a critical step of the differentiation process under nitrogen stress (Mitschke et al. 2011b).

Photosynthetic cyanobacteria have developed specific adaptations that allow them to withstand different environmental conditions, especially different nutrient levels and changes in light quantity and quality. The above studies suggest that cyanobacteria mount adaptive responses to these stresses involving regulatory sRNAs. Further studies of these model organisms will benefit from cataloging their responses to environmentally relevant stresses and improving their genome annotations based on the studies described above.

3 sRNAs Involved in Stress Response Pathways

Regulatory sRNAs play crucial roles in the adaptive responses of bacteria to changes in the environment. These RNAs help alter cellular metabolism, optimizing the utilization of available nutrients and improving the chances of survival. In diverse cyanobacteria, many stress-responsive sRNAs have been experimentally verified under environmental stress conditions, and their functions have been studied in detail (Table 1).

Table 1 Experimental stress-responsive sRNA screens in diverse cyanobacteria

sRNA	Species	Function	Related stress factor	Source
PsrR1	*Synechocystis* sp. PCC6803, conserved in cyanobacteria	Controlling photosynthetic functions	Light-dependent stress	Georg et al. (2014) and Kadowaki et al. (2016)
IsrR	*Synechocystis* sp. PCC6803	Protecting PSI against photoinduced damage	Iron limitation, high light, and oxidative stress	Dühring et al. (2006)
As1_flv	*Synechocystis* sp. PCC6803	Protecting PSII under low-carbon conditions	Inorganic carbon supply	Eisenhut et al. (2012)
PsbA2R, PsbA3R	*Synechocystis* sp. PCC6803	Achieving a maximum level of D1 synthesis	Light-dependent stress	Sakurai et al. (2012)
RblR	*Synechocystis* sp. PCC6803	Regulating photosynthetic machinery	Light-dependent and carbon deficiency stress	Hu et al. (2017)
NsiR1	*Anabaena* sp. PCC7120	Controlling the switch to nitrogen fixation and heterocyst formation	Nitrogen stress	Ionescu et al. (2010) and Muro-Pastor (2014)
NsiR4	*Synechocystis* sp. PCC6803	Regulating nitrogen metabolism	Nitrogen stress	Klahn et al. (2015a, b, 2018)
NsrR1	*Anabaena* sp. PCC 7120	Regulating nitrogen metabolism	Nitrogen stress	Alvarez-Escribano et al. (2018)
α-furA RNA	*Anabaena* sp. PCC 7120, conserved in cyanobacteria	Regulating photosynthetic machinery	Iron stress	Hernández et al. (2006)
IsaR1	*Synechocystis* sp. PCC6803, conserved in cyanobacteria	Acclimating photosynthetic apparatus to iron starvation	Iron stress	Georg et al. (2017) and Rubsam et al. (2018)

3.1 Light-Dependent Stress

Cyanobacteria use sunlight as their major energy source. Therefore, they are particularly vulnerable to the damaging effects of excess light or UV irradiation. Little is known about RNA regulators of photosynthesis in plants, algae, and cyanobacteria. The sRNA PsrR1 (photosynthesis regulatory RNA1), initially named SyR1 (*Synechocystis* RNA1), was first identified by biocomputational prediction and experimental validation in the unicellular model cyanobacteria *Synechocystis* PCC6803, *Synechococcus* elongatus PCC6301, and *Thermosynechococcus elongatus* BP1 and the toxic cyanobacterium *Microcystis aeruginosa* NIES843 (Voss et al. 2009). Complementary microarray analysis and dRNA-Seq revealed two highly expressed ncRNAs, including SyR1 (which is strongly expressed in high light) and ncr0700 (which is strongly expressed in darkness). Overexpressing PsrR1 driven by an inducible promoter led to bleaching accompanied by the loss of photosynthetic pigmentation under moderate light conditions (Mitschke et al. 2011a). During the diurnal cycle, PsrR1 expression peaks early in the morning (Beck et al. 2014). These findings suggest that PsrR1 functions in the adaptation of cyanobacteria to dark/light conditions.

To further explore the regulatory mechanism involving PsrR1, microarray analysis upon pulse overexpression of PsrR1 combined with recently developed advanced computational target prediction (Wright et al. 2014) was performed, yielding 26 possible target mRNAs. The interaction between PsrR1 and the ribosome-binding regions of *psaL*, *psaJ*, *chlN*, and *cpcA* mRNAs was confirmed by mutational analysis in a heterologous reporter system. *psaL* mRNA, a promising direct target of PsrR1, is processed by RNase E only in the presence of PsrR1 (Georg et al. 2014). RNA gel blot analysis together with chromatin affinity purification suggested that PSI genes are activated and *psrR1* is repressed by the binding of RpaB under low-light conditions. A decrease in the DNA-binding affinity of RpaB occurs within 5 min after the shift from low-light to high-light conditions, leading to a prompt decrease in PSI promoter activity together with the derepression of *psrR1* gene expression. The accumulating PsrR1 molecules prevent translation from preexisting PSI transcripts. This dual repression at the transcriptional and posttranscriptional levels ensures the rapid downregulation of PSI expression under high-light conditions. These findings support the role of PsrR1 in regulating photosynthesis-related gene expression when cells are exposed to high light intensities.

At least four important regulators of photosynthetic gene expression have been identified (IsrR, As1-Flv4, RblR, PsbA2R, and PsbA3R) among the antisense RNAs of *Synechocystis* 6803 (Sakurai et al. 2012; Eisenhut et al. 2012; Dühring et al. 2006; Hu et al. 2017). These asRNAs appear to repress (IsrR and As1-Flv4) or activate (RblR, PsbA2R, and PsbA3R) gene expression.

3.2 Nitrogen Stress

Like all bacteria, photosynthetic bacteria must adapt to limited nutrient availability. In response to nitrogen deficiency, some cyanobacteria develop heterocysts, a terminally differentiated cell type specialized for fixing atmospheric nitrogen. In *Nostocales*, this differentiation process requires the sequential activation of many genes involved in regulatory, structural, or enzymatic aspects of heterocyst differentiation and function whose expression is controlled by two major regulators, NtcA and HetR. NsiR1 (nitrogen stress-inducible RNA1) is the first reported bacterial ncRNA that is specifically upregulated in response to nitrogen step-down

(Ionescu et al. 2010). The observation that this ncRNA is dependent on NtcA and HetR established the notion that NsiR1 belongs to the regulatory network that leads to heterocyst differentiation.

The robust heterocyst-specific *nsiR1* promoter is induced during the early stages of differentiation, before any morphological change is detectable (Muro-Pastor 2014). However, the targets of NsiR1 are unknown. In addition to NsiR1, a recent study identified and validated three other sRNAs (Ncr1071, Syr6, and NsiR7) that are responsive to nitrogen depletion in *Synechocystis* PCC6803 and activated by the global transcription factor NtcA (Giner-Lamia et al. 2017). NsiR7 is also induced under conditions of carbon limitation and is thought to represent an NAD(P)H dehydrogenase transcriptional regulator (NdhR)-regulated sRNA (Klahn et al. 2015a). Thus, NsiR7 might integrate signals related to the intracellular nitrogen and carbon status. Another NtcA-regulated sRNA NsrR1 (nitrogen stress-repressed RNA1) has been reported in *Nostoc* PCC7120 (Brenes-Alvarez et al. 2016; Alvarez-Escribano et al. 2018). RNA blot analysis of the *ntcA* mutant and binding assays with purified NtcA protein revealed that repression of NsrR1 transcription is at least partially operated by NtcA. These observations greatly expand the repertoire of NtcA-regulated ncRNAs.

Glutamine synthetase plays a key role in nitrogen assimilation. The activity of this enzyme is regulated in multiple ways in response to various species and concentrations of nitrogen. Cyanobacteria inactivate glutamine synthetase via a unique mechanism involving the specific inhibitory proteins IF7 (inactivating factor 7) and IF17, which are encoded by *gifA* and *gifB*, respectively (Saelices et al. 2011; Garcia-Dominguez et al. 1999). Klahn et al. reported that the sRNA NsiR4 (nitrogen stress-induced RNA 4) and a glutamine riboswitch are two key elements that regulate glutamine synthetase activity during the nitrogen stress response in cyanobacteria (Klahn et al. 2015b, 2018). In silico target prediction, transcriptome profiling following pulse overexpression, and site-directed mutagenesis using a heterologous reporter system showed that NsiR4 interacts with the 5′UTR of *gifA* mRNA and that there is an inverse relationship between the levels of NsiR4 and IF7 in *Synechocystis* 6803. This NsiR4-dependent modulation of *gifA* mRNA accumulation influences the glutamine pool and thus NH^{+}_{4} assimilation via glutamine synthetase (Klahn et al. 2015b). Another ncRNA functions as a glutamine riboswitch within the *gifB* 5′UTR of *Synechocystis* 6803 and is critical for the expression of IF17 (Klahn et al. 2018). These elements, comprising parts of the untranslated regions of bacterial mRNAs, control gene expression via ligand-induced structural modulation (Mandal and Breaker 2004). A mutagenesis experiment showed that the riboswitch mechanism plays at least as great a role in controlling gene expression as promoter-mediated transcriptional regulation (Klahn et al. 2018). Together with the sRNA NsiR4 described above, these results indicate that ncRNA is an indispensable component in the control of nitrogen assimilation in cyanobacteria.

3.3 *Iron Homeostasis*

In aquatic ecosystems, particularly in the ocean, iron is a major limiting factor for photosynthetic activity and phytoplankton growth (Bibby et al. 2001; Behrenfeld 1996). Thus, bacterial iron homeostasis must be tightly regulated. Cyanobacteria, a major class of phytoplankton, respond to iron deficiency by expressing iron stress-induced protein A (IsiA). In terms of both sequence and structure, IsiA is closely related to the photosystem II (PS II) antenna protein CP43 and is therefore also referred to as CP43′. IsrR (iron stress-repressed RNA) was the first RNA shown to regulate a photosynthetic component via the coupled degradation of IsrR/*isiA* mRNA duplexes (Dühring et al. 2006). The artificial overexpression of IsrR in cyanobacteria under iron stress led to a strongly reduced number of IsiA-photosystem I supercomplexes, whereas IsrR depletion resulted in the premature expression of IsiA.

The activation of IsiA expression under iron limitation is mediated by the ferric uptake regulator Fur (Desai et al. 2012). Fur proteins are the principal regulators responsible for maintaining iron homeostasis in prokaryotes by acting as a finely tuned rheostat of genes involved in iron incorporation and storage (Andrews et al. 2003). In the nitrogen-fixing cyanobacterium *Anabaena* sp. PCC 7120, an asRNA (*a-furA* RNA) was identified that covers the complete *furA* transcript and extends upstream of its 5′ end spanning the iron boxes of P_{furA} (Hernández et al. 2006). The expression of *furA* is tuned by *a-furA* RNA, which is co-transcribed with the outer membrane protein Alr1690. Disrupting *a-furA-alr1690* led to enhanced expression of FurA and important changes to the photosynthetic apparatus, leading to lower photosynthetic performance indices (Hernandez et al. 2010). These findings indicate that *a-furA-alr1690* expression is required to maintain the proper arrangement of the thylakoid membrane, efficient regulation of iron uptake, and optimal yield of the photosynthetic machinery.

Statistical and clustering analyses identified 10 sRNAs, 62 asRNAs, 4 5′UTRs, and 7 intragenic elements as potential novel components of the iron regulatory network in *Synechocystis* (Hernandez-Prieto et al. 2012). IsaR1 (iron-stress activated RNA1) acts on the photosynthetic apparatus in multiple ways and controls a complex network that functions in acclimation to low iron levels (Georg et al. 2017). In an experiment involving different environmental perturbations, IsaR1 slowed the osmotic adaptation process in cells subjected to parallel iron starvation (Rubsam et al. 2018).

4 Conclusions and Perspectives

All bacteria have developed extensive regulatory systems that allow them to adapt to limiting nutrient concentrations and both abiotic and biotic stress conditions. Noncoding RNAs are an essential component of bacterial regulatory systems that primarily function at the posttranscriptional level (Storz et al. 2011). As cyanobacteria use sunlight as their major energy source, they are exposed to a particular set of additional regulatory challenges distinct from those of other bacteria. A number of studies have uncovered a complex, sophisticated regulatory sRNA system that helps cyanobacteria adapt to changes in the environment.

The discovery of the intricate mechanisms employed by regulatory sRNAs has spurred a boom in genetic manipulation based on these molecules (Sun et al. 2018; Zess et al. 2016). RNA turnover plays an important role in the regulation of genes in microorganisms and influences the speed at which microorganisms adapt to environmental changes (Steglich et al. 2010). The effectiveness of genetic regulation by sRNAs depends on their rate of production relative to that of target mRNAs. The appropriate balance between these two types of RNA may allow a single sRNA-encoding gene to regulate many genes, as has indeed been observed experimentally (Georg et al. 2014, 2017). Tunable base-pair complementation exhibited by sRNAs allows them to regulate target genes throughout the genome and thereby fine-tune metabolic flux (Rodrigo et al. 2012; Na et al. 2013). In addition, sRNA-related regulatory tools could be used to precisely knock out essential genes in the host cell, in contrast to the traditional method involving lethal knockout of these genes (Nakashima and Tamura 2009). Based on interactions between sRNAs and their target mRNAs, artificial sRNA tools could be used to alter various physiological and metabolic pathways in cyanobacteria under environmental stress conditions (Hu et al. 2017; Dühring et al. 2006). It is worth noting that sRNA regulatory tools generally do not impose any metabolic burden on host cells (Gaida et al. 2013). sRNAs elicit faster responses than proteins when these regulators are produced in response to a stimulus, constituting a more effective strain engineering strategy (Shimoni et al. 2007). Thus, for stimuli that require rapid responses in a short time, sRNA-mediated regulation may be advantageous under transient stress conditions.

Acknowledgments This work was supported jointly by the National Natural Science Foundation of China (31700055, 31770128, and 91851103), the Shaanxi Provincial Natural Science Foundation (2018JQ3019), and Hubei Provincial Natural Science Foundation of China (2017CFA021).

References

Alvarez-Escribano, I., Vioque, A., & Muro-Pastor, A. M. (2018). NsrR1, a nitrogen stress-repressed sRNA, contributes to the regulation of nblA in Nostoc sp. PCC 7120. *Frontiers in Microbiology, 9*, 2267.

Andrews, S. C., Robinson, A. K., & Rodriguez-Quinones, F. (2003). Bacterial iron homeostasis. *FEMS Microbiology Reviews, 27*, 215–237.

Axmann, I. M., Kensche, P., Vogel, J., Kohl, S., Herzel, H., & Hess, W. R. (2005). Identification of cyanobacterial non-coding RNAs by comparative genome analysis. *Genome Biology, 6*, R73.

Babitzke, P., & Romeo, T. (2007). CsrB sRNA family: Sequestration of RNA-binding regulatory proteins. *Current Opinion in Microbiology, 10*, 156–163.

Barrick, J. E., Sudarsan, N., Weinberg, Z., Ruzzo, W. L., & Breaker, R. R. (2005). 6S RNA is a widespread regulator of eubacterial RNA polymerase that resembles an open promoter. *RNA, 11*, 774–784.

Beck, C., Hertel, S., Rediger, A., Lehmann, R., Wiegard, A., Kolsch, A., Heilmann, B., Georg, J., Hess, W. R., & Axmann, I. M. (2014). Daily expression pattern of protein-encoding genes and small noncoding RNAs in synechocystis sp. strain PCC 6803. *Applied and Environmental Microbiology, 80*, 5195–5206.

Behrenfeld, M. J. (1996). Confirmation of iron limitation of phytoplankton photosynthesis in the equatorial Pacific Ocean. *Nature, 383*(6600), 508–511.

Bibby, T. S., Nield, J., & Barber, J. (2001). Iron deficiency induces the formation of an antenna ring around trimeric photosystem I in cyanobacteria. *Nature, 412*, 743–745.

Billis, K., Billini, M., Tripp, H. J., Kyrpides, N. C., & Mavromatis, K. (2014). Comparative transcriptomics between *Synechococcus* PCC 7942 and *Synechocystis* PCC 6803 provide insights into mechanisms of stress acclimation. *PLoS One, 9*, e109738.

Brenes-Alvarez, M., Olmedo-Verd, E., Vioque, A., & Muro-Pastor, A. M. (2016). Identification of conserved and potentially regulatory small RNAs in heterocystous cyanobacteria. *Frontiers in Microbiology, 7*, 48.

Demmig-Adams, B., & Adams Iii, W. (1992). Photoprotection and other responses of plants to high light stress. *Annual Review of Plant Biology, 43*, 599–626.

Desai, D. K., Desai, F. D., & Laroche, J. (2012). Factors influencing the diversity of iron uptake systems in aquatic microorganisms. *Frontiers in Microbiology, 3*, 362.

Dühring, U., Axmann, I. M., Hess, W. R., & Wilde, A. (2006). An internal antisense RNA regulates expression of the photosynthesis gene *isiA*. *Proceedings of the National Academy of Sciences of the United States of America, 103*, 7054–7058.

Eisenhut, M., Georg, J., Klahn, S., Sakurai, I., Mustila, H., Zhang, P., Hess, W. R., & Aro, E. M. (2012). The antisense RNA As1_flv4 in the cyanobacterium *Synechocystis* sp. PCC 6803 prevents premature expression of the flv4-2 operon upon shift in inorganic carbon supply. *The Journal of Biological Chemistry, 287*, 33153–33162.

Gaida, S. M., Al-Hinai, M. A., Indurthi, D. C., Nicolaou, S. A., & Papoutsakis, E. T. (2013). Synthetic tolerance: Three noncoding small RNAs, DsrA, ArcZ and RprA, acting supra-additively against acid stress. *Nucleic Acids Research, 41*, 8726–8737.

Garcia-Dominguez, M., Reyes, J. C., & Florencio, F. J. (1999). Glutamine synthetase inactivation by protein-protein interaction. *Proceedings of the National Academy of Sciences of the United States of America, 96*, 7161–7166.

Georg, J., Voss, B., Scholz, I., Mitschke, J., Wilde, A., & Hess, W. R. (2009). Evidence for a major role of antisense RNAs in cyanobacterial gene regulation. *Molecular Systems Biology, 5*, 305.

Georg, J., Dienst, D., Schurgers, N., Wallner, T., Kopp, D., Stazic, D., Kuchmina, E., Klahn, S., Lokstein, H., Hess, W. R., & Wilde, A. (2014). The small regulatory RNA SyR1/PsrR1 controls photosynthetic functions in cyanobacteria. *Plant Cell, 26*, 3661–3679.

Georg, J., Kostova, G., Vuorijoki, L., Schon, V., Kadowaki, T., Huokko, T., Baumgartner, D., Muller, M., Klahn, S., Allahverdiyeva, Y., Hihara, Y., Futschik, M. E., Aro, E. M., & Hess, W. R. (2017). Acclimation of oxygenic photosynthesis to Iron starvation is controlled by the sRNA IsaR1. *Current Biology, 27*, 1425–1436.e7.

Gierga, G., Voss, B., & Hess, W. R. (2009). The Yfr2 ncRNA family, a group of abundant RNA molecules widely conserved in cyanobacteria. *RNA Biology, 6*, 222–227.

Gierga, G., Voss, B., & Hess, W. R. (2012). Non-coding RNAs in marine *Synechococcus* and their regulation under environmentally relevant stress conditions. *The ISME Journal, 6*, 1544–1557.

Giner-Lamia, J., Robles-Rengel, R., Hernandez-Prieto, M. A., Muro-Pastor, M. I., Florencio, F. J., & Futschik, M. E. (2017). Identification of the direct regulon of NtcA during early acclimation to nitrogen starvation in the cyanobacterium Synechocystis sp. PCC 6803. *Nucleic Acids Research, 45*, 11800–11820.

Hengge-Aronis, R. (2002). Recent insights into the general stress response regulatory network in Escherichia coli. *Journal of Molecular Microbiology and Biotechnology, 4*, 341–346.

Hernández, J. A., Muro-Pastor, A. M., Flores, E., Bes, M. T., Peleato, M. L., & Fillat, M. F. (2006). Identification of a *furA cis* antisense RNA in the Cyanobacterium *Anabaena* sp. PCC 7120. *Journal of Molecular Biology, 355*, 325–334.

Hernandez, J. A., Alonso, I., Pellicer, S., Luisa Peleato, M., Cases, R., Strasser, R. J., Barja, F., & Fillat, M. F. (2010). Mutants of *Anabaena* sp. PCC 7120 lacking *alr1690* and *alpha-furA* antisense RNA show a pleiotropic phenotype and altered photosynthetic machinery. *Journal of Plant Physiology, 167*, 430–437.

Hernandez-Prieto, M. A., Schon, V., Georg, J., Barreira, L., Varela, J., Hess, W. R., & Futschik, M. E. (2012). Iron deprivation in Synechocystis: Inference of pathways, non-coding RNAs, and regulatory elements from comprehensive expression profiling. *G3 (Bethesda), 2*, 1475–1495.

Holmqvist, E., & Wagner, E. G. H. (2017). Impact of bacterial sRNAs in stress responses. *Biochemical Society Transactions, 45*, 1203–1212.

Hong, W., Wu, Y. E., Fu, X., & Chang, Z. (2012). Chaperone-dependent mechanisms for acid resistance in enteric bacteria. *Trends in Microbiology, 20*, 328–335.

Hu, J., Li, T., Xu, W., Zhan, J., Chen, H., He, C., & Wang, Q. (2017). Small antisense RNA RblR positively regulates RuBisCo in Synechocystis sp. PCC 6803. *Frontiers in Microbiology, 8*, 231.

Hüttenhofer, A., & Vogel, J. (2006). Experimental approaches to identify non-coding RNAs. *Nucleic Acids Research, 34*, 635–646.

Huttenhofer, A., Brosius, J., & Bachellerie, J. P. (2002). RNomics: Identification and function of small, non-messenger RNAs. *Current Opinion in Microbiology, 6*, 835–843.

Ionescu, D., Voss, B., Oren, A., Hess, W. R., & Muro-Pastor, A. M. (2010). Heterocyst-specific transcription of NsiR1, a non-coding RNA encoded in a tandem array of direct repeats in cyanobacteria. *Journal of Molecular Biology, 398*, 177–188.

Kadowaki, T., Nagayama R., Georg, J., Nishiyama, Y., Wilde, A., Hess, W. R., & Hihara, Y. (2016). A feed-forward loop consisting of the response regulator RpaB and the small RNA PsrR1 controls light acclimation of photosystem I gene expression in the cyanobacterium Synechocystis sp. PCC 6803. *Plant Cell Physiol, 57*(4), 813–823.

Klahn, S., Orf, I., Schwarz, D., Matthiessen, J. K., Kopka, J., Hess, W. R., & Hagemann, M. (2015a). Integrated Transcriptomic and Metabolomic characterization of the low-carbon response using an ndhR mutant of Synechocystis sp. PCC 6803. *Plant Physiology, 169*, 1540–1556.

Klahn, S., Schaal, C., Georg, J., Baumgartner, D., Knippen, G., Hagemann, M., Muro-Pastor, A. M., & Hess, W. R. (2015b). The sRNA NsiR4 is involved in nitrogen assimilation control in cyanobacteria by targeting glutamine synthetase inactivating factor IF7. *Proceedings of the National Academy of Sciences of the United States of America, 112*, E6243–E6252.

Klahn, S., Bolay, P., Wright, P. R., Atilho, R. M., Brewer, K. I., Hagemann, M., Breaker, R. R., & Hess, W. R. (2018). A glutamine riboswitch is a key element for the regulation of glutamine synthetase in cyanobacteria. *Nucleic Acids Research, 46*(19), 10082–10094.

Kopf, M., & Hess, W. R. (2015). Regulatory RNAs in photosynthetic cyanobacteria. *FEMS Microbiology Reviews, 39*, 301–315.

Kopf, M., Klahn, S., Scholz, I., Matthiessen, J. K., Hess, W. R., & Voss, B. (2014). Comparative analysis of the primary transcriptome of *Synechocystis* sp. PCC 6803. *DNA Research, 21*, 527–539.

Lambrecht, S. J., Wahlig, J. M. L., & Steglich, C. (2018). The GntR family transcriptional regulator PMM1637 regulates the highly conserved cyanobacterial sRNA Yfr2 in marine picocyanobacteria. *DNA Research, 25*, 489–497.

Landt, S. G., Abeliuk, E., Mcgrath, P. T., Lesley, J. A., Mcadams, H. H., & Shapiro, L. (2008). Small non-coding RNAs in *Caulobacter crescentus*. *Molecular Microbiology, 68*, 600–614.

Lee, E. J., & Groisman, E. A. (2010). An antisense RNA that governs the expression kinetics of a multifunctional virulence gene. *Molecular Microbiology, 76*, 1020–1033.

Mandal, M., & Breaker, R. R. (2004). Gene regulation by riboswitches. *Nature Reviews Molecular Cell Biology, 5*, 451–463.

Mitschke, J., Georg, J., Scholz, I., Sharma, C. M., Dienst, D., Bantscheff, J., Voss, B., Steglich, C., Wilde, A., Vogel, J., & Hess, W. R. (2011a). An experimentally anchored map of transcriptional start sites in the model cyanobacterium *Synechocystis* sp. PCC6803. *Proceedings of the National Academy of Sciences of the United States of America, 108*, 2124–2129.

Mitschke, J., Vioque, A., Haas, F., Hess, W. R., & Muro-Pastor, A. M. (2011b). Dynamics of transcriptional start site selection during nitrogen stress-induced cell differentiation in *Anabaena* sp. PCC7120. *Proceedings of the National Academy of Sciences of the United States of America, 108*, 20130–20135.

Mizuno, T., Chou, M. Y., & Inouye, M. (1984). A unique mechanism regulating gene expression: Translational inhibition by a complementary RNA transcript (micRNA). *Proceedings of the National Academy of Sciences of the United States of America, 81*, 1966–1970.

Muro-Pastor, A. M. (2014). The heterocyst-specific NsiR1 small RNA is an early marker of cell differentiation in cyanobacterial filaments. *MBio, 5*, e01079–e01014.

Na, D., Yoo, S. M., Chung, H., Park, H., Park, J. H., & Lee, S. Y. (2013). Metabolic engineering of Escherichia coli using synthetic small regulatory RNAs. *Nature Biotechnology, 31*, 170–174.

Nakashima, N., & Tamura, T. (2009). Conditional gene silencing of multiple genes with antisense RNAs and generation of a mutator strain of Escherichia coli. *Nucleic Acids Research, 37*, e103.

Pei, G., Sun, T., Chen, S., Chen, L., & Zhang, W. (2017). Systematic and functional identification of small non-coding RNAs associated with exogenous biofuel stress in cyanobacterium Synechocystis sp. PCC 6803. *Biotechnology for Biofuels, 10*, 57.

Puente-Sanchez, F., Arce-Rodriguez, A., Oggerin, M., Garcia-Villadangos, M., Moreno-Paz, M., Blanco, Y., Rodriguez, N., Bird, L., Lincoln, S. A., Tornos, F., Prieto-Ballesteros, O., Freeman, K. H., Pieper, D. H., Timmis, K. N., Amils, R., & Parro, V. (2018). Viable cyanobacteria in the deep continental subsurface. *Proceedings of the National Academy of Sciences of the United States of America, 115*, 10702–10707.

Raghavan, R., Groisman, E. A., & Ochman, H. (2011). Genome-wide detection of novel regulatory RNAs in *E. coli*. *Genome Research, 21*, 1487–1497.

Ravasi, T., Suzuki, H., Pang, K. C., Katayama, S., Furuno, M., Okunishi, R., Fukuda, S., Ru, K. L., Frith, M. C., Gongora, M. M., Grimmond, S. M., Hume, D. A., Hayashizaki, Y., & Mattick, J. S. (2006). Experimental validation of the regulated expression of large numbers of non-coding RNAs from the mouse genome. *Genome Research, 16*, 11–19.

Rodrigo, G., Landrain, T. E., & Jaramillo, A. (2012). De novo automated design of small RNA circuits for engineering synthetic riboregulation in living cells. *Proceedings of the National Academy of Sciences of the United States of America, 109*, 15271–15276.

Rubsam, H., Kirsch, F., Reimann, V., Erban, A., Kopka, J., Hagemann, M., Hess, W. R., & Klahn, S. (2018). The iron-stress activated RNA 1 (IsaR1) coordinates osmotic acclimation and iron starvation responses in the cyanobacterium Synechocystis sp. PCC 6803. *Environmental Microbiology, 20*, 2757–2768.

Saelices, L., Galmozzi, C. V., Florencio, F. J., & Muro-Pastor, M. I. (2011). Mutational analysis of the inactivating factors, IF7 and IF17 from Synechocystis sp. PCC 6803: Critical role of arginine amino acid residues for glutamine synthetase inactivation. *Molecular Microbiology, 82*, 964–975.

Sakurai, I., Stazic, D., Eisenhut, M., Vuorio, E., Steglich, C., Hess, W. R., & Aro, E. M. (2012). Positive regulation of *psbA* gene expression by cis-encoded antisense RNAs in *Synechocystis* sp. PCC 6803. *Plant Physiology, 160*, 1000–1010.

Savakis, P., & Hellingwerf, K. J. (2015). Engineering cyanobacteria for direct biofuel production from CO_2. *Current Opinion in Biotechnology, 33*, 8–14.

Shimoni, Y., Friedlander, G., Hetzroni, G., Niv, G., Altuvia, S., Biham, O., & Margalit, H. (2007). Regulation of gene expression by small non-coding RNAs: A quantitative view. *Molecular Systems Biology, 3*, 138.

Sittka, A., Lucchini, S., Papenfort, K., Sharma, C. M., Rolle, K., Binnewies, T. T., Hinton, J. C. D., & Vogel, J. (2008). Deep sequencing analysis of small noncoding RNA and mRNA targets of the global post-transcriptional regulator, Hfq. *PLoS Genetics, 4*, e1000163.

Soo, R. M., Hemp, J., Parks, D. H., Fischer, W. W., & Hugenholtz, P. (2017). On the origins of oxygenic photosynthesis and aerobic respiration in cyanobacteria. *Science, 355*, 1436–1439.

Steglich, C., Futschik, M. E., Lindell, D., Voss, B., Chisholm, S. W., & Hess, W. R. (2008). The challenge of regulation in a minimal photoautotroph: Non-coding RNAs in *Prochlorococcus*. *PLoS Genetics, 4*, e1000173.

Steglich, C., Lindell, D., Futschik, M., Rector, T., Steen, R., & Chisholm, S. W. (2010). Short RNA half-lives in the slow-growing marine cyanobacterium *Prochlorococcus*. *Genome Biology, 11*, R54.

Storz, G. (2002). An expanding universe of noncoding RNAs. *Science, 296*, 1260–1263.

Storz, G., Opdyke, J. A., & Zhang, A. (2004). Controlling mRNA stability and translation with small, noncoding RNAs. *Current Opinion in Microbiology, 7*, 140–144.

Storz, G., Vogel, J., & Wassarman, K. M. (2011). Regulation by small RNAs in bacteria: Expanding frontiers. *Molecular Cell, 43*, 880–891.

Sun, T., Pei, G., Wang, J., Chen, L., & Zhang, W. (2017). A novel small RNA CoaR regulates coenzyme A biosynthesis and tolerance of Synechocystis sp. PCC6803 to 1-butanol possibly via promoter-directed transcriptional silencing. *Biotechnology for Biofuels, 10*, 42.

Sun, T., Li, S., Song, X., Pei, G., Diao, J., Cui, J., Shi, M., Chen, L., & Zhang, W. (2018). Re-direction of carbon flux to key precursor malonyl-CoA via artificial small RNAs in photosynthetic Synechocystis sp. PCC 6803. *Biotechnology for Biofuels, 11*, 26.

Tan, X., Hou, S., Song, K., Georg, J., Klahn, S., Lu, X., & Hess, W. R. (2018). The primary transcriptome of the fast-growing cyanobacterium Synechococcus elongatus UTEX 2973. *Biotechnology for Biofuels, 11*, 218.

Tomizawa, J., Itoh, T., Selzer, G., & Som, T. (1981). Inhibition of ColE1 RNA primer formation by a plasmid-specified small RNA. *Proceedings of the National Academy of Sciences of the United States of America, 78*, 1421–1425.

Tous, C., Vega-Palas, M. A., & Vioque, A. (2001). Conditional expression of RNase P in the cyanobacterium *Synechocystis* sp. PCC6803 allows detection of precursor RNAs. *Journal of Biological Chemistry, 276*, 29059–29066.

Voigt, K., Sharma, C. M., Mitschke, J., Lambrecht, S. J., Voss, B., Hess, W. R., & Steglich, C. (2014). Comparative transcriptomics of two environmentally relevant cyanobacteria reveals unexpected transcriptome diversity. *The ISME Journal, 8*, 2056–2068.

Volpe, T. A., Kidner, C., Hall, I. M., Teng, G., Grewal, S. I., & Martienssen, R. A. (2002). Regulation of heterochromatic silencing and histone H3 lysine-9 methylation by RNAi. *Science, 297*, 1833–1837.

Voss, B., Georg, J., Schon, V., Ude, S., & Hess, W. R. (2009). Biocomputational prediction of non coding RNAs in model cyanobacteria. *BMC Genomics, 10*, 123.

Wassarman, K. M., Zhang, A., & Storz, G. (1999). Small RNAs in Escherichia coli. *Trends in Microbiology, 7*, 37–45.

Watanabe, T., Sugiura, M., & Sugita, M. (1997). A novel small stable RNA, 6Sa RNA, from the cyanobacterium *Synechococcus* sp. strain PCC6301. *FEBS Letters, 416*, 302–306.

Watanabe, T., Sugita, M., & Sugiura, M. (1998). Identification of 10Sa RNA (tmRNA) homologues from the cyanobacterium *Synechococcus* sp. strain PCC6301 and related organisms. *Biochimica et Biophysica Acta, Biomembranes, 1396*, 97–104.

Waters, L. S., & Storz, G. (2009). Regulatory RNAs in bacteria. *Cell, 136*, 615–628.

Weinberg, Z., Barrick, J. E., Yao, Z., Roth, A., Kim, J. N., Gore, J., Wang, J. X., Lee, E. R., Block, K. F., Sudarsan, N., Neph, S., Tompa, M., Ruzzo, W. L., & Breaker, R. R. (2007). Identification of 22 candidate structured RNAs in bacteria using the CMfinder comparative genomics pipeline. *Nucleic Acids Research, 35*, 4809–4819.

Wright, P. R., Georg, J., Mann, M., Sorescu, D. A., Richter, A. S., Lott, S., Kleinkauf, R., Hess, W. R., & Backofen, R. (2014). CopraRNA and IntaRNA: Predicting small RNA targets, networks and interaction domains. *Nucleic Acids Research, 42*, W119–W123.

Xu, W., Chen, H., He, C.-L., & Wang, Q. (2014). Deep sequencing-based identification of small regulatory RNAs in *Synechocystis* sp. PCC 6803. *PLoS One, 9*, e92711.

Zess, E. K., Begemann, M. B., & Pfleger, B. F. (2016). Construction of new synthetic biology tools for the control of gene expression in the cyanobacterium *Synechococcus* sp. strain PCC 7002. *Biotechnology and Bioengineering, 113*, 424–432.

Part III
Artificial Photosynthesis and Light-driven Biofactory

Mimicking the Mn_4CaO_5-Cluster in Photosystem II

Yang Chen, Ruoqing Yao, Yanxi Li, Boran Xu, Changhui Chen, and Chunxi Zhang

Abstract The oxygen-evolving center (OEC) in photosystem II (PSII) is a unique biocatalyst that splits water into electrons, protons, and dioxygen. Recent crystallographic studies of PSII have revealed that the structure of the OEC is an asymmetric Mn_4CaO_5-cluster, while the detailed mechanism for the O=O bond formation is still elusive mainly due to the complexity of the large protein environment and structural uncertainty of the OEC during the catalytic reaction. To understand the structure-function relationship and the catalytic mechanism of this natural Mn_4CaO_5-cluster, as well as to develop efficient man-made water-splitting catalysts in artificial photosynthesis, precise mimics of the OEC are highly required. It is of a great challenge to precisely mimic the structure and function of the OEC in the laboratory. However significant advances have recently been achieved. One of the most important advances is the synthesis of a series of the artificial Mn_4CaO_4-clusters that closely mimics both the geometric and electronic structures of the OEC, which provides a structurally well-defined chemical model to investigate the structure-function relationship of the natural Mn_4CaO_5-cluster, and sheds new insights into the mechanism of the water-splitting reaction in PSII. The artificial Mn_4CaO_4-cluster and its variants may open new avenues to develop efficient artificial catalysts for the water-splitting reaction by using earth-abundant and nontoxic chemical elements in the future.

Keywords Photosystem II · Oxygen-evolving center · Mn_4Ca-cluster · Water-splitting reaction · Artificial photosynthesis

Y. Chen · R. Yao · Y. Li · B. Xu
Laboratory of Photochemistry, Institute of Chemistry, Chinese Academy of Sciences, Beijing, China

University of Chinese Academy of Sciences, Beijing, China

C. Chen · C. Zhang (✉)
Laboratory of Photochemistry, Institute of Chemistry, Chinese Academy of Sciences, Beijing, China
e-mail: chunxizhang@iccas.ac.cn

Q. Wang (ed.), *Microbial Photosynthesis*,
https://doi.org/10.1007/978-981-15-3110-1_13

1 Introduction

Photosynthetic oxygen evolution is a unique function of the photosystem II (PSII), which is a pigment-binding, multi-subunit protein complex embedded in the thylakoid membranes of plants, algae, and cyanobacteria (Barber 2009; Dau and Zaharieva 2009; Junge 2019; Lubitz et al. 2019; Nelson and Yocum 2006; Satoh and Wydrzynski 2005; Shen 2015). The reaction center and the key cofactors of PSII involved to perform this function are shown in Fig. 1. Upon photoexcitation, the primary electron donor (P_{680}) donates one electron to the primary electron acceptor (Pheo), producing the $P_{680}^{+\bullet}$ and $Pheo^{-\bullet}$ charge pair in a few picoseconds (Cardona et al. 2012; Diner and Rappaport 2002; Holzwarth et al. 2006; Renger 2012). $Pheo^{-\bullet}$ then transfers the electron to the primary and secondary plastoquinones (Q_A and Q_B) in sequence through the non-heme iron at the acceptor side (Petrouleas and Crofts 2005; Saito et al. 2013). The high redox potential of $P_{680}^{+\bullet}$ obtains one electron from the secondary electron donor (Tyr_Z), leading to the formation of a neutral radical ($Tyr_Z^{\bullet}$) (Diner and Britt 2005; Styring et al. 2012; Zhang 2007), which then drives the water oxidation at the oxygen-evolving center (OEC), in milliseconds at the donor side (Lubitz et al. 2019; McEvoy and Brudvig 2006; Perez-Navarro et al. 2016; Yano and Yachandra 2014).

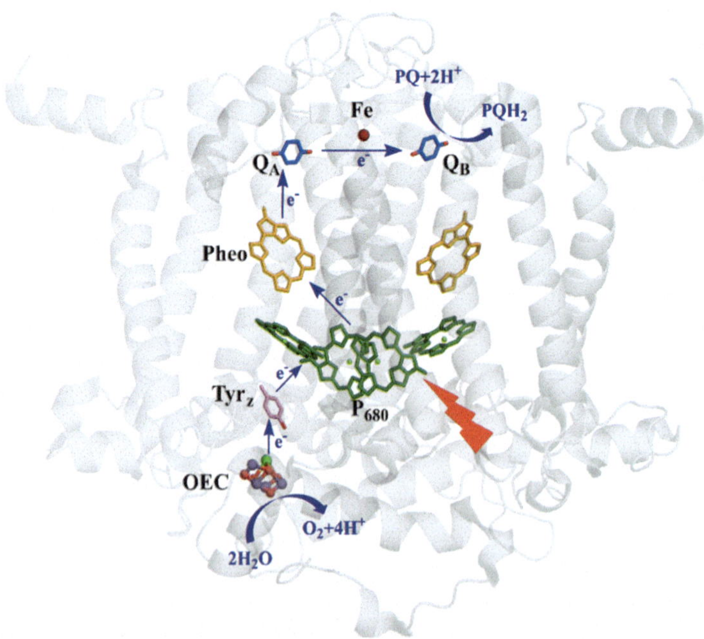

Fig. 1 The reaction center and main cofactors involved for electron transfers and catalytic reactions in PSII (Umena et al. 2011)

The catalytic oxygen-evolving reaction involves five different redox states (S_n, n = 0–4) of the OEC as shown in Fig. 2 (Dau and Haumann 2007; Kok et al. 1970), wherein the S_0 state is the initial and most reduced state and the S_1 state is the dark-stable state. The S_2 and S_3 states are metastable and decay eventually to the dark-stable S_1 state, whereas the S_4 state is a transient state that releases dioxygen and decays to the S_0 state. Changes of the oxidation valences of the four manganese ions occur during the turnover. The oxidation valences for the four manganese ions are S_0 (III, III, III, IV) or (II, III, IV, IV), S_1 (III, III, IV, IV), S_2 (III, IV, IV, IV), and S_3 (IV, IV, IV, IV) (Dau et al. 2008; Krewald et al. 2015; Sauer et al. 2008; Yano and Yachandra 2014). The calcium is an indispensable cofactor for the function of the OEC, and its depletion results in the complete loss of the water oxidation capability of PSII, and it can only be functionally replaced by strontium (Koua et al. 2013; van Gorkom and Yocum 2005; Yocum 2008). Due to the broad interests in fundamental research, as well as potential applications in artificial photosynthesis, the structure and catalytic mechanism of the OEC have attracted extensive studies during the last three decades (Junge 2019; Lubitz et al. 2019; Pantazis 2018).

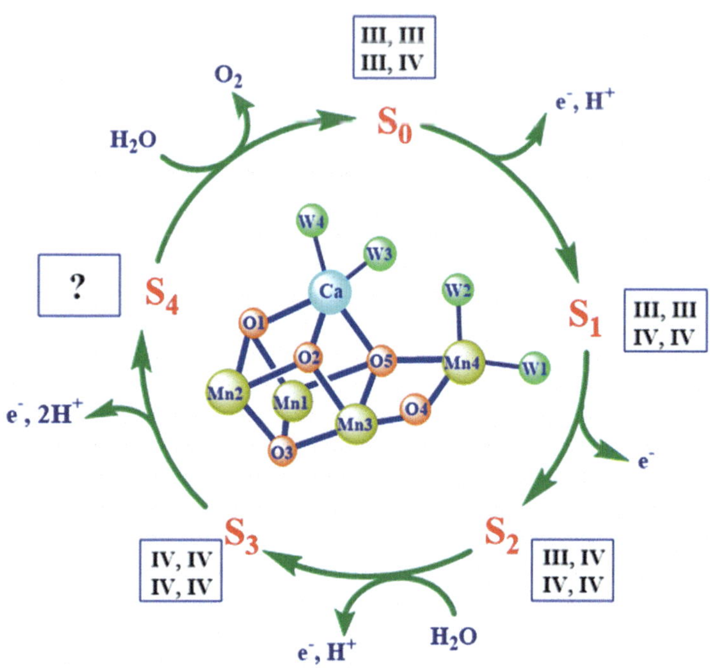

Fig. 2 The turnover of the OEC in PSII (Krewald et al. 2015). The oxidation valences of the four manganese ions in each S state are given in Roman numerals in squares. The structure and the assignment of each atom of the OEC are provided in the middle of this scheme

2 Structure of the OEC

Revealing the structure of the OEC has been a long-standing issue in the field of the photosynthetic research. Before the appearance of the crystal structure of PSII, most structural information of the OEC came from extended X-ray absorption fine structure (EXAFS) (Dau et al. 2008; Yano and Yachandra 2014) and electron paramagnetic resonance (EPR) (Cox et al. 2013; Lubitz et al. 2019; Peloquin and Britt 2001) investigations and theoretical calculations (Siegbahn 2009). Different structural models were proposed to explain the properties of the OEC in PSII (Cinco et al. 1998; Peloquin et al. 2000; Tommos and Babcock 1998; Vrettos et al. 2001). Figure 3a shows the structural model proposed by Zhang et al. in 1999 (Zhang 2016; Zhang et al. 1999), in which the key component of calcium was suggested to be located in the middle of the OEC and be connected with four manganese ions through three oxide bridges and two carboxylate groups. This model was the only model that successfully predicted the binding mode of calcium in OEC of PSII (see below).

The crystal structure data of the PSII have emerged since the beginning of this century (Ferreira et al. 2004; Guskov et al. 2009; Hellmich et al. 2014; Kamiya and Shen 2003; Umena et al. 2011; Zouni et al. 2001). In 2001, Zouni et al. (2001) reported the first crystal structure of PSII from thermophilic cyanobacterium at a resolution of 3.8 Å. In 2004, based on the 3.5 Å resolution structure data, Ferreira et al. (2004), for the first time, proposed that the OEC could be comprised of a Mn_3CaO_4 cubane attached with a "dangler" Mn ion via one bridging oxide, forming a Mn_4CaO_4-cluster. However, the detailed peripheral ligands of the OEC were still elusive due to the low resolution and the possible reduction induced by X-ray radiation during the crystallographic structural determination (Askerka et al. 2017; Grabolle et al. 2006; Yano et al. 2005, 2006).

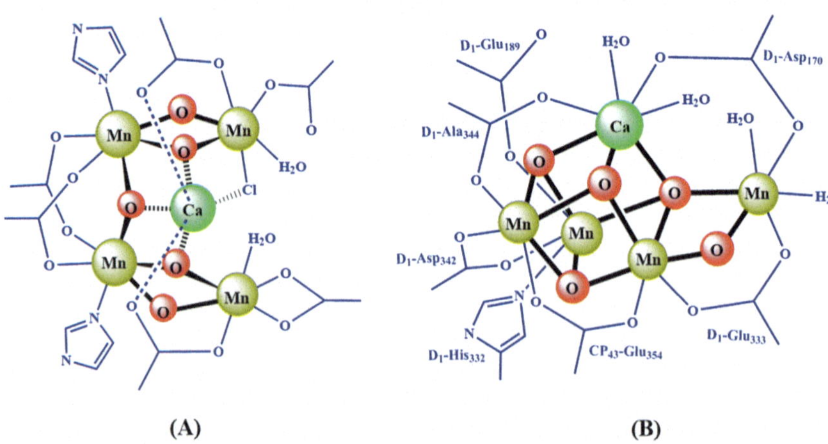

(A) (B)

Fig. 3 Scheme for the structural model (Zhang 2016; Zhang et al. 1999) (**a**) and structure (Umena et al. 2011) (**b**) of the OEC

The detailed structure of the OEC was revealed by the crystal structure of PSII at a resolution of 1.9 Å reported by Umena et al. in 2011. All binding ligands of the OEC have been clearly resolved, including four water molecules, one imidazole group from D_1-His332, and six carboxylate groups from D_1-Asp170, D_1-Glu189, D_1-Glu333, D_1-Asp342, D_1-Ala344, and CP_{43}-Glu354, respectively. One additional bridged oxygen (O4) between Mn4 and Mn3 in the OEC has been observed (Fig. 3b). The whole structure of the OEC is an asymmetric Mn_4CaO_5-cluster. In this structure, Ca^{2+} is located in the middle of the OEC and is connected to the Mn_4-cluster through three oxide bridges and two carboxylate groups, which is consistent with our previous proposal (Fig. 3a) published in 1999 (Zhang 2016; Zhang et al. 1999).

The structure of the OEC (Fig. 3b) was further confirmed by the 1.95 Å resolution data obtained by an X-ray free-electron laser (XFEL) reported by Shen's group (Suga et al. 2015, 2017, 2019) and other groups (Kern et al. 2018; Young et al. 2016). It was also supported by the 2.44 Å resolution reported by Hellmich et al. (2014) and 1.87 Å resolution reported by Tanaka et al. by using conventional synchrotron radiation source at extremely low X-ray doses (0.03 MGy) (Tanaka et al. 2017).

Recently, the structure of the OEC in higher plants (spinach and *Pisum sativum*) has also been revealed by single-particle cryoelectron microscopy (cryo-EM) at the resolution of 3.2~2.7 Å (Su et al. 2017; Wei et al. 2016).

The structures of the S_2 and S_3 states OEC were also reported recently by using XFEL. It has been suggested that the structure revealed by XFEL could be corresponding to the native structure of the OEC (Suga et al. 2015). However, consensus of the atomic positions of the S_1 state OEC revealed by XFEL is still not fully reached for all structures with the results of EXAFS spectroscopy studies on active sample (Askerka et al. 2017). To evaluate the oxidation valences of the four manganese ions in the structure of the OEC revealed by XFEL, we have carried out bond-valence sum (BVS) calculations. BVS method is a popular method in coordination chemistry to estimate the oxidation valences of atoms (Brown 2009; Pauling 1929). It is derived from the bond-valence model (Pauling 1929), which is a simple yet robust model for validating chemical structures with localized bonds or used to predict some of their properties. This method has been used extensively to estimate the oxidation state of active site in various metalloenzymes, as well (Gatt et al. 2012; Liu and Thorp 1993). Table 1 lists the results of the BVS calculations on the XFEL's structures of the OEC in the so-called "native" S_1, S_2, and S_3 states, respectively. Surprisingly, the oxidation valences of the four manganese ions of all these states are remarkably lower than that of widely adopted in the field of photosynthetic research (Dau and Haumann 2008; Krewald et al. 2015; Peloquin and Britt 2001; Yano and Yachandra 2014). It is likely that the reduction of the high oxidation valences of manganese ions in the OEC could still take place during the XFEL structural determination. If it was the case, one would expect that the XFEL structures of the OEC would be different from the native structure of these intermediate states during the catalytic cycle. Recently, Amin et al. found that structural modifications of the OEC would take place due to the radiation damage induced by XFEL (Amin et al. 2016); particularly, they found that the position of the μ_4-oxide bridge (O5, Fig. 2) can be significantly disturbed by XFEL.

Table 1 Bond-valence sum (BVS) calculations on the structures of the OEC revealed by X-ray free-electron laser (XFEL) at different resolutions in different S states. Roman numerals in parentheses indicate the assignment of the possible oxidation valences of four manganese ions in the OEC based on BVS calculations. All atomic coordinates were taken from the first monomer of photosystem II in the crystal structure data with PDB-IDs: 4UB6, 6DHF, 6JLK, 6DHO, and 6JLL, respectively

Resolution of PSII	S_1 (1.95 Å) (4UB6)	S_2 (2.08 Å) (6DHF)	S_2 (2.15 Å) (6JLK)	S_3 (2.07 Å) (6DHO)	S_3 (2.15 Å) (6JLL)
Mn1	3.075 (III)	3.232 (III)	3.204 (III)	3.901 (IV)	4.300 (IV)
Mn2	3.237 (III)	4.316 (IV)	3.775 (IV)	4.193 (IV)	3.852 (IV)
Mn3	2.980 (III)	3.784 (IV)	3.347 (III)	3.232 (III)	3.243 (III)
Mn4	2.318 (II)	3.139 (III)	2.597 (III)	2.932 (III)	2.531 (III)

3 Mechanism for the Water-Splitting Reaction in the OEC

Based on recent crystallographic studies (Ferreira et al. 2004; Kern et al. 2018; Suga et al. 2017, 2019; Umena et al. 2011), spectroscopic investigations (Cox et al. 2014), as well as theoretical calculations (Siegbahn 2013; Yamaguchi et al. 2018), different proposals for the O=O bond formation have been suggested. Figures 4, 5, and 6 show three typical models proposed by different groups.

The mechanism in Fig. 4 was suggested by Barber's group, in which the two oxygen of two water molecules (W2 and W3) were proposed to serve as the oxygen sources for the formation of the O=O bond. The key feature of this mechanism is that the O–O bond is formed by a nucleophilic attack of a calcium-ligated hydroxyl group onto an electrophilic oxo of $Mn^{V} \equiv O$ or Mn^{IV}-$O^{\bullet}$, derived from the deprotonation of the second substrate water molecule. Similar proposals have been suggested by other groups (Chen et al. 2015; Vinyard et al. 2015). However, this proposal was not supported by recent theoretical calculation reported by Siegbahn (Siegbahn 2017).

The second proposal for the mechanism of the water-splitting reaction was suggested by Ishikita's group as shown in Fig. 5 (Kawashima et al. 2018). In this mechanism, the μ_2-oxide bridge (O4) and one water molecule (W1) were suggested to provide the oxygen atoms to form the O=O bond. The key feature of this proposal is that the O-O bond could be formed through the coupling of a bridged oxo and an Mn(IV)-$O^{\bullet}$ oxyl radical. However, the oxidation valences (III, IV, IV, IV) of the four manganese ions in the S_3 state were not consistent with the widely accepted values of (IV, IV, IV, IV).

The third mechanism was suggested by Siegbahn based on theoretical calculations (Siegbahn 2013)(Fig. 6). The main feature of this proposal is that the μ_4-oxide bridge (O5) serves as the active site for the O=O bond formation. Some recent crystallographic results (Suga et al. 2019) and spectroscopic studies (Cox et al. 2014) show experimental evidence to support this proposal in some degree; however, it is still an open question whether the Mn1 is the active site for the binding site of the second substrate water molecule. According to this mechanism, remarkably, the release of O_2 from the S_4 state would result in the formation of four unsaturated

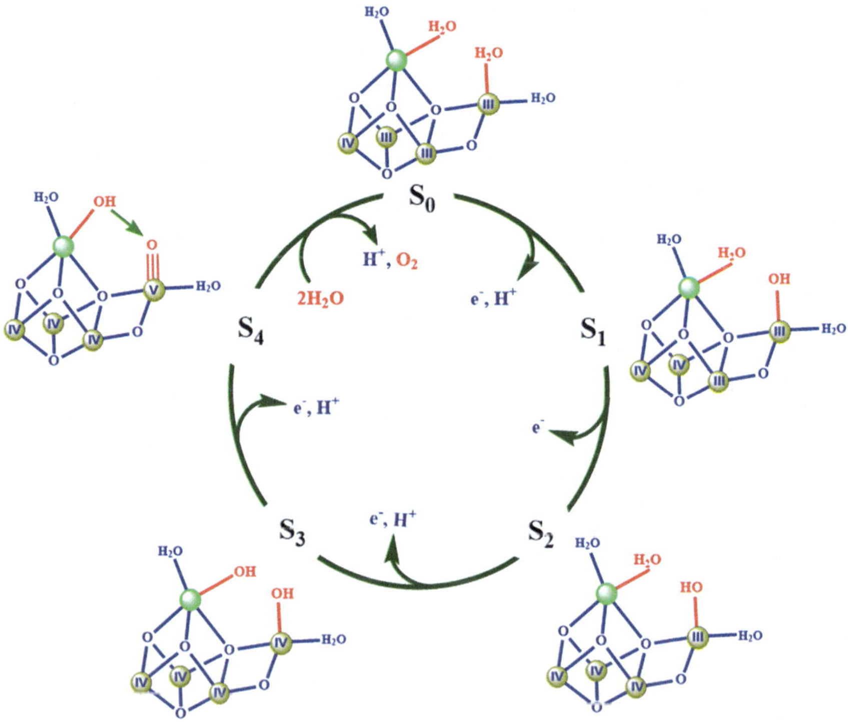

Fig. 4 One possible mechanism for the water-splitting reaction by OEC suggested by Barber's group. Significant changes during the catalytic cycle are given in red color. Mn and Ca are shown in brown and green, respectively. Roman numerals indicate the oxidation states of manganese ions

metal ions, namely, three five-coordinated manganese (i.e., Mn1, Mn3, Mn4) and one six-coordinated calcium (Zhang and Kuang 2018), which could certainly require much high activation energy. Thus, one would expect that the dioxygen release could be the rate-limited step during the catalytic cycle, which is inconsistent with the fast release of the O_2 observed in natural system (Haumann et al. 2005; Junge 2019). It is also noted that the suggestion of the protonated μ_4-oxide bridge (i.e., OH) for O5 has not been supported by the theoretical studies from other group (Ishikita 2019).

As mentioned above, although various mechanisms for the water-splitting reaction have been proposed (Barber 2017; Chen et al. 2015; Isobe et al. 2012; Kawashima et al. 2018; Krewald et al. 2016; Siegbahn 2013; Vinyard et al. 2015), the detailed mechanism for the O=O bond formation is still fully unknown (Junge 2019; Pantazis 2018) due to the complexity of the huge protein environment and the dynamic structural changes of the OEC during the water-splitting reaction. In this regard, therefore, precise structural information for different S states of the OEC is highly required in the future (Kern et al. 2018; Suga et al. 2017, 2019; Young et al. 2016; Zhang and Kuang 2018).

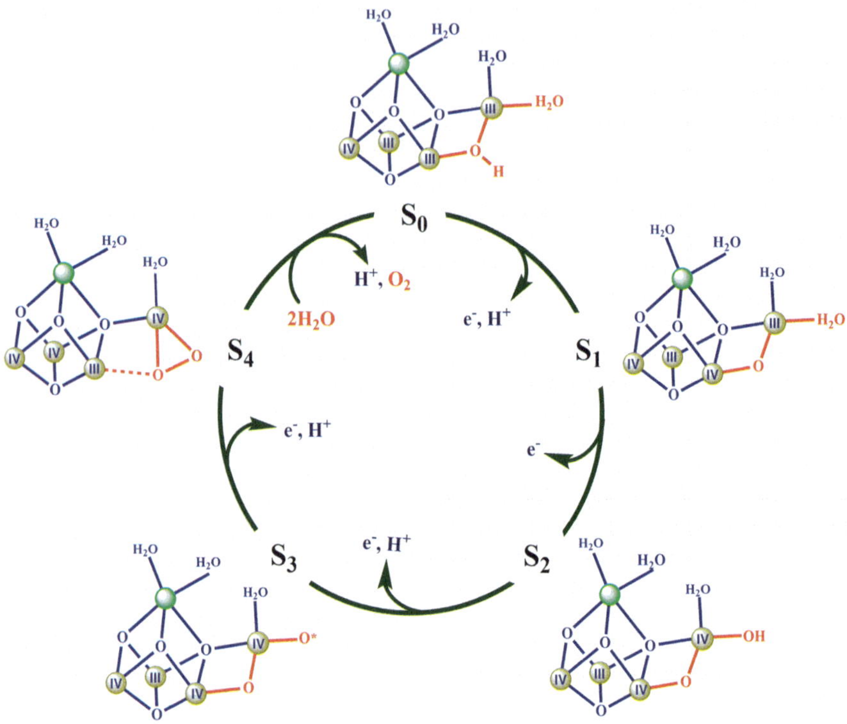

Fig. 5 The second proposal for the O=O bond formation involving the μ_2-oxide bridge (O4) and W1. Significant changes during the catalytic cycle are given in red color. Mn and Ca are shown in brown and green, respectively. Roman numerals indicate the oxidation states of manganese ions

4 Challenge for the Synthesis of the OEC in Laboratory

In order to better understand the structure and properties of the OEC, as well as to develop highly efficient and cheap artificial catalysts for the water-splitting reaction to overcome the bottleneck of artificial photosynthesis, many groups have tried to synthesize the OEC in laboratory since the end of last century. However, it was a great challenge for chemists to synthesize the entire structure of the OEC in laboratory due to several reasons (Zhang 2015, 2016): (1) It is very difficult to incorporate Ca^{2+} into the Mn_4-cluster through μ-oxo bridges, because the affinity of Ca^{2+} to μ-oxo is significantly weaker than that of Mn ion. In general, only a multi-manganese cluster, instead of the manganese-calcium heterometallic cluster, can be synthesized. (2) The core structure of the OEC is an asymmetric Mn_4Ca-cluster. It was unknown whether such a structure could be synthesized in a chemical system. (3) The ligands of the biological OEC are mainly composed of carboxylate groups and water molecules, which are drastically different from the multi-pyridine ligands used in most previous chemical model systems (Kärkäs et al. 2014; Limburg et al. 1999; Mullins and Pecoraro 2008; Tsui et al. 2013). (iv) The redox potential of the

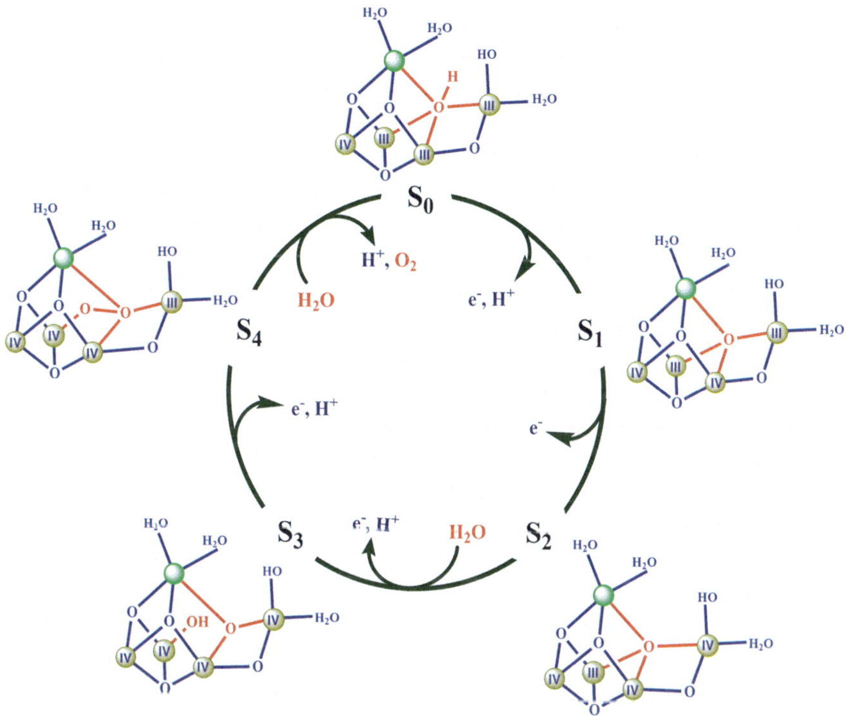

Fig. 6 The third proposal for the O=O bond formation involved the μ_4-oxide bridge (O5). Significant changes during the catalytic cycle are given in red color. Mn and Ca are shown in brown and green, respectively. Roman numerals indicate the oxidation states of manganese ions

OEC is very high (+0.8 ~ +1.0 V vs. NHE) (Dau and Zaharieva 2009; Vass and Styring 1991) due to the presence of the high oxidation states of the manganese ions.

A large number of artificial Mn complexes have been reported in the literature (Dismukes et al. 2009; Gerey et al. 2016; Limburg et al. 1999; Mukhopadhyay et al. 2004; Mullins and Pecoraro 2008; Najafpour et al. 2016; Paul et al. 2017). Among them, tetra-manganese complexes containing Mn_4O_4-cubane (Chakov et al. 2003, 2016; Dismukes et al. 2009; Kanady et al. 2011; Ruettinger et al. 1997) (Fig. 7) are attractive, because the oxidation states of the four manganese ions are (III, III, IV, IV) similar to that of the OEC in the S_1 state (Fig. 2). The peripheral ligands in these artificial complexes, however, it is remarkably different to that of the OEC composing mainly of carboxylate groups (Umena et al. 2011).

Significant advances for the synthesis of the OEC have emerged since 2011. Agapie's group reported the first artificial Mn_3CaO_4-cluster using a multi-pyridylalkoxide ligand (Fig. 8a, b) (Kanady et al. 2011). The same group reported a series of analogues or derivatives (Tsui and Agapie 2013), for example, the artificial Mn_3CaAgO_4-complex (Kanady et al. 2014) (Fig. 8e, f). In 2012, Christou's group reported the $Mn_3Ca_2O_4$-complex with one Ca^{2+} attached to the Mn_3CaO_4 cubane (Mukherjee et al. 2012) (Fig. 8c, d). Distinct from previous Mn_3CaO_4-complexes,

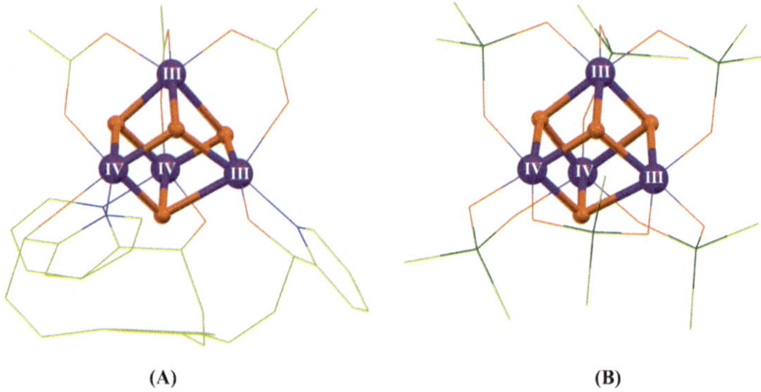

Fig. 7 Structures of two artificial clusters containing Mn_4O_4-cubane. (**a**) $Mn_4O_4L(CH_3CO_2)_3$ complex (L = 1,3,5-tris(2-di(2′-pyridyl)oxymethylphenyl)benzene) (Kanady et al. 2011); (**b**) $Mn_4O_4[(CH_3)_2AsO_2]_6$ complex (Chakov et al. 2016). Mn, O, N, As, and C are shown in purple, orange, blue, dark green, and yellow, respectively. For clarity, all the hydrogen atoms are not shown. Distances are given in Å

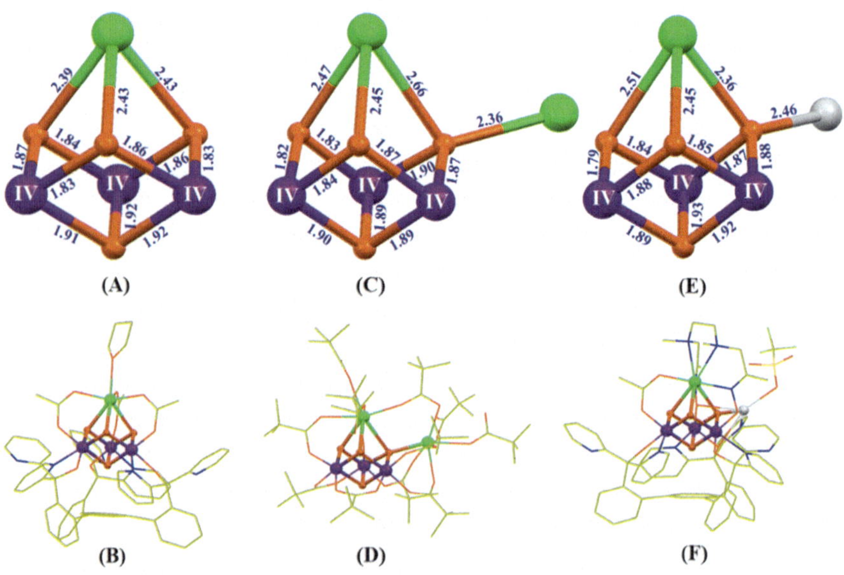

Fig. 8 Structures of artificial complexes containing Mn_3CaO_4 cubane. (**a**) Core structure of the Mn_3CaO_4-complex; (**b**) whole structure of the Mn_3CaO_4-complex (Kanady et al. 2011); (**c**) core structure of the $Mn_3Ca_2O_4$-complex; (**d**) whole structure of the $Mn_3Ca_2O_4$-complex (Mukherjee et al. 2012); (**e**) core structure of the Mn_3CaAgO_4-complex; (**f**) whole structure of the Mn_3CaAgO_4-complex (Kanady et al. 2014). Distances are given in Å units; Mn, Ca, Ag, O, N, F, S, and C are shown in purple, green, gray, orange, blue, green yellow, bright yellow, and yellow, respectively. For clarity, all hydrogen atoms are not shown

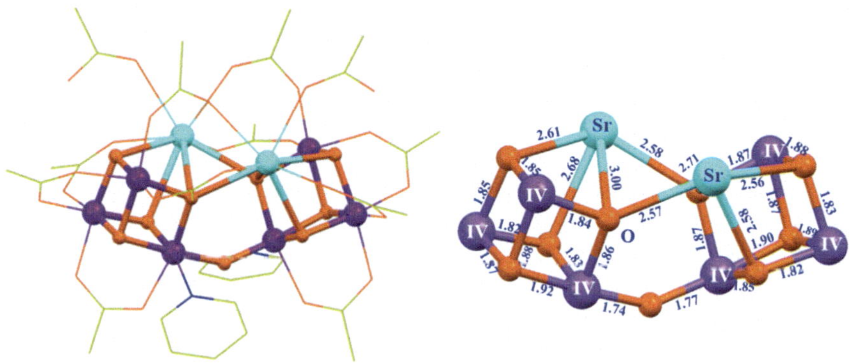

A B

Fig. 9 Structure of the $Mn_6Sr_2O_9$-complex (Chen et al. 2014). (**a**) Whole structure of the $Mn_6Sr_2O_9$-complex; (**b**) core structure of the $Mn_6Sr_2O_9$-complex; Mn, Sr, O, N, and C are shown in purple, cyan, orange, blue, and yellow, respectively. For clarity, all the methyl groups and all the hydrogen atoms are not shown

the peripheral ligands of the $Mn_3Ca_2O_4$-complex are pivalic anions or neutral pivalic acid, closely mimicking the peripheral carboxylate ligands of the OEC in PSII. In these artificial complexes, all the manganese ions are in IV oxidation state, and the typical bond lengths for Mn-O and Ca-O are in the range of 1.8–1.9 Å and 2.4–2.7 Å, respectively. The distances of the Mn…Mn and Mn…Ca are in the range of 2.7–2.8 Å and 3.2–3.5 Å, respectively. In 2014, Zhang's group reported a heterometallic cluster containing two $Mn^{IV}_3SrO_4$-cluster linked by one μ_2-oxide bridge (Chen et al. 2014), which mimics the three types of oxide bridges (μ_2-oxide, μ_3-oxide, and μ_4-oxide) and the Mn_3SrO_4 cubane of the Sr^{2+}-containing OEC (Koua et al. 2013) at the same time.

It should be pointed out that although all complexes shown in Figs. 7, 8, and 9 have mimicked some key structural motifs of the Mn_3CaO_4 or Mn_3SrO_4 cubane in the OEC, until recently, it remains a great challenge to synthesize the entire Mn_4Ca-cluster with similar ligands as seen in the OEC of PSII.

5 Closer Mimicking of the OEC

Inspired by the assembly processes of the OEC in PSII (Dasgupta et al. 2008; Zhang et al. 2017), we successfully prepared the first artificial Mn_4Ca-cluster (Zhang et al. 2015) (Fig. 10b, d). It was synthesized through a two-step procedure using inexpensive commercial chemicals. The first step was to synthesize a precursor through a reaction of $Bu^n_4NMnO_4$ (Bu^n = n-butyl), $Mn(CH_3CO_2)_2$•$(H_2O)_4$, and $Ca(CH_3CO_2)_2$•H_2O (molar ratio of 4:1:1) in boiling acetonitrile in the presence of an excess of pivalic acid. The second step was to treat the precursor with organic base (pyridine) in ethyl acetate, leading to the formation of final product.

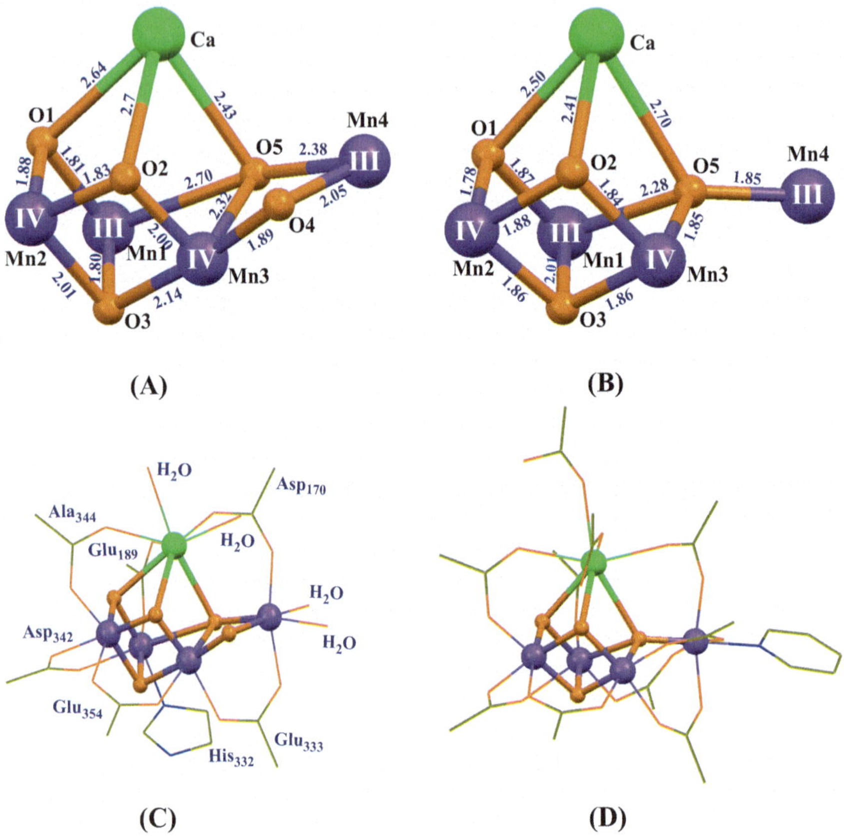

Fig. 10 Structures of the natural OEC (Suga et al. 2015) (**a**, **c**) and the artificial Mn_4Ca-cluster (Zhang et al. 2015) (**b**, **d**). Distances are given in Å units; Mn, Ca, O, N, and C are shown in purple, green, orange, blue, and yellow, respectively. For clarity, all hydrogen atoms are not shown

The core structure of the artificial Mn_4CaO_4-complex contains a Mn_3CaO_4 cubane attached with a dangler Mn ion forming an asymmetric Mn_4CaO_4-cluster, which is exactly the same as the OEC's structure proposed by Ferreira et al. in 2004. The surrounding ligands of the Mn_4CaO_4-cluster are provided by eight $(CH_3)_3CCO_2^-$ anions and three exchangeable neutral ligands (two pivalic acid and one pyridine molecules), similar to the peripheral ligands of the OEC (Fig. 10c).

It should be pointed out that the structure of the artificial Mn_4CaO_4-cluster is well-defined and effect from the X-ray radiation reduction is limited mainly due to the absence of water as solvent in the crystal of the artificial Mn_4CaO_4-complex. BVS calculations have clearly shown that the oxidation states of the four manganese ions of the artificial Mn_4CaO_4-complex are (III, III, IV, IV), which are essentially the same as that proposal for the S_1 state (Fig. 2).

The similarity between the artificial Mn_4CaO_4-cluster and OEC was further supported by the observation of four redox transitions revealed by cyclic voltammo-

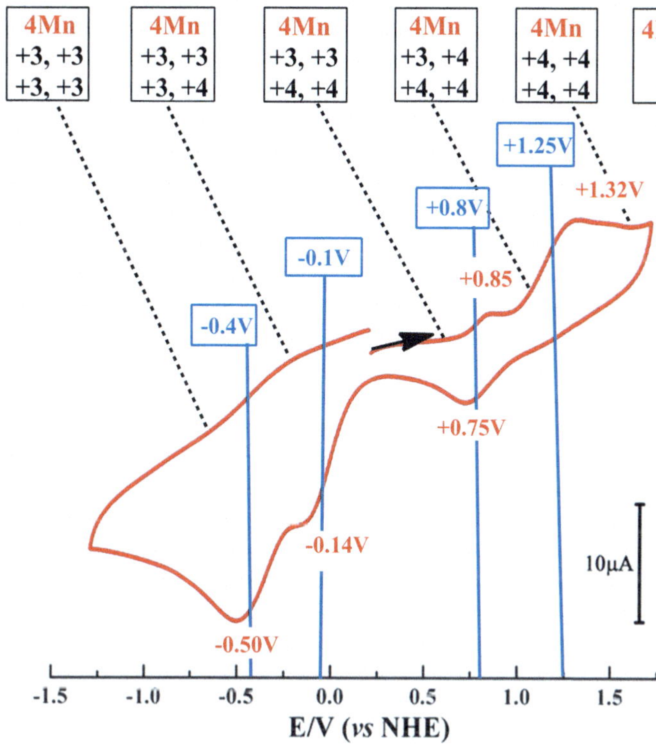

Fig. 11 Cyclic voltammogram (CV) of artificial Mn_4CaO_4-complex in Fig. 10 (Zhang et al. 2015). Potentials vs. NHE; scan rate of 100 mV/s. Peak positions of CV waves are in red and estimated midpoint potentials are in blue. The likely oxidation states of the four Mn ions in various oxidation states of artificial complex are indicated using black numbers

gram (CV) measurements (Fig. 11). The redox potential of ~ 0.8 eV (vs. NHE) for the $S_1 \rightarrow S_2$ transition of artificial Mn_4CaO_4-complex is close to the estimated potential of the corresponding OEC redox transition ($\geq$0.9 V) (Dau and Zaharieva 2009; Vass and Styring 1991) but is remarkably different from that of the previously Mn_3CaO_4-complex without a dangling Mn ion (Kanady et al. 2011). This result indicates that the dangler manganese ion could play a crucial role in tuning the redox potential of Mn_4CaO_4-cluster.

Moreover, the one-electron oxidation of the artificial Mn_4Ca-complex gave rise to two distinct EPR signals ($g = 4.9$ and $g = 2.0$) (Zhang et al. 2015) (Fig. 12a), similar to the $g \approx 4$ and $g = 2.0$ EPR signals (Fig. 12b) observed in PSII for the OEC in the S_2 state (Boussac and Rutherford 2000; Boussac et al. 2018; Dismukes and Siderer 1981; Pantazis et al. 2012; Peloquin and Britt 2001). In the field of photosynthetic research, the latter two EPR signals have been considered as fingerprint spectroscopic characteristics to evaluate the structure and function of the OEC. Therefore, the experimental observation of two EPR signals strongly suggested that the artificial Mn_4Ca-cluster could have similar electronic structures as that of the OEC.

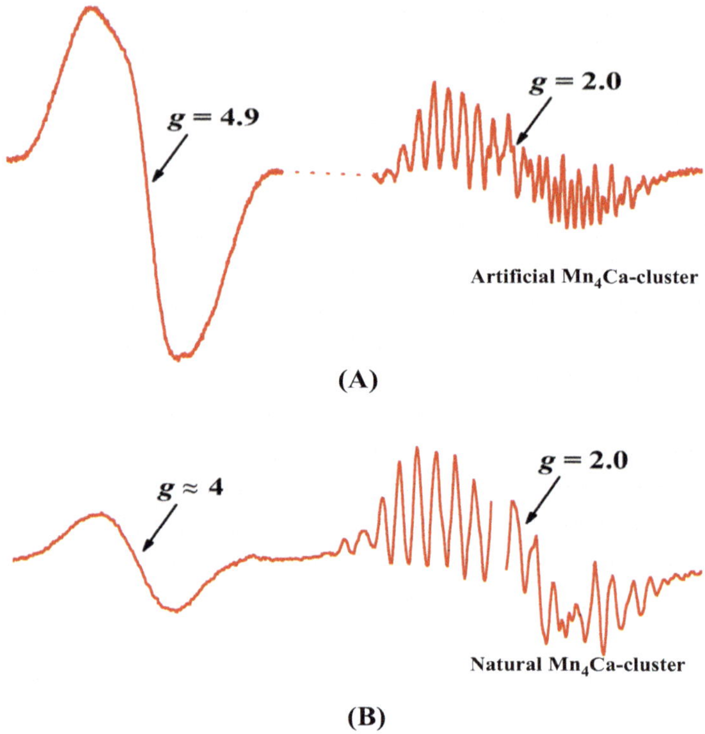

Fig. 12 EPR signals induced from the one-electron oxidation of the artificial Mn_4CaO_4-cluster (Zhang et al. 2015) (**a**) and the OEC in S_2 states (Peloquin and Britt 2001) (**b**). Catalytic activity of the artificial Mn4CaO4-cluster

Considering the close mimicking of both the atomic and electronic structures as well as the redox properties of the OEC, we expect that the artificial Mn_4CaO_4-cluster should be able to serve as a catalyst for the water-splitting reaction. Indeed, a remarkable catalytic current was observed during the CV measurement in the presence of artificial Mn_4CaO_4-complex and 1% water in acetonitrile (Zhang et al. 2015) (Fig. 13a). The artificial Mn_4CaO_4-complexes can also catalyze oxygen-evolving reaction efficiently by using $(CH_3)_3COOH$ (tert-butyl hydroperoxide) as an oxidant in acetonitrile with the presence of 4.5% water (Chen et al. 2017) (Fig. 13b). However, these artificial complexes are very sensitive to the experimental conditions. Quantification of the catalytic reaction has been difficult due to the rapid degradation of the catalyst in the presence of water in solution, making it difficult to quantify the catalytic reaction. The calcium in the artificial complex has been found to be easily dissociated, leading to the formation of multi-manganese complexes.

To improve the stability of the artificial Mn_4CaO_4-cluster, recently, we have prepared two new Mn_4CaO_4-clusters with exchangeable solvent, acetonitrile and N,N-dimethylformamide (DMF) (Chen et al. 2019), on the calcium as shown in Fig. 14. Remarkably, the replacement of one or two ligands on the calcium by organic sol-

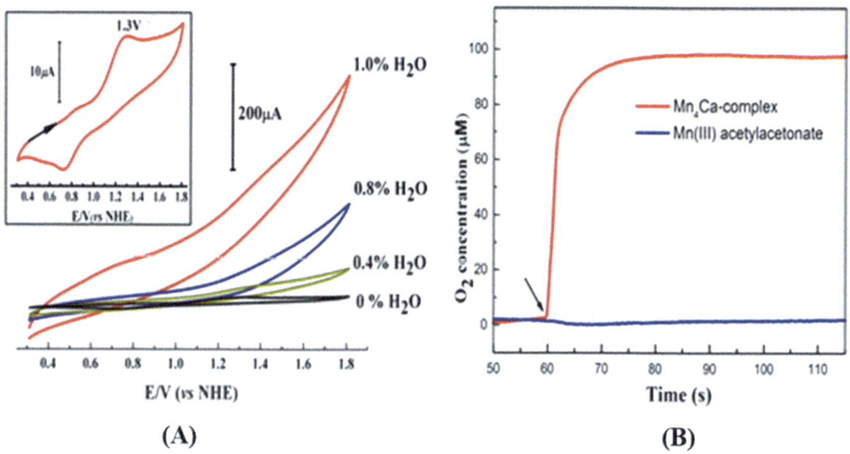

Fig. 13 Catalytic properties of the artificial Mn_4CaO_4-cluster. (**a**) CV measurement of artificial Mn_4CaO_4-complex in acetonitrile with different amounts of H_2O (Zhang et al. 2015). The inset shows the CV without H_2O (0% H_2O). Potentials are referenced to NHE; scan rate, 100 mV/s. (**b**) Oxygen-evolving reaction catalyzed by artificial Mn_4Ca-complex (50 μM) in acetonitrile with the presence of $(CH_3)_3COOH$ (1 M) and 4.5% water in acetonitrile at 25 °C (Chen et al. 2017). A Clark-type O_2 electrode was used to detect the release of dioxygen

vent molecules does not modify the core structure and oxidation valences of the four manganese ions, as well as main peripheral environment ligands. More importantly, these new Mn_4CaO_4-clusters become much stable in the polar solvent, which may provide great opportunity to investigate the catalytic activity of the artificial Mn_4CaO_4-cluster in the future.

6 Implications for the Mechanism of the Water-Splitting Reaction in OEC

Artificial Mn_4CaO_4-clusters shown in Figs. 10 and 14 have mimicked the main structural motifs of the OEC, which could provide distinct structural insight into understanding the principle of the OEC. Firstly, the successful isolation of different artificial Mn_4CaO_4-complexes demonstrates that the Mn_4CaO_4-cluster is thermodynamically stable, supporting the proposal that the Mn_4CaO_4-cluster could be an evolutionary origin of the natural OEC by Barber previously (Barber 2016). Secondly, from the structure of artificial Mn_4CaO_4-cluster, one can clearly see that the μ_4-O^{2-} bridge (O5) is tightly bound to four metal ions (one calcium and three manganese ions). This is a position less prone to be removed or replaced, indicating that the similar μ_4-O^{2-} bridge (O5) in OEC could not be directly involved in serving as an oxygen source to form the O-O bond. Interestingly, Kawashima et al. recently have proposed that the μ_2-O^{2-} bridge (O4) in OEC (Fig. 5), instead, may play a role as the

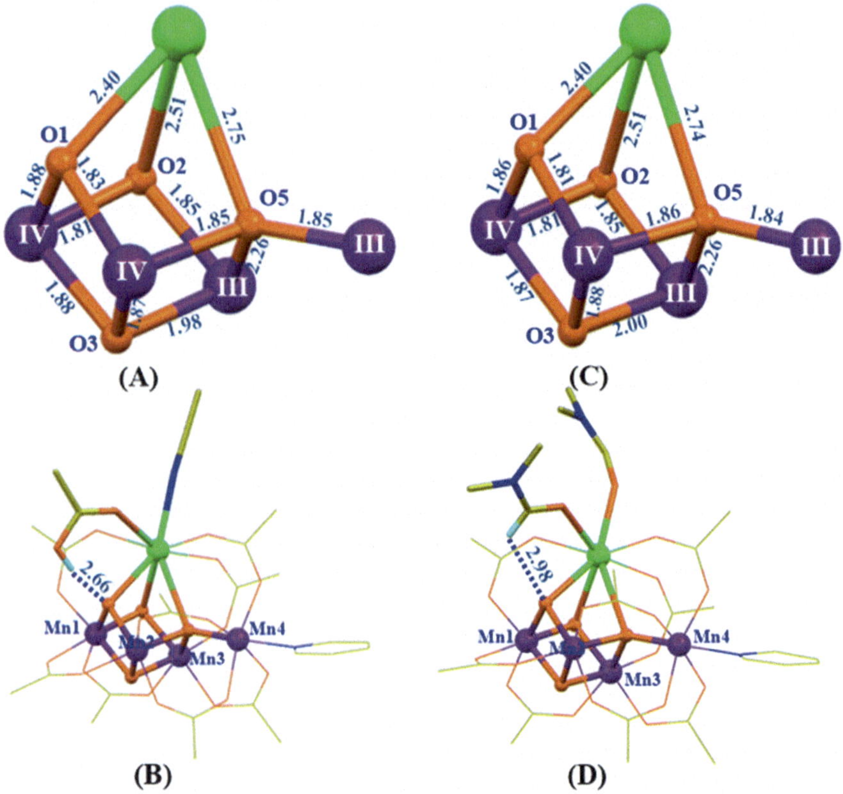

Fig. 14 Crystal structures of two new artificial Mn_4CaO_4-complexes with exchangeable solvent, acetonitrile (**a**, **b**) and N,N-dimethylformamide (**c**, **d**) on the calcium (Chen et al. 2019). Dashed lines indicate the hydrogen bonds. Mn, Ca, O, N, C, and H are shown in purple, green, orange, blue, yellow, and cyan, respectively. Distances are given in Å. The oxidation states of four manganese ions are obtained by the BVS calculation. For clarity, all the methyl groups of pivalic acid are omitted, and only the H atoms involved to form hydrogen bonds are shown

substrate binding site to form the O-O bond (Kawashima et al. 2018). In this proposal and some previous suggestions (Barber 2017; Chen et al. 2015; Vinyard et al. 2015), the Mn_4CaO_4 fragment does not undertake significant changes during the water-splitting reaction, which is consistent with the isolation of the stable artificial Mn_4CaO_4-cluster described here. It should be pointed out that the μ_2-O^{2-} (O4) in OEC is absent in the artificial Mn_4CaO_4-cluster, which is replaced by a bridging carboxyl group in the latter. Obviously, the further investigation of this missing μ_2-O^{2-} bridge in artificial Mn_4Ca-cluster would provide new insights into the mechanism for the O-O bond formation during the OEC turnover.

7 Conclusion

Although the crystallographic studies of PSII have revealed the structure of the OEC, the detailed mechanism for the water-splitting reaction is still elusive due to the complexity of the large protein environment and structural uncertainty of the intermediate states of the OEC during its turnover. To understand the structure-function relationship and the catalytic mechanism of this natural Mn_4CaO_5-cluster, as well as to develop efficient man-made water-splitting catalysts in artificial photosynthesis, many artificial complexes have been synthesized to mimic the OEC in the laboratory. One of the most important advances is the synthesis of a series of the artificial Mn_4CaO_4-clusters that closely mimics both geometric and electronic structures of the OEC, which provides a structurally well-defined chemical model to investigate the structure-function relationship of the natural Mn_4CaO_5-cluster, and sheds new insights into the mechanism of the water-splitting reaction in PSII. This new advance may open new avenues to develop efficient artificial catalysts for the water-splitting reaction by using earth-abundant and nontoxic chemical elements in the future.

Acknowledgment This work was supported by the National Key Research and Development Program of China (No. 2017YFA0503704), the Strategic Priority Research Program of Chinese Academy of Sciences (No. XDA21010212; XDB17030600), and National Natural Science Foundation of China (No. 31770258).

References

Amin, M., Badawi, A., & Obayya, S. S. (2016). Radiation damage in XFEL: Case study from the oxygen-evolving complex of photosystem II. *Scientific Reports, 6*, 36492.

Askerka, M., Brudvig, G. W., & Batista, V. S. (2017). The O_2-evolving complex of photosystem II: Recent insights from quantum mechanics/molecular mechanics (QM/MM), extended X-ray absorption fine structure (EXAFS), and femtosecond X-ray crystallography data. *Accounts of Chemical Research, 50*, 41–48.

Barber, J. (2009). Photosynthetic energy conversion: Natural and artificial. *Chemical Society Reviews, 38*, 185–196.

Barber, J. (2016). Mn_4Ca cluster of photosynthetic oxygen-evolving center: Structure, function and evolution. *Biochemistry, 55*, 5901–5906.

Barber, J. (2017). A mechanism for water splitting and oxygen production in photosynthesis. *Nature Plants, 3*, 17041.

Boussac, A., & Rutherford, A. W. (2000). Comparative study of the g=4.1 EPR signals in the S_2 state of photosystem II. *Biochimica et Biophysica Acta, 1457*, 145–156.

Boussac, A., Ugur, I., Marion, A., Sugiura, M., Kaila, V. R. I., & Rutherford, A. W. (2018). The low spin-high spin equilibrium in the S_2-state of the water oxidizing enzyme. *Biochimica et Biophysica Acta, 1859*, 342–356.

Brown, I. D. (2009). Recent developments in the methods and applications of the bond valence model. *Chemical Reviews, 109*, 6858–6919.

Cardona, T., Sedoud, A., Cox, N., & Rutherford, A. W. (2012). Charge separation in photosystem II: A comparative and evolutionary overview. *Biochimica et Biophysica Acta, 1817*, 26–43.

Chakov, N. E., Abboud, K. A., Zakharov, L. N., Rheingold, A. L., Hendrickson, D. N., & Christou, G. (2003). Reaction of $[Mn_{12}O_{12}(O_2CR)_{16}(H_2O)_4]$ single-molecule magnets with non-carboxylate ligands. *Polyhedron, 22*, 1759–1763.

Chakov, N. E., Thuijs, A. E., Wernsdorfer, W., Rheingold, A. L., Abboud, K. A., & Christou, G. (2016). Unusual $Mn(III/IV)_4$ cubane and $Mn(III)_{16}M_4$ (M = Ca, Sr) looplike clusters from the use of dimethylarsinic acid. *Inorganic Chemistry, 55*, 8468–8477.

Chen, C., Zhang, C., Dong, H., & Zhao, J. (2014). A synthetic model for the oxygen-evolving complex in Sr^{2+}-containing photosystem II. *Chemical Communications, 50*, 9263–9265.

Chen, C., Zhang, C., Dong, H., & Zhao, J. (2015). Artificial synthetic Mn^{IV}Ca-oxido complexes mimic the oxygen-evolving complex in photosystem II. *Dalton Transactions, 44*, 4431–4435.

Chen, C., Li, Y., Zhao, G., Yao, R., & Zhang, C. (2017). Natural and artificial Mn_4Ca cluster for the water splitting reaction. *ChemSusChem, 10*, 4403–4408.

Chen, C., Chen, Y., Yao, R., Li, Y., & Zhang, C. (2019). Artificial Mn_4Ca clusters with exchangeable solvent molecules mimicking the oxygen-evolving center in photosynthesis. *Angewandte Chemie, International Edition, 58*, 3939–3942.

Cinco, R. M., Robblee, J. H., Rompel, A., Fernandez, C., Yachandra, V. K., Sauer, K., & Klein, M. P. (1998). Strontium EXAFS reveals the proximity of calcium to the manganese cluster of oxygen-evolving photosystem II. *The Journal of Physical Chemistry. B, 102*, 8248–8256.

Cox, N., Pantazis, D. A., Neese, F., & Lubitz, W. (2013). Biological water oxidation. *Accounts of Chemical Research, 46*, 1588–1596.

Cox, N., Retegan, M., Neese, F., Pantazis, D. A., Boussac, A., & Lubitz, W. (2014). Electronic structure of the oxygen-evolving complex in photosystem II prior to O-O bond formation. *Science, 345*, 804–808.

Dasgupta, J., Ananyev, G. M., & Dismukes, G. C. (2008). Photoassembly of the water-oxidizing complex in photosystem II. *Coordination Chemistry Reviews, 252*, 347–360.

Dau, H., & Haumann, M. (2007). Eight steps preceding O–O bond formation in oxygenic photosynthesis—A basic reaction cycle of the photosystem II manganese complex. *Biochimica et Biophysica Acta, 1767*, 472–483.

Dau, H., & Haumann, M. (2008). The manganese complex of photosystem II in its reaction cycle—Basic framework and possible realization at the atomic level. *Coordination Chemistry Reviews, 252*, 273–295.

Dau, H., & Zaharieva, I. (2009). Principles, efficiency and blueprint character of solar-energy conversion in photosynthetic water oxidation. *Accounts of Chemical Research, 42*, 1861–1870.

Dau, H., Grundmeier, A., Loja, P., & Haumann, M. (2008). On the structure of the manganese complex of photosystem II: Extended-range EXAFS data and specific atomic-resolution models for four S-states. *Philosophical Transactions of the Royal Society of London B, 363*, 1237–1244.

Diner, B. A., & Britt, R. D. (2005). The redox-active tyrosines Y_Z and Y_D. In T. J. Wydrzynski & K. Satoh (Eds.), *Photosystem II: The light-driven water: Plastoquinone oxidoreductase* (pp. 207–233). Dordrecht: Springer.

Diner, B. A., & Rappaport, F. (2002). Structure, dynamics, and energetics of the primary photochemistry of photosystem II of oxygenic photosynthesis. *Annual Review of Plant Biology, 53*, 551–580.

Dismukes, G. C., & Siderer, Y. (1981). Intermediates of a polynuclear manganese center involved in photosynthetic oxidation of water. *Proceedings of the National Academy of Sciences of the United States of America, 78*, 274–278.

Dismukes, G. C., Brimblecombe, R., Felton, G. A. N., Pryadun, R. S., Sheats, J. E., Spiccia, L., & Swiegers, G. F. (2009). Development of bioinspired Mn_4O_4-cubane water oxidation catalysts: Lessons from photosynthesis. *Accounts of Chemical Research, 42*, 1935–1943.

Ferreira, K. N., Iverson, T. M., Maghlaoui, K., Barber, J., & Iwata, S. (2004). Architecture of the photosynthetic oxygen-evolving center. *Science, 303*, 1831–1838.
Gatt, P., Petrie, S., Stranger, R., & Pace, R. J. (2012). Rationalizing the 1.9 Å crystal structure of photosystem II—A remarkable Jahn-Teller balancing act induced by a single proton transfer. *Angewandte Chemie, International Edition, 51*, 12025–12028.
Gerey, B., Goure, E., Fortage, J., Pecaut, J., & Collomb, M. N. (2016). Manganese-calcium/strontium heterometallic compounds and their relevance for the oxygen-evolving center of photosystem II. *Coordination Chemistry Reviews, 319*, 1–24.
Grabolle, M., Haumann, M., Müller, C., Liebisch, P., & Dau, H. (2006). Rapid loss of structural motifs in the manganese complex of oxygenic photosynthesis by X-ray irradiation at 10-300K. *The Journal of Biological Chemistry, 281*, 4580–4588.
Guskov, A., Kern, J., Gabdulkhakov, A., Broser, M., Zouni, A., & Saenger, W. (2009). Cyanobacterial photosystem II at 2.9-Å resolution and the role of quinones, lipids, channels and chloride. *Nature Structural & Molecular Biology, 16*, 334–342.
Haumann, M., Liebisch, P., Muller, C., Barra, M., Grabolle, M., & Dau, H. (2005). Photosynthetic O_2 formation tracked by time-resolved x-ray experiments. *Science, 310*, 1019–1021.
Hellmich, J., Bommer, M., Burkhardt, A., Ibrahim, M., Kern, J., Meents, A., Müh, F., Dobbek, H., & Zouni, A. (2014). Native-like photosystem II superstructure at 2.44Å resolution through detergent extraction from the protein crystal. *Structure, 22*, 1607–1615.
Holzwarth, A. R., Müller, M. G., Reus, M., Nowaczyk, M., Sander, J., & Rögner, M. (2006). Kinetics and mechanism of electron transfer in intact photosystem II and in the isolated reaction center: Pheophytin is the primary electron acceptor. *Proceedings of the National Academy of Sciences of the United States of America, 103*, 6895–6890.
Ishikita, H. (2019). Protein environment that facilitates proton transfer and electron transfer in photosystem II. In J. Barber, A. V. Ruban, & P. J. Nixon (Eds.), *Oxygen production and reduction in artificial and natural systems* (pp. 191–208). Singapore: World Scientific Publishing Co. Pte Ltd.
Isobe, H., Shoji, M., Yamanaka, S., Umena, Y., Kawakami, K., Kamiya, N., Shen, J. R., & Yamaguchi, K. (2012). Theoretical illumination of water-inserted structures of the $CaMn_4O_5$-cluster in the S_2 and S_3 states of oxygen-evolving complex of photosystem II: Full geometry optimizations by B3LYP hybrid density functional. *Dalton Transactions, 41*, 13727–13740.
Junge, W. (2019). Oxygenic photosynthesis: History, status and perspective. *Quarterly Reviews of Biophysics, 52*, e1.
Kamiya, N., & Shen, J. R. (2003). Crystal structure of oxygen-evolving photosystem II from *Thermosynechococcus vulcanus* at 3.7 Å resolution. *Proceedings of the National Academy of Sciences of the United States of America, 100*, 98–103.
Kanady, J. S., Tsui, E. Y., Day, M. W., & Agapie, T. (2011). A synthetic model of the Mn_3Ca subsite of the oxygen-evolving complex in photosystem II. *Science, 333*, 733–736.
Kanady, J. S., Lin, P. H., Carsch, K. M., Nielsen, R. J., Takase, M. K., Goddard, W. A., & Agapie, T. (2014). Toward models for the full oxygen-evolving complex of photosystem II by ligand coordination to lower the symmetry of the Mn_3CaO_4 cubane: Demonstration that electronic effects facilitate binding of a fifth metal. *Journal of the American Chemical Society, 136*, 14373–14376.
Kärkäs, M. D., Verho, O., Johnston, E. V., & Åkermark, B. (2014). Artificial photosynthesis: Molecular systems for catalytic water oxidation. *Chemical Reviews, 114*, 11863–12001.
Kawashima, K., Takaoka, T., Kimura, H., Saito, K., & Ishikita, H. (2018). O_2 evolution and recovery of the water-oxidizing enzyme. *Nature Communications, 9*, 1247.
Kern, J., Chatterjee, R., Young, I. D., Fuller, F. D., Lassalle, L., Ibrahim, M., Gul, S., Fransson, T., Brewster, A. S., Alonso-Mori, R., et al. (2018). Structures of the intermediates of Kok's photosynthetic water oxidation clock. *Nature, 563*, 421–425.
Kok, B., Forbush, B., & McGloin, M. (1970). Cooperation of charges in photosynthetic O_2 evolution. I. A linear four step mechanism. *Photochemistry and Photobiology, 11*, 457–475.

Koua, F. H. M., Umena, Y., Kawakami, K., & Shen, J. R. (2013). Structure of Sr-substituted photosystem II at 2.1 Å resolution and its implications in the mechanism of water oxidation. *Proceedings of the National Academy of Sciences of the United States of America, 110*, 3889–3894.

Krewald, V., Retegan, M., Cox, N., Messinger, J., Lubitz, W., DeBeer, S., Neese, F., & Pantazis, D. A. (2015). Metal oxidation states in biological water splitting. *Chemical Science, 6*, 1676–1695.

Krewald, V., Retegan, M., Neese, F., Lubitz, W., Pantazis, D. A., & Cox, N. (2016). Spin state as a marker for the structural evolution of nature's water splitting catalyst. *Inorganic Chemistry, 55*, 488–501.

Limburg, J., Vrettos, J. S., Liable-Sands, L. M., Rheingold, A. L., Crabtree, R. H., & Brudvig, G. W. (1999). A functional model for O-O bond formation by the O_2-evolving complex in photosystem II. *Science, 283*, 1524–1527.

Liu, W., & Thorp, H. H. (1993). Bond valence sum analysis of metal-ligand bond lengths in metalloenzymes and model complexes. 2. Refined distances and other enzymes. *Inorganic Chemistry, 32*, 4102–4105.

Lubitz, W., Chrysina, M., & Cox, N. (2019). Water oxidation in photosystem II. *Photosynthesis Research, 142*, 105–125.

McEvoy, J. P., & Brudvig, G. W. (2006). Water-splitting chemistry of photosystem II. *Chemical Reviews, 106*, 4455–4483.

Mukherjee, S., Stull, J. A., Yano, J., Stamatatos, T. C., Pringouri, K., Stich, T. A., Abboud, K. A., Britt, R. D., Yachandra, V. K., & Christou, G. (2012). Synthetic model of the asymmetric [Mn_3CaO_4] cubane core of the oxygen-evolving complex of photosystem II. *Proceedings of the National Academy of Sciences of the United States of America, 109*, 2257–2262.

Mukhopadhyay, S., Mandal, S. K., Bhaduri, S., & Armstrong, W. H. (2004). Manganese clusters with relevance to photosystem II. *Chemical Reviews, 104*, 3981–4026.

Mullins, C. S., & Pecoraro, V. L. (2008). Reflections on small molecule manganese models that seek to mimic photosynthetic water oxidation chemistry. *Coordination Chemistry Reviews, 252*, 416–443.

Najafpour, M. M., Renger, G., Hołyńska, M., Moghaddam, A. N., Aro, E. M., Carpentier, R., Nishihara, H., Eaton-Rye, J. J., Shen, J. R., & Allakhverdiev, S. I. (2016). Manganese compounds as water-oxidizing catalysts: From the natural water-oxidizing complex to nanosized manganese oxide structures. *Chemical Reviews, 116*, 2886–2936.

Nelson, N., & Yocum, C. F. (2006). Structure and function of photosystem I and II. *Annual Review of Plant Biology, 57*, 521–565.

Pantazis, D. A. (2018). Missing pieces in the puzzle of biological water oxidation. *ACS Catalysis, 8*, 9477–9507.

Pantazis, D. A., Ames, W., Cox, N., Lubitz, W., & Neese, F. (2012). Two interconvertible structures that explain the spectroscopic properties of the oxygen-evolving complex of photosystem II in the S_2 state. *Angewandte Chemie, International Edition, 51*, 9935–9940.

Paul, S., Neese, F., & Pantazis, D. A. (2017). Structural models of the biological oxygen-evolving complex: Achievements, insights, and challenges for biomimicry. *Green Chemistry, 19*, 2309–2325.

Pauling, L. (1929). The principles determining the structure of complex ionic crystals. *Journal of the American Chemical Society, 51*, 1010–1026.

Peloquin, J. M., & Britt, R. D. (2001). EPR/ENDOR characterization of the physical and electronic structure of the OEC Mn-cluster. *Biochimica et Biophysica Acta, 1503*, 96–111.

Peloquin, J. M., Campbell, K. A., Randall, D. W., Evanchik, M. A., Pecoraro, V. L., Armstrong, W. H., & Britt, R. D. (2000). ^{55}Mn ENDOR of the S_2-state multiline EPR signal of photosystem II: Implications on the structure of the tetranuclear Mn cluster. *Journal of the American Chemical Society, 122*, 10926–10942.

Perez-Navarro, M., Neese, F., Lubitz, W., Pantazis, D. A., & Cox, N. (2016). Recent developments in biological water oxidation. *Current Opinion in Chemical Biology, 31*, 113–119.

Petrouleas, V., & Crofts, A. R. (2005). The iron-quinone acceptor complex. In T. J. Wydrzynski & K. Satoh (Eds.), *Photosystem II: The light-driven water: Plastoquinone oxidoreductase* (pp. 177–206). Springer: Dordrecht.

Renger, G. (2012). Mechanism of light induced water splitting in photosystem II of oxygen evolving photosynthetic organisms. *Biochimica et Biophysica Acta, 1817*, 1164–1176.

Ruettinger, W. F., Campana, C., & Dismukes, G. C. (1997). Synthesis and characterization of $Mn_4O_4L_6$ complexes with cubane-like core structure: A new class of models of the active site of the photosynthetic water oxidase. *Journal of the American Chemical Society, 119*, 6670–6671.

Saito, K., Rutherford, A. W., & Ishikita, H. (2013). Mechanism of proton-coupled quinone reduction in photosystem II. *Proceedings of the National Academy of Sciences of the United States of America, 110*, 954–959.

Satoh, K., Wydrzynski, T. J., & Govindjee. (2005). Introduction to photosystem II. In T. J. Wydrzynski & K. Satoh (Eds.), *Photosystem II: The light-driven water: Plastoquinone oxidoreductase* (pp. 11–22). Springer: Dordrecht.

Sauer, K., Yano, J., & Yachandra, V. K. (2008). X-ray spectroscopy of the photosynthetic oxygen-evolving complex. *Coordination Chemistry Reviews, 252*, 318–335.

Shen, J. R. (2015). The structure of photosystem II and the mechanism of water oxidation in photosynthesis. *Annual Review of Plant Biology, 66*, 23–48.

Siegbahn, P. E. M. (2009). Structures and energetics for O_2 formation in photosystem II. *Accounts of Chemical Research, 42*, 1871–1880.

Siegbahn, P. E. M. (2013). Water oxidation mechanism in photosystem II, including oxidations, proton release pathways, O-O bond formation and O_2 release. *Biochimica et Biophysica Acta, 1827*, 1003–1019.

Siegbahn, P. E. M. (2017). Nucleophilic water attack is not a possible mechanism for O-O bond formation in photosystem II. *Proceedings of the National Academy of Sciences of the United States of America, 114*, 4966–4968.

Styring, S., Sjöholm, J., & Mamedov, F. (2012). Two tyrosines that changed the world: Interfacing the oxidizing power of photochemistry to water splitting in photosystem II. *Biochimica et Biophysica Acta, 1817*, 76–87.

Su, X., Ma, J., Wei, X., Cao, P., Zhu, D., Chang, W., Liu, Z., Zhang, X., & Li, M. (2017). Structure and assembly mechanism of plant $C_2S_2M_2$-type PSII-LHCII supercomplex. *Science, 357*, 815–820.

Suga, M., Akita, F., Hirata, K., Ueno, G., Murakami, H., Nakajima, Y., Shimizu, T., Yamashita, K., Yamamoto, M., Ago, H., et al. (2015). Native structure of photosystem II at 1.95Å resolution revealed by a femtosecond X-ray laser. *Nature, 517*, 99–103.

Suga, M., Akita, F., Sugahara, M., Kubo, M., Nakajima, Y., Nakane, T., Yamashita, K., Umena, Y., Nakabayashi, M., Yamane, T., et al. (2017). Light-induced structural changes and the site of O=O bond formation in PSII caught by XFEL. *Nature, 543*, 131–135.

Suga, M., Akita, F., Yamashita, K., Nakajima, Y., Ueno, G., Li, H., Yamane, T., Hirata, K., Umena, Y., Yonekura, S., et al. (2019). An oxyl/oxo mechanism for oxygen-oxygen coupling in PSII revealed by an x-ray free-electron laser. *Science, 366*, 334–338.

Tanaka, A., Fukushima, Y., & Kamiya, N. (2017). Two different structures of the oxygen-evolving complex in the same polypeptide frameworks of photosystem II. *Journal of the American Chemical Society, 139*, 1718–1721.

Tommos, C., & Babcock, G. T. (1998). Oxygen production in nature: A light-driven metalloradical enzyme process. *Accounts of Chemical Research, 31*, 18–25.

Tsui, E. Y., & Agapie, T. (2013). Reduction potentials of heterometallic manganese–oxido cubane complexes modulated by redox-inactive metals. *Proceedings of the National Academy of Sciences of the United States of America, 110*, 10084–10088.

Tsui, E. Y., Kanady, J. S., & Agapie, T. (2013). Synthetic cluster models of biological and heterogeneous manganese catalysts for O_2 evolution. *Inorganic Chemistry, 52*, 13833–13848.

Umena, Y., Kawakami, K., Shen, J. R., & Kamiya, N. (2011). Crystal structure of oxygen-evolving photosystem II at a resolution of 1.9Å. *Nature, 473*, 55–60.

van Gorkom, H. J., & Yocum, C. F. (2005). The calcium and chloride cofactor. In T. J. Wydrzynski & K. Satoh (Eds.), *Photosystem II: The light-driven water: Plastoquinone oxidoreductase* (pp. 307–328). Springer: Dordrecht.

Vass, I., & Styring, S. (1991). pH-dependent charge equilibria between tyrosine-D and the S states in photosystem II. Estimation of relative midpoint redox potentials. *Biochemistry, 30*, 830–839.

Vinyard, D. J., Khan, S., & Brudvig, G. W. (2015). Photosynthetic water oxidation: Binding and activation of substrate water for O-O bond formation. *Faraday Discussions, 185*, 37–50.

Vrettos, J. S., Limburg, J., & Brudvig, G. W. (2001). Mechanism of photosynthetic water oxidation: Combining biophysical studies of photosystem II with inorganic model chemistry. *Biochimica et Biophysica Acta, 1503*, 229–245.

Wei, X., Su, X., Cao, P., Liu, X., Chang, W., Li, M., Zhang, X., & Liu, Z. (2016). Structure of spinach photosystem II–LHCII supercomplex at 3.2Å resolution. *Nature, 534*, 69–74.

Yamaguchi, K., Shoji, M., Isobe, H., Yamanaka, S., Kawakami, T., Yamada, S., Katouda, M., & Nakajima, T. (2018). Theory of chemical bonds in metalloenzymes XXI. Possible mechanisms of water oxidation in oxygen evolving complex of photosystem II. *Molecular Physics, 116*, 717–745.

Yano, J., & Yachandra, V. (2014). Mn_4Ca-cluster in photosynthesis: Where and how water is oxidized to dioxygen. *Chemical Reviews, 114*, 4175–4205.

Yano, J., Kern, J., Irrgang, K. D., Latimer, M. J., Bergmann, U., Glatzel, P., Pushkar, Y., Biesiadka, J., Loll, B., Sauer, K., et al. (2005). X-ray damage to the Mn_4Ca complex in single crystals of photosystem II: A case study for metalloprotein crystallography. *Proceedings of the National Academy of Sciences of the United States of America, 102*, 12047–12052.

Yano, J., Kern, J., Sauer, K., Latimer, M. J., Pushkar, Y., Biesiadka, J., Loll, B., Saenger, W., Messinger, J., Zouni, A., et al. (2006). Where water is oxidized to dioxygen: Structure of the photosynthetic Mn_4Ca cluster. *Science, 314*, 821–825.

Yocum, C. F. (2008). The calcium and chloride requirements of the O_2 evolving complex. *Coordination Chemistry Reviews, 252*, 296–305.

Young, I. D., Ibrahim, M., Chatterjee, R., Gul, S., Fuller, F. D., Koroidov, S., Brewster, A. S., Tran, R., Alonso-Mori, R., Kroll, T., et al. (2016). Structure of photosystem II and substrate binding at room temperature. *Nature, 540*, 453–457.

Zhang, C. (2007). Low-barrier hydrogen bond plays key role in active photosystem II – A new model for photosynthetic water oxidation. *Biochimica et Biophysica Acta, 1767*, 493–499.

Zhang, C. (2015). The first artificial Mn_4Ca-cluster mimicking the oxygen-evolving center in photosystem II. *Science China. Life Sciences, 58*, 816–817.

Zhang, C. (2016). From natural photosynthesis to artificial photosynthesis. *Scientia Sinica Chimica, 46*, 1101–1109.

Zhang, C., & Kuang, T. (2018). A new milestone for photosynthesis. *National Science Review, 5*, 444–445.

Zhang, C., Pan, J., Li, L., & Kuang, T. (1999). New structure model of oxygen-evolving center and mechanism for oxygen evolution in photosynthesis. *Chinese Science Bulletin, 44*, 2209–2215.

Zhang, C., Chen, C., Dong, H., Shen, J. R., Dau, H., & Zhao, J. (2015). A synthetic Mn_4Ca-cluster mimicking the oxygen-evolving center of photosynthesis. *Science, 348*, 690–693.

Zhang, M., Bommer, M., Chatterjee, R., Hussein, R., Yano, J., Dau, H., Kern, J., Dobbek, H., & Zouni, A. (2017). Structural insights into the light-driven auto-assembly process of the water-oxidizing Mn_4CaO_5-cluster in photosystem II. *eLife, 6*, e26933.

Zouni, A., Witt, H. T., Kern, J., Fromme, P., Kraub, N., Saenger, W., & Orth, P. (2001). Crystal structure of photosystem II from synechococcus elongatus at 3.8 Å resolution. *Nature, 409*, 739–743.

Photosynthetic Improvement of Industrial Microalgae for Biomass and Biofuel Production

Hyun Gi Koh, Ae Jin Ryu, Seungjib Jeon, Ki Jun Jeong, Byeong-ryool Jeong, and Yong Keun Chang

Abstract Photosynthesis is a process of assimilating carbon dioxide into organic carbons utilizing energy from the sun, during which oxygen is generated. This process thus supports all life on earth by providing food and oxygen. More importantly, it produces biomass that is converted to fossil fuels, which is recapitulated to provide renewable and sustainable biofuels and other chemicals. Microalgae are considered as feedstocks for this purpose, since they possess efficient photosynthetic apparatus, and with their simple body plan, their photosynthetic productivities surpass any crop plants. It should also be noted that photosynthesis also provides carbons and energy for biosynthesis of other molecules, and its improvement should be considered before engineering downstream pathways. One such example would be lipid biosynthesis, and interestingly, certain types of lipids are required for improving photosynthesis, which will be discussed in this review. There have been successful attempts to improve photosynthesis in plants and the model microalgae

H. G. Koh
Advanced Biomass R&D Center, Daejeon, Republic of Korea

Carl R. Woese Institute for Genomic Biology, University of Illinois at Urbana-Champaign, Urbana, IL, USA

A. J. Ryu · K. J. Jeong (✉)
Department of Chemical and Biomolecular Engineering, KAIST, Daejeon, Republic of Korea
e-mail: kjjeong@kaist.ac.kr

S. Jeon · Y. K. Chang (✉)
Advanced Biomass R&D Center, Daejeon, Republic of Korea

Department of Chemical and Biomolecular Engineering, KAIST, Daejeon, Republic of Korea
e-mail: ychang@kaist.ac.kr

B.-r. Jeong (✉)
Department of Chemical and Biomolecular Engineering, KAIST, Daejeon, Republic of Korea

School of Energy and Chemical Engineering, Ulsan National Institute of Science and Technology (UNIST), Ulsan, South Korea

Single-Cell Center, Qingdao Institute of BioEnergy and Bioprocess Technology (QIBEBT), Qingdao, Shandong, China
e-mail: bjeong@unist.ac.kr

Q. Wang (ed.), *Microbial Photosynthesis*,
https://doi.org/10.1007/978-981-15-3110-1_14

Chlamydomonas; however, only a handful reports are available for industrial microalgae including *Chlorella* and *Nannochloropsis*. This review will introduce strategies of improving photosynthesis in the industrial systems, including light-harvesting antenna and photosynthetic pigments, followed by functional lipids relevant to photosynthesis.

Keywords Microalgae · Photosynthesis · Light-harvesting complex (LHC) · Chlorophylls · Tla mutants

1 Introduction

Photosynthesis is employed by photoautotrophic organisms to fix carbon dioxide into organic carbons using the sunlight as the energy source, which is the most efficient sustainable and renewable way of producing organic carbons. The fixed carbons provide not only food for all heterotrophic organisms but also many other chemicals for human needs including fuels, cosmetics, pharmaceuticals, and so on. Therefore, photosynthesis is the most fundamental metabolic pathway for survival of all organisms on earth. Photosynthesis is carried out mostly by plants and algae, and plants have been extensively engineered to provide food for human. In fact, algae are simpler and are more efficient in photosynthesis than plants and are emerging as renewable and sustainable feedstocks for chemicals and fuels, as well as food for human consumption.

Photosynthesis is a complex process mainly comprising light harvest, electron transfer, and Calvin cycle, which is influenced by associated processes such as non-photochemical quenching (NPQ) and carbon-concentrating mechanism (CCM) (Mackinder 2018). These can be targeted for improving photosynthesis, and in fact, efforts have been made to improve the Calvin cycle (Yang et al. 2017; Wei et al. 2017), reduce antenna via random mutagenesis (Mussgnug et al. 2007; Shin et al. 2016c, 2017a), disrupt antenna assembly leading to reduced antenna via targeted mutations (Beckmann et al. 2009a; Jeong et al. 2017; Shin et al. 2016a), and manipulate NPQ (Perozeni et al. 2018; Kromdijk et al. 2016) for improved photosynthetic efficiency. These work have been done mainly in plants and model microalga *Chlamydomonas*; however, pioneering work on unexplored areas of photosynthesis in industrial microalgae, including CCM in *Nannochloropsis* (Wei et al. 2017; Gee and Niyogi 2017), antenna size in *Chlorella* (Shin et al. 2016c; Shin et al. 2017a), and introduction of photosynthetic pigments in *Nannochloropsis* (Koh et al. 2019a), has been reported. There have also been other attempts to manipulate photosynthetic machinery for biohydrogen mostly in *Chlamydomonas* (Batyrova and Hallenbeck 2017), as well as swapping photosynthetic components between different organisms, including cyanobacteria, algae, and plants (Yang et al. 2017; Koh et al. 2019a; Mackinder 2018).

Efficiency of photosynthesis in general is in the range of 12% in terms of energetics for conversion of CO_2 into organic carbon (Stephenson et al. 2011), which can be improved by engineering of carbon-concentrating mechanism (CCM) and other photosynthesis-associated metabolic pathways. Table 1 summarized efforts of genetic and molecular engineering of photosynthesis in industrial microalgae but also included some examples in model microalgae *Chlamydomonas* and plants. There have been excellent reviews regarding genetic and molecular engineering of photosynthesis in the model microalga *Chlamydomonas* and plants (Dehigaspitiya et al. 2019; Mackinder 2018; Work et al. 2012). This chapter will be focused on similar efforts in industrial microalgae and possible applications for large-scale production of biomass that can be converted to biofuels and other high-value chemicals. Overall strategies of genetic engineering employing classic genetic approaches ("bottom-up") and reverse genetics ("top-down" starting with known target genes or chemicals) are summarized in Fig. 1.

2 Genetic and Biological Engineering of Photosynthesis in Microalgae

2.1 Photoprotection Mechanisms and Antenna Size

Ever since primary photosynthetic organism has emerged on earth 3.4 billion years ago, diverse photosynthetic organisms (e.g., cyanobacteria, algae, plants, green-sulfur bacteria) have evolved a variety of light-harvesting pigments/proteins, specialized for their specific habitats (Grossman et al. 1995; Mussgnug et al. 2007). Although each species developed photosynthetic systems in different directions depending on environmental factors such as light wavelength, temperature, and CO_2 availability, the size of the light-harvesting antenna was increased in every species as a means of increasing the chance for exposure to the sunlight. However, excessive amount of light that cannot be converted into energy can induce photooxidative damages, which may cause pigment bleaching or even death in extreme cases (Muller et al. 2001). Hence diverse photoprotection mechanisms have been developed to protect themselves from oxidative stress while remaining competitive in nature.

One of the strategies for photoprotection is to regulate the antenna size upon outer circumstances by controlling the gene expressions or proteolysis. Several genes including *NAB1*, *CAO*, and diverse light-harvesting proteins are known to regulate the antenna size depending on light intensities (Mussgnug et al. 2005; Jia et al. 2016; Dall'osto et al. 2015; Masuda et al. 2003a). In general, the antenna size become larger as the light intensity gets weaker and are often represented by decrease in Chl *a*/*b* ratio. However, the response to light irradiation is species-

Table 1 Genetic engineering of photosynthesis and associated metabolic pathways

Target genes/ mutations	Engineering schemes/host organism	Mechanisms involved in engineering scheme	Improvements	References
Aldolase	Heterologous expression of cyanobacterial fructose 1,6-bisphosphate aldolase/*Chlorella vulgaris*	Enhancement of the Calvin cycle by promoting regeneration of RuBP	Improved photosynthetic capacity and growth	Yang et al. (2017)
CAH1	Knockout of CAH1 by HR/*Nannochloropsis oceanica*	Defective CCM	Reduced growth under ambient CO_2	Gee and Niyogi (2017)
CAO	Heterologous expression *Chlamydomonas* CAO/*Nannochloropsis salina*	Novel production of chlorophyll b for improved photosynthesis under low light	Improved growth, biomass, and lipid productivity	Koh et al. (2019a)
CpSRP43	Random EMS mutation and complementation/*Chlorella vulgaris*	Reduced antenna size for better photosynthesis under high light	Improved growth and biomass	Shin et al. (2016c, 2017a)
	Deletion mutation/*Arabidopsis*	Reduced antennal proteins with normal PS cores	Reduced chlorophylls with increased ratio of Chl a/b	Amin et al. (1999) and Klimyuk et al. (1999)
	Knockout with CRISPR equivalent to TLA3/*Chlamydomonas*	Reduced antenna assembly and photosynthetic pigments	Not tested	Shin et al. (2016a)
CpSRP54	Knockout with CRISPR equivalent to TLA4/*Chlamydomonas*	Reduced assembly of antenna leading to efficient photosynthesis under bright light	Improved photosynthesis	Jeong et al. (2017)
LHCSR3	Overexpression of LHCSR3 in lhcsr mutant/*Chlamydomonas*	Reduced NPQ for improved photosynthetic efficiency	Higher biomass production	Perozeni et al. (2018)
NAB1*, constitutively active	Constitutively active NAB1* mutations/*Chlamydomonas*	Reduced antenna size for improved photosynthetic efficiency with saturating light	Improved growth and photosynthesis	Beckmann et al. (2009a)

(continued)

Table 1 (continued)

Target genes/ mutations	Engineering schemes/host organism	Mechanisms involved in engineering scheme	Improvements	References
PsbS	Overexpression of PsbS together with ZEP and VDE/ Tobacco	Faster NPQ relaxation	Improved crop yield	Kromdijk et al. (2016)
RuBisCO activase	Overexpression of nNoRCA/*Nannochloropsis oceanica*	Improve performance of RuBisCO under low CO_2	Improved growth and photosynthesis, leading to lipid productivity	Wei et al. (2017)
Stm3LR3 mutant	Reduced antennal proteins (LHCI and LHCII)/*Chlamydomonas*	Reduced antenna size is beneficial for efficient photosynthesis under saturating light	Reduced NPQ and improved growth	Mussgnug et al. (2007)

CAH carbonic anhydrase, *CAO* chlorophyll a oxidase, *CpSRP* chloroplast signal recognition particle, *HR* homologous recombination, *LHCSR* light-harvesting complexes stress related, *NAB* nucleic acid-binding protein, *NPQ* non-photochemical quenching, *RuBP* ribulose 1,5-bisphosphate

specific, and the photosynthetic capacity or Chl *a*/*b* ratio does not always correlate with the antenna size (Bailey et al. 2001). The regulation antenna size upon light condition is a slow but a fundamental means of process that actually controls the amount of light captured by the light-harvesting antenna complexes (LHCs).

NPQ is the other strategy for photoprotection, which dissipates the excessive amount of energy already absorbed in a harmless manner (Elrad et al. 2002). When LHCs absorb light exceeding the capacity, the single-state-excited Chl *a* molecule (Chl∗) may decay to the ground state (Chl) via triplet state (^{3}Chl∗), transferring energy to ground-state oxygen O_2 to generate singlet reactive oxygen ($^1O_2$∗).

$$\text{Chlorophyll} + \text{light} \rightarrow {}^1\text{Chlorophyll}^*$$

$${}^1\text{Chlorophyll}^* \rightarrow \text{Chlorophyll} + \text{fluorescence} + \text{qP} + \text{NPQ}$$

$${}^1\text{Chlorophyll}^* \rightarrow {}^3\text{Chlorophyll}^* + \text{O}_2 \rightarrow \text{Chorophyll} + {}^3\text{O}_2^*$$

Hence, to prevent the formation of reactive oxygen, NPQ dissipates excitation energy into three different components: qE, qT, and qI (Muller et al. 2001). Among these components, qE is the main component of NPQ and refers to an energy-dependent quenching. It occurs upon the proton gradient (ΔpH) in thylakoid lumen generated by the excessive light absorption, and the decrease in pH induces activation of xanthophyll cycle (Muller et al. 2001).

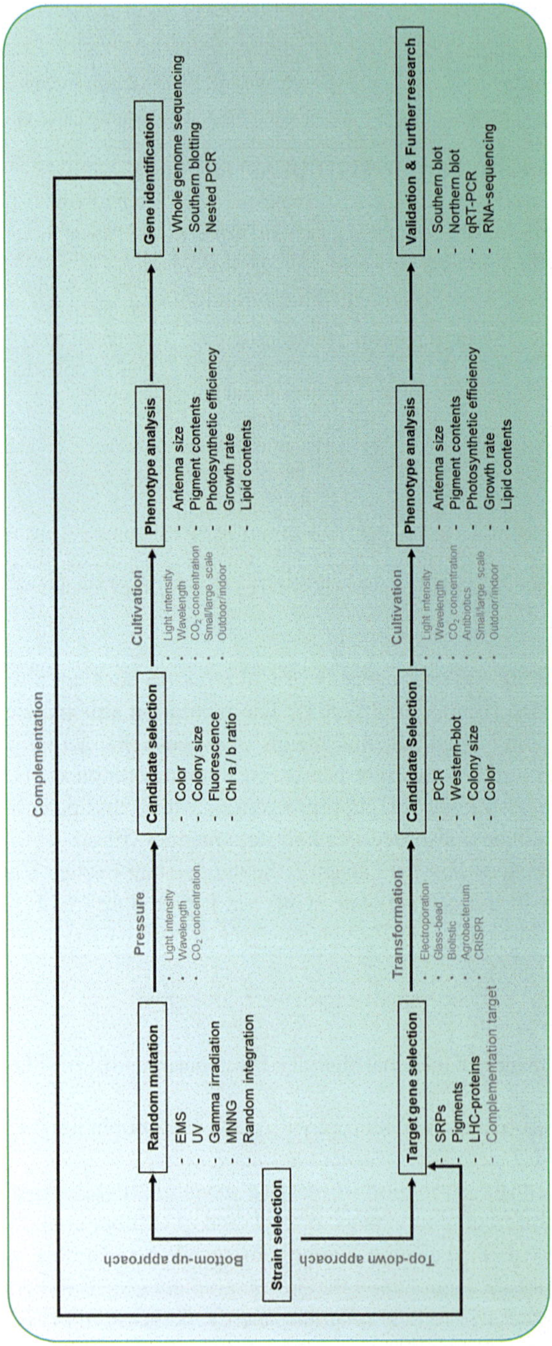

Fig. 1 Scheme of photosynthetic engineering in microalgae

NPQ is an essential photoprotection mechanism especially under intense light stress conditions. However, large size of antennas with energy dissipation through NPQ lowers the entire photosynthesis efficiency, which may result in decrease of biomass productivity. This negative effect prevails particularly when monoculture of microalgae reaches to high density, which leads to shading effects. For this reason, many studies have been conducted on photosynthetic antenna size and NPQ as primary targets for improving the photosynthetic efficiency. In addition, as the cultivation for industrial uses is made in a completely different condition to natural environment, it is necessary to manipulate the LHCs for artificial conditions such as photo-bioreactors or raceway ponds.

2.2 *Manipulation of Antenna and Its Size*

The light-limited condition of underwater environment has forced algal species to develop photosystems with bigger antenna size compared to land plants during their long years of evolutionary process (Perrine et al. 2012; Schenk et al. 2008). This survival strategy may maximize the survival rate of each cell in nature, but from an industrial point of view, the excess amount of light absorption causes energy loss by up to 80% and lowers the overall cultivation efficiency (Polle et al. 2002). Thus, manipulation of antenna size can reduce the energy loss from the shading effects by allowing cells to absorb only the adequate amount of lights that they can utilize (Kirst and Melis 2013; Melis 2009). This concept of antenna size reduction in algae was first suggested by Lien and San Pietro in 1975 (Lien and San Pietro 1976), which were further discussed in detail by Benemann in 1989 (Benemann 1989). After actual evaluation of antenna size reduction was conducted in *Chlorella pyrenoidosa* (Nakajima and Ueda 1997) and *Dunaliella salina* (Melis et al. 1998) in the 1990s, this approach of antenna size reduction became popular, aiming for photosynthetic improvement and higher growth rates.

Most of the research on antenna size manipulation in microalgae was performed in the model strain *Chlamydomonas* sp. owing to its fast-growing traits and relatively well-known genome. The bottom-up approach by random mutation was preferred by many researchers due to limited knowledge on genes encoding key components in photosynthesis. Unlike general random mutations, the selection of truncated light-harvesting antenna (*tla*) mutants is rather simple and accurate as the coloration or fluorescence of *tla* colonies can be distinguished from the parental strain (Perin et al. 2015; Shin et al. 2016d).

Various *tla* mutants have been reported as summarized in Table 2. Based on these research, chloroplast signal recognition particle (CpSRP), which is composed of CpFTSY, CpSRP43, CpSRP53, and ALB3, is found to be a promising target for antenna size reduction (Gohre et al. 2006; Ossenbuhl et al. 2004; Levine and Goodenough 1970; Kirst et al. 2012a, b; Shin et al. 2016b, d, b; Jeong et al. 2017; Bellafiore et al. 2002). As CpSRP proteins are involved in proper folding of light-harvesting proteins and translocation of these proteins to the lipid bilayer of thyla-

Table 2 Reported algae with truncated light-harvesting antennas

Organisms	Target	Referral	Engineering tool	PS II antenna size	Chl a/b	Description on productivity/growth	Reference
Chlamydomonas reinhardtii	TLA1	*tla1*	Insertional random mutation (glass bead)	50% ↓	8.1	Improved productivity at light intensities greater than 1500 μmol photons m^{-2} s^{-1}	Polle et al. (2003) and Tetali et al. (2007)
		tla1-comp	Complementation (glass bead)	n.d.	2.8–3.0	Complementation of TLA1 gene may rescue the mutation	Tetali et al. (2007)
		tla1OE	Overexpression (glass bead)	1–4% ↑	2.2–2.5	n.s.	Mitra et al. (2012)
		tla1Ri	Knockdown (RNAi, glass bead)	11–26% ↓	3.0–6.0	n.s.	
	CpFTSY	*tla2*	Insertional random mutation (glass bead)	35% ↓	8.0–10	Improved productivity in high-density culture at light intensity greater than 800 μmol photons m^{-2} s^{-1}	Kirst et al. (2012a)
		tla2C	Complementation (electroporation)	n.d.	2.7–2.9	Complementation of CpFTSY gene may rescue the mutation	
	CpSRP43	*tla3*	Insertional random mutation (glass bead)	60% ↓	13	Tla3 mutants outperformed WT by 100% in photosynthetic productivity under 2000 μmol photons m^{-2} s^{-1}	Kirst et al. (2012b)
		tla3C	Complementation (glass bead)	40% ↓	2.7–3.0	n.s.	
		CpSRP43	Knockout (CRISPR)	n.d.	8.0–10	n.s.	Shin et al. (2016b)

	CpSRP54	*tla4*	Insertional random mutation (glass bead)	27% ↓	3.3–3.7	The growth rate of *tla4* mutants were 15% higher than WT under 450 μmol photons m^{-2} s^{-1}	Jeong et al. (2017)
		tla4C	Complementation (glass bead)	n.d.	2.5–3.0	Complementation of CpSRP54 gene may rescue the mutation	
	NAB1	*Stm3*	Insertional random mutation (glass bead)	↑	2.21	Growth was reduced by 43% under 400 μmol photons m^{-2} s^{-1}	Mussgnug et al. (2005)
		Stm6GlcT7	Overexpression (glass bead)	10–17% ↓	2.21	The exponential growth rates were improved by 53% in 200 mL scale flask under 700 μmol photons m^{-2} s^{-1}	Beckmann et al. (2009b)
						2.7 folds of biomass were harvested in 2.4 L bioreactor under 1000 μmol photons m^{-2} s^{-1}	
	Alb3	*ac29*	n.d.	↓	n.d.	n.s.	Bellafiore et al. (2002) and Levine and Goodenough (1970)
	Alb3.1p	*ac29*	n.d.	↓	n.d.	n.s.	Ossenbuhl et al. (2004), Levine and Goodenough (1970), and Bellafiore et al. (2002)
	Alb3.2	*ac29*	Knockdown (RNAi)	↓	n.d.	n.s.	Gohre et al. (2006) and Levine and Goodenough (1970)

(continued)

Table 2 (continued)

Organisms	Target	Referral	Engineering tool	PS II antenna size	Chl a/b	Description on productivity/growth	Reference
Chlamydomonas perigranulata	n.d.	*LHC-1*	UV-random mutagenesis	15% ↓	2.7–2.9	Photosynthetic productivity was double in LHC-1 in high-cell density culture	Nakajima et al. (2001)
Chlorella vulgaris	CpSRP43	E5	EMS-random mutation	↓	6.45	19% decreased growth rate under 50 μmol photons m^{-2} s^{-1}	Shin et al. (2016d) and Shin et al. (2017b)
						Similar growth rate under 100 μmol photons m^{-2} s^{-1}	
						44.5% increased growth rate under 400 μmol photons m^{-2} s^{-1}	
		C2–5, C6	Complementation (electroporation)	n.d.	~ 4 ~	Complementation of CpSRP43 gene may rescue the mutation	Shin et al. (2017b)
Chlorella sorokiniana	–	TAM	UV-random mutagenesis	45% ↓	3.3–3.4	Increase biomass productivity by 32% under 450 μmol photons m^{-2} s^{-1} in TAM-2	Cazzaniga et al. (2014)
						No difference observed in TAM-4	
Dunaliella salina	–	*dcd1*	Insertional random mutation (glass bead)	↓	9.9 ~	Similar growth rate under 100 and 2000 μmol photons m^{-2} s^{-1}	Jin et al. (2001)
Nannochloropsis gaditana	–	E2,23,30 I2,3,15,29	EMS-random mutation, insertional random mutation (electroporation)	↓	n.d.	Mutants with decreased chlorophyll contents showed both increased or decreased growth rates in each strains	Perin et al. (2015)

koid membrane through the CpSRP pathway, the mutation in CpSRP-encoding gene often results in antenna size reduction (Kirst and Melis 2014). The CpSRP mutants showed reduction in PS II antenna size and increase in the chlorophyll *a*/*b* ratio, which exhibited improved photosynthetic efficiency and growth rates under intense light conditions. However, decrease in photosynthetic efficiency and growth rates was also observed in *Chlorella* CpSRP43 mutants under low-light conditions (Shin et al. 2016d, 2017b).

In contrast to most of *tla* mutants generated by insertional knockout of photosynthesis-related genes, *C. reinhardtii Stm6GlcT7* transformants, which also have smaller antenna size, were generated by overexpression of an LHC translation repressor NAB1 in a permanent form. It is known that the binding of NAB1 to *LHCBM* mRNA stabilizes mRNA at the preinitiation level and thereby represses translation (Mussgnug et al. 2005). Accordingly, expression of NAB1 in a permanent active form by exchanging two cysteines to serines resulted in reduction of PS II antenna size by 10–17% (Beckmann et al. 2009b). In addition, there are newly discovered genes involved in light-harvesting antenna, such as *tla1*; however, still a lot of *tla* mutants remain unknown for their underlying genes.

In summary, an adequate amount of antenna size reduction often results in improvement of photosynthetic efficiency and biomass production. However, in some cases, the antenna size reduction may not affect (Jin et al. 2001) or adversely affect photosynthetic efficiency depending on light conditions (Shin et al. 2016d). Unfortunately, main reasons for these effects are not yet known but are most likely due to the modification of genes that play a key role in photosynthetic activity.

2.3 Engineering of PS Pigments

Photosynthetic (PS) pigments are one of the most fundamental elements that comprise the photosystems by forming pigment-protein complexes in phototrophic organisms. As PS pigments take roles in light absorption, energy transfer, energy degradation, and stabilization/regulation of photosynthetic apparatus (Scheer 2003), the engineering of PS pigments can also induce considerable improvement in photosynthetic efficiency. Because light-harvesting pigments and antenna size are in an intimate relationship, manipulation of antenna size also results in the changes of PS pigments in most of cases as such summarized in Table 2. Hence, in this chapter, we will only discuss the research which affected the PS pigments in direct ways.

The manipulation of PS pigments in photosynthetic organisms began in 1998, when chlorophyll *b*-synthesizing enzyme "chlorophyllide *a* oxygenase (CAO)" was first discovered from *C. reinhardtii* by Tanaka et al. (1998a). CAO is a unique enzyme found on chloroplast envelope and thylakoid membranes (Eggink et al. 2004) that is responsible for chlorophyll *b* production by a successive hydroxylation of chlorophyll *a* or chlide *a* (Koh et al. 2019b). There are two or three domains in CAO depending on species (Kunugi et al. 2013), where only the C-terminal domain is responsible for the catalytic reaction of chlorophyll *b* synthesis (Yamasato et al.

2005). Because N-terminal domain senses the existing chlorophyll *b* in plastid and regulates the amount by affecting the stability of CAO (Yamasato et al. 2005), sole expression of CAO's C-terminal domain is known to produce higher amount of chlorophyll *b*, which may cause photodamages in plants (Yamasato et al. 2008).

Regulation of chlorophyll *b* by overexpression or knockdown/knockout of CAO results in various changes in LHCs including antenna size, light absorption spectrum, and protein stability (Satoh et al. 2001; Voitsekhovskaja and Tyutereva 2015; Tanaka et al. 2001; Xu et al. 2001; Koh et al. 2019c). In plants, many advantageous effects of CAO overexpression such as retarded senescence in *Arabidopsis* (Sakuraba et al. 2012) or improved photosynthetic efficiency/growth rate/starch contents in tobacco (Biswal et al. 2012) have been discovered. However, in cyanobacteria, expression of CAO did not have much effect on growth rate, even though a marginal increase in light absorption spectrum was shown (Satoh et al. 2001). In addition, a novel chlorophyll species [7-formyl]-Chl *d*p was produced by introduction of CAO in *Acaryochloris marina* (Tsuchiya et al. 2012b), which actually showed functional activities in PS II (Tsuchiya et al. 2012a). As CAO was first discovered in a green alga, *C. reinhardtii*, quite a few research has been done in several algal species including *Chlamydomonas* sp. (Tanaka et al. 1998b; Eggink et al. 2004; Bujaldon et al. 2017), *Dunaliella* sp. (Masuda et al. 2003b), *Micromonas* sp. (Kunugi et al. 2013), and *Nannochloropsis* sp. (Koh et al. 2019b). Although most of research was focused on functional activity of CAO enzyme and not much have been studied on effects of chlorophyll *b* accumulation yet, the decreased photosynthetic efficiency in chlorophyll *b*-less *C. reinhardtii* implies the importance of the pigment. In addition, a heterologous synthesis of chlorophyll *b* in *Nannochloropsis* sp., which does not produce chlorophyll *b* in nature, showed an improved photosynthetic efficiency as well as enhanced growth rate under low-light conditions (Koh et al. 2019b).

Compared to the various works done for the regulation of antenna size in plants and algae, relatively less has been studied on the effects of light-harvesting pigments on growth or photosynthetic efficiencies. This is due to limited knowledge on chlorophyll synthase genes as only genes responsible for chlorophyll *b* and *f* have been discovered yet. Chlorophyll *f* synthase, which oxidizes chlorophyll *a* to synthesize chlorophyll *f*, the far-red light (> 700 nm) absorbing pigment, was first identified in *Chlorogloeopsis fritschii* PCC 9212 and *Synechococcus* sp. PCC 7335 by Ho et al. (2016). So far, chlorophyll *f* has been only found or expressed in cyanobacteria (Ho et al. 2016; Shen et al. 2019), and no successful work of synthesizing chlorophyll *f* in plants or algae has been reported. Hence, it is expected to be worthwhile to express chlorophyll *f* in plants or microalgae as it may not only allow absorption of infrared region of lights but also may result in unexpected effects like the case of CAO expression in cyanobacteria. One of the most important factors to be considered in the heterologous synthesis of light-harvesting pigments is the presence of functional pigment-binding proteins. In the absence of functional pigment-binding proteins, the synthesized pigments remain unstable and become unable to transfer energy to the reaction centers after excitation (Chl*), synthesizing singlet reactive oxygen ($^1O_2*$) by transferring energy to ground-state O_2. However, heterologous synthesis of chlorophyll *b* in a chlorophyll *b*-less cyanobacteria

Synechocystis sp. PCC6803 has revealed that the expressed chlorophyll *b* can function as a light-harvesting antenna in PS I through flexibility of chlorophyll *a*-binding proteins (Satoh et al. 2001).

2.4 Delivery of Heterologous Proteins to the Plastids of Target Species

The translocation of proteins into the chloroplast from cytosol requires a special sequence called transit peptides at the N-terminus. Generally, the transit peptides carry two transport signals in tandem to pass through the prokaryotic double-membrane envelope, which originated from the ancestral cyanobacteria (Day and Theg 2018; Aldridge et al. 2009). Each signal is necessary for passing through the translocons of the outer membrane of chloroplast (TOC) and inner membrane of chloroplast (TIC), which are cleaved off by a zinc-binding metallopeptidase after each translocation. Unfortunately, transit peptide sequences are poorly conserved, and the length can vary from 13 to 146 amino acids long (Zhang and Glaser 2002). Accordingly, the prediction of transit peptide is regarded as one of the most difficult sequences to predict, and although there exist several tools for helping transit peptide prediction such as ChloroP (Emanuelsson et al. 1999) (http://www.cbs.dtu.dk/services/ChloroP/), TargetP (Emanuelsson et al. 2000) (http://www.cbs.dtu.dk/services/TargetP/), Localizer (Sperschneider et al. 2017) (http://localizer.csiro.au/), iPSORT (Bannai et al. 2002) (http://ipsort.hgc.jp/how.html), Predotar (Small et al. 2004) (https://urgi.versailles.inra.fr/predotar/), and HECTAR (Gschloessl et al. 2008) (https://webtools.sb-roscoff.fr/), the results from each tool may conflict to others, and the prediction must be taken as suggestions. Especially for particular proteins, the localization should be predicted in species-dependent manner, as the same sequences sometimes result in different localizations to mitochondria, to chloroplast, or even to the both organelles (Huang et al. 1990; Hurt et al. 1986; Liu et al. 2016). There have been several transit peptides identified in green algae especially in *Chlamydomonas* in which the translocation of nucleus-encoded proteins into chloroplast was verified (Rasala et al. 2014).

In general, transit peptide at the N-terminus is enough for the translocation of nucleus-encoded proteins to the chloroplast. However, there are several species (heterokonts, cryptomonads, haptophytes, dinoflagellates, and apicomplexans) with "complex plastids" (or secondary plastids) that require an additional signal peptide next to the transit peptide at the N-terminus (Lang et al. 1998; Apt et al. 2002). This complex plastid, which originated from the secondary endosymbiosis of red algae (rhodophyte), is enveloped by chloroplast endoplasmic reticulum (cER) and thus surrounded by four membranes (Facchinelli and Weber 2011). *Nannochloropsis* and *Phaeodactylum* are good examples of algae with complex plastid, and both signal and transit peptide sequences required for chloroplast translocation have been identified (Moog et al. 2015a, 2015b; Koh et al. 2019b; Gruber et al. 2007).

It is also possible to express proteins in the chloroplast without using any leader sequences by directly manipulating the chloroplast genome. Although it has been reported that the chloroplast genome editing is more difficult and laborious compared to nucleus genome editing, there are several advantages of chloroplast integration as follows (Bock 2014; Kwon et al. 2018; Gan et al. 2018): (i) the expression level of chloroplast genome is much higher that it can be used for production of pharmaceutical or value-added products, (ii) chloroplastic proteins can be directly synthesized and utilized without translocation, (iii) gene silencing occurs at a lower chance, and (iv) multi-genes can be controlled by an artificial operon. For these reasons, chloroplast transgenics were intensively applied in the model strain *Chlamydomonas* for production of therapeutic proteins such as growth factors (Rasala et al. 2010; Wannathong et al. 2016), antibodies (Tran et al. 2013; Mayfield et al. 2003), antigens (Dreesen et al. 2010; Jones et al. 2013), and antitoxins (Barrera et al. 2015). Majority of the previous studies on chloroplast genome manipulation were performed by biolistic, polyethylene glycol (PEG)-mediated, or glass-bead methods, where there exist several restrictions to overcome for more extensive usage. The biolistic transformation requires gold particles (diameter > 0.6 μm) to penetrate through the chloroplast of algae, which makes it only applicable for cells with relatively large size such as *Chlamydomonas*, *Porphyridium*, and *Platymonas*. For PEG- or glass bead-mediated transformation, cell walls must be removed for protoplast preparation, which is a difficult process especially in species with thick, complex cell walls. Fortunately, many methods for construction of protoplast have been discovered (Hwang et al. 2018; Noda et al. 2017), and also there exist several cell wall-less mutants that can be readily used for transformation (Wannathong et al. 2016; Zhang and Robinson 1990). In addition, recent works by Zhang et al. (2014), Li et al. (2016), and Gan et al. (2018) have revealed the possibility of electroporation-mediated plastid transformation in *Chlamydomonas* and *Nannochloropsis*.

3 Photosynthesis and Lipids

3.1 Classification of Lipids

Lipids are vital cellular constituents in microalgae. Many types of lipids are produced in microalgae, such as neutral lipids, polar lipids, wax esters, sterols, and hydrocarbons, and also phenyl derivatives such as tocopherols, carotenoids, terpenes, quinines, and pyrrole derivatives such as the chlorophylls (Guschina and Harwood 2013). Microalgal lipids are classified two main types: one for storage such as triacylglycerol (TAG), steryl esters, and wax esters and the other for structure as lipid bilayer such as polar lipids and sterols. The reason for classifying lipids in this way is that lipids have been recognized as biomarcromolecules in order to provide the energy stock and to divide the compartments providing a structural basis of cell membranes. In addition, microalgal lipids not only play the roles of membrane structure and energy storage but also perform various important functions in the cell metabolism. These lipids, which have a specific functional role in addition to structure and storage, can be referred to as functional lipids. These functional lipids play an important role as a photosynthetic complex (Kobayashi 2016), as a signaling molecule in the process of defense against the external environment in the cell, or as a signaling molecule specifically expressed in a stress environment. For the improvement of industrial microalgae, it is important to understand the metabolism and function of lipids to selectively obtain the right kind of lipids. Some of important functional lipids from microalgae and other photosynthetic organisms are summarized in Table 3, together with schematic metabolic map of lipids and related chemicals in Fig. 2.

3.2 *Storage Lipids in Microalgae: Triacylglycerol*

Triacylglycerol (TAG), which has a triple ester form in which three long-chain fatty acids are linked to glycerol, accounts for the largest portion of the microalgae in storage lipids (Prabandono and Amin 2015). The fatty acids (FAs) constituting the TAG are predominantly composed of saturated fatty acid (SFA), and some high-accumulation microalgae are composed of a specific long-chain polyunsaturated

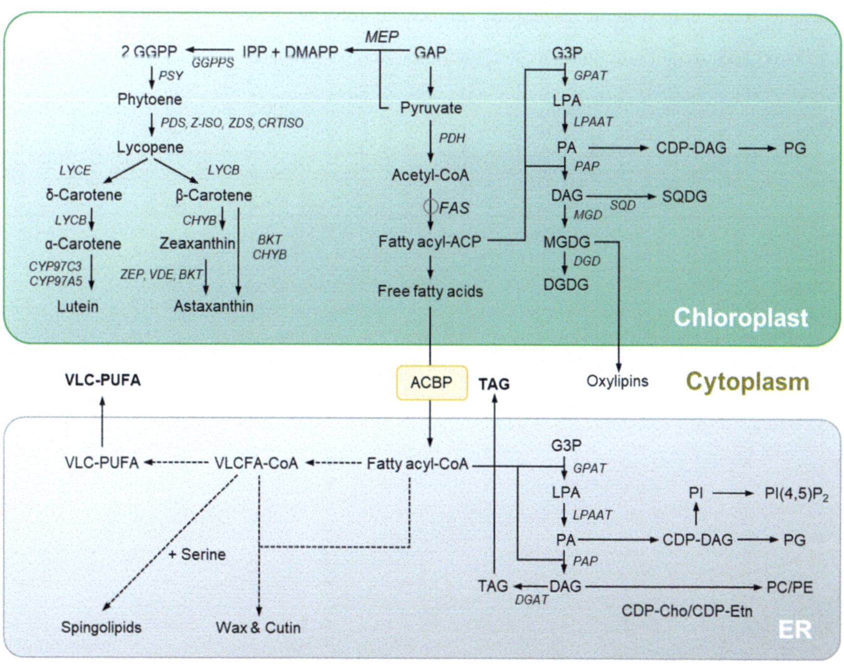

Fig. 2 Lipid metabolic pathways in microalgae

Table 3 Functional lipids and their roles in microalgae

Lipids	Organisms	Functions	Accumulation/production scheme	References
Monogalactosyldiacylglycerol (MGDG)	*Arabidopsis*	The 40% reduction in MGDG levels increases the conductivity of the thylakoid membrane, which inhibits the activation and photoprotection of the thylakoid membrane. The 80% reduction of MGDG with partial DGDG reduction causes disruption of PSII reaction and energy coupling between reaction centers and antenna complexes. The critical loss (95%) of MGDG causes disruption of the formation of PS complexes and loss of PSII activity completely	Overexpression of MGD and other thylakoid membrane lipid-producing enzymes make balance of functional lipids in thylakoid and increase the stability of the thylakoid membrane	Aronsson et al. (2008), Kobayashi et al. (2014), Kobayashi et al. (2013), Schaller et al. (2011), Jahns et al. (2009), and Schaller et al. (2010)
	Tobacco	The reduction of MGDG reduces the cytochrome b^6f complex level and blocked intersystem electron transport		Wu et al. (2013)
Digalactosyldiacylglycerol (DGDG)	*Arabidopsis*	DGDG is closely related to structure and function of PSII and PSI. Decrease of DGDG level increases photoinhibition in high-light condition and could not maintain PSII functionality and trimerization of the LHCII complex	Overexpression of DGD and other thylakoid membrane lipid-producing enzymes make balance of functional lipids in thylakoid and increase the stability of the thylakoid membrane	Steffen et al. (2005), Reifarth et al. (1997), Hölzl et al. (2009), Hartel et al. (1997), and Hölzl et al. (2006)
	Synechocystis	Loss of DGDG causes dissociation of PSII proteins and destabilization of oxygen-evolving complex and increases the sensitivity to photoinhibition in high-temperature and high-light condition by reduction of the repair cycle of PSII		Sakurai et al. (2007b), Mizusawa et al. (2009a), Mizusawa et al. (2009b), Guo et al. (2005), and Ivanov et al. (2006)

Sulfoquinovosyldiacylglycerol (SQDG)	*Chlamydomonas*	Necessity of SQDG in the activity of PSII (not PSI)	Overexpression of SQD and other thylakoid membrane lipid-producing enzymes make balance of functional lipids in thylakoid and increase the stability of the thylakoid membrane	Minoda et al. (2002, 2003) and Sato et al. (1995)
	Arabidopsis	Significant reduction of PSII quantum efficiency is caused by decrease of SQDG level. SQDG maintains the total amount of anionic lipids in membranes in the case of decreased PG in P-deficient condition		Essigmann et al. (1998), Yu et al. (2002), Yu and Benning (2003)
	Synechocystis	Necessity of SQDG in activity of PSII (not PSI), not important for photosynthesis in *Synechococcus* sp. PCC 7942		Aoki et al. (2004) and Guler et al. (1996)

(continued)

Table 3 (continued)

Lipids	Organisms	Functions	Accumulation/production scheme	References
Phosphatidylglycerol (PG)	Pea, spinach	Reduction of PG causes disruption of electron transport in PSII without decreasing PSI activity	Overexpression of PGP and other thylakoid membrane lipid-producing enzymes make balance of functional lipids in thylakoid and increase the stability of the thylakoid membrane	Jordan et al. (1983) and Droppa et al. (1995)
	Chlamydomonas	Necessity of PG in synthesis of core proteins of PSII, formation of LHCII trimer, state transition ability, oxygen evolution ability		El Maanni et al. (1998), Dubertret et al. (1994), and Pineau et al. (2004)
	Arabidopsis	Necessity for photochemical efficiency of PSII, electron transfer in PSII, and cyclic electron transport in PSI		Kobayashi et al. (2015), Kobayashi et al. (2016), Xu et al. (2002), and Yu and Benning (2003)
	Synechocystis	Necessity of PG in electron transport of plastoquinone in PSII, reactivation of PSII, maintenance of the activity of oxygen-evolving complex, activity of PSI complex		Endo et al. (2015), Gombos et al. (2002), Hagio et al. (2000), Itoh et al. (2012), Sakurai et al. (2007a), Sakurai et al. (2003), and Domonkos et al. (2004)
Linoleic acid (C18:2)	Avocado, tomato, bean	Enhancement of resistance against fungal pathogen	Overexpression of delta 12 fatty acid desaturase	Madi et al. (2003), Ongena et al. (2004), and Yaeno et al. (2004a)
	Arabidopsis	Contribution in SAR as a precursor of 9-oxo nonanoic acid which is the substrate of SAR chemical inducer, azelaic acid		Wittek et al. (2014)

Linolenic acid (C18:3)	Avocado, tomato, bean	Enhancement of resistance against fungal pathogen	Overexpression of omega 3 fatty acid desaturase	Madi et al. (2003), Ongena et al. (2004), and Yaeno et al. (2004a)
	Arabidopsis	Contribution in SAR as a precursor of 9-oxo nonanoic acid which is the substrate of SAR chemical inducer, azelaic acid		Wittek et al. (2014)
	Pseudomonas syringae pv. *maculicola*	Regulation of intracellular ROS by affecting NADPH oxidase		Yaeno et al. (2004b)
	Arabidopsis, tobacco	Contribution as a substrate of jasmonic acid which is a major oxylipin in plant		Farmer et al. (2003)
Violaxanthin Zeaxanthin	*Arabidopsis*	Accessory of PS and antioxidant for xanthophyll cycle for non-photochemical quenching	Overexpression of carotenoid metabolic enzymes	Ramel et al. (2012)
Astaxanthin	*Haematococcus*	Reducing the penetration of blue light into plant cell for reduce photooxidative damage		Hagen et al. (1994)
	Haematococcus, *Heliobacteria*	Protection of biomacromolecules against ROS		Kobayashi and Sakamoto (1999), Hagen et al. (1993), and Takaichi et al. (1997)

fatty acid (PUFA) (Prabandono and Amin 2015). These TAGs are used as biodiesel after transesterification with small alcohols. The microalgal TAGs usually have acyl chains having 16 or 18 carbons esterified to glycerol. These fatty acyl chains in microalgae are chemically similar to commercial diesel fuel components that have 10–15 carbons per molecule, and biodiesel from microalgae can be compatible with current diesel engines.

TAG, a typical storage lipid of microalgae, is biosynthesized as cytosolic lipid in the light and normal growth conditions and degraded to be used as an energy source in the dark condition. TAG production and biomass production of microalgae compete for energy use resulting from photosynthesis, and microalgae mainly use energy from photosynthesis in stressful environments. In the stressful environment, microalgae mainly regulate their metabolism to endure the external environment by accumulating TAG type lipid. On the other hand, in the case of PUFA-rich TAGs, they are catabolized in stress condition because of the supplement of specific fatty acids. When the biosynthesis of PUFA is inefficient in unexpected environmental change, PUFA-rich TAG can be a reservoir of PUFA and donate them to monogalactosyldiacylglycerol (MGDG) and other polar lipids for membrane reorganization.

3.3 Functional Lipids in Microalgae

3.3.1 Membrane Lipids for Photosynthesis

The chloroplasts present in plants and microalgae cells are places where they make energy through photosynthesis from carbon dioxide. In the chloroplast, functional lipids, which play an important role in photosynthesis, can be classified into two groups. First groups consist of glycerolipids and phospholipids that make up the structure of thylakoid membrane; the other group is for carotenoids that harvest light for photosynthesis. Among these functional lipids, glycolipids and phospholipid are mentioned first, and then carotenoids are mentioned.

The thylakoid membranes of chloroplasts in higher plant and microalgae cells are composed of glycerolipids and phospholipids (Kobayashi 2016). The lipid composition of the thylakoid membrane is different for each photosynthetic organism, and the special composition is highly conserved. In thylakoid membrane, there are four major lipids found such as monogalactosyldiacylglycerol (MGDG), digalactosyldiacylglycerol (DGDG), sulfoquinovosyldiacylglycerol (SQDG), and phosphatidylglycerol (PG). This membrane lipid plays an important role not only in the formation of lipid bilayers of the thylakoid membrane but also in the biogenesis of the thylakoid membrane (Kobayashi 2016). Second, it plays an important role in photosynthesis by participating in folding and assembly of protein subunits in the photosynthetic complex (Kobayashi 2016). In addition, thylakoid membrane lipid biosynthesis induces the expression of photosynthesis-related genes in both the nucleus and plastid and activates the formation of photosynthetic machinery and development of chloroplast (Kobayashi 2016). Of these, we focus only on the lipid of the thylakoid membrane associated with photosynthesis enhancement.

The four representative lipids consisting the thylakoid, in general, cause the photosystem I (PSI) and the photosystem II (PSII) to malfunction in a number of ways if their level decreases. Studies reported so far have reported more of the role of thylakoid lipids on PSII than on PSI. In *Arabidopsis*, the reduction of MGDG is closely related to the activity of PSII by disturbing the activation of the photoprotection of the thylakoid membrane, energy coupling between reaction centers and antenna complexes, and formation of photosystem complexes (Aronsson et al. 2008; Kobayashi et al. 2014; Kobayashi et al. 2013; Schaller et al. 2011; Jahns et al. 2009; Schaller et al. 2010). It is also known that decrease of DGDG level causes destabilization of PSII in *Arabidopsis* (Steffen et al. 2005; Reifarth et al. 1997; Hölzl et al. 2006, 2009; Hartel et al. 1997) and *Synechocystis* (Sakurai et al. 2007b; Mizusawa et al. 2009a, b; Guo et al. 2005; Ivanov et al. 2006). In the case of SQDG, the influence of the level of SQDG is quite different from many photosynthetic organisms. In *Arabidopsis*, decrease of SQDG level caused significant reduction of PSII quantum efficiency (Essigmann et al. 1998; Yu et al. 2002; Yu and Benning 2003). Among thylakoid lipids, PG is the most closely related to the photosystem, and decrease of level of PG reduces the activity of PSI and PSII in *Arabidopsis* (Kobayashi et al. 2015, 2016; Xu et al. 2002; Yu and Benning 2003).

The results of the loss of photosystem activity following decrease of the level of these thylakoid lipids provide a clue to the study of the increase of photosystem activity or stabilization of the photosystem by proper control of the level of thylakoid lipids. Previous reports have shown that the disruption of metabolic enzymes producing thylakoid lipids leads to a decrease in thylakoid lipid levels and consequent structural destabilization of PS and a reduction in photosynthetic activity, and, on the contrary, it can be considered that increased levels of thylakoid lipids through high expression of these enzymes. The overexpression of MGDG-producing enzyme MGDG synthase (MGD) has been performed on tobacco. It has been reported that when the enzyme is highly expressed, it has a high tolerance by stabilizing the membrane structure by effectively controlling the DGDG-MGDG ratio in the high-salt or high-aluminum condition (Zhang et al. 2016; Wang et al. 2014). These results suggest that proper level of functional lipid in the thylakoid membrane may play an important role in the stability of the chloroplast membrane, and it is also important to maintain the ratio of each functional lipid as well as overexpression. In order to enhance photosynthesis for biofuel production, the regulation of the ratio of these functional lipids in the thylakoid can be considered, and it can be performed by expression and deletion of various genes related to their expression.

3.3.2 Polyunsaturated Fatty Acids (PUFAs) for Plant Defense

PUFA refers to fatty acids containing two or more double bonds, which are produced from palmitic acid as a substrate by various desaturases and elongases from palmitic acid as a substrate in endoplasmic reticulum of plant cells or microalgae cells. Several plants and microalgae are known to produce high levels of PUFAs such as docosahexaenoic acid (DHA) and eicosapentaenoic acid (EPA) which are high-value lipids for nutrients (Adarme-Vega et al. 2012). These value-added PUFAs are not only costly products themselves but also have the advantage of being able to produce more cost-effective biofuel processes by producing each from biomass when utilizing biomass for biofuel production.

The PUFAs are stored in TAGs or stored as MGDG or DGDG to constitute membrane lipids. For TAGs composed of large amounts of PUFAs, they serve as donors to provide PUFAs to MGDG and DGDG in a stressed environment. PUFAs stored in the form of MGDG and DGDG play an important role in plant defense. Also, PUFAs remove reactive oxygen species and weaken cellular damage in stress condition. Among the PUFAs, C18 PUFAs are closely related to defense of fungal pathogen. High-level accumulation of linoleic acid ($C18:2^{\Delta9,12}$) and linolenic acid ($C18:3^{\Delta9,12,15}$) enhances the resistance against various fungal pathogens in avocado, tomato, and bean (Madi et al. 2003; Ongena et al. 2004; Yaeno et al. 2004a).

Including oleic acid ($C18:1^{\Delta9}$), a monounsaturated fatty acid, C18 unsaturated fatty acids such as linoleic acid and linolenic acid play an important role in plants' systemic acquired resistance (SAR) response (Durrant and Dong 2004). The C18 unsaturated fatty acids mentioned above have a double bond at carbon 9 and are the precursor for producing monocarboxylic acid 9-oxo nonanoic acid (ONA) by oxidative cleavage of the double bond at this position (Wittek et al. 2014). ONA is also the precursor of nine-carbon dicarboxylic acid, called azelaic acid (AzA), and it is one of the chemical inducers of SAR and is a mobile signal that activates defense response in plant.

Also, the desaturated fatty acid C18:3 is related to the accumulation of reactive oxygen species (ROS) which are the regulators of SA. C18:3 may affect NADPH oxidase that produces superoxide, and then they can be converted to ROS. For this reason, C18:3 can regulate the ROS production and thereby contribute to the defense response (Yaeno et al. 2004b).

Moreover, PUFAs can be substrates for the production of oxylipins in plant cells. Oxylipin is an oxygenated PUFA produced by oxidation from a polyunsaturated acyl group of galactolipids of the chloroplast membrane (Andreou et al. 2009). Jasmonic acid (JA), the major defense phytohormone in the plant, is produced from hexadecanoic acid (C16:3) or octadecanoic acid (C18:3) precursors by autoxidation or enzymatic oxidation depending on the type of photosynthetic organisms (Farmer et al. 2003).

3.3.3 Carotenoids for Stress Response and Photosynthesis

Carotenoids, the natural lipid-soluble pigments, are compounds composed of isoprene units and belong to the isoprenoid compound group (Frank and Cogdell 1996). In addition, carotenoids can be a good product in microalgae production along with biofuels because they are considered important production materials in the food, cosmetic, medical, and pharmaceutical industries (Mezzomo and Ferreira 2016). There are two types of carotenoids: unoxygenated carotenoids such as carotene and oxygen-containing carotenoids such as xanthophylls (Prof and Gupta 2014). Carotenoids usually have from 3 to 13 conjugated double bonds and absorb light whose wavelength is from 450 nm to 570 nm (Havaux 2014). This light-absorbing carotenoid plays a major role in harvesting light for photosynthesis by consisting light-harvesting complex in plant and microalgae. In addition, carotenoids also perform photoprotection functions in plants and microalgae cells with their antioxidant activity. Microalgae and plant cells have a carotenoid-based protective mechanism called xanthophyll cycle for non-photochemical quenching. This is a mechanism to prevent photosynthetic organisms from photodamage and oxidative stress caused by the photoinhibition and inactivation of photosystem in excess light conditions and the production of ROS and free radicals. Among the carotenoids, violaxanthin and zeaxanthin dissipate excess energy by converting from an epoxidized form to a de-epoxidized form under excessive light conditions (Ramel et al. 2012).

In addition, as well as the primary carotenoids mentioned above, astaxanthin, which is classified as secondary carotenoid, also has 13 conjugated double bonds, so it has a very powerful ability as an antioxidant and effectively removes ROS. Also, astaxanthin plays a role in plants and microalgae cells than other carotenoids. It is known that astaxanthin acts to reduce the photooxidative damage of PS by reducing blue light entering chloroplasts under strong light conditions (Hagen et al. 1994). In addition, astaxanthin also protects biomacromolecules such as DNA, RNA, enzymes, and membrane lipids from being attacked by ROS (Kobayashi and Sakamoto 1999; Hagen et al. 1993; Takaichi et al. 1997).

In summary, various functional lipids are involved in photosynthesis of plants and microalgae. Several glycerolipids and phospholipids present in thylakoid membrane play an essential role in the structure and activity of PSI and PSII. PUFAs contribute to plant defense for photosynthesis inhibition by ROS in stress conditions, and linolenic acid is directly related to intracellular ROS concentration. Carotenoids act as an accessory in the PS and contribute to photosynthesis by helping to absorb or quenching light from plants and microalgae and as an antioxidant to prevent photosynthesis inhibition by ROS under stress conditions.

Acknowledgments This work was supported by the Advanced Biomass R&D Center (ABC) of the Global Frontier Project funded by the Ministry of Science and ICT (ABC-2010-0029728 and 2011-0031350).

Competing Financial Interests The authors declare no competing financial or nonfinancial interests.

References

Adarme-Vega, T. C., Lim, D. K. Y., Timmins, M., Vernen, F., Li, Y., & Schenk, P. M. (2012). Microalgal biofactories: A promising approach towards sustainable omega-3 fatty acid production. *Microbial Cell Factories, 11*, 96–96.

Aldridge, C., Cain, P., & Robinson, C. (2009). Protein transport in organelles: Protein transport into and across the thylakoid membrane. *FEBS Journal, 276*, 1177–1186.

Amin, P., Sy, D. A., Pilgrim, M. L., Parry, D. H., Nussaume, L., & Hoffman, N. E. (1999). Arabidopsis mutants lacking the 43- and 54-kilodalton subunits of the chloroplast signal recognition particle have distinct phenotypes. *Plant Physiology, 121*, 61–70.

Andreou, A., Brodhun, F., & Feussner, I. (2009). Biosynthesis of oxylipins in non-mammals. *Progress in Lipid Research, 48*, 148–170.

Aoki, M., Sato, N., Meguro, A., & Tsuzuki, M. (2004). Differing involvement of sulfoquinovosyl diacylglycerol in photosystem II in two species of unicellular cyanobacteria. *European Journal of Biochemistry, 271*, 685–693.

Apt, K. E., Zaslavkaia, L., Lippmeier, J. C., Lang, M., Kilian, O., Wetherbee, R., Grossman, A. R., & Kroth, P. G. (2002). In vivo characterization of diatom multipartite plastid targeting signals. *Journal of Cell Science, 115*, 4061–4069.

Aronsson, H., Schöttler, M. A., Kelly, A. A., Sundqvist, C., Dörmann, P., Karim, S., & Jarvis, P. (2008). Monogalactosyldiacylglycerol deficiency in Arabidopsis affects pigment composition in the prolamellar body and impairs thylakoid membrane energization and photoprotection in leaves. *Plant Physiology, 148*, 580–592.

Bailey, S., Walters, R. G., Jansson, S., & Horton, P. (2001). Acclimation of Arabidopsis thaliana to the light environment: The existence of separate low light and high light responses. *Planta, 213*, 794–801.

Bannai, H., Tamada, Y., Maruyama, O., Nakai, K., & Miyano, S. (2002). Extensive feature detection of N-terminal protein sorting signals. *Bioinformatics, 18*, 298–305.

Barrera, D. J., Rosenberg, J. N., Chiu, J. G., Chang, Y. N., Debatis, M., Ngoi, S. M., Chang, J. T., Shoemaker, C. B., Oyler, G. A., & Mayfield, S. P. (2015). Algal chloroplast produced camelid VH H antitoxins are capable of neutralizing botulinum neurotoxin. *Plant Biotechnology Journal, 13*, 117–124.

Batyrova, K., & Hallenbeck, P. C. (2017). Hydrogen production by a *Chlamydomonas reinhardtii* strain with inducible expression of photosystem II. *International Journal of Molecular Sciences, 18*(3), 647.

Beckmann, J., Lehr, F., Finazzi, G., Hankamer, B., Posten, C., Wobbe, L., & Kruse, O. (2009a). Improvement of light to biomass conversion by de-regulation of light-harvesting protein translation in *Chlamydomonas reinhardtii*. *Journal of Biotechnology, 142*, 70–77.

Beckmann, J., Lehr, F., Finazzi, G., Hankamer, B., Posten, C., Wobbe, L., & Kruse, O. (2009b). Improvement of light to biomass conversion by de-regulation of light-harvesting protein translation in Chlamydomonas reinhardtii. *Journal of Biotechnology, 142*, 70–77.

Bellafiore, S., Ferris, P., Naver, H., Gohre, V., & Rochiax, J. D. (2002). Loss of Albino3 leads to the specific depletion of the light-harvesting system. *Plant Cell, 14*, 2303–2314.

Benemann, J. R. (1989). The future of microalgal biotechnology. *Algal and Cyanobacterial Biotechnology*, 317–337.

Biswal, A. K., Pattanayak, G. K., Pandey, S. S., Leelavathi, S., Reddy, V. S., Govindjee, & Tripathy, B. C. (2012). Light intensity-dependent modulation of chlorophyll b biosynthesis and photosynthesis by overexpression of chlorophyllide a oxygenase in tobacco. *Plant Physiology, 159*, 433–449.

Bock, R. (2014). Genetic engineering of the chloroplast: Novel tools and new applications. *Current Opinion in Biotechnology, 26*, 7–13.

Bujaldon, S., Kodama, N., Rappaport, F., Subramanyam, R., DE Vitry, C., Takahashi, Y., & Wollman, F. A. (2017). Functional accumulation of antenna proteins in chlorophyll b-less mutants of *Chlamydomonas reinhardtii*. *Molecular Plant, 10*, 115–130.

Cazzaniga, S., Dall'osto, L., Szaub, J., Scibilia, L., Ballottari, M., Purton, S., & Bassi, R. (2014). Domestication of the green alga *Chlorella sorokiniana*: Reduction of antenna size improves light-use efficiency in a photobioreactor. *Biotechnology for Biofuels, 7*, 157.

Dall'osto, L., Bressan, M., & Bassi, R. (2015). Biogenesis of light harvesting proteins. *Biochimica et Biophysica Acta, 1847*, 861–871.

Day, P. M., & Theg, S. M. (2018). Evolution of protein transport to the chloroplast envelope membranes. *Photosynthesis Research, 138*, 315–326.

Dehigaspitiya, P., Milham, P., Ash, G. J., Arun-Chinnappa, K., Gamage, D., Martin, A., Nagasaka, S., & Seneweera, S. (2019). Exploring natural variation of photosynthesis in a site-specific manner: Evolution, progress, and prospects. *Planta, 250*, 1033–1050.

Domonkos, I., Malec, P., Sallai, A., Kovács, L., Itoh, K., Shen, G., Ughy, B., Bogos, B., Sakurai, I., Kis, M., Strzalka, K., Wada, H., Itoh, S., Farkas, T., & Gombos, Z. (2004). Phosphatidylglycerol is essential for oligomerization of photosystem I reaction center. *Plant Physiology, 134*, 1471–1478.

Dreesen, I. A., Charpin-El Hamri, G., & Fussenegger, M. (2010). Heat-stable oral alga-based vaccine protects mice from Staphylococcus aureus infection. *Journal of Biotechnology, 145*, 273–280.

Droppa, M., Horváth, G., Hideg, É., & Farkas, T. J. P. R. (1995). The role of phospholipids in regulating photosynthetic electron transport activities: Treatment of thylakoids with phospholipase C. *Photosynthesis Research, 46*, 287–293.

Dubertret, G., Mirshahi, A., Mirshahi, M., Gerard-Hirne, C., & Trémolieres, A. (1994). Evidence from in vivo manipulations of lipid composition in mutants that the Δ3-trans-hexadecenoic acid-containing ohosphatidylglycerol is involved in the biogenesis of the light-harvesting chlorophyll a/b-protein complex of Chlamydomonas reinhardtii. *European Journal of Biochemistry, 226*, 473–482.

Durrant, W. E., & Dong, X. (2004). Systemic acquired resistance. *Annual Review of Phytopathology, 42*, 185–209.

Eggink, L. L., Lobrutto, R., Brune, D. C., Brusslan, J., Yamasato, A., Tanaka, A., & Hoober, J. K. (2004). Synthesis of chlorophyll *b*: Localization of chlorophyllide *a* oxygenase and discovery of a stable radical in the catalytic subunit. *BMC Plant Biology, 4*, 5.

EL Maanni, A., Dubertret, G., Delrieu, M.-J., Roche, O., & Trémolières, A. (1998). Mutants of Chlamydomonas reinhardtii affected in phosphatidylglycerol metabolism and thylakoid biogenesis. *Plant Physiology and Biochemistry, 36*, 609–619.

Elrad, D., Niyogi, K. K., & Grossman, A. R. (2002). A major light-harvesting polypeptide of photosystem II functions in thermal dissipation. *The Plant Cell Online, 14*, 1801–1816.

Emanuelsson, O., Nielsen, H., & VON Heijne, G. (1999). ChloroP, a neural network-based method for predicting chloroplast transit peptides and their cleavage sites. *Protein Science, 8*, 978–984.

Emanuelsson, O., Nielsen, H., Brunak, S., & VON Heijne, G. (2000). Predicting subcellular localization of proteins based on their N-terminal amino acid sequence. *Journal of Molecular Biology, 300*, 1005–1016.

Endo, K., Mizusawa, N., Shen, J.-R., Yamada, M., Tomo, T., Komatsu, H., Kobayashi, M., Kobayashi, K., & Wada, H. J. P. R. (2015). Site-directed mutagenesis of amino acid residues of D1 protein interacting with phosphatidylglycerol affects the function of plastoquinone QB in photosystem II. *Photosynthesis Research,* 126, 385–397.

Essigmann, B., Güler, S., Narang, R. A., Linke, D., & Benning, C. (1998). Phosphate availability affects the thylakoid lipid composition and the expression of SQD1, a gene required for sulfolipid biosynthesis in Arabidopsis thaliana. *Proceedings of the National Academy of Sciences of the United States of Americ, 95*, 1950–1955.

Facchinelli, F., & Weber, A. P. M. (2011). The metabolite transporters of the plastid envelope: An update. *Frontiers in Plant Science, 2*, 50.

Farmer, E. E., Alméras, E., & Krishnamurthy, V. (2003). Jasmonates and related oxylipins in plant responses to pathogenesis and herbivory. *Current Opinion in Plant Biology, 6*, 372–378.

Frank, H. A., & Cogdell, R. J. (1996). Carotenoids in Photosynthesis. *Photochemistry and Photobiology, 63*, 257–264.

Gan, Q., Jiang, J., Han, X., Wang, S., & Lu, Y. (2018). Engineering the chloroplast genome of oleaginous marine microalga Nannochloropsis oceanica. *Frontiers in Plant Science, 9*, 439.

Gee, C. W., & Niyogi, K. K. (2017). The carbonic anhydrase CAH1 is an essential component of the carbon-concentrating mechanism in *Nannochloropsis oceanica*. *Proceedings of the National Academy of Sciences of the United States of America, 114*, 4537–4542.

Gohre, V., Ossenbuhl, F., Crevecoeur, M., Eichacker, L. A., & Rochaix, J. D. (2006). One of two Alb3 proteins is essential for the assembly of the photosystems and for cell survival in Chlamydomonas. *Plant Cell, 18*, 1454–1466.

Gombos, Z., Várkonyi, Z., Hagio, M., Iwaki, M., Kovács, L., Masamoto, K., Itoh, S., & Wada, H. (2002). Phosphatidylglycerol requirement for the function of electron acceptor Plastoquinone QB in the photosystem II reaction center. *Biochemistry, 41*, 3796–3802.

Grossman, A. R., Bhaya, D., Apt, K. E., & Kehoe, D. M. (1995). Light-harvesting complexes in oxygenic photosynthesis: Diversity, control, and evolution. *Annual Review of Genetics, 29*, 231–288.

Gruber, A., Vugrinec, S., Hempel, F., Gould, S. B., Maier, U. G., & Kroth, P. G. (2007). Protein targeting into complex diatom plastids: Functional characterisation of a specific targeting motif. *Plant Molecular Biology, 64*, 519–530.

Gschloessl, B., Guermeur, Y., & Cock, J. M. (2008). HECTAR: A method to predict subcellular targeting in heterokonts. *BMC Bioinformatics, 9*, 393.

Guler, S., Seeliger, A., Hartel, H., Renger, G., & Benning, C. (1996). A null mutant of Synechococcus sp. PCC7942 deficient in the sulfolipid sulfoquinovosyl diacylglycerol. *The Journal of Biological Chemistry, 271*, 7501–7507.

Guo, J., Zhang, Z., Bi, Y., Yang, W., Xu, Y., & Zhang, L. (2005). Decreased stability of photosystem I in dgd1 mutant of Arabidopsis thaliana. *FEBS Letters, 579*, 3619–3624.

Guschina, I. A., & Harwood, J. L. (2013). Algal lipids and their metabolism. In M. A. Borowitzka & N. R. Moheimani (Eds.), *Algae for biofuels and energy*. Dordrecht: Springer.

Hagen, C., Braune, W., & Greulich, F. (1993). Functional aspects of secondary carotenoids in Haematococcus lacustris [Girod] Rostafinski (Volvocales) IV. Protection from photodynamic damage. *Journal of Photochemistry and Photobiology B: Biology, 20*, 153–160.

Hagen, C., Braune, W., & Björn, L.O. (1994). Functional aspects of secondary carotenoids in Haematococcus lacustris (volvocales). III. Action as A "sunshade" 1. *Journal of Phycology, 30*, 241–248.

Hagio, M., Gombos, Z., Várkonyi, Z., Masamoto, K., Sato, N., Tsuzuki, M., & Wada, H. (2000). Direct evidence for requirement of phosphatidylglycerol in photosystem II of photosynthesis. *American Society of Plant Physiologists, 124*, 795–804.

Hartel, H., Lokstein, H., Dormann, P., Grimm, B., & Benning, C. (1997). Changes in the composition of the photosynthetic apparatus in the Galactolipid-deficient dgd1 mutant of Arabidopsis thaliana., *Science, 115*, 1175–1184.

Havaux, M. (2014). Carotenoid oxidation products as stress signals in plants. *The Plant Journal, 79*, 597–606.

Ho, M. Y., Shen, G. Z., Canniffe, D. P., Zhao, C., & Bryant, D. A. (2016). Light-dependent chlorophyll f synthase is a highly divergent paralog of PsbA of photosystem II. *Science, 353*(6302), aaf9178.

Hölzl, G., Witt, S., Kelly, A. A., Zähringer, U., Warnecke, D., Dörmann, P., & Heinz, E. (2006). Functional differences between galactolipids and glucolipids revealed in photosynthesis of higher plants. *Proceedings of the National Academy of Sciences of the United States of America, 103*, 7512–7517.

Hölzl, G., Witt, S., Gaude, N., Melzer, M., Schöttler, M. A., & Dörmann, P. (2009). The role of Diglycosyl lipids in photosynthesis and membrane lipid homeostasis in Arabidopsis. *Plant Physiology, 150*, 1147–1159.

Huang, J., Hack, E., Thornburg, R. W., & Myers, A. M. (1990). A yeast mitochondrial leader peptide functions in vivo as a dual targeting signal for both chloroplasts and mitochondria. *Plant Cell, 2*, 1249–1260.

Hurt, E. C., Soltanifar, N., Goldschmidt-Clermont, M., Rochaix, J. D., & Schatz, G. (1986). The cleavable pre-sequence of an imported chloroplast protein directs attached polypeptides into yeast mitochondria. *The EMBO Journal, 5*, 1343–1350.

Hwang, H. J., Kim, Y. T., Kang, N. S., & Han, J. W. (2018). A simple method for removal of the Chlamydomonas reinhardtii cell wall using a commercially available Subtilisin (Alcalase). *Journal of Molecular Microbiology and Biotechnology, 28*, 169–178.

Itoh, S., Kozuki, T., Nishida, K., Fukushima, Y., Yamakawa, H., Domonkos, I., Laczkó-Dobos, H., Kis, M., Ughy, B., & Gombos, Z. (2012). Two functional sites of phosphatidylglycerol for regulation of reaction of plastoquinone QB in photosystem II. *Biochimica et Biophysica Acta (BBA) – Bioenergetics, 1817*, 287–297.

Ivanov, A. G., Hendrickson, L., Krol, M., Selstam, E., Öquist, G., Hurry, V., & Huner, N. P. A. (2006). Digalactosyl-diacylglycerol deficiency impairs the capacity for photosynthetic intersystem Electron transport and state transitions in Arabidopsis thaliana due to photosystem I acceptor-side limitations. *Plant and Cell Physiology, 47*, 1146–1157.

Jahns, P., Latowski, D., & Strzalka, K. (2009). Mechanism and regulation of the violaxanthin cycle: The role of antenna proteins and membrane lipids. *Biochimica et Biophysica Acta (BBA) – Bioenergetics, 1787*, 3–14.

Jeong, J., Baek, K., Kirst, H., Melis, A., & Jin, E. (2017). Loss of CpSRP54 function leads to a truncated light-harvesting antenna size in Chlamydomonas reinhardtii. *Biochimica et Biophysica Acta-Bioenergetics, 1858*, 45–55.

Jia, T., Ito, H., & Tanaka, A. (2016). Simultaneous regulation of antenna size and photosystem I/II stoichiometry in Arabidopsis thaliana. *Planta, 244*, 1041–1053.

Jin, E. S., Polle, J. E. W., & Melis, A. (2001). Involvement of zeaxanthin and of the Cbr protein in the repair of photosystem II from photoinhibition in the green alga Dunaliella salina. *Biochimica et Biophysica Acta-Bioenergetics, 1506*, 244–259.

Jones, C. S., Luong, T., Hannon, M., Tran, M., Gregory, J. A., Shen, Z., Briggs, S. P., & Mayfield, S. P. (2013). Heterologous expression of the C-terminal antigenic domain of the malaria vaccine candidate Pfs48/45 in the green algae Chlamydomonas reinhardtii. *Applied Microbiology and Biotechnology, 97*, 1987–1995.

Jordan, B. R., Chow, W.-S., & Baker, A. J. (1983). The role of phospholipids in the molecular organisation of pea chloroplast membranes. Effect of phospholipid depletion on photosynthetic activities. *Biochimica et Biophysica Acta (BBA) – Bioenergetics, 725*, 77–86.

Kirst, H., & Melis, A. (2013). The chloroplast signal recognition particle (CpSRP) pathway as a tool to minimize chlorophyll antenna size and maximize photosynthetic productivity. *Biotechnology Advances, 32*, 66–72.

Kirst, H., & Melis, A. (2014). The chloroplast signal recognition particle (CpSRP) pathway as a tool to minimize chlorophyll antenna size and maximize photosynthetic productivity. *Biotechnology Advances, 32*, 66–72.

Kirst, H., Garcia-Cerdan, J. G., Zurbriggen, A., & Melis, A. (2012a). Assembly of the light-harvesting chlorophyll antenna in the green alga Chlamydomonas reinhardtii requires expression of the TLA2-CpFTSY gene. *Plant Physiology, 158*, 930–945.

Kirst, H., Garcia-Cerdan, J. G., Zurbriggen, A., Ruehle, T., & Melis, A. (2012b). Truncated photosystem chlorophyll antenna size in the green microalga Chlamydomonas reinhardtii upon deletion of the TLA3-CpSRP43 gene. *Plant Physiology, 160*, 2251–2260.

Klimyuk, V. I., Persello-Cartieaux, F., Havaux, M., Contard-David, P., Schuenemann, D., Meiherhoff, K., Gouet, P., Jones, J. D., Hoffman, N. E., & Nussaume, L. (1999). A chromodomain protein encoded by the *Arabidopsis* CAO gene is a plant-specific component of the chloroplast signal recognition particle pathway that is involved in LHCP targeting. *Plant Cell, 11*, 87–99.

Kobayashi, K. (2016). Role of membrane glycerolipids in photosynthesis, thylakoid biogenesis and chloroplast development. *Journal of Plant Research, 129*, 565–580.

Kobayashi, M., & Sakamoto, Y. J. B. L. (1999). Singlet oxygen quenching ability of astaxanthin esters from the green alga Haematococcus pluvialis. *Biotechnology Letters, 21*, 265–269.

Kobayashi, K., Narise, T., Sonoike, K., Hashimoto, H., Sato, N., Kondo, M., Nishimura, M., Sato, M., Toyooka, K., Sugimoto, K., Wada, H., Masuda, T., & Ohta, H. (2013). Role of galactolipid biosynthesis in coordinated development of photosynthetic complexes and thylakoid membranes during chloroplast biogenesis in Arabidopsis. *Biotechnology Letters, 73*, 250–261.

Kobayashi, K., Fujii, S., Sasaki, D., Baba, S., Ohta, H., Masuda, T., & Wada, H. (2014). Transcriptional regulation of thylakoid galactolipid biosynthesis coordinated with chlorophyll biosynthesis during the development of chloroplasts in Arabidopsis. *Frontiers in Plant Science, 5*, 272.

Kobayashi, K., Fujii, S., Sato, M., Toyooka, K., & Wada, H. J. P. C. R. (2015). Specific role of phosphatidylglycerol and functional overlaps with other thylakoid lipids in Arabidopsis chloroplast biogenesis. *Plant Cell Reports, 34*, 631–642.

Kobayashi, K., Endo, K., & Wada, H. (2016). Multiple Impacts of Loss of Plastidic Phosphatidylglycerol Biosynthesis on Photosynthesis during Seedling Growth of Arabidopsis. *Frontiers in Plant Science, 7*.

Koh, H. G., Kang, N. K., Jeon, S., Shin, S.-E., Jeong, B.-R., & Chang, Y. K. (2019a). Heterologous synthesis of chlorophyll b in *Nannochloropsis salina* enhances growth and lipid production by increasing photosynthetic efficiency. *Biotechnology for Biofuels, 12*, 122.

Koh, H. G., Kang, N. K., Jeon, S., Shin, S. E., Jeong, B. R., & Chang, Y. K. (2019b). Heterologous synthesis of chlorophyll b in Nannochloropsis Salina enhances growth and lipid production by increasing photosynthetic efficiency. *Biotechnology for Biofuels, 12*, 122.

Koh, H. G., Kang, N. K., Jeon, S., Shin, S. E., Jeong, B. R., & Chang, Y. K. (2019c). Heterologous synthesis of chlorophyll b in Nannochloropsis Salina enhances growth and lipid production by increasing photosynthetic efficiency. *Biotechnology for Biofuels, 12*, 122.

Kromdijk, J., Glowacka, K., Leonelli, L., Gabilly, S. T., Iwai, M., Niyogi, K. K., & Long, S. P. (2016). Improving photosynthesis and crop productivity by accelerating recovery from photoprotection. *Science, 354*, 857–861.

Kunugi, M., Takabayashi, A., & Tanaka, A. (2013). Evolutionary changes in Chlorophyllide a Oxygenase (CAO) structure contribute to the acquisition of a new light-harvesting complex in Micromonas. *Journal of Biological Chemistry, 288*, 19330–19341.

Kwon, Y. M., Kim, K. W., Choi, T. Y., Kim, S. Y., & Kim, J. Y. H. (2018). Manipulation of the microalgal chloroplast by genetic engineering for biotechnological utilization as a green biofactory. *World Journal of Microbiology and Biotechnology, 34*, 183.

Lang, M., Apt, K. E., & Kroth, P. G. (1998). Protein transport into "complex" diatom plastids utilizes two different targeting signals. *Journal of Biological Chemistry, 273*, 30973–30978.

Levine, R. P., & Goodenough, U. W. (1970). The genetics of photosynthesis and of the chloroplast in Chlamydomonas reinhardi. *Annual Review of Genetics, 4*, 397–408.

Li, X., Zhang, R., Patena, W., Gang, S. S., Blum, S. R., Ivanova, N., Yue, R., Robertson, J. M., Lefebvre, P. A., Fitz-Gibbon, S. T., Grossman, A. R., & Jonikas, M. C. (2016). An indexed, mapped mutant library enables reverse genetics studies of biological processes in Chlamydomonas reinhardtii. *Plant Cell, 28*, 367–387.

Lien, S., & San Pietro, A. (1976). *Inquiry into biophotolysis of water to produce hydrogen*. Indiana University, Bloomington (USA). Department of Plant Sciences.

Liu, X. Y., Ouyang, L. L., & Zhou, Z. G. (2016). Phospholipid: Diacylglycerol acyltransferase contributes to the conversion of membrane lipids into triacylglycerol in Myrmecia incisa during the nitrogen starvation stress. *Scientific Reports, 6*.

Mackinder, I. C. M. (2018). The *Chlamydomonas* CO_2-concentrating mechanism and its potential for engineering photosynthesis in plants. *The New Phytologist, 217*, 54–61.

Madi, L., Wang, X., Kobiler, I., Lichter, A., & Prusky, D. (2003). Stress on avocado fruits regulates Δ9-stearoyl ACP desaturase expression, fatty acid composition, antifungal diene level and resistance to Colletotrichum gloeosporioides attack. *Physiological and Molecular Plant Pathology, 62*, 277–283.

Masuda, T., Tanaka, A., & Melis, A. (2003a). Chlorophyll antenna size adjustments by irradiance in Dunaliella salina involve coordinate regulation of chlorophyll a oxygenase (CAO) and Lhcb gene expression. *Plant Molecular Biology, 51*, 757–771.

Masuda, T., Tanaka, A., & Melis, A. (2003b). Chlorophyll antenna size adjustments by irradiance in Dunaliella salina involve coordinate regulation of chlorophyll a oxygenase (CAO) and Lhcb gene expression. *Plant Molecular Biology, 51*, 757–771.

Mayfield, S. P., Franklin, S. E., & Lerner, R. A. (2003). Expression and assembly of a fully active antibody in algae. *Proceedings of the National Academy of Sciences of the United States of America, 100*, 438–442.

Melis, A. (2009). Solar energy conversion efficiencies in photosynthesis: Minimizing the chlorophyll antennae to maximize efficiency. *Plant Science, 177*, 272–280.

Melis, A., Neidhardt, J., & Benemann, J. R. (1998). Dunaliella salina (Chlorophyta) with small chlorophyll antenna sizes exhibit higher photosynthetic productivities and photon use efficiencies than normally pigmented cells. *Journal of Applied Phycology, 10*, 515–525.

Mezzomo, N., & Ferreira, S. (2016). Carotenoids functionality. *Sources, and Processing by Supercritical Technology: A Review, 2090-9063*(2016), 1–16.

Minoda, A., Sato, N., Nozaki, H., Okada, K., Takahashi, H., Sonoike, K., & Tsuzuki, M. (2002). Role of sulfoquinovosyl diacylglycerol for the maintenance of photosystem II in Chlamydomonas reinhardtii. *European Journal of Biochemistry, 269*, 2353–2358.

Minoda, A., Sonoike, K., Okada, K., Sato, N., & Tsuzuki, M. (2003). Decrease in the efficiency of the electron donation to tyrosine Z of photosystem II in an SQDG-deficient mutant of Chlamydomonas. *FEBS Letters, 553*, 109–112.

Mitra, M., Kirst, H., Dewez, D., & Melis, A. (2012). Modulation of the light-harvesting chlorophyll antenna size in Chlamydomonas reinhardtii by TLA1 gene over-expression and RNA interference. *Philosophical Transactions of the Royal Society B-Biological Sciences, 367*, 3430–3443.

Mizusawa, N., Sakata, S., Sakurai, I., Sato, N., & Wada, H. (2009a). Involvement of digalactosyldiacylglycerol in cellular thermotolerance in Synechocystis sp. PCC 6803. *Archives of Microbiology, 191*, 595–601.

Mizusawa, N., Sakurai, I., Sato, N., & Wada, H. (2009b). Lack of digalactosyldiacylglycerol increases the sensitivity of Synechocystis sp. PCC 6803 to high light stress. *FEBS Letters, 583*, 718–722.

Moog, D., Rensing, S. A., Archibald, J. M., Maier, U. G., & Ullrich, K. K. (2015a). Localization and evolution of putative triose phosphate translocators in the diatom Phaeodactylum tricornutum. *Genome Biology and Evolution, 7*, 2955–2969.

Moog, D., Stork, S., Reislohner, S., Grosche, C., & Maier, U. G. (2015b). In vivo localization studies in the stramenopile alga *Nannochloropsis oceanica. Protist, 166*, 161–171.

Muller, P., Li, X. P., & Niyogi, K. K. (2001). Non-photochemical quenching. A response to excess light energy. *Plant Physiology, 125*, 1558–1566.

Mussgnug, J. H., Wobbe, L., Elles, I., Claus, C., Hamilton, M., Fink, A., Kahmann, U., Kapazoglou, A., Mullineaux, C. W., Hippler, M., Nickelsen, J., Nixon, P. J., & Kruse, O. (2005). NAB1 is an RNA binding protein involved in the light-regulated differential expression of the light-harvesting antenna of *Chlamydomonas reinhardtii. Plant Cell, 17*, 3409–3421.

Mussgnug, J. H., Thomas-Hall, S., Rupprecht, J., Foo, A., Klassen, V., Mcdowall, A., Schenk, P. M., Kruse, O., & Hankamer, B. (2007). Engineering photosynthetic light capture: Impacts on improved solar energy to biomass conversion. *Plant Biotechnology Journal, 5*, 802–814.

Nakajima, Y., & Ueda, R. (1997). Improvement of photosynthesis in dense microalgal suspension by reduction of light harvesting pigments. *Journal of Applied Phycology, 9*, 503–510.

Nakajima, Y., Tsuzuki, M., & Ueda, R. (2001). Improved productivity by reduction of the content of light-harvesting pigment in Chlamydomonas perigranulata. *Journal of Applied Phycology, 13*, 95–101.

Noda, J., Muhlroth, A., Bucinska, L., Dean, J., Bones, A. M., & Sobotka, R. (2017). Tools for biotechnological studies of the freshwater alga Nannochloropsis limnetica: Antibiotic resistance and protoplast production. *Journal of Applied Phycology, 29*, 853–863.

Ongena, M., Duby, F., Rossignol, F., Fauconnier, M. L., Dommes, J., & Thonart, P. (2004). Stimulation of the lipoxygenase pathway is associated with systemic resistance induced in bean by a nonpathogenic Pseudomonas strain. *Molecular Plant-Microbe Interactions, 17*, 1009–1018.

Ossenbuhl, F., Gohre, V., Meurer, J., Krieger-Liszkay, A., Rochaix, J. D., & Eichacker, L. A. (2004). Efficient assembly of photosystem II in Chlamydomonas reinhardtii requires Alb3.1p, a homolog of Arabidopsis ALBINO3. *Plant Cell, 16*, 1790–1800.

Perin, G., Bellan, A., Segalla, A., Meneghesso, A., Alboresi, A., & Morosinotto, T. (2015). Generation of random mutants to improve light-use efficiency of *Nannochloropsis gaditana* cultures for biofuel production. *Biotechnology for Biofuels, 8*, 161.

Perozeni, F., Stella, G. R., & Ballottari, M. (2018). LHCSR expression under HSP70/RBCS2 promoter as a strategy to increase productivity in microalgae. *International Journal of Molecular Sciences, 19*, 155.

Perrine, Z., Negi, S., & Sayre, R. T. (2012). Optimization of photosynthetic light energy utilization by microalgae. *Algal Research-Biomass Biofuels and Bioproducts, 1*, 134–142.

Pineau, B., Girard-Bascou, J., Eberhard, S., Choquet, Y., Trémolières, A., Gérard-Hirne, C., Bennardo-Connan, A., Decottignies, P., Gillet, S., & Wollman, F.-A. (2004). A single mutation that causes phosphatidylglycerol deficiency impairs synthesis of photosystem II cores in Chlamydomonas reinhardtii. *European Journal of Biochemistry, 271*, 329–338.

Polle, J. E. W., Kanakagiri, S., Jin, E., Masuda, T., & Melis, A. (2002). Truncated chlorophyll antenna size of the photosystems - a practical method to improve microalgal productivity and hydrogen production in mass culture. *International Journal of Hydrogen Energy, 27*, 1257–1264.

Polle, J. E. W., Kanakagiri, S. D., & Melis, A. (2003). tla1, a DNA insertional transformant of the green alga Chlamydomonas reinhardtii with a truncated light-harvesting chlorophyll antenna size. *Planta, 217*, 49–59.

Prabandono, K., & Amin, S. (2015). Chapter 10 – Biofuel production from microalgae. In S.-K. Kim (Ed.), *Handbook of marine microalgae*. Boston: Academic Press.

Prof, D., & Gupta, C. (2014). Carotenoids: Chemistry and health benefits. *Phytochemicals of Nutraceutical Importance*, 181–195.

Ramel, F., Birtic, S., Cuiné, S., Triantaphylidès, C., Ravanat, J.-L., & Havaux, M. (2012). Chemical quenching of singlet oxygen by carotenoids in plants. *Plant Physiology, 158*, 1267–1278.

Rasala, B. A., Muto, M., Lee, P. A., Jager, M., Cardoso, R. M., Behnke, C. A., Kirk, P., Hokanson, C. A., Crea, R., Mendez, M., & Mayfield, S. P. (2010). Production of therapeutic proteins in algae, analysis of expression of seven human proteins in the chloroplast of Chlamydomonas reinhardtii. *Plant Biotechnology Journal, 8*, 719–733.

Rasala, B. A., Chao, S. S., Pier, M., Barrera, D. J., & Mayfield, S. P. (2014). Enhanced genetic tools for engineering multigene traits into green algae. *Plos One, 9*, e94028.

Reifarth, F., Christen, G., Seeliger, A. G., Dörmann, P., Benning, C., & Renger, G. (1997). Modification of the water oxidizing complex in leaves of the dgd1 mutant of Arabidopsis thaliana deficient in the Galactolipid Digalactosyldiacylglycerol. *Biochemistry, 36*, 11769–11776.

Sakuraba, Y., Balazadeh, S., Tanaka, R., Mueller-Roeber, B., & Tanaka, A. (2012). Overproduction of Chl b retards senescence through transcriptional reprogramming in Arabidopsis. *Plant and Cell Physiology, 53*, 505–517.

Sakurai, I., Hagio, M., Gombos, Z., Tyystjärvi, T., Paakkarinen, V., Aro, E.-M., & Wada, H. (2003). Requirement of phosphatidylglycerol for maintenance of photosynthetic machinery. *Plant Physiology, 133*, 1376–1384.

Sakurai, I., Mizusawa, N., Ohashi, S., Kobayashi, M., & Wada, H. (2007a). Effects of the lack of phosphatidylglycerol on the donor side of photosystem II. *Plant Physiology, 144*, 1336–1346.

Sakurai, I., Mizusawa, N., Wada, H., & Sato, N. (2007b). Digalactosyldiacylglycerol is required for stabilization of the oxygen-evolving complex in photosystem II. *Plant Physiology -Rockville Pike Bethesda, 145*, 1361–1370.

Sato, N., Sonoike, K., Tsuzuk, M., & Kawaguchi, A. (1995). Impaired photosystem II in a mutant of chlamydomonas reinhardtii. *Defective in Sulfoquinovosyl Diacylglycerol, 234*, 16–23.

Satoh, S., Ikeuchi, M., Mimuro, M., & Tanaka, A. (2001). Chlorophyll b expressed in cyanobacteria functions as a light harvesting antenna in photosystem I through flexibility of the proteins. *Journal of Biological Chemistry, 276*, 4293–4297.

Schaller, S., Latowski, D., Jemioła-Rzemińska, M., Wilhelm, C., Strzałka, K., & Goss, R. (2010). The main thylakoid membrane lipid monogalactosyldiacylglycerol (MGDG) promotes the deepoxidation of violaxanthin associated with the light-harvesting complex of photosystem II (LHCII). *Biochimica et Biophysica Acta (BBA) – Bioenergetics, 1797*, 414–424.

Schaller, S., Latowski, D., Jemioła-Rzemińska, M., Dawood, A., Wilhelm, C., Strzałka, K., & Goss, R. (2011). Regulation of LHCII aggregation by different thylakoid membrane lipids. *Biochimica et Biophysica Acta (BBA) – Bioenergetics, 1807*, 326–335.

Scheer, H. (2003). *Light harvesting antenna in photosynthesis*. Netherlands: Kluwer Academic Publishers.

Schenk, P. M., Thomas-Hall, S. R., Stephens, E., Marx, U. C., Mussgnug, J. H., Posten, C., Kruse, O., & Hankamer, B. (2008). Second generation biofuels: High-efficiency microalgae for biodiesel production. *Bioenergy Research, 1*, 20–43.

Shen, G., Canniffe, D. P., Ho, M. Y., Kurashov, V., Van Der Est, A., Golbeck, J. H., & Bryant, D. A. (2019). Characterization of chlorophyll f synthase heterologously produced in Synechococcus sp. PCC 7002. *Photosynthesis Research, 140*, 77–92.

Shin, S. E., Lim, J. M., Koh, H. G., Kim, E. K., Kang, N. K., Jeon, S., Kwon, S., Shin, W. S., Lee, B., Hwangbo, K., Kim, J., Ye, S. H., Yun, J. Y., Seo, H., Oh, H. M., Kim, K. J., Kim, J. S., Jeong, W. J., Chang, Y. K., & Jeong, B.-R. (2016a). CRISPR/Cas9-induced knockout and knock-in mutations in *Chlamydomonas reinhardtii*. *Scientific Reports, 6*, 27810.

Shin, S. E., Lim, J. M., Koh, H. G., Kim, E. K., Kang, N. K., Jeon, S., Kwon, S., Shin, W. S., Lee, B., Hwangbo, K., Kim, J., Ye, S. H., Yun, J. Y., Seo, H., Oh, H. M., Kim, K. J., Kim, J. S., Jeong, W. J., Chang, Y. K., & Jeong, B. R. (2016b). CRISPR/Cas9-induced knockout and knock-in mutations in Chlamydomonas reinhardtii. *Scientific Reports, 6*.

Shin, W.-S., Lee, B., Jeong, B.-R., Chang, Y. K., & Kwon, J.-H. (2016c). Truncated light-harvesting chlorophyll antenna size in *Chlorella vulgaris* improves biomass productivity. *Journal of Applied Phycology, 28*, 3193–3202.

Shin, W. S., Lee, B., Jeong, B. R., Chang, Y. K., & Kwon, J. H. (2016d). Truncated light-harvesting chlorophyll antenna size in Chlorella vulgaris improves biomass productivity. *Journal of Applied Phycology, 28*, 3193–3202.

Shin, W.-S., Lee, B., Kang, N. K., Kim, Y.-U., Jeong, W.-J., Kwon, J.-H., Jeong, B.-R., & Chang, Y. K. (2017a). Complementation of a mutation in CpSRP43 causing partial truncation of light-harvesting chlorophyll antenna in *Chlorella vulgaris*. *Scientific Reports, 7*, 17929.

Shin, W. S., Lee, B., Kang, N. K., Kim, Y. U., Jeong, W. J., Kwon, J. H., Jeong, B. R., & Chang, Y. K. (2017b). Complementation of a mutation in CpSRP43 causing partial truncation of light-harvesting chlorophyll antenna in Chlorella vulgaris. *Scientific Reports, 7*, 17929.

Small, I., Peeters, N., Legeai, F., & Lurin, C. (2004). Predotar: A tool for rapidly screening proteomes for N-terminal targeting sequences. *Proteomics, 4*, 1581–1590.

Sperschneider, J., Catanzariti, A. M., Deboer, K., Petre, B., Gardiner, D. M., Singh, K. B., Dodds, P. N., & Taylor, J. M. (2017). LOCALIZER: Subcellular localization prediction of both plant and effector proteins in the plant cell. *Scientific Reports, 7*, 44598.

Steffen, R., Kelly, A. A., Huyer, J., Dörmann, P., & Renger, G. (2005). Investigations on the reaction pattern of photosystem II in leaves from Arabidopsis thaliana wild type plants and mutants with genetically modified lipid content. *Biochemistry, 44*, 3134–3142.

Stephenson, P. G., Moore, C. M., Terry, M. J., Zubkov, M. V., & Bibby, T. S. (2011). Improving photosynthesis for algal biofuels: Toward a green revolution. *Trends in Biotechnology, 29*, 615–623.

Takaichi, S., Inoue, K., Akaike, M., Kobayashi, M., Oh-Oka, H., & Madigan, M. T. (1997). The major carotenoid in all known species of heliobacteria is the C30 carotenoid 4,4′-diaponeurosporene, not neurosporene. *Archives of Microbiology, 168*, 277–281.

Tanaka, A., Ito, H., Tanaka, R., Tanaka, N. K., Yoshida, K., & Okada, K. (1998a). Chlorophyll *a* oxygenase (*CAO*) is involved in chlorophyll *b* formation from chlorophyll *a*. *Proceedings of the National Academy of Sciences of the United States of America, 95*, 12719–12723.

Tanaka, A., Ito, H., Tanaka, R., Tanaka, N. K., Yoshida, K., & Okada, K. (1998b). Chlorophyll a oxygenase (CAO) is involved in chlorophyll b formation from chlorophyll a. *Proceedings of the National Academy of Sciences of the United States of America, 95*, 12719–12723.

Tanaka, R., Koshino, Y., Sawa, S., Ishiguro, S., Okada, K., & Tanaka, A. (2001). Overexpression of chlorophyllide a oxygenase (CAO) enlarges the antenna size of photosystem II in Arabidopsis thaliana. *Plant Journal, 26*, 365–373.

Tetali, S. D., Mitra, M., & Melis, A. (2007). Development of the light-harvesting chlorophyll antenna in the green alga Chlamydomonas reinhardtii is regulated by the novel Tla1 gene. *Planta, 225*, 813–829.

Tran, M., Henry, R. E., Siefker, D., Van, C., Newkirk, G., Kim, J., Bui, J., & Mayfield, S. P. (2013). Production of anti-cancer immunotoxins in algae: Ribosome inactivating proteins as fusion partners. *Biotechnology and Bioengineering, 110*, 2826–2835.

Tsuchiya, T., Akimoto, S., Mizoguchi, T., Watabe, K., Kindo, H., Tomo, T., Tamiaki, H., & Mimuro, M. (2012a). Artificially produced [7-formyl]-chlorophyll d functions as an antenna pigment in the photosystem II isolated from the chlorophyllide a oxygenase-expressing Acaryochloris marina. *Biochimica et Biophysica Acta-Bioenergetics, 1817*, 1285–1291.

Tsuchiya, T., Mizoguchi, T., Akimoto, S., Tomo, T., Tamiaki, H., & Mimuro, M. (2012b). Metabolic engineering of the Chl d-dominated cyanobacterium Acaryochloris marina: Production of a novel Chl species by the introduction of the Chlorophyllide a oxygenase gene. *Plant and Cell Physiology, 53*, 518–527.

Voitsekhovskaja, O. V., & Tyutereva, E. V. (2015). Chlorophyll *b* in angiosperms: Functions in photosynthesis, signaling and ontogenetic regulation. *Journal of Plant Physiology, 189*, 51–64.

Wang, S., Uddin, M. I., Tanaka, K., Yin, L., Shi, Z., Qi, Y., Mano, J., Matsui, K., Shimomura, N., Sakaki, T., Deng, X., & Zhang, S. (2014). Maintenance of chloroplast structure and function by overexpression of the rice MONOGALACTOSYLDIACYLGLYCEROL SYNTHASE gene leads to enhanced salt tolerance in tobacco. *Plant Physiology, 165*, 1144–1155.

Wannathong, T., Waterhouse, J. C., Young, R. E., Economou, C. K., & Purton, S. (2016). New tools for chloroplast genetic engineering allow the synthesis of human growth hormone in the green alga Chlamydomonas reinhardtii. *Applied Microbiology and Biotechnology, 100*, 5467–5477.

Wei, L., Wang, Q., Xin, Y., Lu, Y., & Xu, J. (2017). Enhancing photosynthetic biomass productivity of industrial oleaginous microalgae by overexpression of RuBisCO activase. *Algal Research, 27*, 366–375.

Wittek, F., Hoffmann, T., Kanawati, B., Bichlmeier, M., Knappe, C., Wenig, M., Schmitt-Kopplin, P., Parker, J. E., Schwab, W., & Vlot, A. C. (2014). Arabidopsis ENHANCED DISEASE SUSCEPTIBILITY1 promotes systemic acquired resistance via azelaic acid and its precursor 9-oxo nonanoic acid. *Journal of Experimental Botany, 65*, 5919–5931.

Work, V. H., D'adamo, S., Radakovits, R., Jinkerson, R. E., & Posewitz, M. C. (2012). Improving photosynthesis and metabolic networks for the competitive production of phototroph-derived biofuels. *Current Opinion in Biotechnology, 23*, 290–297.

Wu, W., Ping, W., Wu, H., Li, M., Gu, D., & Xu, Y. (2013). Monogalactosyldiacylglycerol deficiency in tobacco inhibits the cytochrome b6f-mediated intersystem electron transport process and affects the photostability of the photosystem II apparatus. *Biochimica et Biophysica Acta (BBA) – Bioenergetics, 1827*, 709–722.

Xu, H., Vavilin, D. & Vermaas, W. 2001. Chlorophyll b can serve as the major pigment in functional photosystem II complexes of cyanobacteria. *Proceedings of the National Academy of Sciences of the United States of America, 98*, 14168–14173.

Xu, C., Härtel, H., Wada, H., Hagio, M., Yu, B., Eakin, C. & Benning, C. (2002). The pgp1 mutant locus of Arabidopsis encodes a Phosphatidylglycerolphosphate synthase with impaired activity. *Plant Physiology, 129*, 594–604.

Yaeno, T., Matsuda, O., & Iba, K. (2004a). Role of chloroplast trienoic fatty acids in plant disease defense responses. *The Plant Journal, 40*, 931–941.

Yaeno, T., Matsuda, O., & Iba, K. (2004b). Role of chloroplast trienoic fatty acids in plant disease defense responses. *The Plant Journal, 40*, 931–941.

Yamasato, A., Nagata, N., Tanaka, R., & Tanaka, A. (2005). The N-terminal domain of chlorophyllide *a* oxygenase confers protein instability in response to chlorophyll *b* accumulation in Arabidopsis. *Plant Cell, 17*, 1585–1597.

Yamasato, A., Tanaka, R., & Tanaka, A. (2008). Loss of the N-terminal domain of chlorophyllide a oxygenase induces photodamage during greening of Arabidopsis seedlings. *BMC Plant Biology, 8*, 64.

Yang, B., Liu, J., Ma, X., Guo, B., Liu, B., Wu, T., Jiang, Y., & Chen, F. (2017). Genetic engineering of the Calvin cycle toward enhanced photosynthetic CO_2 fixation in microalgae. *Biotechnology for Biofuels, 10*, 229.

Yu, B., & Benning, C. (2003). Anionic lipids are required for chloroplast structure and function in Arabidopsis. *The Plant Journal, 36*, 762–770.

Yu, B., Xu, C., & Benning, C. (2002). Arabidopsis disrupted in SQD2 encoding sulfolipid synthase is impaired in phosphate-limited growth. *Proceedings of the National Academy of Sciences of the United States of America, 99*, 5732–5737.

Zhang, X. P., & Glaser, E. (2002). Interaction of plant mitochondrial and chloroplast signal peptides with the Hsp70 molecular chaperone. *Trends in Plant Science, 7*, 14–21.

Zhang, Y. H., & Robinson, D. G. (1990). Cell-wall synthesis in Chlamydomonas reinhardtii: An immunological study on the wild type and wall-less mutants cw2 and cw15. *Planta, 180*, 229–236.

Zhang, R., Patena, W., Armbruster, U., Gang, S. S., Blum, S. R., & Jonikas, M. C. (2014). High-throughput genotyping of green algal mutants reveals random distribution of mutagenic insertion sites and endonucleolytic cleavage of transforming DNA. *Plant Cell, 26*, 1398–1409.

Zhang, M., Deng, X., Yin, L., Qi, L., Wang, X., Wang, S. & Li, H. 2016. Regulation of galactolipid biosynthesis by overexpression of the rice MGD gene contributes to enhanced aluminum tolerance in tobacco. *Frontiers in Plant Science, 7*.

Self-Assembly, Organisation, Regulation, and Engineering of Carboxysomes: CO_2-Fixing Prokaryotic Organelles

Yaqi Sun, Fang Huang, and Lu-Ning Liu

Abstract Carboxysomes are a group of bacterial microcompartments (BMCs) that encapsulate Rubisco and carbonic anhydrase to enhance CO_2 fixation in cells. Through self-assembly of hundreds of proteins into a virus-like icosahedral organelle, carboxysomes provide all cyanobacteria and some chemoautotrophs with the ability to utilise limited environmental CO_2 and function as a key component of CO_2-concentrating mechanisms. In this chapter, we will summarise recent advances in understanding the composition, biogenesis, structural and functional regulation of carboxysomes, as well as synthetic engineering of carboxysomes.

Keywords Carbon fixation · Bacterial microcompartment · Carboxysome · Self-assembly · Metabolic organelle · CO_2-concentrating mechanisms

1 Bacterial Microcompartments

Organelle compartmentalisation within cells provides the structural foundation for physiological optimisation and modulation of metabolic reactions in space and time (Kerfeld et al. 2010; Bobik et al. 2015). In eukaryotic cells, the specialised subcellular organelles (such as chloroplasts, mitochondria, and lysosomes) encapsulate specific enzymes and metabolic pathways from the intricate cytosolic environment without mutual interference (Satori et al. 2013; Gabaldón and Pittis 2015). Similar compartmentalisation has been observed in prokaryotic cells, and one example is

Y. Sun · F. Huang
Institute of Integrative Biology, University of Liverpool, Liverpool, UK

L.-N. Liu (✉)
Institute of Integrative Biology, University of Liverpool, Liverpool, UK

College of Marine Life Sciences, Ocean University of China, Qingdao, China
e-mail: luning.liu@liverpool.ac.uk

Q. Wang (ed.), *Microbial Photosynthesis*,
https://doi.org/10.1007/978-981-15-3110-1_15

the bacterial microcompartments (BMCs) that are widespread among bacterial phyla (Axen et al. 2014).

Unlike eukaryotic organelles, BMCs are proteinaceous mega-complexes assembled with hundreds of different proteins. The BMC consists of a virus-like shell and encapsulated cargo enzymes (Fig. 1) (Bobik 2006; Kerfeld et al. 2010; Yeates et al. 2011). Through encapsulation of enzymes, concentrating substrates, or avoiding leakage of toxic intermediates that are to be consumed sequentially, BMCs enhance diverse enzymatic reactions within the subcellular "micro-factories" and make notable contributions to the metabolic diversities of bacteria accommodating specific habitats (Havemann et al. 2002; Yeates et al. 2008; Huang et al. 2001; Chowdhury et al. 2014). Since the first BMC has been discovered in 1956 (Drews and Niklowitz 1956), tremendous studies have been carried out in the past 60 years to provide advanced knowledge about the structures, biosynthesis, and functionality of BMCs.

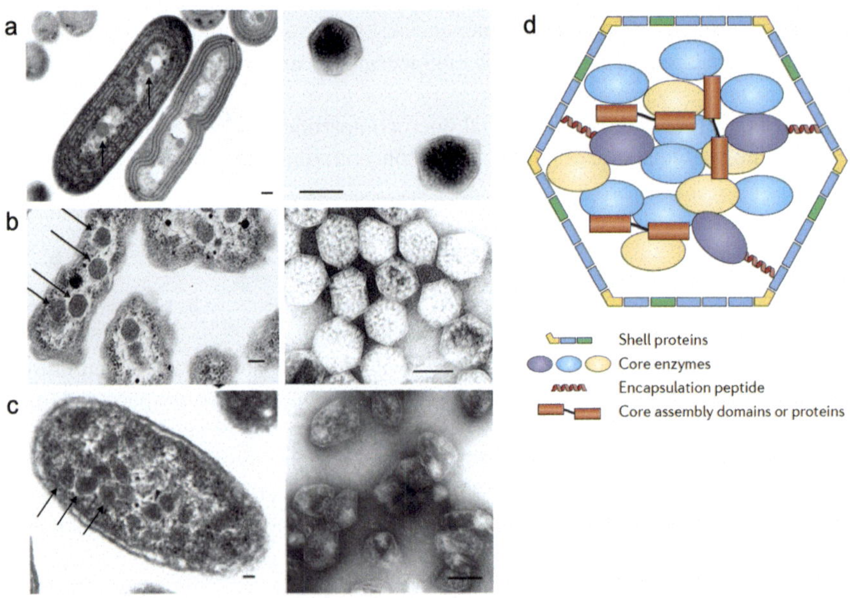

Fig. 1 Structures of BMCs. (**a**–**c**) Electron microscopy images showing BMCs within native host cells and isolated BMCs from the cyanobacterium *Synechococcus elongatus PCC7942* (**a**, β-carboxysome), *Halothiobacillus neapolitanus* (**b**, α-carboxysome), and *Salmonella enterica* (**c**, propanediol utilisation microcompartment). Images were modified from (Tsai et al. 2007; Crowley et al. 2008; Faulkner et al. 2017). Scale bar, 100 nm. (**d**) Schematic of BMC showing the composition of proteins, modified from (Kerfeld et al. 2018)

1.1 The BMC Shells

The BMC shell is made of thousands of polypeptides that belong to multiple protein paralogs (Yeates et al. 2010). The single-layer shell acts as a physical barrier to concentrate and protect enzymes within the BMC lumen and is semipermeable to specific metabolic molecules (Kerfeld et al. 2005; Tsai et al. 2007; Tanaka et al. 2008; Kerfeld and Erbilgin 2015; Faulkner et al. 2017; Fan et al. 2012). Three conserved protein paralogs form the shell: hexamers (BMC-H), trimers (BMC-T), and pentamers (BMC-P) (Kerfeld and Erbilgin 2015). The BMC-H protein peptides self-assemble into hexameric units and construct the majority of the shell facets (Kerfeld et al. 2010; Dryden et al. 2009). The BMC-T proteins, with a tandem BMC-H domain, function as pseudo-hexamers that make up a smaller fraction of the shells; it has a large pore that was suggested to be the potential gateways for molecule passage through the shell (Klein et al. 2009; Cai et al. 2013). The overall assembly model of BMC-H, BMC-P, and BMC-T is believed to be of icosahedral symmetry (Tanaka et al. 2008), which has been confirmed recently by the structure analysis of a synthetic shell comprising all three types of protein paralogs (Sutter et al. 2017). Based on the icosahedral geometry, the BMC-P proteins were suggested to occupy the vertices of the shell (Iancu et al. 2007; Tanaka et al. 2008; Sutter et al. 2013; Wheatley et al. 2013). A diagram of the BMC shell organisation is illustrated in Fig. 2. All shell proteins possess a central pore, with distinctive

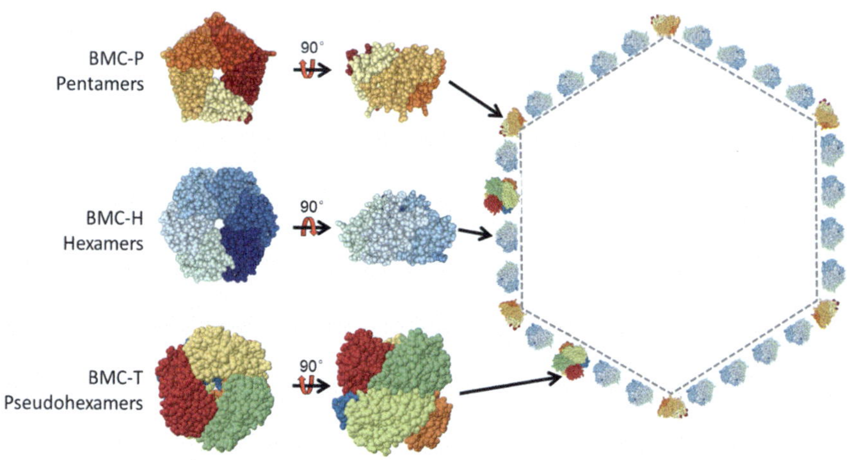

Fig. 2 Schematic diagram of BMC shell structure by pentameric and (pseudo)-hexameric proteins. BMC-P proteins are deduced to locate on the vertexes of the polyhedron shells, whereas BMC-H and BMC-T proteins form the surfaces of the facet. Assembly was shown for illustration only; numbers and ratios of building blocks do not represent the real stoichiometry. Space fill models were adopted from PDB entries (BMC-T PDB ID, 5LT5; BMC-H PDB ID, 2A1B; BMC-P PDB ID, 2QW7) reported in (Sutter et al. 2017; Tanaka et al. 2008; Kerfeld et al. 2005)

electrostatic properties on the concave and convex sides (Sutter et al. 2017; Tanaka et al. 2008; Kerfeld et al. 2005). The pores of shell proteins are speculated to mediate the passages of substrates and products to enter or exit from BMCs (Chowdhury et al. 2015; Crowley et al. 2010; Park et al. 2017).

1.2 The BMC Cargo Enzymes

Although BMC shells are structurally alike and phylogenetically related, they encapsulate a variety of enzymes responsible for diverse biological functions, allowing the host bacteria to survive in different environmental niches and play important roles in CO_2 fixation, pathogenesis, and microbial ecology (Yeates et al. 2010; Bobik et al. 2015; Yeates et al. 2008; Kerfeld et al. 2018). These enzymes require specific linker proteins to form defined assemblies and interact with shell proteins (Ryan et al. 2019; Wang et al. 2019b).

Three major types of BMCs have been identified according to the biological reactions encapsulated inside the BMCs: the carboxysomes for CO_2 fixation, the Pdu metabolosomes for vitamin B12-dependent 1,2-propanediol utilisation, and the Eut metabolosome for ethanolamine utilisation (Kerfeld et al. 2018). The anabolic carboxysomes were found in all phototrophic cyanobacteria and some chemoautotrophic organisms (Shively et al. 1973; Price et al. 1998). The catabolic BMCs, such as Pdu and Eut metabolosomes, provide competitive advantages in the metabolism of organic compounds of pathogenic bacteria (Bobik et al. 1999; Kofoid et al. 1999). In addition, bioinformatic studies have revealed uncharacterised BMCs with a variety of functions in the bacteria kingdom (Axen et al. 2014; Ravcheev et al. 2019), including a new type of Pdu microcompartment that is vitamin B12-independent (Zarzycki et al. 2017; Levin and Balskus 2018), the Planctomycetes and Verrucomicorbia microcompartments that are capable of degrading the L-fucose and L-rhamnose (Erbilgin et al. 2014), as well as the choline-utilisation BMCs found in human gut bacteria (Martinez-Del Campo et al. 2015; Craciun et al. 2016; Backman et al. 2017). With the progressively growing database of bacteria genomics, more uncharacterised BMCs are entering our views that await further investigations (Axen et al. 2014; Kerfeld et al. 2018). The expanding range of BMC structures and functions suggests the adaptive nature of BMCs to cope with specific environmental niches (Chowdhury et al. 2014; Ravcheev et al. 2019).

2 CO_2-Concentrating Mechanisms and CO_2 Uptake Systems

In cyanobacteria, carboxysomes coordinate with inorganic carbon (Ci) transporters in plasma membranes and CO_2 uptake complexes in thylakoid membranes, forming CO_2-concentrating mechanisms (CCMs) to ensure efficient carbon assimilation (Badger and Price 2003) (Fig. 3). Ci transporters are capable of transport HCO_3^- passively into the cytoplasm (Price et al. 2008). So far, multiple types of bicarbonate

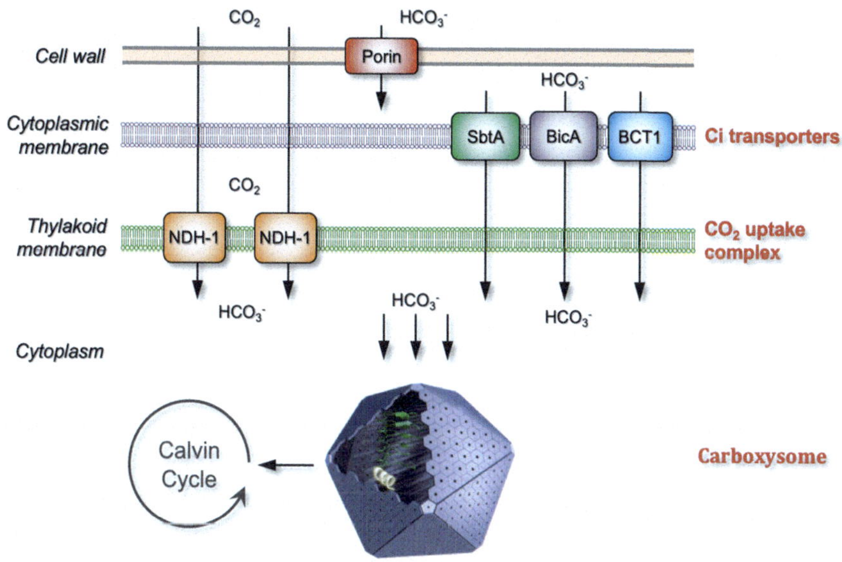

Fig. 3 Schematic model of CO_2-concentrating mechanisms in cyanobacteria

transporters have been identified, including an ATP-dependent high-affinity transporter BCT1 (Omata et al. 1999) and electrochemical Na^+-coupled medium affinity transporters SbtA and BicA (Price et al. 2004; Shibata et al. 2002). Recently, the structure and function of BicA, existing as a dimer, in the cyanobacterium *Synechocystis* sp. PCC 6803 have been studied (Wang et al. 2019a). Likewise, recent studies have revealed the presence of bicarbonate transporters (DAB1 and DAB2) in the chemoautotroph *Halothiobacillus neapolitanus* (Desmarais et al. 2019). The DAB homologs also appear in human pathogens, as well as a diverse range of bacteria and archaea.

In addition, two isoforms of CO_2 uptake NDH-I complexes located in thylakoid membranes convert CO_2 that diffuse in cells across membranes into HCO_3^-, further elevating the concentration of HCO_3^- in cytoplasm (Burnap et al. 2015; Maeda et al. 2002; Shibata et al. 2001; Liu et al. 2012). Overall, thanks to the Ci uptake systems, the concentration of Ci pools in the cytoplasm could reach a strikingly 1000-fold higher than that of the environmental Ci level (Woodger et al. 2005; Price et al. 1998), therefore remarkably enhancing CO_2 fixation (Schwarz et al. 1995; Eisenhut et al. 2008; Hagemann et al. 2013).

3 Carboxysome: Anabolic BMCs

Carboxysomes were the first BMC discovered (Drews and Niklowitz 1956) and have been extensively characterised as a model BMC system. Unlike the Pdu and Eut microcompartments which are only required by host cells to utilise one replicable source of nutrients, carboxysomes are physiologically vital for cell survival in the natural environment. Removing or impairing carboxysomes could render cells incapable of CO_2 fixation in ambient air and thus lead to cell death (Price and Badger 1989). Additionally, carboxysomes contain the smallest set of shell and enzyme proteins, therefore suitable for studying the biogenesis, function, and structures of BMCs.

3.1 Encapsulated Enzymes in the Carboxysome: Rubisco and Carbonic Anhydrase

Two types of enzymes are encased within the carboxysomes lumen: the Ribulose-1,5-bisphosphate carboxylase/oxygenase (Rubisco) (Shively et al. 1973) and the carbonic anhydrase (CA) (Mcgurn et al. 2016). Rubisco has both carboxylation and oxidation activities by binding with CO_2 and O_2 (Stec 2012). The carboxylation of Rubisco is the rate-limiting step of the Calvin-Benson-Bassham (CBB) cycle (Bassham et al. 1950). The fixation of O_2 leads to the "wasteful" process of photorespiration (Peterhansel et al. 2010). Rubisco exhibits natural catalytic diversity in the evolutionary context. In cyanobacteria, green algae, and C_4 plants, Rubisco enzymes have higher carboxylation rates (V_{CO2}) but lower CO_2 affinities (higher K_m for CO_2) and lower specificities to distinguish CO_2 and O_2 ($S_{C/O}$). By contrast, Rubisco in most C_3 crop plants that lack any CCM possess slower carboxylation rates but higher CO_2 affinities and CO_2/O_2 specificities. Due to the relatively low specificity for CO_2 over O_2 (Fig. 4), free forms of cyanobacterial Rubiscos are inca-

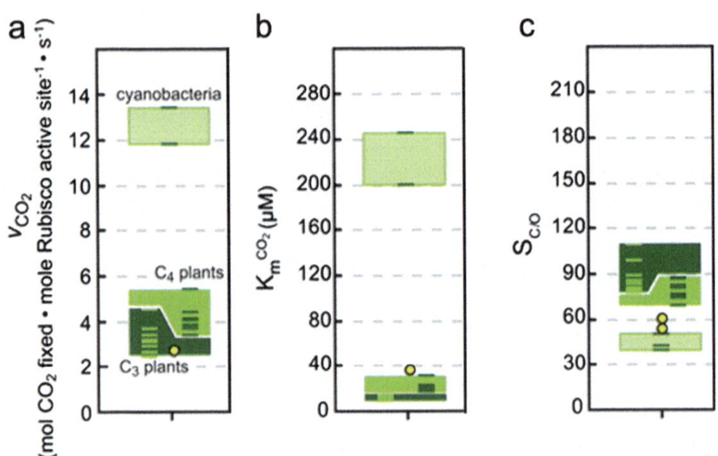

Fig. 4 Comparison of the catalytic activities of cyanobacterial and plant Rubisco enzymes. The carbon-fixation rate as V_{CO2} (**a**), CO_2 affinity as K_m^{CO2} (**b**), and specificity of CO_2 over O_2 as $S_{C/O}$ calculated from $(V_{CO2}/K_m^{O2})/(V_{O2}/K_m^{CO2})$ (**c**). Circles coloured in yellow correspond to values for green algae. The figure was adapted from (Whitney et al. 2011)

pable to fulfil carboxylation activities at ambient levels of CO_2 (Galmes et al. 2014; Mcnevin et al. 2006). To compensate it, carboxysomal Rubisco possess a higher catalytic rate in CO_2 conversion, at a rate of 12–14 CO_2 molecules per Rubisco per second, compared with 2–6 for plant Rubisco (Whitney et al. 2011).

The enzymatic properties of Rubisco are optimised for carboxylase activities with the assistance of CA (Fig. 5). Inside the carboxysome, CA catalyses the conversion of HCO_3^- to CO_2, providing high levels of CO_2 around Rubisco to facilitate CO_2 fixation. Cyanobacterial cells lacking CA exhibited a high-CO_2 requirement phenotype, indicating the significance of CA in cyanobacterial CO_2 fixation (So et al. 2002). Some cyanobacterial species do not have the carboxysomal β-CA homologs; instead, the N-terminal domain of CcmM functions as an active γ-CA (Pena et al. 2010).

3.2 *Shell Permeability*

The selective permeability of carboxysome shell is crucial for internal enzyme activity (Griffiths 2006; Frey et al. 2016). The carboxysome shell is permeable to charged HCO_3^-, 3-phosphoglyceric acid (3-PGA), and Ribulose 1,5-bisphosphate (RuBP) and allows the entry/exit of these substrates/products across the shell through the central pores of shell proteins (Fig. 5). Meanwhile, O_2 diffusion through the shell was believed to be precluded; this feature, together with the elevated CO_2 levels in the carboxysome lumen, facilitates Rubisco carboxylation activities and minimises photorespiration (Rae et al. 2013; Dou et al. 2008; Tanaka et al. 2009; Turmo et al. 2017).

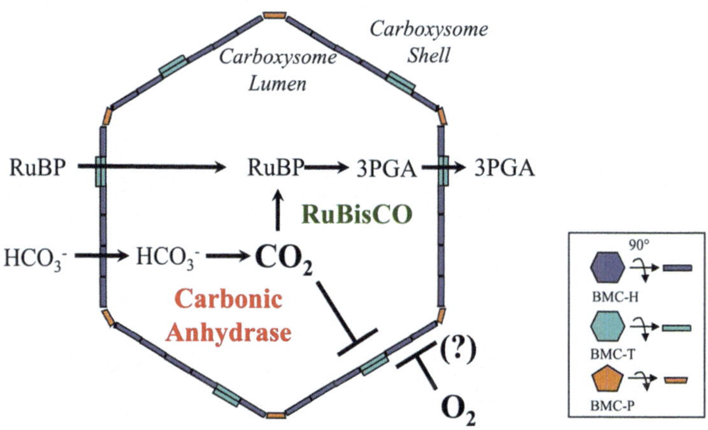

Fig. 5 Diagrams of carboxysome functions. Carboxysomes are capable of converting CO_2 and RuBP into 3PGA, supported by permeable shells that allow influx/efflux of substrates/products. The figure was adapted from (Turmo et al. 2017). *RuBP* Ribulose 1,5-bisphosphate, *3PGA* 3-phosphoglyceric acid

4 Two Types of Carboxysomes

According to the types of Rubisco (form 1A and form 1B) encapsulated, carboxysomes can be categorised into two groups: α-carboxysomes and β-carboxysomes. The independently evolved α- and β-carboxysomes were found mostly in oceanic and freshwater cyanobacterial strains, respectively (Rae et al. 2013).

4.1 Model Organisms for Studying Carboxysomes

The chemoautotroph *Halothiobacillus neapolitanus* C2 (*H. neapolitanus*) and the cyanobacterium *Synechococcus elongatus* PCC7942 (Syn7942) are currently the best-studied model organisms (Rae et al. 2012, 2013). *H. neapolitanus*, a bacterium that fix CO_2 by their energy derived through oxidation of reduced sulphur compounds, has been widely adopted as a model strain for α-carboxysome study. Syn7942, previously known as *Anacystis nidulans* R2, has been extensively selected as a model strain to study carbon assimilation (Tchernov et al. 2001), acclimation to environmental changes (Bustos and Golden 1992; Schwarz and Grossman 1998; Tsinoremas et al. 1994), as well as bacterial circadian clocks (Cohen and Golden 2015; Golden 2003; Swan et al. 2018). This rod-shaped organism is superior for microscopic studies over β-carboxysomes due to its distinctive carboxysome contents (three to four carboxysomes per cell) in ordered cytosolic spaces and well-separated subcellular localisations (Yokoo et al. 2015; Savage et al. 2010). Additionally, Syn7942 is reliably transformable by exogenous DNA naturally (Shestakov and Khyen 1970), providing easy genetic modification.

4.2 Gene Organisation of Carboxysomes

The genetic composition and organisation of α- and β-carboxysomes vary, as displayed in Fig. 6. For α-carboxysomes, all genes required for carboxysome assembly and function (9 out of 11 genes) are located over a single *cso* operon (Bonacci et al. 2012). The additional satellite carboxysomal genes *csoS1D* (Klein et al. 2009) and *cbbQ/cbbO* (Sutter et al. 2015; Tsai et al. 2015) were shown to play roles in enhancing carbon fixation. By contrast, the genetic content of β-carboxysomes has higher complexity than that of α-carboxysomes. Besides the major *ccm* and *rbc* operons that encode carboxysomal genes from *ccmK2* to *rbcS*, four distant loci have been identified encoding BMC-T and BMC-H shell proteins, as well as the crucial CA (CcaA) and Rubisco chaperone RbcX (Fig. 6).

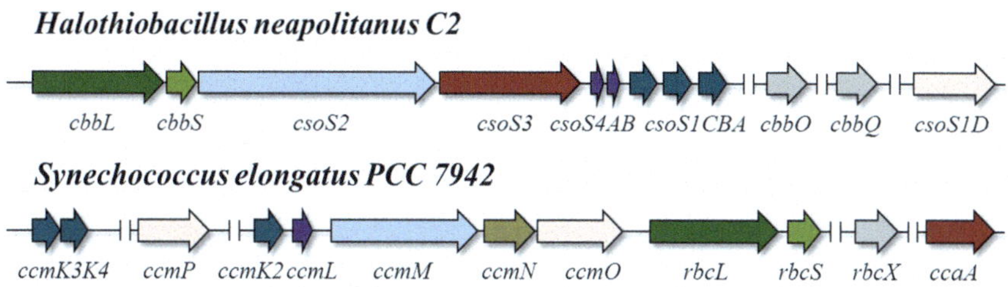

Fig. 6 The organisation of α- and β-carboxysome genes in the genomes of *H. neapolitanus* and Syn7942. Double-slash lines indicate gaps between separated carboxysome gene operons

4.3 Protein Composition of Carboxysomes

In contrast to the diverse gene organisation, α- and β-carboxysomes share high degrees of similarities in protein composition and overall structure assemblies. The protein compositions of *H. neapolitanus* α-carboxysomes and Syn7942 β-carboxysomes are listed in Table 1.

Form IA and IB Rubisco share a hexadecameric (L_8S_8) structure (Schneider et al. 1992). The Rubisco holoenzyme is composed of eight large subunits coded by *cbbL* or *rbcL* and eight small subunits coded by *cbbS* or *rbcS* for α- and β-carboxysomes, respectively. The reaction core is formed through tetramerisation of large subunit dimmers in the form of L_8. Eight small subunits dock at the top and bottom of the L_8 cylinder, resulting in the formation of an L_8S_8 complex (Spreitzer 2003).

Table 1 Protein information of typical α- and β-carboxysomes in model organisms *H. neapolitanus* C2 and *Syn7942*

Description	α-carboxysome	β-carboxysome
Rubisco large subunit	CbbL	RbcL
Rubisco small subunit	CbbS	RbcS
Carbonic anhydrase	CsoS3	CcaA
Shell hexamers	CsoS1A	CcmK2
	CsoS1B	CcmK3
	CsoS1C	CcmK4
Pentamers, putative vertices of shell	CsoS4A	CcmL
	CsoS4B	
Structural protein	CsoS2	CcmM
		CcmN
Pseudo-hexamers; putative shell edge proteins		CcmO
Pseudo- hexamers, minor shell protein	CsoS1D	CcmP
Rubisco assembly chaperon		RbcX
Rubisco activase	CbbQ/CbbO	

The CAs coded by *csoS3* in the α-carboxysome and *ccaA* in the β-carboxysome (Long et al. 2007) are ε-class and β-class CAs, respectively (So et al. 2004; Mcgurn et al. 2016). These CAs have no significant sequential similarity but share similar catalytic properties in HCO_3^-/CO_2 conversion (So et al. 2004). In many cyanobacterial organisms, additional CA domains appear in the core assembly protein, such as CcmM in β-carboxysomes (Pena et al. 2010); in a subset of β-carboxysome-containing cyanobacteria, the CA function could be completely replaced by CcmM, leading to the loss of the *ccaA* gene in the genome (Pena et al. 2010). The convergent evolution of CAs supports the independent evolution of carboxysomes.

The shell proteins CcmK2-K4 in the β-carboxysome (Kerfeld et al. 2005; Tanaka et al. 2008; Samborska and Kimber 2012; Tanaka et al. 2009) and CsoS1ABC in the α-carboxysome (Tsai et al. 2007) belong to the BMC-H family and cover the majority of shell facets. The co-existence of these homologs likely lays the foundation for the structural integrity of the shell and the regulation of shell permeability. Recently, it was reported that CcmK3 and CcmK4 can form heterohexamers in a pH-dependent manner, which may lead to the capping/uncapping transition of the carboxysome shell (Sommer et al. 2019).

CcmO and CcmP in the β-carboxysome (Cai et al. 2013; Marco et al. 1994; Larsson et al. 2017) and CsoS1D in the α-carboxysome (Klein et al. 2009) belong to the BMC-T family and are minor components in the shell facets. Structural analysis of CcmP and CsoS1D revealed a larger pore compared with BMC-H proteins, suggesting a potential metabolite channel, with the open/close capability, for passage of large molecules (Klein et al. 2009; Cai et al. 2013). Further investigation is needed to determine what triggers the pore formation changes of BMC-T proteins.

CcmL in the β-carboxysome (Tanaka et al. 2008) and CsoS4AB in the α-carboxysome (Cai et al. 2009) from the BMC-P family were deduced to cap the vertices of the polyhedron shells. Lacking BMC-P proteins leads to defects in carboxysomes shape, and typical "rod-like" carboxysome structures were commonly observed (Rae et al. 2013; Cameron et al. 2013; Long et al. 2018).

There are also associated proteins that are required for the carboxysome formation and functionality, such as GroEL/GroES for RbcL folding (Hayer-Hartl et al. 2016), RbcX/CbbX and Rubisco accumulation factor 1 (Raf1) for $RbcL_8$ assembly (Emlyn-Jones et al. 2006; Occhialini et al. 2016; Saschenbrecker et al. 2007), and CbbO/CbbQ for Rubisco activation. The genes encoding these proteins are not always located in the operons of carboxysomes and, therefore, were conventionally considered not to be integrated into carboxysomes (Tanaka et al. 2007; Emlyn-Jones et al. 2006; Zarzycki et al. 2013). However, recent studies have revealed that RbcX is part of the mature β-carboxysome structure and loss of RbcX could affect the carboxysome organisation (Huang et al. 2019).

4.4 Assembly Pathways of Carboxysomes

The strategies for nucleation of Rubisco enzymes to form the interior matrix and the overall assembly pathways of carboxysomes were proposed to differ in the α- and β-carboxysomes. In α-carboxysomes, the enzyme cores were observed to be encapsulated partially by shells (Iancu et al. 2010; Dai et al. 2018), suggesting a concomitant assembly of Rubisco assemblies with shells (Fig. 7). The Rubisco aggregation is driven by the structural protein CsoS2, which could associate with shell proteins to recruit shell encapsulation (Cai et al. 2015a; Oltrogge et al. 2019). Interestingly, protein interactions have also been found between the N-terminus of CbbL and all shell hexamers (Liu et al. 2018), which supports the concomitant assembly model of α-carboxysomes.

In β-carboxysomes, the interior assembly is mediated by structural proteins CcmM and CcmN (Fig. 7b). CcmN serves as a linker to bind with shell proteins CcmK through the C-terminus encapsulation peptide (Aussignargues et al. 2015) and associate with CcmM by the N-terminal domain (Kinney et al. 2012). CcmM exists in two forms: a short isoform CcmM35 and a long isoform CcmM58 (Long et al. 2007). CcmM58 processes an additional N-terminal domain that binds CcmN and, therefore, enables encapsulation of shells (Long et al. 2007, 2010, 2011). Both CcmM35 and CcmM58 contain three RbcS-like homolog domains (SSUL) which were previously believed to replace three RbcS in the Rubisco L_8S_8 structure, leading to the $L_8S_5M_3$ stoichiometry of Rubisco-CcmM supercomplexes that is vital for compact-packing paracrystalline of Rubisco assemblies (Long et al. 2011). Recently, Cryo-EM analysis of the reconstituted Rubisco-CcmM structure revealed that the SSUL of CcmM binds to the side of Rubisco L_8S_8 complex instead of replacing RbcS, leading to the formation of a liquid-like matrix of β-carboxysome core components (Wang et al. 2019b). It remains still unclear what results in the reduction in RbcS content against RbcL, which has been determined previously (Long et al. 2011; Faulkner et al. 2017; Sun et al. 2019).

Recent studies documented that de novo assembly of β-carboxysomes exploits the "inside out" model: Rubisco and CcmM forming the core first, followed by encapsulation of shell proteins that is mediated by CcmN and CcmK (Cameron et al. 2013; Chen et al. 2013). This model could explain the higher Rubisco packing density in β-carboxysome (Sun et al. 2019; Rae et al. 2013) and the greater carboxysome diameters ranging from 100 to 400 nm (Sun et al. 2019; Cai et al. 2015a). The dense core could also contribute to the structural robustness and integrity of β-carboxysomes (Kerfeld and Melnicki 2016; Rae et al. 2013). By contrast, a smaller diameter of the α-carboxysome, ranging from 100 to 160 nm (Heinhorst et al. 2014), limits the enzyme loading per carboxysome compared with β-carboxysomes. Through computational simulations, both concomitant and "inside out" assembly modes have been examined (Perlmutter et al. 2016; Rotskoff and Geissler 2018).

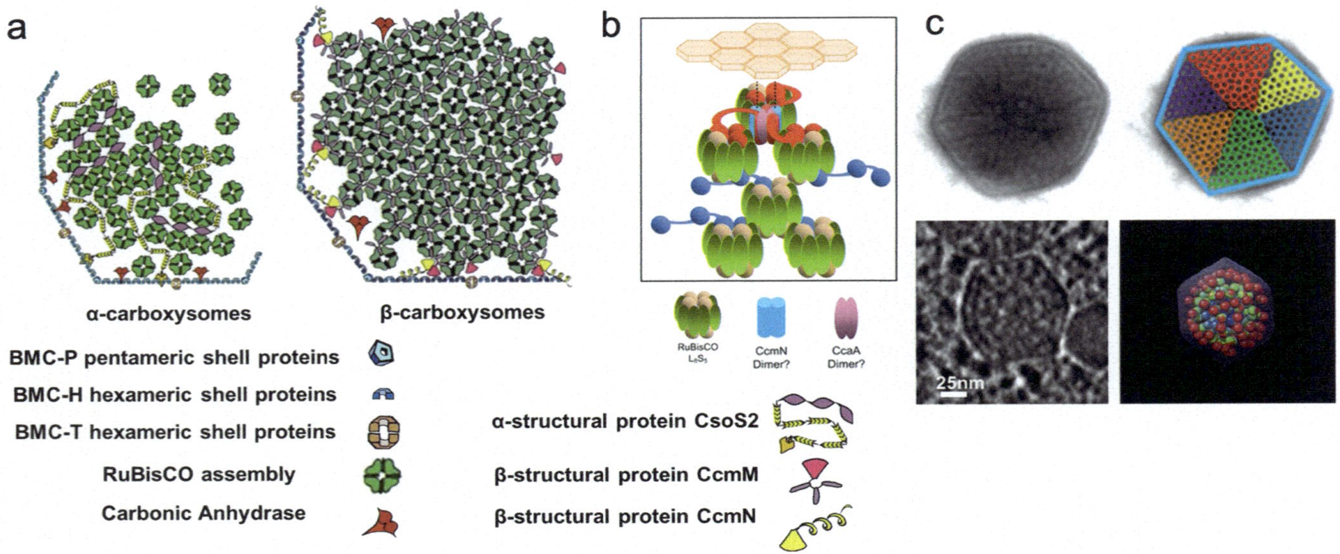

Fig. 7 Models of the assembly pathways of α- and β-carboxysomes, adapted from (Kerfeld and Melnicki 2016; Rae et al. 2013; Faulkner et al. 2017; Dai et al. 2018). (**a**) Diagrams of the assembly pathways of α- and β-carboxysome. (**b**) A model of $RbcL_8RbcS_5CcmM_3$ formation in the β-carboxysome. (**c**) Electron microscopy images of α- and β-carboxysomes (bottom and top, respectively). Circles in different colour indicate clusters of Rubiscos with less ordered/ordered formation for α- and β-carboxysomes

5 Structural Plasticity of Carboxysomes

The formation and structures of carboxysomes in living cells are highly regulated in the physiological context in response to environmental/growth conditions. Transcriptomic studies have reported the alternation of carboxysomal transcriptional profiles within different growth environments such as light intensities and CO_2 availabilities (Gill et al. 2002; Hihara et al. 2001; Huang et al. 2002; McGinn et al. 2004). The multi-operon organisation of β-carboxysome genes could be a strategy to modulate carboxysome protein expression and self-assembly to form functional entities in freshwater habitats (Whitehead et al. 2014). Moreover, photosynthetic electron flow has been reported to be a key regulatory factor in β-carboxysome formation and location in Syn7942 cells (Sun et al. 2016). Likewise, the composition of individual protein within the β-carboxysome and the overall β-carboxysome structure were demonstrated to be adaptively alternated depending on light intensity and CO_2 availabilities (Sun et al. 2019) (Fig. 8).

The structural plasticity of carboxysomes was also shown in protein-protein interactions. It was found that CcmK3 and CcmK4 of Syn7942 could form heterohexamers in a pH-dependent manner, with a 2:4 stoichiometry (Sommer et al. 2019). The combination variability of shell hexamers potentially permits tune of shell permeability in carboxysomes. The formation of CcmK3 and CcmK4 heterohexamers has also been observed in *Synechocystis* sp. PCC 6803 (Garcia-Alles et al. 2019), probably indicating a general principle underlying the shell organisation and permeability regulation of cyanobacterial carboxysomes. Moreover, capping of hexamers and pentamers in the carboxysome shell (Sommer et al. 2019) and other BMC shells (Hagen et al. 2018) has been reported, providing alternative means to alter metabolite flux across the shell.

The variability of shell organisation and composition appears as a common feature of carboxysomes and other BMCs, as well as reconstituted shells (Greber et al. 2019). Atomic force microscopy (AFM) has revealed the self-assembly dynamics of BMC shell proteins in seconds and the formation process of shell facets that is dependent upon environmental conditions (Sutter et al. 2016; Faulkner et al. 2019) (Fig. 9). In addition, AFM nano-indentation analysis demonstrated that the β-carboxysome structure is mechanically softer than virus capsids, likely a physical nature of this bacterial organelle (Faulkner et al. 2017; Rodriguez-Ramos et al. 2018).

6 Subcellular Positioning of Carboxysomes

The number and subcellular localisation of carboxysomes are of physiological importance for cell metabolism and growth. In Syn7942, three to four β-carboxysomes were observed to be evenly spaced along the centreline of the longitudinal axis of cells (Fig. 10a), ensuring equal segregation of β-carboxysomes between daughter cells (Savage et al. 2010). Moreover, the numbers of carboxysomes per Syn7942

a

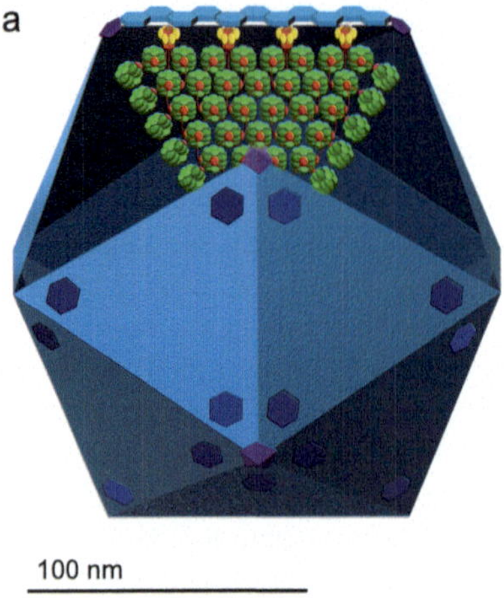

	Copy of functional units			
	Air/ML	CO_2/ML	LL	HL
Rubisco*	853	550	367	1507
CcaA	14	21	11	20
CcmM35 + CcmM58	719	468	483	1176
CcmN	74	52	51	82
CcmL	7.4	13.2	6.8	13.8
CcmK4	52	94	52	51
CcmK3	15	29	14	14

b

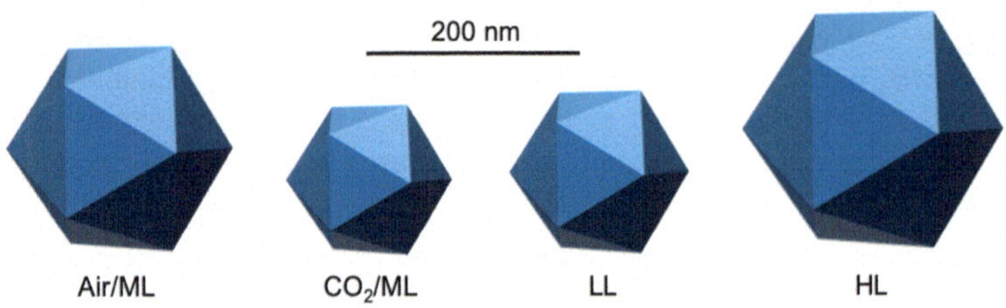

Fig. 8 Model of the β-carboxysome structure and protein stoichiometry. (**a**) Diagram of an icosahedral carboxysome structure and organization of building components. The stoichiometry of each building component within the carboxysome and its variations in response to changes in CO_2 and light intensity. ∗Rubisco content was estimated from RbcS stoichiometry based on the $RbcL_8S_8$ Rubisco structure. The majority of shell facets shown in light blue is tiled by the major shell protein $CcmK_2$. The total abundance of CcmM58 and CcmM35 was estimated. The components RbcL, $CcmK_2$, CcmO, and CcmP were not directly determined in this work and thus are not shown in this model. (**b**) The carboxysome diameter is variable in response to changes in the CO_2 level and light intensity (Sun et al. 2019)

cells vary depending on environmental conditions (Sun et al. 2016; Sun et al. 2019). The spatial organisation of carboxysomes in the cell correlates with the redox state of the photosynthetic electron transport chain (Sun et al. 2016), as shown in Fig. 10b. Additionally, previous studies have indicated that carboxysomes could also locate close to thylakoid membranes (Mckay et al. 1993), which may potentially have the advantages in Ci utilisation when carboxysomes are closer to Ci uptake systems.

The specific organisation of carboxysomes within cyanobacterial cells was proposed to be mediated by interactions between carboxysomes and cellular components, such as cytoskeletons (Savage et al. 2010). The cytoskeletal proteins such as ParA and MreB have been identified in the fraction of isolated carboxysomes (Faulkner et al. 2017). Another model that has recently been proposed identified a self-organising ParA-type ATPase system, McdAB, which modulate the positioning of carboxysome in nucleoid body through a non-cytoskeletal system in Syn7942 using a Brownian-ratchet mechanism (Maccready et al. 2018).

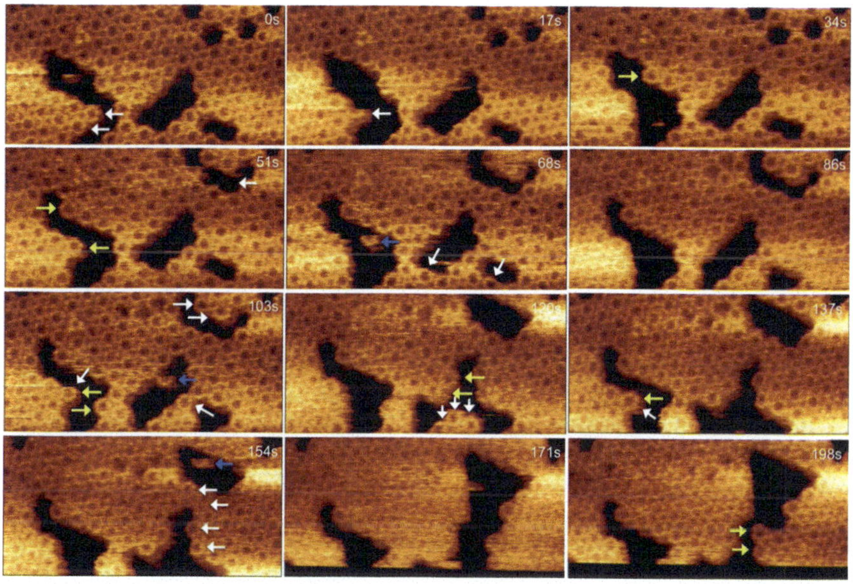

Fig. 9 Aligned time-lapse images reveal the dynamics of BMC-H sheet formation by atomic force microscopy (AFM) (Sutter et al. 2016). Hexamers are both removed from (white arrows) and incorporated into the sheet (yellow arrows) during the course of scanning. Blue arrows depict hexamers not associated with the sheet that is translocating across the mica surface. Frame interval: 17 s

7 Synthetic Engineering of Carboxysomes

The self-assembly and modularity nature, as well as metabolic enhancement of carboxysomes, make them an ideal engineering objective. The first attempt was to express α-carboxysome operon from *H. neapolitanus* and has led to the production of recombinant carboxysome-like structures with CO_2 fixation activity in *Escherichia coli* (Bonacci et al. 2012). The operon has also been used to reconstitute α-carboxysomes in a Gram-positive organism *Corynebacterium glutamicum* (Baumgart et al. 2017). By contrast, reconstructing β-carboxysome structures in prokaryotic organisms is far behind the α-carboxysome engineering, due to the multiple operons encoding β-carboxysome proteins. Recently, it has been shown that carboxysome-like structures could be formed when expressing all the carboxysome proteins in *E. coli*, with the detectable CO_2-fixing activity (Fang et al. 2018), representing a step toward constructing functional β-carboxysomes in eukaryotic organisms, such as plants.

The shared structural features and high sequence similarity of carboxysome proteins from different hosts raise the possibility of constructing chimeric carboxysomes. It has been reported that the form 1A Rubisco large subunit CbbL and shell protein CsoS1A of α-carboxysomes from *H. neapolitanus* could be incorporated into the natural β-carboxysomes in Syn7942 individually (Fang et al. 2018; Cai et al. 2015b), verifying the interchangeability of carboxysome building components.

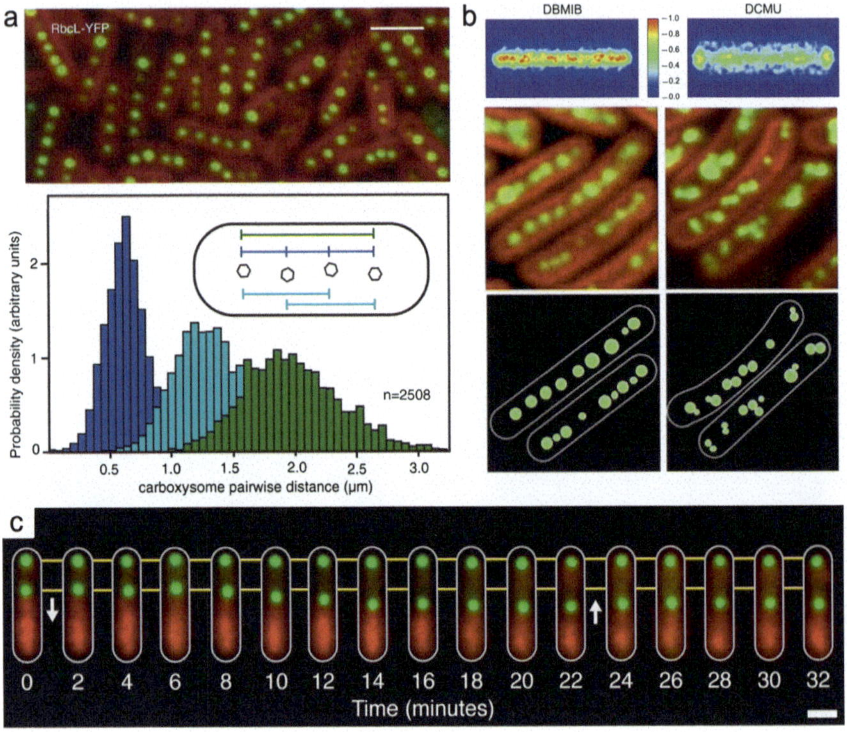

Fig. 10 Spatial localisation of β-carboxysomes in cyanobacteria cells. (**a**) Even distribution of β-carboxysomes in vivo. (**b**) The spatial organisation of β-carboxysomes is related to the redox states of the photosynthetic electron transport chain when treated with inhibitors DCMU (3-(3,4-dichlorophenyl)-1,1-dimethylurea) and DBMIB (dibromothymoquinone). (**c**) Time-lapse confocal imaging revealed the interactions between carboxysomes (green) and McdA (red). White arrow indicates the moving direction of the central carboxysome. (Images were adapted from (MacCready et al. 2018; Sun et al. 2019; Savage et al. 2010))

Design and engineering of functional carboxysome structures with simplified composition are necessary for biotechnology applications. Expression of *ccmK1*, *ccmK2*, *ccmL*, and *ccmO* from the cyanobacterium *Halothece* sp. PCC 7418 in *E. coli* could generate synthetic β-carboxysome shells (Cai et al. 2016). The shell proteins self-assemble predominantly into two families of shells with pseudo-T = 3 and pseudo-T = 4 icosahedral symmetry with diameters of 21 and 24.5 nm, respectively (Sutter et al. 2019). The empty carboxysome shells, with fewer shell proteins and without encapsulated enzymes, were observed to be notably smaller than the native β-carboxysomes (Faulkner et al. 2017). Moreover, β-carboxysome structures were formed in Syn7942 by expressing a chimeric hybrid CcmC, which replaced native CcmM, CcaA, and CcmN (Gonzalez-Esquer et al. 2015). The resulting carboxysomes exhibited reduced diameter and uneven Rubisco loading, demonstrating the necessity of optimising carboxysome composition and assembly for its structural and functional integrity.

The capacity of carboxysomes in enhancing carbon assimilation has attracted tremendous interest in installing carboxysomes into plants to improve photosynthesis and crop yields (Hanson et al. 2016). Attempts have been made to express carboxysome components in the chloroplasts of the model crop tobacco *Nicotiana benthamiana*, including expressing Syn7942 Rubisco with a fast carboxylation rate to replace endogenous Rubisco (Lin et al. 2014b; Occhialini et al. 2016) and transiently expressing β-carboxysome shell proteins (Lin et al. 2014a). Recently, a simplified α-carboxysome structure, including CbbL, CbbS, CsoS1A, and CsoS2 from *Cynaobium* sp. PCC 7001, was reconstituted in tobacco chloroplasts (Long et al. 2018).

8 Concluding Remarks

Carboxysomes are prokaryotic organelles made of purely proteins. Functionally resembling eukaryotic organelles, carboxysomes encapsulate the key CO_2-fixing Rubisco to elevate Rubisco concentration and CO_2 levels around Rubisco, thereby improving carbon assimilation. Despite that the carboxysome studies have been performed for more than 60 years, the exponential growth of the research outcomes in the field has only been seen in the past decade. Significant research on the structural and functional nature of carboxysomes, in particular, their structures, protein stoichiometry, assembly pathways, encapsulation principles, positioning, and physiological adaptation, has shed light on the building and regulatory principles of carboxysomes, extendable to diverse bacterial microcompartments. Moreover, the emergence of synthetic biology has spurred ambitions to engineer functional carboxysome modules and repurpose carboxysome structures to generate new nanoreactors and scaffolding materials for use in biotechnology. Further systematic studies require a combination of multidisciplinary techniques, ranging from nanoscale structural characterisation and manipulation to high-throughput "-omics," computational analysis and design, as well as machine learning.

References

Aussignargues, C., Paasch, B. C., Gonzalez-Esquer, R., Erbilgin, O., & Kerfeld, C. A. (2015). Bacterial microcompartment assembly: The key role of encapsulation peptides. *Communicative & Integrative Biology, 8*, e1039755.

Axen, S. D., Erbilgin, O., & Kerfeld, C. A. (2014). A taxonomy of bacterial microcompartment loci constructed by a novel scoring method. *PLoS Computational Biology, 10*, e1003898.

Backman, L. R. F., Funk, M. A., Dawson, C. D., & Drennan, C. L. (2017). New tricks for the glycyl radical enzyme family. *Critical Reviews in Biochemistry and Molecular Biology, 52*, 674–695.

Badger, M. R., & Price, G. D. (2003). CO2 concentrating mechanisms in cyanobacteria: Molecular components, their diversity and evolution. *Journal of Experimental Botany, 54*, 609–622.

Bassham, J. A., Benson, A. A., & Calvin, M. (1950). The path of carbon in photosynthesis. *The Journal of Biological Chemistry, 185*, 781–787.

Baumgart, M., Huber, I., Abdollahzadeh, I., Gensch, T., & Frunzke, J. (2017). Heterologous expression of the *Halothiobacillus neapolitanus* carboxysomal gene cluster in *Corynebacterium glutamicum*. *Journal of Biotechnology, 258*, 126–135.

Bobik, T. A. (2006). Polyhedral organelles compartmenting bacterial metabolic processes. *Applied Microbiology and Biotechnology, 70*, 517–525.

Bobik, T. A., Havemann, G. D., Busch, R. J., Williams, D. S., & Aldrich, H. C. (1999). The propanediol utilization (pdu) operon of Salmonella enterica serovar Typhimurium LT2 includes genes necessary for formation of polyhedral organelles involved in coenzyme B(12)-dependent 1, 2-propanediol degradation. *Journal of Bacteriology, 181*, 5967–5975.

Bobik, T. A., Lehman, B. P., & Yeates, T. O. (2015). Bacterial microcompartments: Widespread prokaryotic organelles for isolation and optimization of metabolic pathways. *Molecular Microbiology, 98*, 193–207.

Bonacci, W., Teng, P. K., Afonso, B., Niederholtmeyer, H., Grob, P., Silver, P. A., & Savage, D. F. (2012). Modularity of a carbon-fixing protein organelle. *Proceedings of the National Academy of Sciences of the United States of America, 109*, 478–483.

Burnap, R. L., Hagemann, M., & Kaplan, A. (2015). Regulation of CO(2) concentrating mechanism in cyanobacteria. *Life, 5*, 348–371.

Bustos, S. A., & Golden, S. S. (1992). Light-regulated expression of the psbD gene family in Synechococcus sp. strain PCC 7942: Evidence for the role of duplicated psbD genes in cyanobacteria. *Molecular & General Genetics, 232*, 221–230.

Cai, F., Menon, B. B., Cannon, G. C., Curry, K. J., Shively, J. M., & Heinhorst, S. (2009). The pentameric vertex proteins are necessary for the icosahedral carboxysome shell to function as a CO_2 leakage barrier. *PLoS One, 4*, e7521.

Cai, F., Sutter, M., Cameron, J. C., Stanley, D. N., Kinney, J. N., & Kerfeld, C. A. (2013). The structure of CcmP, a tandem bacterial microcompartment domain protein from the beta-carboxysome, forms a subcompartment within a microcompartment. *The Journal of Biological Chemistry, 288*, 16055–16063.

Cai, F., Dou, Z., Bernstein, S. L., Leverenz, R., Williams, E. B., Heinhorst, S., Shively, J., Cannon, G. C., & Kerfeld, C. A. (2015a). Advances in understanding carboxysome assembly in *Prochlorococcus* and *Synechococcus* implicate CsoS2 as a critical component. *Life (Basel), 5*, 1141–1171.

Cai, F., Sutter, M., Bernstein, S. L., Kinney, J. N., & Kerfeld, C. A. (2015b). Engineering bacterial microcompartment shells: Chimeric shell proteins and chimeric carboxysome shells. *ACS Synthetic Biology, 4*, 444–453.

Cai, F., Bernstein, S. L., Wilson, S. C., & Kerfeld, C. A. (2016). Production and characterization of synthetic carboxysome shells with incorporated luminal proteins. *Plant Physiology, 170*, 1868–1877.

Cameron, J. C., Wilson, S. C., Bernstein, S. L., & Kerfeld, C. A. (2013). Biogenesis of a bacterial organelle: The carboxysome assembly pathway. *Cell, 155*, 1131–1140.

Chen, A. H., Robinson-Mosher, A., Savage, D. F., Silver, P. A., & Polka, J. K. (2013). The bacterial carbon-fixing organelle is formed by shell envelopment of preassembled cargo. *PLoS One, 8*, e76127.

Chowdhury, C., Sinha, S., Chun, S., Yeates, T. O., & Bobik, T. A. (2014). Diverse bacterial microcompartment organelles. *Microbiology and Molecular Biology Reviews, 78*, 438–468.

Chowdhury, C., Chun, S., Pang, A., Sawaya, M. R., Sinha, S., Yeates, T. O., & Bobik, T. A. (2015). Selective molecular transport through the protein shell of a bacterial microcompartment organelle. *Proceedings of the National Academy of Sciences, 112*, 2990–2995.

Cohen, S. E., & Golden, S. S. (2015). Circadian rhythms in cyanobacteria. *Microbiology and Molecular Biology Reviews, 79*, 373–385.

Craciun, S., Marks, J. A., & Balskus, E. P. (2016). Correction to characterization of choline trimethylamine-lyase expands the chemistry of glycyl radical enzymes. *ACS Chemical Biology, 11*, 2068.

Crowley, C. S., Sawaya, M. R., Bobik, T. A., & Yeates, T. O. (2008). Structure of the PduU shell protein from the Pdu microcompartment of Salmonella. *Structure, 16*, 1324–1332.

Crowley, C. S., Cascio, D., Sawaya, M. R., Kopstein, J. S., Bobik, T. A., & Yeates, T. O. (2010). Structural insight into the mechanisms of transport across the Salmonella enterica Pdu Microcompartment Shell. *The Journal of Biological Chemistry, 285*, 37838–37846.

Dai, W., Chen, M., Myers, C., Ludtke, S. J., Pettitt, B. M., King, J. A., Schmid, M. F., & Chiu, W. (2018). Visualizing individual RuBisCO and its assembly into carboxysomes in marine cyanobacteria by Cryo-Electron Tomography. *Journal of Molecular Biology, 430*, 4156–4167.

Desmarais, J. J., Flamholz, A. I., Blikstad, C., Dugan, E. J., Laughlin, T. G., Oltrogge, L. M., Chen, A. W., Wetmore, K., Diamond, S., Wang, J. Y., & Savage, D. F. (2019). DABs are inorganic carbon pumps found throughout prokaryotic phyla. *Nature Microbiology, 4*(12), 2204–2215.

Dou, Z., Heinhorst, S., Williams, E. B., Murin, C. D., Shively, J. M., & Cannon, G. C. (2008). CO_2 fixation kinetics of *Halothiobacillus neapolitanus* mutant carboxysomes lacking carbonic anhydrase suggest the shell acts as a diffusional barrier for CO_2. *The Journal of Biological Chemistry, 283*, 10377–10384.

Drews, G., & Niklowitz, W. (1956). Cytology of Cyanophycea. II. Centroplasm and granular inclusions of Phormidium uncinatum. *Archiv für Mikrobiologie, 24*, 147–162.

Dryden, K. A., Crowley, C. S., Tanaka, S., Yeates, T. O., & Yeager, M. (2009). Two-dimensional crystals of carboxysome shell proteins recapitulate the hexagonal packing of three-dimensional crystals. *Protein Science, 18*, 2629–2635.

Eisenhut, M., Ruth, W., Haimovich, M., Bauwe, H., Kaplan, A., & Hagemann, M. (2008). The photorespiratory glycolate metabolism is essential for cyanobacteria and might have been conveyed endosymbiontically to plants. *Proceedings of the National Academy of Sciences of the United States of America, 105*, 17199–17204.

Emlyn-Jones, D., Woodger, F. J., Price, G. D., & Whitney, S. M. (2006). RbcX can function as a rubisco chaperonin, but is non-essential in *Synechococcus* PCC7942. *Plant & Cell Physiology, 47*, 1630–1640.

Erbilgin, O., Mcdonald, K. L., & Kerfeld, C. A. (2014). Characterization of a Planctomycetal organelle: A novel bacterial microcompartment for the aerobic degradation of plant saccharides. *Applied and Environmental Microbiology, 80*, 2193–2205.

Fan, C., Cheng, S., Sinha, S., & Bobik, T. A. (2012). Interactions between the termini of lumen enzymes and shell proteins mediate enzyme encapsulation into bacterial microcompartments. *Proceedings of the National Academy of Sciences of the United States of America, 109*, 14995–15000.

Fang, Y., Huang, F., Faulkner, M., Jiang, Q., Dykes, G. F., Yang, M., & Liu, L. N. (2018). Engineering and modulating functional cyanobacterial CO2-fixing organelles. *Frontiers in Plant Science, 9*, 739.

Faulkner, M., Rodriguez-Ramos, J., Dykes, G. F., Owen, S. V., Casella, S., Simpson, D. M., Beynon, R. J., & Liu, L. N. (2017). Direct characterization of the native structure and mechanics of cyanobacterial carboxysomes. *Nanoscale, 9*, 10662–10673.

Faulkner, M., Zhao, L. S., Barrett, S., & Liu, L. N. (2019). Self-assembly stability and variability of bacterial microcompartment shell proteins in response to the environmental change. *Nanoscale Research Letters, 14*, 54.

Frey, R., Mantri, S., Rocca, M., & Hilvert, D. (2016). Bottom-up construction of a primordial carboxysome mimic. *Journal of the American Chemical Society, 138*, 10072–10075.

Gabaldón, T., & Pittis, A. A. (2015). Origin and evolution of metabolic sub-cellular compartmentalization in eukaryotes. *Biochimie, 119*, 262–268.

Galmes, J., Kapralov, M. V., Andralojc, P. J., Conesa, M. A., Keys, A. J., Parry, M. A., & Flexas, J. (2014). Expanding knowledge of the rubisco kinetics variability in plant species: Environmental and evolutionary trends. *Plant, Cell & Environment, 37*, 1989–2001.

Garcia-Alles, L. F., Root, K., Maveyraud, L., Aubry, N., Lesniewska, E., Mourey, L., Zenobi, R., & Truan, G. (2019). Occurrence and stability of hetero-hexamer associations formed by β-carboxysome CcmK shell components. *PLoS One, 14*(10), e0223877.

Gill, R. T., Katsoulakis, E., Schmitt, W., Taroncher-Oldenburg, G., Misra, J., & Stephanopoulos, G. (2002). Genome-wide dynamic transcriptional profiling of the light-to-dark transition in *Synechocystis* sp. strain PCC 6803. *Journal of Bacteriology, 184*, 3671–3681.

Golden, S. S. (2003). Timekeeping in bacteria: The cyanobacterial circadian clock. *Current Opinion in Microbiology, 6*, 535–540.

Gonzalez-Esquer, C. R., Shubitowski, T. B., & Kerfeld, C. A. (2015). Streamlined construction of the cyanobacterial CO_2-fixing organelle via protein domain fusions for use in plant synthetic biology. *Plant Cell, 27*, 2637–2644.

Greber, B. J., Sutter, M., & Kerfeld, C. A. (2019). The plasticity of molecular interactions governs bacterial microcompartment shell assembly. *Structure, 27*, 749–763.e4.

Griffiths, H. (2006). Designs on Rubisco. *Nature, 441*, 940.

Hagemann, M., Fernie, A. R., Espie, G. S., Kern, R., Eisenhut, M., Reumann, S., Bauwe, H., & Weber, A. P. (2013). Evolution of the biochemistry of the photorespiratory C2 cycle. *Plant Biology (Stuttgart, Germany), 15*, 639–647.

Hagen, A., Sutter, M., Sloan, N., & Kerfeld, C. A. (2018). Programmed loading and rapid purification of engineered bacterial microcompartment shells. *Nature Communications, 9*, 2881.

Hanson, M. R., Lin, M. T., Carmo-Silva, A. E., & Parry, M. A. (2016). Towards engineering carboxysomes into C3 plants. *The Plant Journal, 87*, 38–50.

Havemann, G. D., Sampson, E. M., & Bobik, T. A. (2002). PduA is a shell protein of polyhedral organelles involved in coenzyme B(12)-dependent degradation of 1,2-propanediol in *Salmonella enterica* serovar typhimurium LT2. *Journal of Bacteriology, 184*, 1253–1261.

Hayer-Hartl, M., Bracher, A., & Hartl, F. U. (2016). The GroEL-GroES chaperonin machine: A nano-cage for protein folding. *Trends in Biochemical Sciences, 41*, 62–76.

Heinhorst, S., Cannon, G. C., & Shively, J. M. (2014). Carboxysomes and their structural organization in prokaryotes. In L. L. Barton, D. A. Bazylinski, & H. Xu (Eds.), *Nanomicrobiology: Physiological and environmental characteristics*. New York: Springer.

Hihara, Y., Kamei, A., Kanehisa, M., Kaplan, A., & Ikeuchi, M. (2001). DNA microarray analysis of cyanobacterial gene expression during acclimation to high light. *Plant Cell, 13*, 793–806.

Huang, X., Holden, H. M., & Raushel, F. M. (2001). Channeling of substrates and intermediates in enzyme-catalyzed reactions. *Annual Review of Biochemistry, 70*, 149–180.

Huang, L., McCluskey, M. P., Ni, H., & LaRossa, R. A. (2002). Global gene expression profiles of the cyanobacterium *Synechocystis* sp. strain PCC 6803 in response to irradiation with UV-B and white light. *Journal of Bacteriology, 184*, 6845–6858.

Huang, F., Vasieva, O., Sun, Y., Faulkner, M., Dykes, G. F., Zhao, Z., & Liu, L. N. (2019). Roles of RbcX in carboxysome biosynthesis in the cyanobacterium Synechococcus elongatus PCC7942. *Plant Physiology, 179*, 184–194.

Iancu, C. V., Ding, H. J., Morris, D. M., Dias, D. P., Gonzales, A. D., Martino, A., & Jensen, G. J. (2007). The structure of isolated *Synechococcus* strain WH8102 carboxysomes as revealed by electron cryotomography. *Journal of Molecular Biology, 372*, 764–773.

Iancu, C. V., Morris, D. M., Dou, Z., Heinhorst, S., Cannon, G. C., & Jensen, G. J. (2010). Organization, structure, and assembly of alpha-carboxysomes determined by electron cryotomography of intact cells. *Journal of Molecular Biology, 396*, 105–117.

Kerfeld, C. A., & Erbilgin, O. (2015). Bacterial microcompartments and the modular construction of microbial metabolism. *Trends in Microbiology, 23*, 22–34.

Kerfeld, C. A., & Melnicki, M. R. (2016). Assembly, function and evolution of cyanobacterial carboxysomes. *Current Opinion in Plant Biology, 31*, 66–75.

Kerfeld, C. A., Sawaya, M. R., Tanaka, S., Nguyen, C. V., Phillips, M., Beeby, M., & Yeates, T. O. (2005). Protein structures forming the shell of primitive bacterial organelles. *Science, 309*, 936–938.

Kerfeld, C. A., Heinhorst, S., & Cannon, G. C. (2010). Bacterial microcompartments. *Annual Review of Microbiology, 64*, 391–408.

Kerfeld, C. A., Aussignargues, C., Zarzycki, J., Cai, F., & Sutter, M. (2018). Bacterial microcompartments. *Nature Reviews Microbiology, 16*, 277.

Kinney, J. N., Salmeen, A., Cai, F., & Kerfeld, C. A. (2012). Elucidating essential role of conserved carboxysomal protein CcmN reveals common feature of bacterial microcompartment assembly. *Journal of Biological Chemistry, 287*, 17729–17736.

Klein, M. G., Zwart, P., Bagby, S. C., Cai, F., Chisholm, S. W., Heinhorst, S., Cannon, G. C., & Kerfeld, C. A. (2009). Identification and structural analysis of a novel carboxysome shell protein with implications for metabolite transport. *Journal of Molecular Biology, 392*, 319–333.

Kofoid, E., Rappleye, C., Stojiljkovic, I., & Roth, J. (1999). The 17-gene ethanolamine (eut) operon of Salmonella typhimurium encodes five homologues of carboxysome shell proteins. *Journal of Bacteriology, 181*, 5317–5329.

Larsson, A. M., Hasse, D., Valegard, K., & Andersson, I. (2017). Crystal structures of β-carboxysome shell protein CcmP: Ligand binding correlates with the closed or open central pore. *Journal of Experimental Botany, 68*, 3857–3867.

Levin, B. J., & Balskus, E. P. (2018). Characterization of 1,2-Propanediol dehydratases reveals distinct mechanisms for B12-dependent and glycyl radical enzymes. *Biochemistry, 57*, 3222–3226.

Lin, M. T., Occhialini, A., Andralojc, P. J., Devonshire, J., Hines, K. M., Parry, M. A., & Hanson, M. R. (2014a). β-Carboxysomal proteins assemble into highly organized structures in Nicotiana chloroplasts. *The Plant Journal, 79*, 1–12.

Lin, M. T., Occhialini, A., Andralojc, P. J., Parry, M. A. J., & Hanson, M. R. (2014b). A faster rubisco with potential to increase photosynthesis in crops. *Nature, 513*, 547–550.

Liu, L. N., Bryan, S. J., Huang, F., Yu, J. F., Nixon, P. J., Rich, P. R., & Mullineaux, C. W. (2012). Control of electron transport routes through redox-regulated redistribution of respiratory complexes. *Proceedings of the National Academy of Sciences of the United States of America, 109*, 11431–11436.

Liu, Y., He, X., Lim, W., Mueller, J., Lawrie, J., Kramer, L., Guo, J., & Niu, W. (2018). Deciphering molecular details in the assembly of alpha-type carboxysome. *Scientific Reports, 8*, 15062.

Long, B. M., Badger, M. R., Whitney, S. M., & Price, G. D. (2007). Analysis of carboxysomes from *Synechococcus* PCC7942 reveals multiple Rubisco complexes with carboxysomal proteins CcmM and CcaA. *The Journal of Biological Chemistry, 282*, 29323–29335.

Long, B. M., Tucker, L., Badger, M. R., & Price, G. D. (2010). Functional cyanobacterial β-carboxysomes have an absolute requirement for both long and short forms of the CcmM protein. *Plant Physiology, 153*, 285–293.

Long, B. M., Rae, B. D., Badger, M. R., & Price, G. D. (2011). Over-expression of the beta-carboxysomal CcmM protein in Synechococcus PCC7942 reveals a tight co-regulation of carboxysomal carbonic anhydrase (CcaA) and M58 content. *Photosynthesis Research, 109*, 33–45.

Long, B. M., Hee, W. Y., Sharwood, R. E., Rae, B. D., Kaines, S., Lim, Y. L., Nguyen, N. D., Massey, B., Bala, S., Von Caemmerer, S., Badger, M. R., & Price, G. D. (2018). Carboxysome encapsulation of the CO2-fixing enzyme rubisco in tobacco chloroplasts. *Nature Communications, 9*, 3570.

Maccready, J. S., Hakim, P., Young, E. J., Hu, L., Liu, J., Osteryoung, K. W., Vecchiarelli, A. G., & Ducat, D. C. (2018). Protein gradients on the nucleoid position the carbon-fixing organelles of cyanobacteria. *eLife, 7*, e39723.

Maeda, S., Badger, M. R., & Price, G. D. (2002). Novel gene products associated with NdhD3/D4-containing NDH-1 complexes are involved in photosynthetic CO2 hydration in the cyanobacterium, Synechococcus sp. PCC7942. *Molecular Microbiology, 43*, 425–435.

Marco, E., Martinez, I., Ronen-Tarazi, M., Orus, M. I., & Kaplan, A. (1994). Inactivation of ccmO in Synechococcus sp. strain PCC 7942 results in a mutant requiring high levels of CO(2). *Applied and Environmental Microbiology, 60*, 1018–1020.

Martinez-Del Campo, A., Bodea, S., Hamer, H. A., Marks, J. A., Haiser, H. J., Turnbaugh, P. J., & Balskus, E. P. (2015). Characterization and detection of a widely distributed gene cluster that predicts anaerobic choline utilization by human gut bacteria. *MBio, 6*.

McGinn, P. J., Price, G. D., & Badger, M. R. (2004). High light enhances the expression of low-CO_2-inducible transcripts involved in the CO_2-concentrating mechanism in *Synechocystis* sp. PCC6803. *Plant, Cell & Environment, 27*, 615–626.

Mcgurn, L. D., Moazami-Goudarzi, M., White, S. A., Suwal, T., Brar, B., Tang, J. Q., Espie, G. S., & Kimber, M. S. (2016). The structure, kinetics and interactions of the beta-carboxysomal beta-carbonic anhydrase, CcaA. *The Biochemical Journal, 473*, 4559–4572.

Mckay, R. M. L., Gibbs, S. P., & Espie, G. S. (1993). Effect of dissolved inorganic carbon on the expression of carboxysomes, localization of Rubisco and the mode of inorganic carbon transport in cells of the cyanobacterium Synechococcus UTEX 625. *Archives of Microbiology, 159*, 21–29.

Mcnevin, D., Von Caemmerer, S., & Farquhar, G. (2006). Determining RuBisCO activation kinetics and other rate and equilibrium constants by simultaneous multiple non-linear regression of a kinetic model. *Journal of Experimental Botany, 57*, 3883–3900.

Occhialini, A., Lin, M. T., Andralojc, P. J., Hanson, M. R., & Parry, M. A. (2016). Transgenic tobacco plants with improved cyanobacterial Rubisco expression but no extra assembly factors grow at near wild-type rates if provided with elevated CO_2. *The Plant Journal, 85*, 148–160.

Oltrogge, L. M., Chaijarasphong, T., Chen, A. W., Bolin, E. R., Marqusee, S., & Savage, D. F. (2019). α-carboxysome formation is mediated by the multivalent and disordered protein CsoS2. *bioRxiv*, 708164.

Omata, T., Price, G. D., Badger, M. R., Okamura, M., Gohta, S. & Ogawa, T. 1999. Identification of an ATP-binding cassette transporter involved in bicarbonate uptake in the cyanobacterium Synechococcus sp. strain PCC 7942. *Proceedings of the National Academy of Sciences, 96*, 13571–13576.

Park, J., Chun, S., Bobik, T. A., Houk, K. N., & Yeates, T. O. (2017). Molecular dynamics simulations of selective metabolite transport across the propanediol bacterial microcompartment shell. *The Journal of Physical Chemistry. B, 121*, 8149–8154.

Pena, K. L., Castel, S. E., DE Araujo, C., Espie, G. S. & Kimber, M. S. 2010. Structural basis of the oxidative activation of the carboxysomal gamma-carbonic anhydrase, CcmM. Proceedings of the National Academy of Sciences of the United States of America, 107, 2455–2460.

Perlmutter, J. D., Mohajerani, F., & Hagan, M. F. (2016). Many-molecule encapsulation by an icosahedral shell. *eLife, 5*, e14078.

Peterhansel, C., Horst, I., Niessen, M., Blume, C., Kebeish, R., Kurkcuoglu, S., & Kreuzaler, F. (2010). Photorespiration. *Arabidopsis Book, 8*, e0130.

Price, G. D., & Badger, M. R. (1989). Isolation and characterization of high CO(2)-requiring-mutants of the cyanobacterium Synechococcus PCC7942: Two phenotypes that accumulate inorganic carbon but are apparently unable to generate CO(2) within the carboxysome. *Plant Physiology, 91*, 514–525.

Price, G. D., Sültemeyer, D., Klughammer, B., Ludwig, M., & Badger, M. R. (1998). The functioning of the CO2 concentrating mechanism in several cyanobacterial strains: A review of general physiological characteristics, genes, proteins, and recent advances. *Canadian Journal of Botany, 76*, 973–1002.

Price, G. D., Woodger, F. J., Badger, M. R., Howitt, S. M., & Tucker, L. (2004). Identification of a SulP-type bicarbonate transporter in marine cyanobacteria. *Proceedings of the National Academy of Sciences of the United States of America, 101*, 18228–18233.

Price, G. D., Badger, M. R., Woodger, F. J., & Long, B. M. (2008). Advances in understanding the cyanobacterial CO2-concentrating-mechanism (CCM): Functional components, Ci transporters, diversity, genetic regulation and prospects for engineering into plants. *Journal of Experimental Botany, 59*, 1441–1461.

Rae, B. D., Long, B. M., Badger, M. R., & Price, G. D. (2012). Structural determinants of the outer shell of beta-carboxysomes in *Synechococcus elongatus* PCC 7942: Roles for CcmK2, K3-K4, CcmO, and CcmL. *PLoS One, 7*, e43871.

Rae, B. D., Long, B. M., Badger, M. R., & Price, G. D. (2013). Functions, compositions, and evolution of the two types of carboxysomes: Polyhedral microcompartments that facilitate CO_2 fixation in cyanobacteria and some proteobacteria. *Microbiology and Molecular Biology Reviews, 77*, 357–379.

Ravcheev, D. A., Moussu, L., Smajic, S., & Thiele, I. (2019). Comparative genomic analysis reveals novel microcompartment-associated metabolic pathways in the human gut microbiome. *Frontiers in Genetics, 10*, 636.

Rodriguez-Ramos, J., Faulkner, M., & Liu, L. N. (2018). Nanoscale visualization of bacterial microcompartments using atomic force microscopy. *Methods in Molecular Biology, 1814*, 373–383.

Rotskoff, G. M., & Geissler, P. L. (2018). Robust nonequilibrium pathways to microcompartment assembly. *Proceedings of the National Academy of Sciences of the United States of America, 115*, 6341–6346.

Ryan, P., Forrester, T. J. B., Wroblewski, C., Kenney, T. M. G., Kitova, E. N., Klassen, J. S., & Kimber, M. S. (2019). The small RbcS-like domains of the beta-carboxysome structural protein CcmM bind RubisCO at a site distinct from that binding the RbcS subunit. *The Journal of Biological Chemistry, 294*, 2593–2603.

Samborska, B., & Kimber, M. A. T. T. H. E. W. S. (2012). A dodecameric CcmK2 structure suggests β-carboxysomal shell facets have a double-layered organization. *Structure, 20*, 1353–1362.

Saschenbrecker, S., Bracher, A., Rao, K. V., Rao, B. V., Hartl, F. U., & Hayer-Hartl, M. (2007). Structure and function of RbcX, an assembly chaperone for hexadecameric Rubisco. *Cell, 129*, 1189–1200.

Satori, C. P., Henderson, M. M., Krautkramer, E. A., Kostal, V., Distefano, M. M., & Arriaga, E. A. (2013). Bioanalysis of eukaryotic organelles. *Chemical Reviews, 113*, 2733–2811.

Savage, D. F., Afonso, B., Chen, A. H., & Silver, P. A. (2010). Spatially ordered dynamics of the bacterial carbon fixation machinery. *Science, 327*, 1258–1261.

Schneider, G., Lindqvist, Y., & Branden, C. I. (1992). RUBISCO: Structure and mechanism. *Annual Review of Biophysics and Biomolecular Structure, 21*, 119–143.

Schwarz, R., & Grossman, A. R. (1998). A response regulator of cyanobacteria integrates diverse environmental signals and is critical for survival under extreme conditions. *Proceedings of the National Academy of Sciences of the United States of America, 95*, 11008–11013.

Schwarz, R., Reinhold, L., & Kaplan, A. (1995). Low activation state of ribulose-1,5-bisphosphate carboxylase/oxygenase in carboxysome-defective Synechococcus mutants. *Plant Physiology, 108*, 183–190.

Shestakov, S. V., & Khyen, N. T. (1970). Evidence for genetic transformation in blue-green alga Anacystis nidulans. *Molecular & General Genetics, 107*, 372–375.

Shibata, M., Ohkawa, H., Kaneko, T., Fukuzawa, H., Tabata, S., Kaplan, A., & Ogawa, T. (2001). Distinct constitutive and low-CO2-induced CO2 uptake systems in cyanobacteria: Genes involved and their phylogenetic relationship with homologous genes in other organisms. *Proceedings of the National Academy of Sciences of the United States of America, 98*, 11789–11794.

Shibata, M., Katoh, H., Sonoda, M., Ohkawa, H., Shimoyama, M., Fukuzawa, H., Kaplan, A., & Ogawa, T. (2002). Genes essential to sodium-dependent bicarbonate transport in cyanobacteria: Function and phylogenetic analysis. *The Journal of Biological Chemistry, 277*, 18658–18664.

Shively, J. M., Ball, F., Brown, D. H., & Saunders, R. E. (1973). Functional organelles in prokaryotes: Polyhedral inclusions (carboxysomes) of *Thiobacillus neapolitanus*. *Science, 182*, 584–586.

So, A. K. C., Cot, S. S. W, & Espie, G. S. (2002). Characterization of the C-terminal extension of carboxysomal carbonic anhydrase from Synechocystis sp. PCC6803. *Functional Plant Biology, 29*, 183–194.

So, A. K. C., Espie, G. S., Williams, E. B., Shively, J. M., Heinhorst, S., & Cannon, G. C. (2004). A novel evolutionary lineage of carbonic anhydrase (ε class) is a component of the carboxysome shell. *Journal of Bacteriology, 186*, 623–630.

Sommer, M., Sutter, M., Gupta, S., Kirst, H., Turmo, A., Lechno-Yossef, S., Burton, R. L., Saechao, C., Sloan, N. B., Cheng, X., Chan, L.-J. G., Petzold, C. J., Fuentes-Cabrera, M., Ralston, C. Y. & Kerfeld, C. A. 2019. Heterohexamers formed by CcmK3 and CcmK4 increase the complexity of beta carboxysome shells. Plant Physiology, 179(1), 156–167.

Spreitzer, R. J. (2003). Role of the small subunit in ribulose-1,5-bisphosphate carboxylase/oxygenase. *Archives of Biochemistry and Biophysics, 414*, 141–149.

Stec, B. (2012). Structural mechanism of RuBisCO activation by carbamylation of the active site lysine. *Proceedings of the National Academy of Sciences of the United States of America, 109*, 18785–18790.

Sun, Y., Casella, S., Fang, Y., Huang, F., Faulkner, M., Barrett, S., & Liu, L. N. (2016). Light modulates the biosynthesis and organization of cyanobacterial carbon fixation machinery through photosynthetic electron flow. *Plant Physiology, 171*, 530–541.

Sun, Y., Wollman, A. J. M., Huang, F., Leake, M. C., & Liu, L. N. (2019). Single-organelle quantification reveals the stoichiometric and structural variability of carboxysomes dependent on the environment. *Plant Cell, 31*, 1648–1664.

Sutter, M., Wilson, S. C., Deutsch, S., & Kerfeld, C. A. (2013). Two new high-resolution crystal structures of carboxysome pentamer proteins reveal high structural conservation of CcmL orthologs among distantly related cyanobacterial species. *Photosynthesis Research, 118*, 9–16.

Sutter, M., Roberts, E. W., Gonzalez, R. C., Bates, C., Dawoud, S., Landry, K., Cannon, G. C., Heinhorst, S., & Kerfeld, C. A. (2015). Structural characterization of a newly identified component of α-carboxysomes: The AAA+ domain protein CsoCbbQ. *Scientific Reports, 5*, 16243.

Sutter, M., Faulkner, M., Aussignargues, C., Paasch, B. C., Barrett, S., Kerfeld, C. A., & Liu, L. N. (2016). Visualization of bacterial microcompartment facet assembly using high-speed atomic force microscopy. *Nano Letters, 16*, 1590–1595.

Sutter, M., Greber, B., Aussignargues, C., & Kerfeld, C. A. (2017). Assembly principles and structure of a 6.5-MDa bacterial microcompartment shell. *Science, 356*, 1293–1297.

Sutter, M., Laughlin, T. G., Sloan, N. B., Serwas, D., Davies, K. M., & Kerfeld, C. A. (2019). Structure of a synthetic beta-carboxysome shell. *Plant Physiology, 181*, 1050.

Swan, J. A., Golden, S., Liwang, A., & Partch, C. L. (2018). Structure, function, and mechanism of the core circadian clock in cyanobacteria. *Journal of Biological Chemistry, 293*, 5026.

Tanaka, S., Sawaya, M. R., Kerfeld, C. A., & Yeates, T. O. (2007). Structure of the RuBisCO chaperone RbcX from *Synechocystis* sp. PCC6803. *Acta Crystallographica. Section D, Biological Crystallography, 63*, 1109–1112.

Tanaka, S., Kerfeld, C. A., Sawaya, M. R., Cai, F., Heinhorst, S., Cannon, G. C., & Yeates, T. O. (2008). Atomic-level models of the bacterial carboxysome shell. *Science, 319*, 1083–1086.

Tanaka, S., Sawaya, M. R., Phillips, M., & Yeates, T. O. (2009). Insights from multiple structures of the shell proteins from the β-carboxysome. *Protein Science, 18*, 108–120.

Tchernov, D., Helman, Y., Keren, N., Luz, B., Ohad, I., Reinhold, L., Ogawa, T., & Kaplan, A. (2001). Passive entry of CO2 and its energy-dependent intracellular conversion to HCO3- in cyanobacteria are driven by a photosystem I-generated deltamuH+. *The Journal of Biological Chemistry, 276*, 23450–23455.

Tsai, Y., Sawaya, M. R., Cannon, G. C., Cai, F., Williams, E. B., Heinhorst, S., Kerfeld, C. A., & Yeates, T. O. (2007). Structural analysis of CsoS1A and the protein shell of the Halothiobacillus neapolitanus carboxysome. *PLoS Biology, 5*, e144.

Tsai, Y.-C. C., Lapina, M. C., Bhushan, S., & Mueller-Cajar, O. (2015). Identification and characterization of multiple rubisco activases in chemoautotrophic bacteria. *Nature Communications, 6*, 8883.

Tsinoremas, N. F., Schaefer, M. R., & Golden, S. S. (1994). Blue and red light reversibly control psbA expression in the cyanobacterium Synechococcus sp. strain PCC 7942. *The Journal of Biological Chemistry, 269*, 16143–16147.

Turmo, A., Gonzalez-Esquer, C. R., & Kerfeld, C. A. (2017). Carboxysomes: Metabolic modules for CO2 fixation. *FEMS Microbiology Letters, 364*.

Wang, C., Sun, B., Zhang, X., Huang, X., Zhang, M., Guo, H., Chen, X., Huang, F., Chen, T., Mi, H., Yu, F., Liu, L. N., & Zhang, P. (2019a). Structural mechanism of the active bicarbonate transporter from cyanobacteria. *Nature Plants, 5*(11), 1184–1193.

Wang, H., Yan, X., Aigner, H., Bracher, A., Nguyen, N. D., Hee, W. Y., Long, B. M., Price, G. D., Hartl, F. U., & Hayer-Hartl, M. (2019b). Rubisco condensate formation by CcmM in beta-carboxysome biogenesis. *Nature, 566*, 131–135.

Wheatley, N. M., Gidaniyan, S. D., Liu, Y., Cascio, D., & Yeates, T. O. (2013). Bacterial micro-compartment shells of diverse functional types possess pentameric vertex proteins. *Protein Science, 22*, 660–665.

Whitehead, L., Long, B. M., Price, G. D., & Badger, M. R. (2014). Comparing the *in vivo* function of alpha-carboxysomes and beta-carboxysomes in two model cyanobacteria. *Plant Physiology, 165*, 398–411.

Whitney, S. M., Houtz, R. L., & Alonso, H. (2011). Advancing our understanding and capacity to engineer nature's CO2-sequestering enzyme, rubisco. *Plant Physiology, 155*, 27–35.

Woodger, F. J., Badger, M. R., & Price, G. D. (2005). Sensing of inorganic carbon limitation in Synechococcus PCC7942 is correlated with the size of the internal inorganic carbon pool and involves oxygen. *Plant Physiology, 139*, 1959–1969.

Yeates, T. O., Kerfeld, C. A., Heinhorst, S., Cannon, G. C., & Shively, J. M. (2008). Protein-based organelles in bacteria: Carboxysomes and related microcompartments. *Nature Reviews. Microbiology, 6*, 681–691.

Yeates, T. O., Crowley, C. S., & Tanaka, S. (2010). Bacterial microcompartment organelles: Protein shell structure and evolution. *Annual Review of Biophysics, 39*, 185–205.

Yeates, T. O., Thompson, M. C., & Bobik, T. A. (2011). The protein shells of bacterial microcompartment organelles. *Current Opinion in Structural Biology, 21*, 223–231.

Yokoo, R., Hood, R. D., & Savage, D. F. (2015). Live-cell imaging of cyanobacteria. *Photosynthesis Research, 126*, 33–46.

Zarzycki, J., Axen, S. D., Kinney, J. N., & Kerfeld, C. A. (2013). Cyanobacterial-based approaches to improving photosynthesis in plants. *Journal of Experimental Botany, 64*, 787–798.

Zarzycki, J., Sutter, M., Cortina, N. S., Erb, T. J., & Kerfeld, C. A. (2017). In vitro characterization and concerted function of three core enzymes of a glycyl radical enzyme – Associated bacterial microcompartment. *Scientific Reports, 7*, 42757.

Correction to: Adaptive Mechanisms of the Model Photosynthetic Organisms, Cyanobacteria, to Iron Deficiency

Hai-Bo Jiang, Xiao-Hui Lu, Bin Deng, Ling-Mei Liu, and Bao-Sheng Qiu

Correction to:
Chapter 11 in: Q. Wang (ed.), *Microbial Photosynthesis*, https://doi.org/10.1007/978-981-15-3110-1_11

The original version of the book was revised with the following corrections.

1. In the 10th line of Page 6, the word Fe(II)L was changed as Fe(Π)L.
2. In Fig. 1, the figure legend was replaced from "Miethke and Marahiel 2007" to "Shaked and Lis 2012".
3. A list of references was newly added along with its in-text citations as given below:

Trujillo, A. (2011). *The iron hypothesis*. http://www.homepages.ed.ac.uk/shs/Climatechange/Carbon%20sequestration/Martin%20iron.htm

Moeck, G. S., & Coulton, J. W. (1998). TonB-dependent iron acquisition: Mechanisms of siderophore-mediated active transport. *Molecular Microbiology, 28*(4), 675–681.

Teresa, B. M., Hernández, J. A., Luisa, P. M., et al. (2001). Cloning, overexpression and interaction of recombinant Fur from the cyanobacterium *Anabaena* PCC 7119 with isiB and its own promoter. *FEMS Microbiology Letters, 194*(2), 187–192.

The updated online version of this chapter can be found at https://doi.org/10.1007/978-981-15-3110-1_11

Q. Wang (ed.), *Microbial Photosynthesis*,
https://doi.org/10.1007/978-981-15-3110-1_16

The manufacturer's authorised representative in the EU is Springer Nature Customer Service Centre GmbH, Europaplatz 3, 69115 Heidelberg, Germany. If you have any concerns regarding our products, please contact ProductSafety@springernature.com

Printed and bound by CPI Group (UK) Ltd, Croydon, CR0 4YY

30/09/2026

02225248-0009